D1338567

Heavy Metals in Soils

Heavy Metals in Soils

Second edition

edited by

B.J. Alloway

Professor of Soil Science
The University of Reading
Reading
UK

BLACKIE ACADEMIC & PROFESSIONAL

An Imprint of Chapman & Hall

London · Glasgow · Weinheim · New York · Tokyo · Melbourne · Madras

Published by
Blackie Academic and Professional, an imprint of Chapman & Hall, Wester Cleddens Road, Bishopbriggs, Glasgow G64 2NZ, UK

Chapman & Hall, 2–6 Boundary Row, London SE1 8HN, UK

Blackie Academic & Professional, Wester Cleddens Road, Bishopbriggs, Glasgow G64 2NZ, UK

Chapman & Hall GmbH, Pappelallee 3, 69469 Weinheim, Germany

Chapman & Hall USA, One Penn Plaza, 41st Floor, New York NY 10119, USA

Chapman & Hall Japan, ITP-Japan, Kyowa Building, 3F, 2-2-1 Hirakawacho, Chiyoda-ku, Tokyo 102, Japan

DA Book (Aust.) Pty Ltd, 648 Whitehorse Road, Mitcham 3132, Victoria, Australia

Chapman & Hall India, R. Seshadri, 32 Second Main Road, CIT East Madras 600 035, India

First edition 1990
Second edition 1995

Typeset in 10/12pt Times by Julia Stevenson, Hove, East Sussex
Printed in Great Britain by St Edmundsbury Press, Bury St Edmunds, Suffolk

ISBN 0 7514 0198 6

A catalogue record for this book is available from the British Library
Library of Congress Catalogue Card Number: 94–79286

∞ Printed on acid-free text paper, manufactured in accordance with ANSI/NISO Z39.48-1992 (Permanence of Paper)

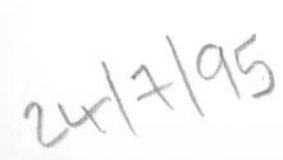

Preface

Heavy metals in soils continue to receive increasing attention due to the greater understanding of their toxicological importance in ecosystems, agriculture and human health, the growing scientific and public awareness of environmental issues and the development of even more sensitive analytical techniques to measure their concentrations accurately. Building on the success and acclaim of the first edition, this book has been thoroughly revised and updated and continues to provide a balanced and comprehensive review of the subject in two sections: the first providing an introduction to the metals' chemistry, sources and methods used for their analysis; and the second containing chapters dealing with individual elements in detail.

This volume is for soil scientists, research chemists, geochemists, agronomists, environmental scientists and professionals who deal with contaminated land.

B.J.A.

Contributors

Professor B.J. Alloway	Department of Soil Science, The University of Reading, Whiteknights, PO Box 233, Reading, Berks RG6 2DW, UK
Professor D.E. Baker	Land Management Decisions Inc., 3048 Research Drive, State College, Pennsylvania 16801, USA
Professor B.E. Davies	Department of Environmental Science, University of Bradford, Richmond Road, Bradford BD7 1DP, UK
Dr. R. Edwards	School of Biological and Earth Sciences, Liverpool John Moores University, Byrom Street, Liverpool L3 3AF, UK
Dr. K.C. Jones	Institute of Environmental and Biological Sciences, University of Lancaster, Bailrigg, Lancaster LA1 4YQ, UK
Professor L. Kiekens	Industriele Hogeschool van het Gemeenschap-sonderwijs, C.T.L. 9000 Gent, Voskenslaan 270, Belgium
Professor N.W. Lepp	School of Biological and Earth Sciences, Liverpool John Moores University, Byrom Street, Liverpool L3 3AF, UK
Professor S.P. McGrath	Soils and Crop Production Division, Rothamsted Experimental Station, Harpenden, Herts. AL5 2JQ, UK
Dr. R.H. Neal	Statewide Air Pollution Research Centre, University of California, Riverside, California 92521, USA
Dr. P.O'Neill	Department of Environmental Science, University of Plymouth, Drake's Circus, Plymouth, Devon PL4 8AA, UK
Dr. J.E. Paterson	Department of Soil Science, Scottish Agricultural College, West Mains Road, Edinburgh EH9 3JG, UK
Dr. J.P. Senft	Land Management Decisions Inc., 3048 Research Drive, State College, Pennsylvania 16801, USA
Dr. K.A. Smith	Department of Soil Science, East of Scotland College of Agriculture, West Mains Road, Edinburgh EH9 3JG, UK

Professor E. Steinnes Department of Chemistry, University of Trondheim, AVH, N-7055 Dravoll, Norway

Dr. A.M. Ure Department of Pure and Applied Chemistry, University of Strathclyde, Thomas Graham Building, 295 Cathedral Street, Glasgow G1 1XL, UK

Contents

Section 1 — General principles

1 Introduction

B.J. ALLOWAY

This book is dedicated to the occurrence and behaviour of heavy metals in the soil. It is important to realise that the soil is both a source of metals and also a sink for metal contaminants. The factors controlling the total and bioavailable concentrations of heavy metals in soils are of great importance with regard to both human toxicology and agricultural productivity. By acting as a sink for metals (and many other contaminants) the soil also functions as a filter protecting the groundwater from inputs of potentially harmful metals.

Heavy metals, or trace metals, is the term applied to a large group of trace elements which are both industrially and biologically important. Although not completely satisfactory from a chemical point of view, 'heavy metals' is the most widely recognised and used term for the large group of elements with an atomic density greater than 6 g/cm^3 [1]. Other names have been used for this group of elements and 'toxic metals' is even less appropriate because all trace elements are toxic to living organisms when present in excess. However, some such as Co, Cr, Cu, Mn, Mo and Zn are essential in small but critical concentrations for the normal healthy growth of either plants, animals or both, although they are toxic at high concentrations. Even 'heavy metal' is often used perjoratively with emphasis on the pollution and toxicity aspects rather than as a collective name for a group of elements of considerable economic and environmental importance. A newer term which is being increasingly used with reference to the harmful properties of these metals is 'potentially toxic element' (or PTE) and this seems to be gaining acceptance. The heavy metals which tend to give rise to the greatest amount of concern with regard to the human health, agriculture and ecotoxicology are As, Cd, Hg, Pb, Tl, and U. On the other hand, agricultural productivity on large areas of land in many parts of the world can be limited by deficiencies of 'essential' trace elements or heavy metals (often called 'micronutrients') such as Zn, Cu and Mn in the case of crops and Co, Mn, Cu and Zn in livestock.

'Pollution' is, in some ways, a term easier to define than 'heavy metals', but it is commonly confused or used interchangeably with the term 'contamination'. Although several interpretations of the terms 'pollution' and 'contamination' exist, the definition given by Holdgate [2] is widely accepted. This states that *pollution* is "the introduction by man into the environment of substances or energy liable to cause hazards to human health, harm to living resources and ecological systems, damage to structures or amenity, or interference with legitimate uses of the environment". Other definitions use the term 'contamination' where the anthropogenic inputs do not appear to cause obvious harmful effects and 'pollution' is applied only to situations where toxicity has occurred. However, this is unsatisfactory because the effects of the 'contaminant' may not be fully understood at the time. In Holdgate's definition, 'pollution' covers any concentration of a potentially harmful substance, whether or not adverse effects are observed. In practice, the terms 'contamination' and 'pollutions' are frequently used interchangeably, although pollution is usually more pejorative. For soils, there is a fairly widely adopted convention to use the term 'contamination' for any situation in which elevated concentrations of a substance occur.

Studies of heavy metals in ecosystems have indicated that many areas near urban complexes, metalliferous mines or major road systems contain anomalously high concentrations of these elements. In particular, soils in such regions have been polluted from a wide range of sources with Pb, Cd, Hg, As and other heavy metals. Nriagu [3] wrote that we may be experiencing a "silent epidemic of environmental metal poisoning" from the ever increasing amounts of metals wasted into the biosphere. The mining, manufacture and disposal of metals and metal-containing materials inevitably cause environmental pollution. The trends in the production of several industrially important metals and estimates of the amounts reaching the soil are shown in Table 1.1.

From Table 1.1 it can be seen that more Cu is produced than any of the other metals listed, but Zn is the highest in terms of global emissions to soil. Over the fifteen year period, the production of Cd, Cu and Zn showed a general increase but Hg gradually declined. This was largely due to its severe mammalian toxicity which has led to its replacement in many uses by less hazardous materials. The decrease in Pb is due to the same reasons, but especially to its reduction and/or removal from petrol (gasoline). Nickel and Sn show a fluctuation in production due to economic factors and substitution in uses. Tin has been replaced by Al in drink cans and the substitution of deep freezing food for canning has led to a reduced demand for Sn-plated steel for making cans. Nevertheless, large amounts of all these metals are produced from ores each year and large amounts are either

Table 1.1 Changes in the primary production of metals and the current rate of global heavy metal emmissions reaching the soil (10^3 t/yr). From World Resources Institute [4] Nriagu and Pacyna [5]

Metal	Year				Global emissions to soil (Nriagu and Pacyna [5])
	1975	1980	1985	1990	1980s
Cd	15.2	18.2	19.1	20.2	22
Cu	6739.0	7204.0	7870.0	8814.0	954
Hg	8.7	6.8	6.1	5.8	8.3
Ni	723.8	658.2	687.3	836.9	325
Pb	3432.2	3448.2	3431.2	3367.2	796
Sn	232.2	247.3	180.7	219.3	—
Zn	3975.4	4030.3	4723.1	5570.9	1372

recycled or dispersed into the environment in a non-recoverable form.

With regard to emissions to soil, it can be seen that those for Cd exceed the tonnage of primary production. This is due to its presence as a guest element or contaminant in many metal ores and to its dispersion as an atmospheric contaminant from metalliferous industries and as a contaminant in some phosphatic fertilisers. Its annual production has more than tripled over the last 40 years (production 6000 t in 1950) as a result of its increasing uses in plastics, steel corrosion prevention and dry cells (Ni–Cd). However, before it was intentionally used, Cd contamination has occurred from ores and phosphatic rocks. There is now a drive to reduce its use and dispersion in the environment owing to its potential toxic risk to humans and ecosystems. Although Zn is generally considered a much lower toxicological risk to animals and humans than Cd, As, Hg or Pb, recent evidence suggests that its dispersion into the environment and accumulation in soils, especially from sewage sludges, may have far reaching soil fertility implications due to its toxicity to some soil microorganisms.

With increasing industrial demand for metals there is a continuous need to locate new reserves of ore minerals. Soils, stream sediments and natural vegetation samples are analysed in large numbers for mineral exploration purposes. Although originally developed for mineral exploration, stream sediment reconnaissance has subsequently been found to be of great value for highlighting geochemical anomalies of importance to agriculture and for showing regional patterns of soil pollution.

In contrast to the problem of heavy metal excesses resulting from pollution, agricultural land in many parts of the world has been found to be deficient in one or more micronutrients. This includes the heavy metals/metalloids: Cu, Mn and Zn which are essential for plants and animals and Co, Cr and Se which are only necessary for animals. Iodine, B and Fe are also important micronutrients but are not classed as heavy metals. Zinc deficiency is probably the most important heavy metal micronutrient problem in crops on a world scale, but deficiencies of Cu and Mn also affect crops in many areas. These crop deficiencies are rectified either by amending the soils with salts of the relevant metal or by the application of salts or chelates directly to the crop in foliar sprays. Livestock deficiencies are usually treated by feed supplements or injections. In certain parts of the world, human health has been shown to have been seriously affected by heavy metal micronutrient deficiencies. These include deficiencies of Se in China and of Zn in the USA.

Being at the interface between the atmosphere and the earth's crust as well as the substrate for natural and agricultural ecosystems, the soil is open to inputs of heavy metals from many sources. Heavy metals occur naturally in soils, usually at relatively low concentrations, as a result of the weathering and other pedogenic processes acting on the rock fragments on which the soils develop (soil parent material). Considerable variations occur between the normal concentration ranges of different elements, from Au (0.001–0.002 mg/kg) to Mn (20–10000 mg/kg). Rocks of different types vary considerably in their mineralogical and elemental composition giving rise to marked differences in concentrations between soils even in the absence of significant inputs from external sources. Heavy metals are also classed as 'trace elements' because they occur in concentrations of less than 1% (frequently below 0.01% or 100 mg/kg) in the rocks of the earth's crust. The heavy metal concentrations inherited from the soil parent material are modified by pedogenic and biochemical processes, by natural inputs such as dust particles derived from soil, rocks and volcanic ash and, most importantly, by anthropogenic inputs, i.e. pollution.

In the five or more years since the first edition of this book was being prepared, several important developments have occurred with regard to heavy metals in soils and the environment as a whole. These include:

i) Soil protection and the sustainability of agricultural production have become increasingly important topics of discussion and research and the heavy metal contamination of soils features prominently in many aspects of the subject as a result of their potentially toxic effects and long-term persistence. Several conferences have been held on this topic and research projects focussing on some of the key isssues have been initiated.

ii) Although not a recent discovery, the inhibition of soil microorganism activity by various heavy metals has received increasing research attention and many important papers on this topic have been published within the last few years. Attention has mainly been focussed on *Rhizobia* species which are involved in symbiotic nitrogen fixation in the roots of clovers because these have been found to be inhibited in some pastures on sewage sludge amended soils. The general order of toxicity to *Rhizobium leguminosum* bv *trifolii* appears to be Cu > Cd > Ni > Zn [6]. However, most concern is being expressed about Zn, which, although less toxic than Cu, Cd and Ni, tends to occur at much higher concentrations. Significant reductions in *Rhizobia* numbers have been found in experimental field plots in both the UK and Germany where the total Zn contents have been within the UK permissible limits for Zn in sludged soils (300 mg/kg). A recent committee set up by the UK Ministry of Agriculture, Fisheries and Food (MAFF) recommended that the statutory maximum Zn content of sludge amended soils in the UK be reduced to 200 mg/kg [7]. From a soil protection point of view, these findings are important because they indicate that normal biomass activity in the soil is significantly inhibited by the presence of some heavy metals reaching the soil in sewage sludge applied at normal agricultural rates and other species of microorganisms involved in soil fertility may also be adversely affected. (See Chapters 2, 6 and 13).

iii) Heavy metals in sewage sludge have received considerable attention in both the USA and the UK. In the USA, the Environmental Protection Agency produced Standards for the Use and Disposal of Sewage Sludge (Part 503 of Section 405 of the Clean Water Act, finalised, November 1992) [8]. This contained tables giving ceiling concentrations of metals in sludges, cumulative loading rates, pollutant concentrations in soils and annual pollutant loading rates (see Chapter 3). It is particularly notewothy that the values adopted and the rationale behind them differs markedly from those in the Directive of the European Community.

In the UK, the Ministry of Agriculture, Fisheries and Food (MAFF) commissioned independent scientific committees to review the rules for sludge application to agricultural land with regard to the soil fertility, food safety and animal health aspects of potentially toxic elements. Reports from these committees were published in November 1993 [7]. (See Chapter 3).

iv) New guideline values for heavy metals in contaminated soils have been introduced in Germany, the Netherlands and in Canada. The Guide Values and Quality Standards for metals (and organic contaminants) in soil used in the Netherlands have been revised [9]. Instead of giving A, B and C values as in the original (1986) version [10], the guidance now comprises a new set of Target Values and the original C values (intervention concentrations). This scheme is still based on the multifunctionality of land use. However, the Interim Environmental Quality Standards introduced in Canada in 1991 give values for different land uses, with the lowest allowable values in agricultural soils and the highest in soils used for industrial premises [11].

v) Many more cases of severe heavy metal pollution of soils and the environment have been reported, especially in the former communist countries of Central and Eastern Europe and Russia. Some of the problems are associated with abandoned military bases and training grounds but most are due to inadequate pollution control in metal mining and smelting operations and metallurgical industries. For example, in Romania, the rural and urban areas around three large non-ferrous metal smelting complexes (Bio Mare, Zlatna and Copsa Mica) have been shown to be heavily contaminated by a range of heavy metals including Cd, Cu, Pb and Zn. In Copsa Mica the pollution has been found to have significantly affected 180,000 ha of surrounding land and the environmental problems have been exacerbated by atmospheric emissions of organic particulate pollutants from a carbon black factory adjacent to the smelter.

However, historical contaminated land problems are still continuing to be discovered in technologically advanced countries in Western Europe, the USA and many other parts of the World. In Denmark, part of the small town of Mundelstrup, in the County of Aarhus, was found to have been heavily polluted with lead from a fertiliser factory operating between 1871 and 1921. Concentrations of up to 67,562 mg/kg Pb were found in the topsoils when the contamination was identified in 1987. In view of the risk to the occupants of houses built on contaminated soil, a carefully planned clean-up and restoration programme was carried out in 1991–92. All soil with Pb concentrations above 40 mg/kg was removed and this amounted to 50,000 m^3 of soil from around 33 houses. The contaminated soil was removed to a safe containment site and the original soil excavated to create this landfill was used to replace the contaminated soil around the houses. The cost of this operation was more than 33×10^6 Dkr (c. £4×10^6) [12].

vi) Lead concentrations in air have been shown to have decreased significantly as a result of the decreases in the amounts of Pb added to petrol (gasoline) and the widescale use of unleaded petrol in many countries. Unleaded petrol was introduced in the USA and some other countries in the 1970s and in Europe mainly in the 1980s; the numbers of cars using this fuel is now very large and the effect on atmospheric Pb levels is quite marked. Although the residence time for Pb in many soils is considered to be hundreds or thousands of years, there are indications that Pb contents are showing significant decreases in some localities as a result of this reduction in the most ubiquitous input of the metal (see Chapter 9).

vii) The concept of 'chemical time bombs' has attracted a lot of interest following a European 'state-of -the-art-conference'on 'Delayed Effects of Chemicals in Soils and Sediments' held in the Netherlands in 1992 [13] and articles in the popular scientific press. Many of the examples of chemical time bombs which have been recognised so far have involved heavy metals in soils. The delayed effects which have been observed have often been marked increases in the solubility and bioavailability of heavy metals resulting from acidification, leakage from landfills and chemical dumps, soil erosion and depletion of organic matter levels in soils.

viii) An atlas of 17 elements in the soils of England and Wales by McGrath and Loveland was published in 1992 [14]. This '*Soil Geochemical Atlas*' was based on the analysis of 5692 topsoil (0–15 cm) samples collected on a 5 km × 5 km cell basis between 1978 and 1982 as part of the UK National Soil Inventory. Maps and summary data are given for 17 elements including the nine heavy metals; Ba, Cd, Cr, Co, Cu, Pb, Mn, Ni and Zn (see Chapter 3).

ix) The use of metal hyperaccumulator plants for the remediation of heavily contaminated soils has been proposed by Baker and McGrath [15] and these and other workers are currently conducting research into the feasibility of this approach. Plant species such as *Thlaspi Caerulescens* have been found to accumulate <7000 mg/kg Zn which accounts for an uptake of 43 kg/ha and it may be possible to use plant is this way as an *in situ* method of reducing the metal contents of contaminated soils.

This book brings together information on the behaviour of antimony, arsenic, cadmium, copper, chromium, cobalt, gold, lead, manganese, mercury, molybdenum, nickel, selenium, silver, thallium, tin, uranium, vanadium, and zinc in the soil, their sources and uptake by plants and their entry into the food chain. Following on from the highly successful First Edition, the chapters in this Second Edition have been revised by their original authors to bring them up to date with the latest developments. Where possible, the authors have used data and information from a wide variety of sources in order to present the global situation. Section 1 comprises a brief introduction to the relevant properties and chemical processes of soils which affect the dynamics of metals (Chapter 2), a summary of the major sources of heavy metals (Chapter 3) and a review of the methods used for the analysis of metals in soils (Chapter 4). This section is intended to complement the detailed chapters on each element in Section 2 and to provide an introduction to non-specialists who may need to deal with problems of heavy metal excess or deficiency. The appendices contain summarised data on the concentrations of metals found in soils, plants and sewage sludges and the critical concentrations which have so far been established. However, practitioners are recommended to refer to regional data for critical concentrations of metals in soils and crops where these are available because significant variations can occur as a result of differences in soil properties, soil test methods and crop varieties in different parts of the world.

The reader is also referred to several other books which complement the coverage of this volume: Bowen [16], Kabata-Pendias and Pendias [17], Adriano [18], the detailed reviews of several elements edited by Nriagu, including Cd [20], Cu [22], Ni [21], Hg [23], Pb [24] and Zn [25] and the SCOPE Report on Pb, Hg, Cd and As in the Environment [26]. Various specialised texts covering the detailed aspects of analytical and soil chemistry, mineralogy and plant physiology are referred to in the subsequent chapters.

References

1. Phipps, D.A., in *Effects of Heavy Metal Pollution on Plants*, ed. Lepp, N.W. Applied Science Publishers, London (1981), 1–54.
2. Holdgate, M.W., *A Perspective of Environmental Pollution*, University Press, Cambridge (1979).
3. Nriagu, J.O., *Environ. Pollut.* **50** (1988) 139–161.
4. World Resources Institute, *World Resources 1992/93*, Oxford University Press, New York (1992).
5. Nriagu, J.O. and Pacyna, J.M. *Nature (London)*, **333** (1988) 134–139.
6. Chaudri, A.M., McGrath, S.P. and Giller, K.E. *Soil Biol. Biochem.* **24** (1992) 625–632.
7. Ministry of Agriculture, Fisheries and Food, *Review of the Rules for Sewage Sludge Application to Agricultural Land: Soil fertility aspects of potentially toxic elements*, MAFF Publications, No. PB 1561, London (1993).
8. US Environmental Protection Agency, 40 CFR Parts 257, 403 and 503 [FRL-4203-3] *Standards for the Use and Disposal of Sewage Sludge*, US EPA, Washington DC (1992).
9. Ministry of Housing, Physical Planning and Environment, Director General for Environmental Protection (Netherlands), *Environmental Standards for Soil and Water*, Leidschendam (1991).
10. Moen, J.E.T., Cornet, J.P. and Evers, C.W.A., in *Contaminated Land*, Assink, J.W. and van der Brink, W.J. ed. Martinus Nijhoff, Dordrecht (1986).
11. Canada Council of Ministers of the Environment, Interim Canadian Environmental Quality Criteria for Contaminated Sites. Report CCME EPC-c534, Winnipeg, Manitoba (1992).
12. Bauman, J. (1992), personal communication.
13. ter Meulen, G.R.B., Stigliani, W.M., Salomons, W., Bridges, E.M. and Ineson, A.C. *Chemical Time Bombs — Proc. of the State of the Art Conference on Delayed Effects of Chemicals in Soils and Sediments*, Foundation for Ecodevelopment, Hoofddorp (1993).
14. McGrath, S.P. and Loveland, P.J. *The Soil Geochemical Atlas of England and Wales*, Blackie Academic and Professional, Glasgow (1992).
15. Baker, A.J.M., McGrath, S.P., Sidoli, C.M.D., Reeves, R.D. *Resources, Conservation and Recycling* (1994) (in the Press).
16. Bowen, H.J.M., *Environmental Chemistry of the Elements*, Academic Press, London (1979).
17. Kabata Pendias, A. and Pendias, H., *Trace Elements in Soils and Plants (2nd edition*, CRC Press, Boca Raton Fla. (1992).
18. Adriano, D.C., *Trace Elements in the Terrestrial Environment*, Springer-Verlag, Heidelberg (1986).
19. Fergussion, J.E. *The Heavy Elements: Chemistry, Environmental Impact and Health Effects*, Pergamon Press, Oxford (1990).
20. Nriagu, J.O ed., *Cadmium in the Environment: Part 1; Ecological Cycling*, John Wiley & Sons, New York (1980).
21. Nriagu, J.O ed., *Nickel in the Environment*, John Wiley & Sons, New York (1980).
22. Nriagu, J.O ed., *Copper in the Environment: Part 1; Ecological Cycling*, John Wiley & Sons, New York (1979).
23. Nriagu, J.O ed., *The Biogenochemistry of Mercury in the Environment*, Elsever/North Holland, Amsterdam (1979).

24. Nriagu, J.O ed., *The Biogeochemistry of Lead in the Environment*, Elsever/North Holland, Amsterdam (1978).
25. Nriagu, J.O ed., *Zinc in the Environment: Part 1; Ecological Cycling*, John Wiley & Sons, New York (1980).
26. Hutchinson, T.C. and Meema, K.M. eds., *Lead, Mercury, Cadmium and Arsenic in the Environment*, SCOPE 31, John Wiley & Sons, Chichester (1987).

2 Soil processes and the behaviour of metals

B.J. ALLOWAY

2.1 Introduction to the soil

The soil is a key component of terrestrial ecosystems, both natural and agricultural, being essential for the growth of plants and the degradation and recycling of dead biomass. It is a complex heterogeneous medium comprising mineral and organic solids, aqueous and gaseous components. The minerals present are usually weathering (chemically decomposing) rock fragments and secondary minerals such as phyllo-silicates or clay minerals, oxides* of Fe, Al, and Mn and sometimes carbonates (usually $CaCO_3$). The organic matter comprises living organisms (mesofauna and microorganisms), dead plant material (litter) and colloidal humus formed by the action of microorganisms on plant litter. These solid components are usually clustered together in the form of aggregates, thus creating a system of interconnected voids (pores) of various sizes filled with either water or air. The solid components have the ability to adsorb ions, but this differs between materials and is strongly influnced by the prevailing pH and redox conditions and the relative concentrations of the ions present in the aqueous soil solution.

This structured heterogeneous mixture of organic and mineral components is the habitat for many organisms, as well as the medium in which plant roots grow, extracting water, oxygen and ions. Roots also release CO_2 and exude organic compounds which are responsible for the intense microbial activity in the interfacial zone between the root and the soil called the 'rhizosphere'. Plant roots modify the chemical and physical properties of the soil around them and thus influence the bioavailability of some chemical elements.

The soil is a dynamic system, subject to short-term fluctuations, such as variations in moisture status, pH and redox conditions and also undergoing gradual alterations in response to changes in management and environmental factors. These changes in soil properties affect the form and bioavailability of metals, and need to be considered in decisions on the management of polluted soils or the use of soils for disposal of waste materials. Soils can show marked spatial variability in physical and chemical properties at the macro- and micro-scales, thus emphasising the need for thorough sampling to include the range of variation in parameters at any site investigated.

* The term 'oxides' includes all forms of oxides including hydrous oxides and oxyhydroxides.

2.2 Key soil properties

2.2.1 Soil pH

The soil reaction is the pre-eminent factor controlling the chemical behaviour of metals and many other important processes in the soil. However, the pH concept is not as precise for soils as it is for solutions *in vitro* because of the heterogeneity of soils, the relatively small proportion of solution present in the pores of the solid soil and the adsorption of (cationic) H^+ (ions) on to solid surfaces [1, 2]. The pH of a soil applies to the H^+ (ion) concentration in the solution present in soil pores which is in dynamic equilibrium with the predominantly negatively charged surfaces of the soil particles. Hydrogen ions are strongly attracted to the surface negative charges, and they have the power to replace most other cations.

The diffuse layer close to a negatively charged surface therefore has a higher concentration of H^+ ions than the bulk soil solution. When the soil solution is diluted the diffuse layer expands, causing the pH of the bulk solution to increase. This has important implications for the measurement of soil pH in the laboratory. This normally involves mixing dry soil with 2–2.5 times its weight of water, shaking and then measuring the pH in the supernatant solution after 30 minutes. The pH value obtained is about 1 to 1.5 units higher than that of the soil solution near to the solid surfaces where the reactions take place. This dilution effect is usually overcome by measuring the pH in a suspension of soil in a solution of a neutral salt, such as $CaCl_2$ or KCl. Variable charge soils may give slightly higher pH values in dilute $CaCl_2$ solution. Normally, pH is measured in suspensions of soil with either distilled water or a dilute solution of a neutral salt such as $CaCl_2$ or KCl. pH values are usually expressed together with the soil:solution ratio and the solvent used. It is assumed that if the solvent is not mentioned the pH was measured in distilled water.

Soil pH is affected by the changes in redox potential which occur in soils that become waterlogged periodically. Reducing conditions generally cause a pH increase, and oxidation brings about a decrease. Variations of up to 2 units can occur over a year in gley soils prone to waterlogging. Oxidation of pyrite (FeS_2) in a soil parent material can cause a marked decrease in pH.

Soils have several mechanisms which serve to buffer pH to varying extents, including hydroxyaluminium ions, CO_2, carbonates and cation exchange reactions [1]. However, even with these buffering mechanisms, soil pH differs significantly due to localised variations within the soil. Diurnal fluctuations of more than 1 unit may occur as well as variations in different parts of a field. Soil pH usually increases with depth in humid regions where bases are leached down the profile, and can decrease with depth in arid environments where evaporation causes salts to accumulate in the surface horizon. As a result of the variations which can occur it is not necessarily meaningful to express soil pH measurements more accurately than to the nearest 0.2 division of a unit.

In general, heavy metal cations are most mobile under acid conditions and increasing the pH by liming usually reduces their bioavailabilty. However, molybdate anions become more available with increasing pH.

Soils generally have pH values within the range 4–8.5, owing to the buffering by Al at the lower end and by $CaCO_3$ at the upper end of the range [2]. Brady [3] states that the normal pH is 5–7 in soils of humid regions, and pH 7–9 in the soils of arid regions. However, the maximum range of pH conditions found in soils is 2–10.5. In a typical temperate environment, such as the UK, soils normally have a pH in the range 4–8. The optimum pH for most arable crops is 6.5 on mineral soils and 5.5 on peaty soils. The optimum pH for grassland in the UK is 6.0 on mineral soils and 5.5 on peaty soils. Soil pH can be raised by liming, but it is impractical to acidify agricultural soils more alkaline than these.

2.2.2 Soil organic matter

The main feature which distinguishes soil from regolith (decomposed rock) is the presence of living organisms, organic debris and humus. All soils contain organic matter, although the amount and type may vary considerably. Colloidal soil organic matter has a major influence on the chemical properties of soils, and can be divided into 'non-humic' and 'humic' substances. The non-humic substances comprise unaltered biochemicals such as amino acids, carbohydrates, organic acids, fats and waxes that have not changed from the form in which they were synthesised by living organsims. Humic substances are a series of acidic, yellow to black coloured polyelectrolytes of moderately high molecular weight. They are formed by secondary synthesis reactions involving microorganisms and have characteristics which are dissimilar to any compounds in living organisms [4]. They have a wide variety of functional groups, including carboxyl, phenolic hydroxyl, carbonyl, ester and possibly quinone and methoxy groups [5, 6]. While predominantly composed of humic substances, soil humus also contains some biochemicals bound to the humic polymers.The elemental composition of humus is typically (on an ash-free basis): 44–53% C, 3.6–5.4% H, 1.8–3.6% N and 40–47% O [7].

Traditionally, humus has been separated in the laboratory into three fractions: (1) humin, which is insoluble in alkali and acid (ii) humic acid, which is soluble in alkali and insoluble in acid and (iii) fulvic acid, which is soluble in both acid and alkali. These substances cannot be regarded as distinctly different, but merely as part of a continuum of compounds varying in molecular weight, C content, O content, acidity and cation exchange capacity (CEC) in the order: humin > humin acid > fulvic acid, with N content decreasing through the same sequence [6]. Less than half the C in humic acids is aromatic; much of the remainder occurs as unsaturated aliphatic chains containing carboxyl groups which also contribute most of the titratable acidity. Humic acids have molecular weights in the range 20000–100000, and the fulvic acid fraction generally consists of lower-molecular-weight compounds. Apart from containing up to 10% polysaccharides, the general composition of fulvic acid is similar to that of humic acid. The fulvic acid fraction may contain precursors and degradation products of the humic acid fraction. Humins

Table 2.1 Typical values for the organic matter contents of soils

Soil	Organic matter content (%)	Reference
Cultivated soils (general)	<10	9
Mineral soils (general)	3–5	10
Arable soil (SE England)	<2	9
Arable soil (N England/S Scotland)	2–6	9
Perm. grassland (S Scotland)	7.9–9.5	9
Prairie grassland (USA)	5–6	3
Poorly drained soil (USA)	10	3
All soils (W Virginia, USA)	0.54–15	3
Tropical soils (S America)	0.5–21.7	11
Tundra (USSR)	73	12
Podzols (USSR)	10	12
Chernozems (USSR)	3.5–10	12

are considered to be humic acid-type compounds adsorbed onto minerals [7].

Methods used to determine the organic matter content of soils include either the percentage loss in weight after ignition in a furnace at 375°C for 16 h or the oxidation of C by acid dichromate followed by the titration of excess dichromate [8]. Organic acid matter contains about 58–60% organic C (% orgC × 1.67 = % OM). Within the soil profile, the organic matter content is always highest in the surface horizon, but podzols and vertisols may have some translocated humic material lower down the profile.

Tables 2.2 and 2.3 give the summarised values for the soil pH and organic matter contents in representative samples of the ranges of soils in the USA and England and Wales, respectively. Table 2.2 also contains the data for the cation exchange capacities of the American soils. These data give a very useful indication of the ranges of values for these parameters in these two countries. It should be noted that there are a much wider range of soil types in the USA compared with England and Wales. The heavy metal concentrations in these soils are given in Chapter 3 (Tables 3.11 and 3.12).

Table 2.2 Summarised results for pH, % organic carbon and cation exchange capacity in agricultural soils of the USA

Parameter	Minimum	Mean	Median	Maximum	CV%Δ
pH	3.9	6.26	6.1	8.9	17
Org. C (%)	0.09	4.18	1.05	63.0	228
CEC (cmolc/kg)	0.6	26.3	14.0	204.0	143

From Holmgren *et al.* [13]. Based on the analysis of 3045 samples of soils from 307 different soil series at sites with healthy crops remote from obvious sources of contamination (pH measured in water).

Table 2.3 Summarised results for pH and % organic carbon in soils in England and Wales

Parameter	No. of samples	Minimum	Median	Maxium
pH	5679	3.1	6.0	9.2
Org. C (%)	5666	0.1	3.6	65.9

From McGrath and Loveland [14]. Samples collected on a 5 × 5 km grid basis to avoid bias (pH measured in water).

2.2.3 Clay minerals

Clay minerals are products of rock weathering and have marked effects on both the physical and chemical properties of the soils. Their contribution to soil chemical properties results from their comparatively large surface area and permanent surface negative charge. The soil textural class is dependent on the percentages of clay, silt and sand-sized particles. The clay fraction comprises the dispersed minerals of diameter less than 2 μm. In most cases, this specifically applies to the mineralogically distinct group of clay minerals, although it can also include finely ground particles of other minerals. The clay minerals consist of continuous two-dimensional tetrahedral sheets of composition T_2O_5 (T = tetrahedral cation, usually Si^{4+} or Al^{3+}), in which individual tetrahedra are linked with neighbouring tetrahedra by sharing three corners each (the basal oxygens) [15]. The fourth tetrahedral corner (the apical oxygen) is in the direction normal to the sheet and also forms part of an immediately adjacent octahedral sheet. The octahedral cations are normally Al, Mg, Fe^{2+} and Fe^{3+}; however, many other medium sized cations such as Li, Ti, V, Cr, Co, Ni, Cu and Zn may also occur in some species. There are two ways of filling the octahedral sites depending on the valence of the cation. If a trivalent cation such as Al^{3+} occupies the sites this arrangement is called dioctahedral and if the dominant cation in the octahedral sheet is divalent, like Mg^{2+} then it is classified as trioctahedral.

The most common types of clay minerals include the kaolinites comprising one silica sheet and one gibbsite sheet referred to as a 1:1 clay mineral; the illites comprising two silica sheets and one gibbsite (2:1), and the smectites (also montmorillonites) with two silica sheets and one gibbsite sheet (2:1). In all the minerals except kaolinites, isomorphous substitution within the mineral lattice gives rise to a net negative charge on the surface of the mineral. This is caused by trivalent Al substituting for tetravalent Si and divalent Mg substituting for Al. In kaolinite the 1:1 units are tightly bonded together by hydrogen bonds between hydrogen and oxygen atoms of adjacent layers. The specific surface of these minerals is relatively small (5–40 m^2/g) in comparison with the other clay minerals, and their cation exchange capacity is low (3–20 $cmols_c$/kg) because very little isomorphous substitution has occurred. Illite minerals have the 2:1 units bonded by K ions, and the specific surface and CEC are larger than those of kaolinite minerals (100–200 m^2/g and 10–40 $cmols_c/kg$). Smectites have the largest specific surfaces (700–800 m^2/g) owing to relatively weak interlayer bonding allowing soil solution to penetrate between the units. The CEC is also quite high (80–120 $cmols_c/kg$) as a result of this large specific surface. Vermiculites have an intermediate specific surface (300–500 m^2/g) and a high CEC (100–150 $cmols_c/kg$) [7].

Clay minerals rarely exist in pure form in soils; they usually have humic colloids and hydrous oxide precipitates linked to them. The combined organo-mineral collidal complex plays a very important role in controlling the concentrations ions in the soil solution. Space does not permit an adequate discussion of soil constituents, but these are covered in soil chemistry texts [2, 3, 7, 12].

2.2.4 Oxides of iron, manganese and aluminium

The oxides of Fe, Al and Mn which are commonly referred to as hydrous oxides, or sesquioxides in the case of Fe and Al, play an important role in the chemical behaviour of metals in soils. They occur in the clay-size (<2 μm) fraction and they are usually mixed with clays and have a disordered structure. However, in the more rigorous weathering environment of the tropics these oxides are often more abundant than clay minerals [2].

In freely drained (oxic) soils, oxides of Fe, Al and Mn precipitate from solution and occur as: (i) coatings on soil particles where they are often intimately mixed with clays, (ii) fillings in voids, and (iii) concentric nodules. Hydrous Fe oxide minerals tend to be the most abundant of all the oxides in soils. Gibbsite is a common form of Al hydroxide but is much less abundant then the Fe oxides except in some tropical soils which have undergone severe weathering [2].

Precipitation of Fe is usually in the form of the gelatinous ferrihydrite ($5Fe_2O_3.9H_2O$) initially and this gradually dehydrates to more stable forms such as goethite (-FeOOH) [2, 21]. Goethite is the most common Fe oxide found in soils and hematite (α-Fe_2O_3) is found mainly in tropical soils. Lepidocrocite (γ-FeOOH) is characteristic of the fluctuating redox conditions in gleyed soils.The common mineral forms of Mn oxides in soils are birnessite, hollandite and lithiophorite [22].

With regard to the dynamics of heavy metals in soils, Fe and Mn oxides co-precipitate and adsorb (scavenge) cations including Co, Cr, Cu, Mn, Mo, Ni, V and Zn, anions, such as HPO_4^{2-} and AsO_4^{3-}, from solution. This is due to a pH-dependent charge which, generally speaking, is negative in alkaline conditions and positive in acid conditions but the pH at which there is no net charge, called the PZC (point of zero charge), varies for different hydrous oxide minerals. In pure form, the PZC values for Fe oxides are in the range 7–10, and those for Al oxides pH 8–9.4; however, when mixed with clays in soils the PZC values are usually much lower. The PZC values for Mn oxides are very low, usually in the range 1.5–4.6.

As discussed in Section 2.2.5, variations in redox conditions have a profound effect on both the quantities of hydrous oxides present in a soil and also on the adsorptive capacity of that soil for a wide range of cations and anions. The onset of reducing conditions induced either by waterlogging or a decrease in the percentage of air-filled pores caused by structural degradation, results in the dissolution of the oxides and the release of their adsorbed ions. Specialised bacteria, such as *Thiobacillus ferrooxidans* and *Metallogenum* spp are also involved in the precipitation of Fe and Mn hydrous oxides, respectively.

2.2.5 Oxidation and reduction in soils

Soils are subject to variations in oxidation–reduction (redox) status and this mainly affects the elements C, N, O, S, Fe and Mn, although Ag, As, Cr, Cu, Hg and Pb can also be affected [16]. Redox equilibria are controlled by the aqueous free-electron activity, which can be expressed as either the pE value (the negative log of the electron activity) or E_h (the millivolt difference in potential between a Pt electrode and the standard H electrode) [17]. The pE unit has the advantage of allowing electrons to be treated like other reactants or products, allowing both chemical and electrochemical equilibria to be expressed with the single equilibrium constant. The conversion factor for the unit is E_h (mV) = 59.2 pE [17]. Large positive values of pE (or E_h) favour the existence of oxidised species, and low or negative values of pE (or E_h) are associated with reduced species [18].

Redox potential E_h is measured using a Pt electrode and a calomel electrode connected to a millivolt meter, but it is difficult to get accurate readings. However, oxic soil conditions usually give values in the range +300 to +800 mV (pE 5.1–13.5) but mostly from +400 to +600 mV (pE 6.8–10.1). Anaerobic soils have values from +118 to −414 mV (pE +2 to −7) [15, 19]. Measurements of E_h can be used to determine whether oxidising or reducing conditions exist, but frequently soil colours provide a good indication of redox status. Red and brown colours indicate oxic conditions, blue-green and grey colours indicate gleyed soils. However, strong-coloured parent materials may mask these colour changes in some soils.

Redox reactions in soils are frequently slow but are catalysed by soil microorganisms which are able to live over the full range of pH and pE conditions normally found in soils (pH 3–10 and pE +12.7 to −6.0) [18]. Respiration by microorganisms, mesofauna and plant roots consumes a relatively large amount of oxygen. If the oxygen in a zone of soil becomes exhausted, as happens with waterlogging or compaction, microorganisms with anaerobic respiration predominate and susceptible elements (Mn, Cr, Hg, Fe, Cu and Mo) are gradually reduced [18, 19]. When Fe (III) is reduced to Fe (II) there is a slight pH increase in acid soils and a slight decrease in alkaline soils. Most waterlogged soils have a pH of 6.7–7.2 [19]. When the pE begins to rise, Fe (II) is oxidised in preference to Mn (II).

The combined effect of E_h (or pE) and pH conditions on the forms Fe and Mn are best illustrated by an E_h/pH diagram, as shown in Figure 2.1. From Figure 2.1 it can be seen that the oxides of both Fe and Mn can be dissolved by either decreasing

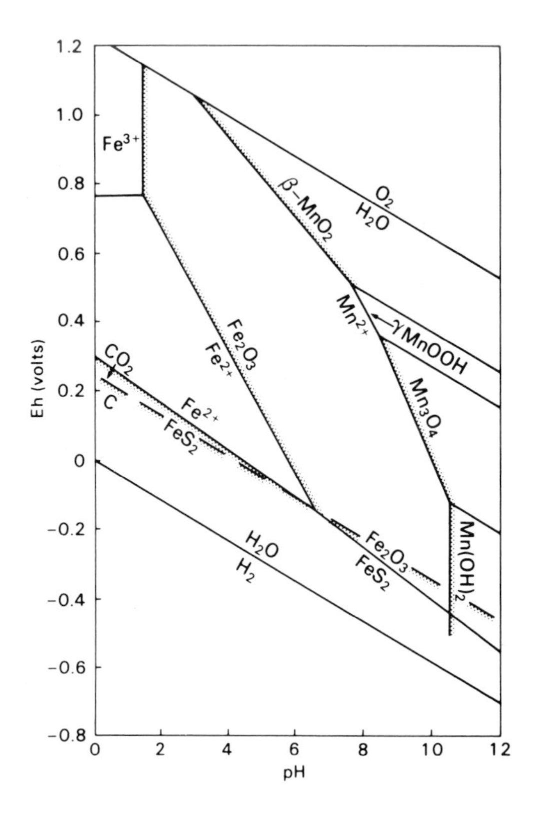

Figure 2.1 An E_h/pH diagram showing the stability of Fe and Mn oxides, pyrite and organic C. (Redrawn and adapted from Rose *et al.* [20], with permission.)

pH or E_h, but Mn oxides are more easily dissolved than Fe oxides. With increasing E_h and pH, Fe oxides precipitate before those of Mn. Small changes in E_h or pH can give rise to either extensive dissolution or precipitation of Fe oxides [20].

Several heavy metals can be affected by the indirect effects of reducing conditions. Sulphate ions are reduced to sulphide below pE –2.0, and this can lead to the precipitation of metal sulphides, such as Fe S_2, HgS, CdS, CuS, MnS and ZnS [18]. When reducing conditions cause the dissolution of hydrous Mn, Al and Fe oxides, their co-precipitated metals are released into the soil solution [16].

2.3 The adsorption of metal ions by soils and their constituents

The most important chemical processes affecting the behaviour and bioavailabilty of metals in soils are those concerned with the adsorption of metals from the liquid phase on to the solid phase. These processes control the concentrations of metal ions and complexes in the soil solution and thus exert a major influence on their uptake by plant roots. Several different mechanisms can be involved in the adsorption of metal ions, including cation exchange (or non-specific adsorption), specific adsorption, co-precipitation and organic complexation. However, although the extent of adsorption can be measured and isotherms derived it is frequently difficult to be precise about which particular process is responsible for the retention of metals in any particular soil. In order to provide an introduction to the adsorption mechanisms referred to in the following chapters, the four processes mentioned above will be discussed briefly.

2.3.1 Cation exchange

Most heavy metals (with certain exceptions, including the metalloids As, Sb and Se and the metals Mo and V) exist mainly as cations in the soil solution, and their adorption therefore depends on the density of negative charges on the surfaces of the soil colloids. In order to maintain electroneutrality, the surface negative charge is balanced by an equal quantity of cations. Ion exchange refers to the exchange between the counter-ions balancing the surface charge on the colloids and the ions in the soil solution [22]. It has the following characteristics: it is reversible, diffusion controlled, stochiometric and, in most cases, there is some selectivity or preference for one ion over another by the adsorbent [23]. This selectivity gives rise to a replacing order amongst the cations, determined by their valency and degree of hydration. The higher the valency of an ion, the greater its replacing power; H^+ ions behave like polyvalent ions and the greater the degree of hydration, the lower its replacing power, other things being equal. Adsorption by cation exchange can also be described as the formation of outer-sphere complexes with the surface functional groups to which they are bound electrostatically [18].

The CEC of mineral soils can range from a few to 60 $cmols_c$/kg but in organic soils it may exceed 200 $cmol_c$/kg [24]. The CEC of soils is far larger than their anion exchange capacity, owing to the greater number of negative charges on the colloid surfaces. These negative charges are of two types (i) permanent charges (independent of pH due to isomorphous substitution), and (ii) pH-dependent charges on the edges of clay minerals, on humus polymers and oxides. The negative charges are due to the dissociation of protons from carboxyl and phenolic groups on humic polymers and O and OH groups on the edges of clay minerals and oxides. As mentioned in Section 2.2.4 there are PZC values at which the net charge is zero for all these materials (freshly deposited oxides of Fe have a PZC c. pH 8.5 and Al c. pH 8.3) but the PZC of mixtures in soils tends to differ from those of pure forms. In general, the oxides tend to contribute little to the CEC of soils below pH 7 but develop an anion exchange capacity in acid soils [12].

In humic polymers, carboxyl groups have pK (dissociation constant) values of 3–5, whereas phenolic groups have values over 7. Some basic NH_3 groups exist at pH values below 3. All of the adsorptive sites on the humic collids are thus pH dependent, with CEC values of between 150 and 300 $cmols_c$/kg. Although the organic matter content of a soil is usually much less than that of clay, it nevertheless makes a major contribution to the CEC of a soil because of its high adsorptive capacity at pHs above 5.

2.3.2 Specific adsorption

Specific adsorption involves the exchange of heavy metal cations and most anions with surface ligands to form partly covalent bonds with lattice ions [26]. It results in metal ions being adsorbed to a far greater extent than would be expected from the CEC of a soil. For example, Brummer [27] showed that the sorptive capacities of amorphous Fe and Al oxides for Zn were 7 and 26 times greater, respectively, than their CECs at pH 7.6. As shown earlier, this value is below the PZC for the pure forms of Al and Fe oxides and so they would not be expected to have a significant CEC. Specific adsorption is strongly pH dependent and is related to the hydrolysis of the heavy metal ions [27]. The metals most able to form hydroxy complexes are specifically adsorbed to the greatest extent. Therefore, the pK (equilibrium constant) values of the reaction $M^{2+} + H_2O = MOH^+ + H^+$ determine the adsorption behaviour of the different metals. Specific adsorption increases with decreasing pK values but, in the case of Cu and Pb which have the same pK value, Pb with the greater ionic size is more strongly adsorbed. Brummer [27] gives the order for increasing specific adsorption as: Cd (pK = 10.1) < Ni (pK = 9.9) < Co (pK = 9.7) < Zn (pK = 9.0) << Cu (pK = 7.7) < Pb (pK = 7.7) < Hg (pK = 3.4). The hydrous oxides of Al, Fe and Mn are thought to be the main soil constituents involved in the specific adsorption reaction.

In addition to being adsorbed on mineral surfaces, heavy metal ions can also diffuse into minerals such as goethite, Mn oxides, illites, smectites and some other minerals [27]. The relative rate of diffusion of the metal ions into minerals increases with pH up to a maximum which is equal to the pK value for the situation when $M^{2+} = MOH^+$ on the mineral surface. Above this pH the $MOH^+ > M^{2+}$ and the relative diffusion rate decreases. For example, the maximum relative diffusion rates for Co, Ni and Zn decrease in the order Ni > Zn > Cd and can be related to their ionic diameters (Ni = 0.69 nm, Zn = 0.74 nm and Cd = 0.97 nm). Adsorption of metals by goethite therefore comprises three different steps: first, surface adsorption; second, diffusion into goethite particles and third, adsorption and fixation at positions within the mineral particles [27].

2.3.3 Co-precipitation

Co-precipitation is defined as the simultaneous precipitation of a chemical agent in conjunction with other elements by any mechanism and at any rate [16]. The types of mixed solid commonly formed include clay minerals, hydrous Fe and Mn oxides and calcite in which isomorphous substitution has occurred (Table 2.4). In addition to co-precipitation, replacement of Ca^{2+} cations by Cd^{2+} can also occur

Table 2.4 Trace metals normally found co-precipitated with secondary minerals in soils (from Sposito [16])

Mineral	Co-precipitated trace metals
Fe oxides	V, Mn, Ni, Cu, Zn, Mo
Mn oxides	Fe, Co, Ni, Zn, Pb
Ca carbonates	V, Mn, Fe, Co, Cd
Clay minerals	V, Ni, Co, Cr, Zn, Cu, Pb, Ti, Mn, Fe

in the surface layer of calcite when it comes into contact with solutions containing Cd. When all the calcite surface has reacted in the solid solution mode, the remaining Cd^{2+} in solution precipitates out as $CdCO_3$ [28].

2.3.4 *Insoluble precipitates of heavy metals in soils*

When the physico-chemical conditions and concentrations of appropriate ions are sufficiently high, many metals can form insoluble precipitates (solid phases) which could play a role in controlling their solubility in the soil solution. The following summarised information for selected metals is taken from Lindsay [17].

Cadmium. Octavite ($CdCO_3$) could be a major factor controlling the solubility of Cd in high pH soils. In strongly gleyed soils (with reducing conditions) the sulphide mineral greenockite (CdS) can form, which explains the low solubility of Cd in flooded paddy soils. However, drainage of these paddy soils and a return to oxidising conditions results in Cd^{2+} and SO_4^{2-} being formed along with a marked decrease in pH which results in an increase in the mobility and bioavailability of Cd.

Copper. Under most physico-chemical conditions encountered in soils the adsorbed forms of Cu (soil-Cu) are more stable than any Cu minerals except in strongly reducing conditions when cuprous ferrite ($Cu_2Fe_2O_4$) is more stable than soil-Cu.

Lead. Several Pb phosphates can occur in soils, including $Pb_5(PO_4)_3OH$, $Pb_3(PO_4)_2$ and $Pb_5(PO_4)_3Cl$. The latter form, chloropyromorphite is the most insoluble of the Pb phosphate minerals and could control the solubility of Pb^{2+} throughout a wide pH range especially in soils of high P status, such as sewage sludge amended soils.

Manganese. Under well oxidised conditions the most stable Mn mineral is pyrolusite β-MnO_2. Manganese generally forms hydrous oxides with mixed valency states but under strong reducing conditions manganite (g-MnOOH) is formed.

Mercury. The halide complexes Hg_2I_2, Hg_2Cl_2 and Hg_2Br_2 are possible mineral forms in soils if the respective anion concentrations are sufficiently high. In reducing conditions the sulphides Hg_2S and HgS can occur. However, methylated forms of mercury can also be formed.

Molybdenum. Controlling minerals are $MnCO_3$ (rhodochrocite) in low redox situations and Fe_3O_4 (magnetite) under oxidising conditions. Ferrimolybdite $Fe_2(MoO_4)_3$ and Pb molybdate $PbMoO_4$ could also be important controlling minerals.

Zinc. The sorbed forms of Zn in soils (soil-Zn) are normally more stable than most Zn minerals except franklinite $ZnFe_2O_4$ which could be an important factor controlling the solubility of Zn depending on the Fe(II) concentrations.

2.3.5 Organic complexation

In addition to being involved in cation exchange reactions, solid-phase humic substances such as humic acids also adsorb metals by forming chelate complexes. The stability constants of chelates with metals tend to be in the following decreasing order:Cu > Fe = Al > Mn = Co > Zn. Low-molecular-weight organic ligands, not necessarily humic in origin, can form soluble complexes with metals and prevent them from being adsorbed or precipitated. Humic compounds with suitable reactive groups, such as hydroxyl, phenoxyl and carboxyl form coordination complexes with metallic ions. Carboxyl groups play a predominant role in metal binding in both humic and fulvic acids. The maxium amount of any given metal that can be bound is found to be approximately equal to the number of carboxyl groups [4].

2.3.6 Selectivity of adsorbents for different metals

Metals vary in the extent to which they are adsorbed by the mechanisms described and the adsorbents show differences in selectivity sequences for metals, as shown in Table 2.5.

It can be seen from Table 2.5 that the order of selectivity for metals differs between adsorbent materials and, in the case of hydrous Fe oxides, variations also occur in sequence order for different oxide minerals. The relative selectivity for metal cations can be largely explained in terms of the Lewis hard–soft acid–base (HSAB) principle. Hard Lewis acids prefer to react or complex with hard Lewis bases, and soft acids prefer soft bases [30]. The term ‘hard’ indicates a high electronegativity, low polarisabilty and small ionic radius, and ‘soft’ implies the opposite of these. Hard Lewis acids include the following cations: Li^+, Na^+, K^+, Rb^+, Cs^+, Be^{2+}, Mg^{2+} Ca^{2+}, Sr^{2+}, Ti^{4+}, Cr^{3+}, Mn^{2+}, Fe^{3+}, Co^{3+}. Soft Lewis acids include Cu^+, Ag^+, As^+, Cd^{2+}, Hg^+, Hg^{2+}, Tl^+, Pd^{2+}, Pt^{2+}. Borderline metals which do not fit so well into the classification include Fe^{2+}, Co^{2+}, Ni2+, Zn^{2+}, Sb^{3+} and Pb^{4+} [37]. Complexing ligands can complicate the application of HSAB principles, as in the case of Cl^- which reacts with Cd^{2+} and reduces the amount of Cd adsorbed [32]. Water is a very hard base, clay minerals behave as soft bases and Fe oxides are hard bases. Consequently, clay minerals show a preference for Cd, relative to Zn or Ni, which are comparatively hard and are preferentially adsorbed by Fe oxides, which are harder bases [30]. The siloxane ditrigonal cavity adsorption sites on montmorillonite are harder than the hydroxy functional groups on the edge of kaolinite crystals which show a preference for the soft acid Cd^{2+} [30].

2.3.7 *Quantitative description of metal ion adsorption*

Traditionally, the adsorption of ions by soils has been quantitatively described by either the *Langmuir* or the *Freundlich* adsorption isotherms. The Langmuir equation has the form

$$\frac{C}{x/m} = \frac{1}{Kb} + \frac{C}{b}$$

where C is the concentration of the ion in the equilibrium solution, x/m is the amount of C adsorbed per unit of adsorbate, K is a constant related to the bonding energy and b is the maximum amount of ions that will be adsorbed by a given sorbent [26]. This equation is convenient to use for soils because both K and b can be easily determined by experiment and thus the amount adsorbed at any input can then be estimated [38]. The adsorption maximum 'b' agrees with experimental findings that metals are held less strongly as adsorption increases. In some polluted soils there is a possibility that adsorption sites could become saturated at high metal concentrations.

The Freundlich adsorption isotherm has the form

$$x = kc^n \quad \text{or} \quad \log x = \log k + n \log c$$

where x is the amount adsorbed per unit of adsorbent at concentration of c of adsorbate, and k and n are constants [26]. This equation does not include an adsorption maximum [38].

Neither of these equations provides information on the adsorption mechanisms involved, and both assume a uniform distribution of adsorption sites and the absence of any reactions between adsorbed ions. Nevertheless, they are still generally considered to be useful and many workers have found their data fitted these isotherms. De Haan and Zwerman [38] have used a simplified sorption equation for specific conditions, based on the Langmuir and Freundlich equations:

$$x/m = K_d c_o$$

where K_d is the distribution constant corresponding to the slope of the isotherm, and c_o is the equilibrium concentration of the adsorbing compound once adsorption has been established. The distribution coefficient K_d is a useful parameter for comparing the sorptive capacities of different soils or materials for any particular ion (when measured under the same experimental conditions):

$$K_d = \frac{\text{amount sorbed per unit weight of soil}}{\text{amount in solution per unit volume of liquid}}$$

Allen [39] has reviewed recent literature concerned with the speciation of metals in sediments and, more briefly, their application to soils. Of particular interest is his reference to work by Lee *et al.* [40] who used K_d values of soils to determine safe maximum concentrations for metals which would prevent groundwater concentrations rising above the Drinking Water Standards. They determined the K_d values of soils for Cd at a range of pH values and then established a soil total concentration that equalled the D.W.S. From this they developed the Soil Quality Criteria value, in mg/kg (*Soil QC*), using the Drinking Water Standard for the metal (*DWS*) and the porosity (n), particle density (D_s) and degree of water saturation (p) of the soils investigated:

$$Soil\ QC = DWS\ K_d + \frac{n \times p}{D_s \times (1 - n)}$$

For a range of New Jersey (USA) soils with pHs between 3.9 and 6.2, organic matter contents up to 2.9%, clay contents up to 37% and CEC values between 0.9 and 9.5 $cmols_c$/kg they obtained *Soil QC* concentrations of Cd of between 0.09 and 4.5 mg/kg.

2.3.8 *Adsorption as surface complexation*

In view of the limitations of the traditional approaches to metal ion adsorption, Sposito and Page [18] suggest an alternative approach in which the adsorption reactions are considered as complexation reactions with functional groups on the solid surfaces creating surface metal species analogous to the main aqueous species which exist in the soil solution. Table 2.5 lists the main metal species in decreasing order of relative abundance, based on predictions from the GEOCHEM model [18]. Many soils are maintained at around pH 7, so the predominant species may differ in some cases from those listed for acid and alkaline soils.

Surface functional groups, including hydroxyl groups on the edges of clays and on hydrous oxides, ditrigonal cavities in the basal planes of clays, and carboxyl, amino and phenoxyl groups on the surfaces of organic matter, react with metal species to form surface complexes [18]. These surface complexes can be of two types: (i) inner-sphere complexes, in which no molecule of the bathing solvent (water) is interposed between the surface groups and the ion or molecule it binds, and (ii) outer-sphere complexes, in which at least one molecule of the solvent comes between the functional group and the ion. These outer-sphere complexes are usually electrostatically bonded and are the equivalent of the cation exchange reaction. They are less stable than inner-sphere complexes, which are equivalent to the specifically adsorbed and the organically chelated metal ions.

In addition to the solid phase, there are interactions with metal-complexing ligands in the soil solution.

2.3.9 *Biological methylation of heavy metals* (From Fergusson [41])

Some elements, including Hg, As, Se, Te, Tl and In can undergo methylation by microorganisms to form volatile molecules, such as CH_3Hg^+, CH_3Se and CH_3As, and this can be a major route for losses of these elements from soils. Methylation is known to be brought about by both aerobic and anaerobic species of bacteria and fungi, but tends to occur more readily in anaerobic sediments in aquatic environments. All biological methylation involves methyl cobalamin, a methylated derivative of B_{12} which contains Co (see Chapter 10). The rate at which biological methylation occurs depends on the ambient conditions, including temperature, redox and pH, but non-biological methylation can also occur. The methylated forms of Hg are $(CH_3)_2Hg$ (most stable in alkaline conditions) and CH_3Hg^+ (stable in neutral to acid soils). Lead is also thought to be methylated in the environment by both biological and abiotic mechanisms but the evidence is not conclusive. However, most organo-Pb compounds in the environment are probably derived from additives in petrol.

2.4 Soil–plant relationships of heavy metals

2.4.1 The soil–plant system

The major interrelationships affecting the dynamics of heavy metals between the soil and the plant are shown in Figure 2.2. The soil–plant system is an open system subject to inputs, such as contaminants, fertilisers and pesticides, and to losses, such as the removal of metals in harvested plant material, leaching, erosion and volatilisation.

2.4.2 Plant uptake of metals

The factors affecting the amounts of metal absorbed by a plant are those controlling: (i) the concentrations and speciation of the metal in the soil solution, (ii) the movement of the metal from the bulk soil to the root surface, (iii) the transport of the metal from the root surface into the root, and (iv) its translocation from the root to the shoot [2, 43]. Plant uptake of mobile ions present in the soil solution is largely determined by the total quantity of this ion in the soil but, in the case of strongly adsorbed ions, absorption is more dependent upon the amount of root produced [2]. Mycorrhizae are symbiotic fungi which effectively increase the absorptive area of the root and can assist in the uptake of nutrient ions, such as orthophosphates and micronutrients. Roots possess a significant CEC, due largely to the presence of carboxyl groups, and this may form part of the mechanism of moving ions through the outer part of the root to the plasmalemma where active absorption occurs.

Absorption of metals by plant roots can be by both passive and active (metabolic) processes. Passive (non-metabolic) uptake involves diffusion of ions in the soil solution into the root endodermis. On the other hand, active uptake takes place against a concentration gradient but requires metabolic energy and can therefore be inhibited by toxins. The mechanisms appear to differ between metals; for instance

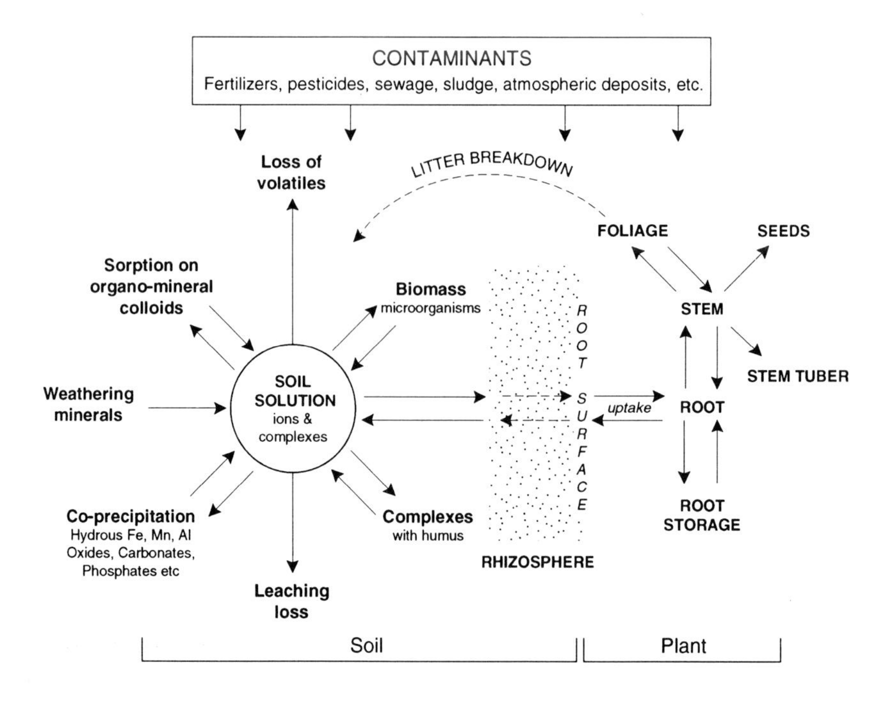

Figure 2.2 The soil–plant system showing the key components concerned with the dynamics of heavy metals (adapted from Peterson and Alloway [42]).

Pb uptake is generally considered to be passive while that of Cu, Mo and Zn, is thought to be either active metabolic uptake, or a combination of both active and passive uptake [44].

Absorption mechanisms can vary for different metal ions, but ions which are absorbed into the root by the same mechanisms are likely to compete with each other. For example, Zn absorption is inhibited by Cu and H^+, but not by Fe and Mn; Cu absorption is inhibited by Zn, NH_4^+, Ca and K [45, 46].

The rhizosphere is the zone about 1–2 mm wide between plant roots and the surrounding soil. It receives appreciable amounts of organic material from the roots, including exudates, mucilage, sloughed-off cells and their lysates [47]. These organic compounds give rise to intense microbiological and biochemical activity in the rhizosphere which enables roots to mobilise some of the metals which are strongly adsorbed in the soil, by acidification, redox changes, or the formation of organic complexes. Phenolic compounds and certain amino acids are known to be involved in the solubilisation of Fe^{3+} and Mn^{4+} [47].

Cereals deficient in micronutrients such as Fe and Zn appear to have root exudates containing substances such as phytosiderophore 2′-deoxymugineic acid which are effective in mobilising these and other metals from sorption sites in the vicinity of the root [44]. Mench and Martin [48] showed that the root exudates with identical

Table 2.5 Soil–plant transfer coefficients of heavy metals

Element	Soil–plant transfer coefficient
Cd	1–10
Co	0.01–0.1
Cr	0.01–0.1
Cu	0.1–10
Hg	0.01–0.1
Ni	0.1–1.0
Pb	0.01–0.1
Tl	1–10
Zn	1–10
As	0.01–0.1
Be	0.01–0.1
Se	0.1–10
Sn	0.01–0.1

From Kloke *et al.* [51]

carbon contents from maize and tobacco extracted amounts of Mn, Cu, Cd and Fe which differed with the plant species. Tobacco root exudates increased the extraction of Cd but decreased that of Fe; those of maize did not affect the concentrations of either of these metals.

The uptake of metals from soils is greater in plants grown in pots of soil in the greenhouse than from the same soil in the field [46, 47]. De Vries and Tiller [49, 50] found the uptake of Cd by lettuce and onion bulbs grown in pots to be 6 and 25 times greater, respectively, than when grown in the same soil in the field. This is probably due to differences in microclimate and soil moisture, and to the roots of container-grown plants growing solely in contaminated soil, whereas those of field-grown plants may reach down to less contaminated soil.

Relative differences in the uptake of metal ions between plant species and cultivars is genetically controlled and can be due to various factors, including: surface area of the root, root CEC, root exudates and the rate of evapotranspiration. The latter mechanism affects the mass flow of the soil solution in the vicinity of the root and thus the movement of ions to the root absorbing surface. Kloke *et al.* [51] gave the general order of the transfer coefficients for most of the biologically important heavy metals which are shown in Table 2.5. The transfer coefficient is the metal concentration in plant tissue above ground divided by the total metal concentration in the soil. Although numerous soil and plant factors can affect the accumulation of metals in plants, the values given are intended as guides to the order of magnitude of the transfer coefficients and not precise values. From the values in Table 2.5, it can be seen that Cd, Tl and Zn have the highest transfer coefficients and are the most readily taken up and translocated of all the metals considered.

Certain species of plants have been found to accumulate very high concentrations of certain heavy metals and these are referred to as 'hyperaccumulator' species. Baker *et al.* [52] reported that some *Thlaspi* species which are naturally adapted to metal-rich soils in areas of Pb–Zn mineralisation in Europe are hyperaccumulators of Zn, Cd and Pb. These *Thlaspi* species have been found to accumulate concentrations of over 3% Zn, 0.01% Cd and 0.8% Pb. *Alyssum* species which are naturally adapted to serpentine soils, can accumulate over 2% Ni. These hyperaccumulator species therefore have potential for being used for the *in situ* cleaning-up of metal contaminated soils. In preliminary experiments, McGrath *et al.* [53] found that *T. caerulescens* showed an accumulation of Zn which was 150 times that of a non-accumulator species and has the potential for extracting twice as much Zn as would be added in sewage sludge in one season under the existing UK guideline values [53]. *A. tenium* could extract around 45% of the guideline addition of Ni in sludged soils.

2.4.3 Foliar absorption

In addtion to root absorption, plants can also derive significant amounts of some elements through foliar absorption. This is exploited in agriculture as a means of supplying plants with micronutrients, such as Mn and Cu, but can also be a significant route for the entry of atmospheric pollutants, such as Cd, into the food chain [54]. Foliar absorption of solutes depends on the plant species, its nutritional status, the thickness of its cuticle, the age of the leaf, the presence of stomata guard cells, the humidity at the leaf surface and the nature of the solutes [47, 55]. Metal antagonisms, such as between Cu and Zn, can occur in foliar absorption as well as in the root, and the accompanying ions also have an effect [55]. Aerosol-deposited Pb particles do not penetrate the cuticle of higher plants, but tend to adhere to the surface of leaves; they can, however, be absorbed through the cuticle of some bryophytes [56].

2.4.4 Translocation of metals within plants

Once the ions have been absorbed through the roots or leaves and have been transported to the xylem vessels, there is the possibility of movement throughout the whole plant. The rate and extent of movement within plants depends on the metal concerned, the plant organ and the age of the plant. Chaney and Giordano [43] classified Mn, Zn, Cd, B, Mo and Se as elements which were readily translocated to the plant tops; Ni, Co and Cu were intermediate, and Cr, Pb and Hg were translocated to the least extent. Work on xylem sap has shown that Mn may be largely present as the free ion, although in rice 35% of Mn is organically bound; Ni and Zn may exist as anionic complexes and Cr exists as a trioxalate-Cr^{3+} anion [57]. Copper may exist in organic complexes with amino acids, or in other anionic complexed forms [58].

In the leaves, metal ions may be incorporated into proteins or translocated around the plant in the phloem with photosynthates. The relative distribution of heavy

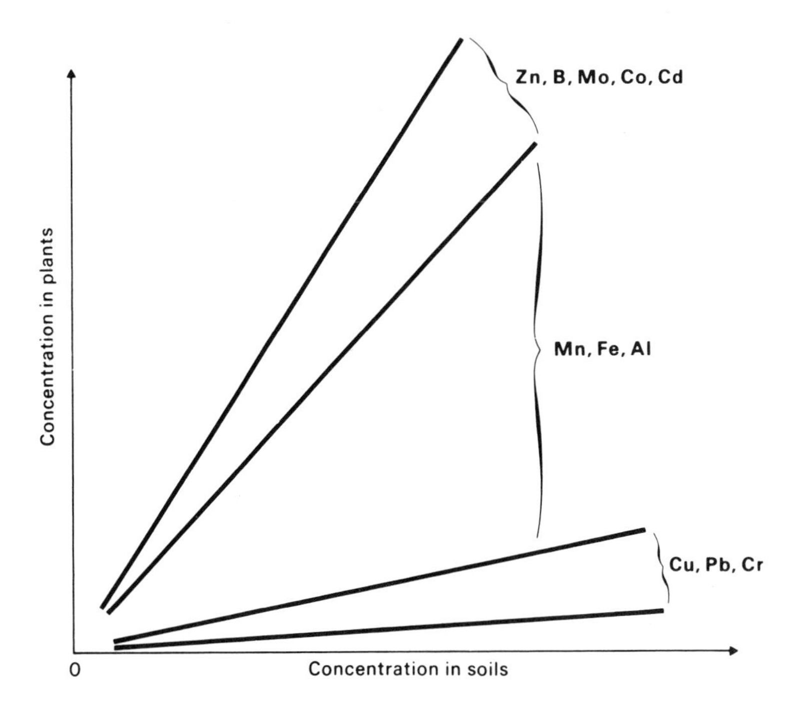

Figure 2.3 Metal uptake into aerial parts of plants as a function of their concentrations in nutrient or soil solutions (after Cottenie *et al.* [76], reproduced by permission of the authors).

metals in plant tops, compared with their concentrations in nutrient or soil solutions is shown in Figure 2.3. Following root absorption, the extent to which elements are translocated decreases in the order Cd > B > Zn > Cu > Pb [44]. It is important to recognise that many species of fungi can accumulate relatively large amounts of metals even from unpolluted forest soils. This is very important in the case of edible species, particularly mushroom, which can accumulate potentially hazardous metals, such as Cd, to significantly high levels, where soils have been significantly contaminated. In many countries, for example Russia and other Central and East European states, mushrooms are collected from forests and preserved in various ways for eating during the winter and could thus comprise a significant contribution to dietary intakes of metals.

2.4.5 Interactions between metals and other elements

Figure 2.4 clearly summarises the known interactions between trace elements within plants and in the soil at the root surface where interactions can affect absorption. Antagonistic and synergistic interactions can also occur between heavy metals and major elements [44].

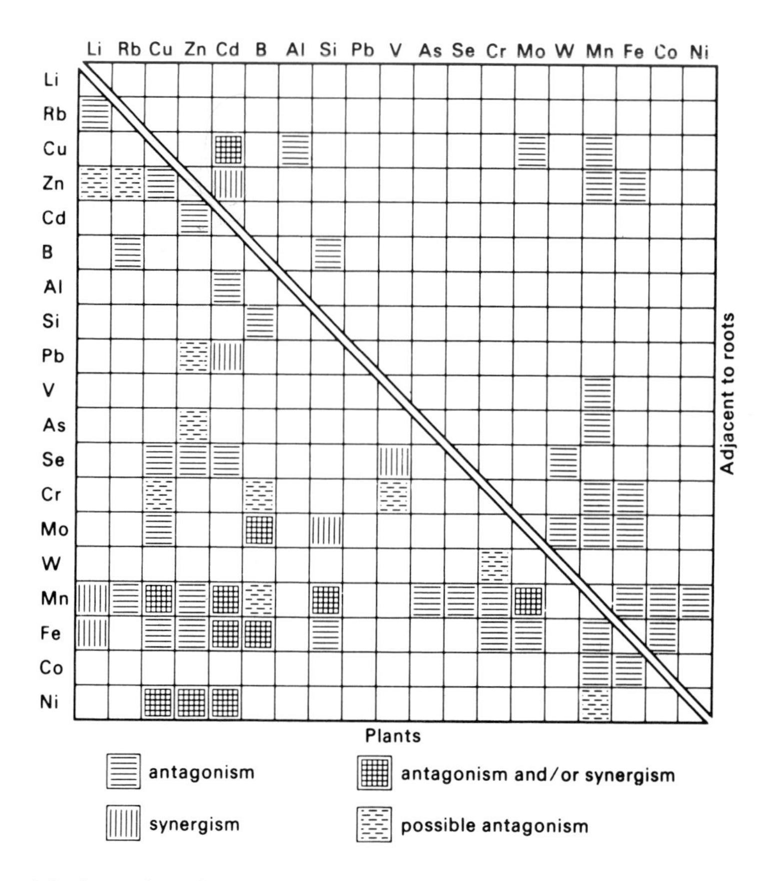

Figure 2.4 Interactions of trace elements within plants and at the root surface (after Kabata-Pendias and Pendias [44], reproduced by permission of the authors and publishers).

2.4.6 The biological essentiality of trace elements

There are three criteria for establishing whether or not a trace element is essential for the normal healthy growth of plants and/or animals:

(i) The organism can neither grow nor complete its life cycle without an adequate supply of the element.
(ii) The element cannot be wholly replaced by any other element.
(iii) The element has a direct influence on the organism and is involved in its metabolism [57].

Apart from C, H, O, N, P, K and S, the elements which have been shown to be essential for plants are:

Al, B, Br (algae), Ca, Cl, Co, Cu, F, Fe, I, K,
Mg, Mn, Mo, Na, Ni, Rb, Si, Ti, V, Zn [44].

However, although these elements fulfil the criteria for essentiality many are unlikely to cause deficiency problems in agricultural crops. The unequivocally essential trace elements which are most likely to give rise to deficiency problems in plants are: B, Cu, Fe, Mn, Mo, and Zn. Cobalt is vitally important for symbiotic bacteria in the roots of legumes. [44, 59].

The elements essential for animals are:

As, Ca, Cl, Cr, Co, Cu, F, Fe, I, K, Mg, Mn, Mo, Na, Ni, Se, Si, Sn, V, and Zn.

In addition, Ba, Cd, Pb, and Sr may be essential at very low concentrations [57].

Several of these elements are essential at very low concentrations and are of little practical importance with regard to deficiencies; Pb is a major toxicity hazard in animals. The unequivocally essential trace elements for animals are Co (in ruminants), Cu, Fe, I, Mg, Mn, Se and Zn [60].

Essential trace elements are frequently referred to as 'micronutrients'. If an organism's supply of a micronutrient is inadequate, its growth is adversely affected. At the other extreme, an excessive supply of a micronutrient will cause toxicity. Typical dose–response curves for micronutrients and for non-essential trace elements are shown in Figure 2.5. From Figure 2.5 (*a*) it can be seen that when the supply of a micronutrient to a plant is inadequate, growth and yield are severely reduced and symptoms of deficiency are manifested. With an increasing supply of micronutrient, the yield reduction becomes progressively lower and the symptoms are less marked. Some micronutrients, such as Cu, give rise to *subclinical* or *hidden* deficiencies where yield reductions of around 20% can occur without the manifestations of obvious symptoms, but not all micronutrients show this

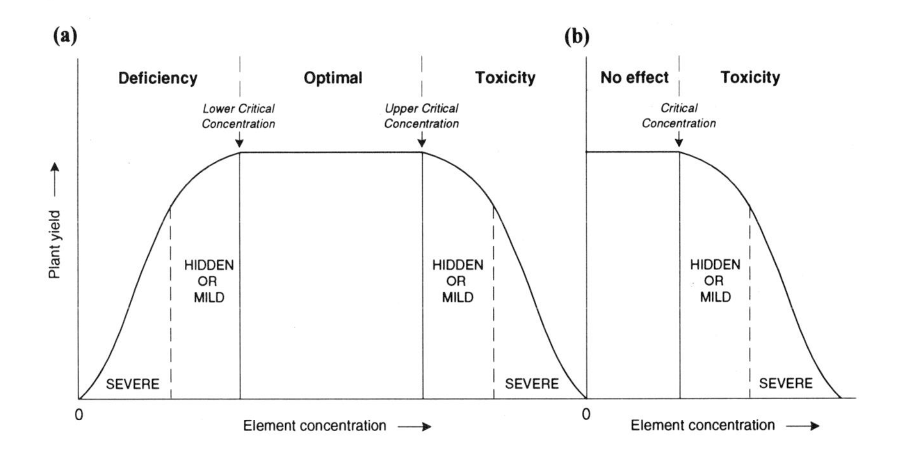

Figure 2.5 Typical dose–response curves for (a) essential trace elements (micronutrients) and (b) non-essential trace elements.

phenomenon. As the supply of the micronutrient increases beyond the lower critical concentration, there is a zone of luxury consumption with no effect on yield. The upper critical concentration heralds the commencement of yield reductions due to toxicity, which becomes more severe until the lethal dose is reached.

The curve for non-essential elements in Figure 2.5 (b) shows that there is no deficiency effect with low concentrations of the element; yield is not affected until the upper concentration limit is reached, after which toxicity occurs in the same way as with an excess of a micronutrient.

2.4.7 *Sensitivity of plants to micronutrient deficiencies*

Plant species and varieties (cultivars) differ greatly in their sensitivity to both deficiencies and toxicities. Table 2.6 shows the differences between species of major crops to deficient supplies of micronutrients. However, some cultivars may differ from this general classification.

2.4.8 *Heavy metal toxicity in plants*

Excessive concentrations of both essential and non-essential metals result in phytotoxicity. Kabata-Pendias and Pendias [44] list the following possible causal mechanisms:

(i) Changes in the permeability of the cell membrane:

Ag, Au, Br, Cd, Cu, F, Hg, I, Pb, UO_2.

(ii) Reactions of sulphydryl (–SH) groups with cations:

Ag, Hg, Pb.

(iii) Competition for sites with essential metabolites:

As, Sb, Se, Te, W, F.

(iv) Affinity for reacting with phosphate groups and active groups of ADP or ATP:

Al, Be, Y, Zr, lanthanides and, possibly, all heavy metals.

(v) Replacement of essential ions (mainly major cations):

Cs, Li, Rb, Se, Sr.

(vi) Occupation of sites for essential groups such as phosphate and nitrate:

arsenate, fluorate, borate, bromate, selenate, tellurate and tungstate.

Although the relative toxicity of different metals to plants can vary with plant genotype and experimental conditions, the metals which, when present in excessive amounts, are the most toxic to higher plants and microorganisms are Hg, Cu, Ni, Pb, Co, Cd and possibly also Ag, Be and Sn [48]. Food plants which tolerate

Table 2.6 Sensitivity of crop plants to deficient supplies of micronutrients (H high, M medium, L low). Adapted from Lucas and Knezek [61] and Shorrocks and Alloway [62]

Crop	Zn	Fe	Mn	Mo	Cu	B
Barley	M	H	M	L	M	L
Corn (maize)	H	M	L	L	M	L
Potato	M	—	M	L	L	L
Rice	M	H	M	L	H	—
Soya bean	M	H	H	M	L	L
Sugar beet	M	H	H	M	M	H
Wheat	L	L	H	L	H	L

relatively high concentrations of these potentially hazardous metals are likely to create a greater health risk than those which are more sensitive and show definite symptoms of toxicity. Logan and Chaney [63] showed that relatively high concentrations of Zn, Cu, Ni and Cd in sludge-amended soils were most toxic to chard, lettuce, carrot, red beet, turnip and peanuts; while maize, sudangrass, smooth bromegrass and 'Merlin' red fescue were very tolerant of that particular combination of metals.

The types of metal tolerance mechanisms in plants include [44].

(i) Selective uptake of ions
(ii) Decreased permeability of membranes or other differences in the structure and function of membranes.
(iii) Immobilisation of ions in roots, foliage and seeds.
(iv) Removal of ions from metabolism by deposition (storage) in fixed and/or insoluble forms.
(v) Alteration in metabolic patterns—increased enzyme system that is inhibited, or increased antagonistic metabolite, or reduced metabolic pathway by-passing an inhibited site.
(vi) Adaptation to toxic metal replacement of a physiological metal in an enzyme.
(vii) Release of ions from plants by leaching from foliage, guttation, leaf shedding and excretion from roots.

Table 2.7 Relative metal accumulations (Cd and Pb in edible portions; Cu, Ni and Zn in leaves). Based on data from Davis and Calton-Smith [65]

Metal	High accumulations	Low accumulations
Cd	Lettuce, spinach, celery, cabbage	Potato, maize, french bean, peas
Pb	Kale, ryegrass, celery	Some barley cvs, potato, maize
Cu	Sugar beet, certain barley cvs	Leek, cabbage, onion
Ni	Sugar beet, ryegrass, mangold, turnip	Maize, leek, barley cv, onion
Zn	Sugar beet, mangold, spinach, beetroot	Potato, leek, tomato, onion

Tolerance is usually specific to one particular metal, although a single plant may possess mechanisms enabling it to tolerate excesses of more than one element. Many species in several plant families have been found in show tolerance to heavy metals. For example, tolerance to excess Cu has been found in species of bacteria, algae, protozoa, fungi, lichens, mosses, liverworts, grasses, herbaceous plants, trees and shrubs [64].

From a study of a wide range of crops on the same sludge-amended soils, Davis and Calton Smith found differences in relative metal accumulation as shown in Table 2.7 [65].

2.4.9 Effects of heavy metals on the soil microbial biomass

Several authors have reported that high concentrations of various heavy metals in soils have had marked inhibitory effects on microbial activity. In Sweden, Tyler [66] reported that the normal decomposition of conifer litter and recycling of plant nutrients was inhibited in the forest surrounding a brass foundry which had emitted large amounts of Cu, Zn and other metals as aerosols over many years. The result of this inhibition of microbial activity was that the growth of trees in the area was retarded due to deficiencies in plant macronutrients. However, other authors, such as Olsen and Thornton [67] have reported that soils from severely metal contaminated sites, such as Shipham in Somerset (see Chapter 6) contained bacteria which showed a marked tolerance to Cd, relative to bacteria in uncontaminated soils. In the Netherlands, Doelman and Haanstra [68] showed that Pb inhibited both microbial respiration and dehydrogenase activity in polluted soils. Although tolerant populations of microorganisms were found in polluted soils, there was a change in the balance of the different types of microorganisms present which could have an impact on soil fertility.

The subject received a lot of attention recently when it was found that soils with a relatively high content of heavy metals from a long term field experiment with sewage sludge had a microbial biomass 50% smaller than in adjacent low metal soils [69]. In particular, it was found that the metals from sewage sludge applied between 1942 and 1961 had a marked inhibitory effect on symbiotic nitrogen fixation in the roots of white clover due to toxicity affecting *Rhizobium leguminosum* bv *trifolii.* Experiments conducted *in vitro* showed a decreasing order of toxicity as being: Cu > Cd > Ni > Zn [70]. Other workers studying *Rhizobia* in more recent field experiments elsewhere in the UK concluded that the low numbers of these bacteria present was probably due to the inhibitory effect of Cd [71]. Investigations on nitrogen fixing bacteria in the USA failed to show the same degree of sensitivity to metal toxicity [72]. Soils contaminated over 90 years by fallout from a Zn smelter were found to have populations of *Rhizobium meliloti* (associated with alfalfa—*Medicago sativa*) which were not correlated with extractable metal concentrations [72]. Studies in Japan on the effects of Cd, Cr, Cu, Ni and Pb on organic decomposition in gley and adosol soils showed that although all the metals inhibited the evolution of CO_2, Cd and Cu had the greatest inhibitory effect and Pb the least [73].

However, although Cu and Cd have been shown by most workers to be the most toxic to soil bacteria, especially *Rhizobia*, it is Zn which constitutes the greatest problem because of its high concentrations in sludges and sludged soils [74]. Although Zn does not appear to be a serious toxicity hazard to crop plants on sludged soils, the fact that numbers of *Rhizobium leguminosum* are significantly reduced in soils within the limits for Zn (300 mg/kg in the UK and EC) is a major cause for concern [74, 75]. The nitrogen fixed by clovers in pastures could be supplied in fertiliser form, but it is an ominous indication of the impact of sludges at 'normal' application rates on soil fertility.

References

1. Bache, B.W., in *The Encyclopaedia of Soil Science*, eds. Fairbridge, R.W. and Finkl, C.W. Dowden, Hutchinson & Ross, Stroudsburg, Pa (1979), 487–492.
2. Wild, A. (ed.), *Russell's Soil Conditions and Plant Growth* 11th edn. Longman, London (1988).
3. Brady, N.C., *The Nature and Properties of Soils*, 9th edn. Collier Macmillan, New York (1984).
4. Chen, Y. and Stevenson, F.J., in *The Role of Organic Matter in Modern Agriculture*, eds. Chen, Y. and Avinmelech, Y. Martinus Nijhoff, Dordrecht (1986), 73–112.
5. Stevenson, F.J., *in Encyclopaedia of Soil Sciences*, eds. Fairbridge, R.W. and Finkl, C.W. Dowden, Hutchinson & Ross, Stroudsburg, Pa (1979), 195–205.
6. Hayes, M.H.B. and Swift, R.S., in *The Chemistry of Soil Constituents*, eds. Greenland D.J. and Hayes, M.H.B. John Wiley, Chichester (1978), 179–320.
7. White, R.E., *Introduction to the Principles and Practice of Soil Science*, 2nd edn. Blackwell, Oxford (1987).
8. Ball, D.F., *J. Soil Sci.* **15** (1964), 84–92.
9. Simpson, K., *Soil.* Longman, London (1983).
10. Finkl, C.W., in *Encyclopaedia of Soil Science*, eds. Fairbridge, R.W. and Finkl, C.W. Dowden, Hutchinson & Ross, Stroudsburg, Pa (1979), 348–349.
11. Sanchez, P.A., *Properties and Management of Soils in the Tropics*. John Wiley, New York (1976).
12. Marshall, C.E., *The Physical Chemistry and Mineralogy of Soils. Vol. 1: Soil Materials.* Robert K. Kreiger Publ., Huntingdon NY (1964).
13. Holmgren, C.G.S., Meyer, M.W., Chaney, R.L. and Daniels, R.B., *J. Environ Qual.* **22** (1993) 335.
14. McGrath, S.P. and Loveland, P.J., *The Soil Geochemical Atlas of England and Wales*, Blackie Academic and Professional, Glasgow (1992).
15. Bailey, S.W., *Crystal Structures of Clay Minerals and their X-ray Identification*, Mineralogical Society, London (1980).
16. Sposito, G., in *Applied Environmental Geochemistry* ed. Thornton, I. Academic Press, London (1983), 123–170.
17. Lindsay, W.L., *Chemical Equilibria in Soils.* John Wiley, New York (1979).
18. Sposito, G. and Page, A.L., in *Metal Ions in Biological Systems Vol. 18 Circulation of Metals in the Environment*, ed. Sigel, H. Marcel Dekker, New York (1985), 287–332.
19. Rowell, D.L., in *The Chemistry of Soil Processes* eds. Greenland, D.J. and Hayes, M.H.B. John Wiley, Chichester (1981), 401–462.
20. Rose, A.W., Hawkes, H.E. and Webb, J.S., *Geochemistry in Mineral Exploration*, 2nd edn. Academic Press, London (1979).
21. O'Neill, P., *Environmental Chemistry*. George Allen & Unwin, London (1985).
22. Brown, G., *J. Soil Sci.* **5** (1954), 145–155.
23. Gast, R.G., in *Encyclopaedia of Soil Science* eds. Fairbridge, R.W. and Finkl, C.W. Dowden, Hutchinson & Ross, Stroudsburg Pa. (1979), 148–152.
24. Wiklander, L., in *Chemistry of the Soil*, 2nd edn. ed. Bear, F.E. Reinhold, New York (1964), 163–205.

25. Talibudeen, O., in *The Chemistry of Soil Processes* eds. Greenland, D.J. and Hayes, M.H.B. John Wiley, Chichester (1981), 115–178.
26. Keeney, D.R., in *Encyclopaedia of Soil Science*, eds. Fairbridge, R.W. and Finkl, C.W. Dowden, Hutchinson & Ross, Stroudsburg, Pa (1979), 8–9.
27. Brummer, G.W., in *The Importance of Chemical Speciation in Environmental Processes* Springer-Verlag, Berlin (1986), 169–192.
28. Papodopoulos, P. and Rowell, D.K., *J. Soil Sci.* **39** (1988), 23–36.
29. Bittel, J.E. and Miller, R.J., *J. Environ. Qual.* **3** (1974), 243–244.
30. Pulls, R.W. and Bohn, H.L., *Soil Sci. Soc. Am. J.* **52** (1988), 1289–1292.
31. Stuanes, A., *Acta Agric. Scand.* **26** (1976), 243–250.
32. Kinniburgh, D.G., Jackson, M.L. and Syers, J.K., *Soil Sci. Soc. Am. J.* **40** (176), 769–799
33. Mackenzie, R.M., *Aust. J. Soil Res.* **18** (1980), 61–73.
34. Forbes, E.A., Posner, A.M. and Quirk, J.P., *J. Soil Sci.* **27** (1976), 154–166.
35. Harmsen, K., *Behaviour of Heavy Metals in Soils.* Centre for Agricultural Publishing and Documentation, Wageningen (1977).
36. Bunzl, K., Schmidt, W. and Sansomi, B., *J. Soil Sci.* **27** (1976), 32–41.
37. Phipps, D.A., in *Effect of Heavy Pollution on Plants* Vol. 1 ed. Lepp, N.W. Applied Science Publishers, London (1981), 1–54.
38. De Haan, F.A.M. and Zweman, P.J., in *Soil Chemistry: A Basic Element* eds. Bolt, G.J. and Bruggenwirt, M.G.M. Elsvier, Amsterdam (1976), 192–271.
39. Allen, H.E., *Sci. Total Environ.* Supplement (1993), 23.
40. Lee, J., Chen, B., Allen, H.E., Huang, C.P., Sparks, D.L. and Sanders, P., in *Proc. of 24th Mid-Atlantic Industrial Waste Conference*, Reed, B.E. and Sack, W.A. ed.
41. Fergusson, J.E. *The Heavy Elements: Chemistry, Environmental Impact and Health Effects*, Pergammon Press, Oxford (1990).
42. Peterson, P.J. and Alloway, B.J., in *The Chemistry, Biochemsitry and Biology of Cadmium* ed. Webb, M. Elsvier/North-Holland. Amsterdam (1979), 45–92.
43. Chaney, R.L. and Giordano, P.M., in *Soils for the Management of Organic Wastes and Waste Waters* eds. Elliot, L.F. and Stevenson, F.J., Soil Sci. Soc. Am., Am. Soc. Agron & Crop Sci. Soc. Am., Madison (1977), 235–279.
44. Kabata Penias, A. and Pendia, H., *Trace Elements in Soils and Plants* (2nd edition) CRC Press, Boca Raton, Fla. (1992).
45. Barber, S., *Soil Nutrient Bioavailability—A Mechanistic Approach.* John Wiley, New York (1984).
46. Graham, R.D., in *Copper in Soils and Plants* eds. Loneragan, J.F., Robson, A.D. and Graham, R.D. Academic Press, Sydney (1981), 141–163.
47. Marschner, H., *Mineral Nutrition in Higher Plants.* Academic Press, London (1986).
48. Mench, M. and Martin, E. *Plant & Soil* **132** (1991) 187.
49. Page, A.L. and Chang, A.C., in *Proceedings 5th National Conf. on Acceptable Sludge Disposal Techniques* Information Transfer Inc., Rockville Md. (1978), 91–96.
50. De Vries, M.P.C. and Tiller, K.FG., *Environ. Pollut.* **16** (1978), 231–240.
51. Kloke, A. Sauerbeck, D.R. and Vetter, H. In *Changing Metal Cycles and Human Health*, Nriagu, J. ed, Springer-Veriag, Berlin (1994) 113.
52. Baker, A.J.M., McGrath, S.P., Sidoli, C.M.D., Reeves, R.D. *Resources, Conservation and Recycling* (1994) (in the Press).
53. McGrath, S.P., Sidoli, C.M.D., Baker, A.J.M., and Reeves, R.D., in *Integrated Soil and Sediment Research: A Basis for Proper Protection.* Eijackers, H.J.P. and Hamers, T. ed, Kluwer Academic Publishers, Dordrecht (1993) 673.
54. Hovmand, M.F., Tjell, J.C. and Mossbaek, H., *Environ. Pollut* (Series A) **30** (1983), 27–38.
55. Chamel, A., in *Foliar Fertilization* ed. Alexader, A. Martinus Nijhoff, Dordrecht (1986), 66–86.
56. Zimdahl, R.L. and Koeppe, D.E., in *Lead in the Environment* ed. Boggess, W.R. National Science Foundation, Washington D.C. (1977), 99–104.
57. Bowen, H.J.M., *Environmental Chemistry of the Elements.* Academic Press, London (1979).
58. Loneragan, J.F., in *Copper in Soils and Plants* eds. Loneragan, J.F., Robson, A.D. and Graham, R.D. Academic Press, Sydney (1979), 165–188.
59. Price, C.A., Clark, H.E. and Funkhauser, E.A., in *Micronutrients in Agriculture* eds. Mortvedt, J.J., Giordano, P.M. and Lindsay, W.L. Soil Sci. Soc. Am., Madison Wisc. (1972), 231–242.

60. Scott, M.L., in *Micronutrients in Agriculture* eds. Mortvedt, J.J. Giordano, P.M. and Lindsay, W.L. Soil Sci. Soc. Am., Madison Wisc, (1972) 265–288.
61. Lucas, R.E. an Knezek, B.D., in *Micronutrients in Agriculture* eds. Mortvedt, J.J. Giordano, P.M. and Lindsay, W.L. Soil Sci. Soc. Am., Madison Wisc, (1972) 265–288.
62. Shorrocks, V.M. and Alloway, B.J., *Copper in Plant, Animal and Human Nutrition* Copper Development Association, Potters Bar, UK (1986).
63. Logan, T.J. and Chaney, R.L., in *Utilization of Municipal Wastewater and Sludge on Land* eds. Page, A.L. Gleason, T.l., Smith, J., Iskandar, I.K and Sommers, L.E. Univ. California Press, Riverside, Ca (1984), 165–267.
64. Lepp, N.W., in *Effect of Heavy Metal Pollution on Plants Vol. 1* Applied Science Publishers, London (1981), 111–143.
65. Davis, R.D. and Calton-Smith C., *Crop as Indicators of the Significance of Contamination of Soil by Heavy Metals* Tech. Rept. 140 Water Research Centre, Stevenage UK (1980).
66. Tyler, G., Balsberg, M.A., Bengtsson, G., Baath, E. and Trannk, L. *Water Air, Soil Pollut.* **47** (1989) 189–215.
67. Olsen, B.H. and Thornton, I.T. *J. Soil Sci.* **33**, 271.
68. Doelman, P. and Haanstra, L., *Soil Biol. Biochem.* **11** (1979), 475.
69. Brooks, P.C. ad McGrath, S.P., *J. Soil Sci.* **35** (1984), 341.
70. Chaudri, A.M., McGrath, S.P. and Giller, K.E., *Soil Biol. Biochem.* **24** (1992) 625.
71. Obbard, J.P and Jones, K.C., *Environ, Pollut* **79** (1993), 105.
72. Angle, J.S. and Chaney, R.L., *Water, Air and Soil Pollut.* **57–58** (1991), 597.
73. Hattori, H., *Soil Sci and Plant Nutr.* **38** (1992) 93.
74. Chaudri, A.M., McGrath, S.P, Giller, K.E., Reitz, E. and Sauerbeck, D.R., *Soil Biol. Biochem.* **25** (1993), 301.
75. Ministry of Agriculture, Fisheries and Food, *Review of the Rules for Sewage Sludge Application to Agricultural Land: Soil Fertility Aspects of Potentially Toxic Elements*, MAFF Publications, No. PB 1561, London (1993).
76. Cottenie, A., Velghe, G., Verloo, M., and Kiekens, L., *Biological & Analytical Aspects of Soil Pollution*, Laboratory of Analytical & Agrochemistry, State University of Ghent, Belgium (1982).

3 The origins of heavy metals in soils

B.J. ALLOWAY

3.1 Geochemical origins of heavy metals

Ten 'major elements', O, Si, Al, Fe, Ca, Na, K, Mg, Ti and P, constitute over 99% of the total element content of the earth's crust. The remainder of the elements in the periodic table are called 'trace' elements' and their individual concentrations do not normally exceed 1000 mg/kg (0.1%), in fact most have average concentrations of less than 100 mg/kg [1]. However, ore minerals containing high concentrations of one or more heavy metals also occur and these constitute the main commercial sources of the particular metals. The important ore minerals of the metals discussed in this book are mentioned in the relevant chapters.

Trace elements occur as trace constituents of primary minerals in igneous rocks (which crystallize from molten magma). They become incorporated into these minerals by isomorphously substituting in the crystal lattice for ions of one of the major elements at the time of crystallisation. This substitution is governed by the ionic charge, ionic radius and electronegativity of the major element and of the trace element replacing it. Substitution can occur when the radii of the major constituent ion and that of the trace metal are within 15% of each other and when the charges differ by not more than one unit [2]. The trace constituents of the common rock-forming primary minerals are given in Table 3.1.

Sedimentary rocks comprise approximately 75% of the rocks outcropping at the earth's surface and are therefore more important than igneous rocks as soil parent materials. They are formed by the lithification of sediments comprising rock fragments or resistant primary minerals, secondary minerals such as clays, or chemical precipitates such as $CaCO_3$. The trace element concentrations in sedimentary rocks are dependent upon the mineralogy and adsorptive properties of the sedimentary material, the matrix and the concentrations of metals in the water in which the sediments were deposited. In general, clays and shales tend to have relatively high concentrations of many elements due to their ability to adsorb metal ions. Black (or bituminous) shales contain high concentrations of several metals and metalloids, including Ag, As, Cd, Cu, Pb, Mo, U, V and Zn [3]. The sediments from which they are formed acts both as an adsorbent for heavy metals and also as a subsubstrate for microorganisms. The latter catalyse the development of reducing conditions which lead to further heavy metal accumulation through the precipitation of metal sulphides. Kim and Thornton [4] reported concentrations of 0.4–46 mg/kg Cd, 0.1–992 mg/kg Mo and <0.1–41 mg/kg Se in the Okchon uraniferous black shales in South Korea. The concentrations of these elements in

Table 3.1 Trace constituents of common rock-forming minerals (adapted from Mitchell [1]).

Mineral	Trace constituents	Susceptibility to weathering
Olivine	Ni, Co, Mn, Li, Zn, Cu, Mo	Easily weathered
Hornblende	Ni, Co, Mn, Sc, Li, V, Zn, Cu, Ga	
Augite	Ni, Co, Mn, Sc, Li, V, Zn, Pb, Cu, Ga	
Biotite	Rb, Ba, Ni, Co, Sc, Li, Mn, V, Zn, Cu, Ga	
Apatite	Rare earths, Pb, Sr	
Anorthite	Sr, Cu, Ga, Mn	
Andesine	Sr, Cu, Ga, Mn	
Oligoclase	Cu, Ga	
Albite	Cu, Ga	
Garnet	Mn, Cr, Ga	Moderately stable
Orthoclase	Rb, Ba, Sr, Cu, Ga	
Muscovite	F, Rb, Ba, Sr, Cu, Ga, V	
Titanite	Rare earths, V, Sn	
Ilmenite	Co, Ni, Cr, V	
Magnetite	Zn, Co, Ni, Cr, V	
Tourmaline	Li, F, Ga	
Zircon	Hf, U	
Quartz	—	Very resistant to weathering

Table 3.2 Mean heavy metal contents of major rock types (mg/kg). Adapted from Krauskopf [2], Rose *et al.* [3] and from this volume

	Earth's crust	Igneous rocks			Sedimentary rocks		
		Ultra-mafic*	Mafic*	Granitic	Lime-stone	Sand-stone	Shales*
Ag	0.07	0.06	0.1	0.04	0.12	0.25	0.07
As	1.5	1	1.5	1.5	1	1	13 (1–900)
Au	0.004	0.003	0.003	0.002	0.002	0.003	0.0025
Cd	0.1	0.12	0.13	0.09	0.028	0.05	0.22 (<240)
Co	20	110	35	1	0.1	0.3	19
Cr	100	2980	200	4	11	35	90 (<500)
Cu	50	42	90	13	5.5	30	39 (<300)
Hg	0.05	0.004	0.01	0.08	0.16	0.29	0.18
Mn	950	1040	1500	400	620	460	850
Mo	1.5	0.3	1	2	0.16	0.2	2.6 (<300)
Ni	80	2000	150	0.5	7	9	68 (<300)
Pb	14	14	3	24	5.7	10	23 (<400)
Sb	0.2	0.1	0.2	0.2	0.3	0.005	1.5
Se	0.05	0.13	0.05	0.05	0.03	0.01	0.5 (<675)
Sn	2.2	0.5	1.5	3.5	0.5	0.5	6
Tl	0.6	0.0005	0.08	1.1	0.14	0.36	1.2
U	2.4	0.03	0.43	4.4	2.2	0.45	3.7 (<1250)
V	160	40	250	72	45	20	130 (<2000)
W	1	0.1	0.36	1.5	0.56	1.6	1.9
Zn	75	58	100	52	20	30	120 (<1000)

* Ultramafic rocks are also called 'ultrabasic', e.g. dunite, peridotite and serpentinite, Mafic rocks are also called 'basic igneous rocks', e.g. basalt. 'Shales'also include clays.

the soils associated with outcrops of these shales were: 0.38.3 mg/kg Cd, 0.1–275 mg/kg Mo and <0.1–24 mg/kg Se. Uranium contents of 2.7–91 mg/kg were found in the soils. These shales and their soils had exceptionally high concentrations of Mo and Se which were also reflected in the composition of the crops grown on them. Similar conditions occurred to a certain extent in the early stages of coal formation, which accounts for the significant concentrations of several metals in coal. Sandstones usually contain only low concentrations of most elements because they consist mainly of quartz grains which have no trace constituents and little ability to adsorb metals. Table 3.2 gives the mean concentrations of heavy metals in a range of representative types of igneous and sedimentary rocks.

The heavy metals considered in this book can be classified according to Goldschmidt's geochemical classification of the elements (Table 3.3). Briefly, this is based on the affinity of elements for iron (siderophile elements), those elements that have a high affinity for sulphur and normally occur in sulphide deposits (Chalcophile elements). The remaining elements which occur in air as gases are classed as atmophile elements. The classification is not very widely used nowadays, except for the chalcophile group which include important metals found in ore deposits.

3.2 Pedogenesis and the translocation of heavy metals in soils

Pedogenesis (soil formation) is the process by which a thin surface layer of soil develops on weathered rock material, gradually increases in thickness and undergoes differentiation to from a soil profile. The soil profile comprises distinct layers (called horizons) differing in colour and/or texture and structure, and is the unit of classification of soils [6]. The main soil profile horizons are shown in Figure 3.1.

With regard to the soil–plant relationships of metals in cultivated soils, the properties and composition of the plough layer (Ap horizon) are of predominant importance because this uppermost part of the profile contains the majority of the root mass. The Ap horizon (also called the 'topsoil') comprises the original organic (O) and the organo-mineral (A) horizons, plus, in some cases, the upper part of the underlying E or B horizons. These will have been mixed together by cultivation over many years, amended by manures and fertilisers and also contaminated by

Table 3.3 The metals/metalloids discussed in this book classified according to Goldschmidt's geochemical classification of the elements. The parentheses indicate that an element belongs primarily to another group but is also related to this group. Adapted from Krauskopf [2]

Siderophile	Chalcophile	Lithophile
Co, Ni, Au Mo, (Pb), (As)	Cu, Ag, (Au), Zn Cd, Hg, Pb, As, Sb, Se, Tl (Mo)	V, Cr, Mn, U, (Tl)

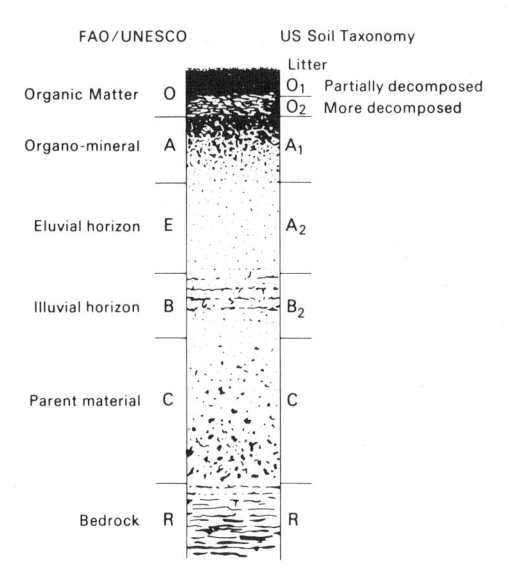

Figure 3.1 A diagrammatic soil profile with FAO/UNESCO and US Soil Taxonomy Horizon nomenclature.

atmospheric deposition from various sources during this time to at least a slight extent.

Within the soil profile, Ag, As, Cd, Cu, Hg, Pb, Sb and Zn are found concentrated in the surface horizons as a result of cycling through vegetation, atmospheric deposition and adsorption by the soil organic matter. The elements found concentrated in the lower horizons of the soil profile include Al, Fe, Ga, Mg, Ni, Sc, Ti, V and Zr, which tend to be associated with accumulations of translocated clays and hydrous oxides [6]. However, recently polluted soils often have higher contents of the pollutant metals in the topsoil because the pedogenic processes have not been operating long enough to effect a redistribution within the profile.

3.2.1 Pedogenesis

Soils form as a result of the interaction of environmental conditions and biological activity on the surface of the weathering rock in the landscape. Jenny [7] expressed the factors of soil formation in an equation:

$$\text{Soil} = f(Cl, O, R, P, T)$$

where f is a function, Cl is climate, O are organisms, R is topography, P is parent material and T is time. These factors control soil formation by determining the intensity of the pedogenic processes operating at any site. The most important aspects of pedogenesis relating to the behaviour of heavy metals in soils are those affecting: (i) the release of metals from the parent material by weathering, and

(ii) the translocation and accumulation of soil constituents which absorb metals, such as clays, hydrous oxides and organic matter.

Several distinct pedogenic processes are responsible for the development of soil horizons and soil profiles. At any site, one of these processes usually predominates and gives rise to the profile characteristics used in the classification of soils (Appendix 5). Weathering, which is the physical disintegration and chemical decomposition of minerals, is an important process in the formation of all soils. However, it is not peculiar to soils since it also occurs in rocks not in direct contact with the biosphere.

The weathering of minerals in the soil parent material involves hydrolysis, hydration, dissolution, oxidation and reduction, ion-exchange and carbonation. These reactions are dependent on water, both for the reaction to proceed, and also for the products to be removed. The rates of the reactions are directly related to temperature, hence rates are high in the humid tropics and low in cold and/or dry environments. Variations in the susceptibilty of minerals and their particle size also affect the rate of weathering. The presence of organic matter and biomass in the soil results in biochemical weathering occurring contemporaneously with the inorganic reactions [7].

The other pedogenic processes which are important in relation to the behaviour of metals are those involving the translocation of soil constituents. These are the processes of leaching, eluviation, salinisation, calcification, podzolisation and ferralitisation together with gleying, which is the development of reducing conditions as a result of waterlogging, and organic matter accumulation [8, 9].

3.3 Sources of heavy metal contaminants in soils

Although heavy metals are ubiquitous in soil parent materials, the major anthropogenic source of metals to soils and the environment are:

a) Metalliferous mining and smelting
b) Agricultural and horticultural materials
c) Sewage sludges
d) Fossil fuel combustion
e) Metallurgical industries — manufacture, use and disposal of metal commodities
f) Electronics — manufacture, use and disposal of electronic commodities
g) Chemical and other manufacturing industries
h) Waste disposal
i) Sports shooting and fishing
j) Warfare and military training

3.3.1 Metalliferous mining and smelting

Metals utilised in manufacturing are obtained from either the mining of ore bodies in the earth's crust, or the recycling of scrap metal. Ores are naturally occurring concentrations of minerals with a sufficiently high concentration of metals to render them economically worthwhile exploiting. With increasing demand for metals and improvements in mineral extraction technology, ore bodies with progressively lower metal contents are being mined. These lower grade ore bodies are usually larger in extent and require a higher proportion of rock to be mined per tonne of metal extracted and consequently much greater quantities of waste are produced, especially tailings (finely milled fragments of rock and residual particles of ore not removed by the separation process). Modern mineral dressing techniques are generally quite efficient and so the tailings produced from them only contain relatively small concentrations of metals. However, historic mine sites which operated during the nineteenth and early twentieth century have tailings with higher metal contents due to much less efficient ore separation. Therefore these tailings particles, which can be transported by either wind or water, constitute a significant source of metal contamination in soils surrounding the mines and in alluvial soils downstream.

Accidental flooding resulting from the failure of dams in tailings lagoons has been responsible for many severe pollution events in several countries. Once in the soil, the ore mineral fragments undergo oxidation and other weathering reactions which result in the metal ions becoming distributed within the soil system and hence more potentially bioavailable. Some of the common ore minerals of non-ferrous metals are shown in Table 3.4.

Table 3.4 Common ore minerals of non-ferrous metals

Metal	Ore minerals	Associated heavy metals
Ag	Ag_2S, PbS	Au, Cu, Sb, Zn, Pb, Se, Te
As	FeAsS, AsS, Cu ores	Au, Ag, Sb, Hg, U, Bi, Mo, Sn, Cu
Au	Au*, $AuTe_2$, $(Au, Ag)Te_2$	Te, Ag, As, Sb, Hg, Se
Ba	$BaSO_4$	Pb, Zn
Cd	ZnS	Zn, Pb, Cu
Cr	$FeCr_2O_4$	Ni, Co
Cu	$CuFeS_2$, Cu_5FeS_4, Cu_2S, Cu_3AsS_4, CuS, Cu*	Zn, Cd, Pb, As, Se, Sb, Ni, Pt, Mo, Au, Te
Hg	HgS, Hg*, Zn ores	Sb, Se, Te, Ag, Zn, Pb
Mn	MnO_2	Various (e.g. Fe, Co, Ni, Zn, Pb)
Mo	MoS_2	Cu, Re, W, Sn
Ni	$(Ni, Fe)_9S_8$, NiAs, $(Co, Ni)_3S_4$	Co, Cr, As, Pt, Se, Te
Pb	PbS	Ag, Zn, Cu, Cd, Sb, Tl Se, Te
Pt	Pt*, $PtAs_2$	Ni, Cu, Cr
Sb	Sb_2S_3, Ag_3SbS_3	Ag, Au, Hg, As
Se	Cu ores	As, Sb, Cu, Ag, Au
Sn	SnO_2, $Cu_2(Fe, Zn)SnS_4$	Nb, Ta, W, Rb
U	U_3O_8	V, As, Mo, Se, Pb, Cu, Co, Ag
V	C_2O_5, VS_4	U
W	WO_3, $CaWO_4$	Mo, Sn, Nb
Zn	ZnS	Cd, Cu, Pb, As, Se, Sb, Ag, Au, In

* Native metal deposits
(Information from refs [3] and [9])

Many of the non-ferrous metalliferous ores are sulphide minerals whose oxidation in soils will produce SO_4^{2-} and lead to increased acidification, unless soils have a high buffering capacity. Large amounts of SO_2 are emitted during the smelting of these sulphides and this can cause severe acid deposition in large areas downwind of smelters if there is inadequate control of the gaseous and particulate emissions. This acid deposition can kill vegetation and cause severe soil erosion as well as widespread metal pollution and this has occurred in many areas including the Cu–Ni smelter at Sudbury in Ontario, Canada and the Pb–Zn smelter at Zlatna in Romania. In technologically advanced countries, smelter flue gases are normally scrubbed to remove SO_2 but at long established works there is likely to be a legacy of acidification and metal pollution from before the installation of pollution control measures.

It can be seen from Table 3.4 that most ore minerals contain several other metals as minor inclusions and the soils around mines and smelters will have been significantly contaminated by these metals as well the major metals in the ores, for example, although Cd has only been widely used during the last 50 years, Pb–Zn smelters will have been emitting this metal into the surrounding environment because it was present in the ores. Pacyna [10] estimated that primary non-ferrous metal production gave rise to atmospheric emissions of 1630 t Cd/yr. Copper smelters are important sources of As contamination. Specific examples relating to pollution from mining and smelting are given in Chapter 5 (As), 6 (Cd), 8 (Cu), 9 (Pb), 13 (Zn) and 14 (Sb and Sn).

3.3.2 Agricultural and horticultural materials

Agricultural practices constitute very important non-point sources of metals which make significant contributions to their total concentrations in soils in many parts of the world, especially in regions of intensive farming. The main sources are [11]:

- impurities in fertilisers: Cd, Cr, Mo, Pb, U, V, Zn
- sewage sludge: especially Cd, Ni, Cu, Pb, Zn (and many other elements)
- manures from intensive animal production, especially pigs and poultry: Cu, As, Zn
- pesticides: Cu, As, Hg, Pb, Mn, Zn
- refuse derived composts (not widely used in agriculture): Cd, Cu, Ni, Pb, Zn
- dessicants: As
- wood preservatives: As, Cu, Cr
- corrosion of metal objects (galvanised metal roofs and wire fences: Zn, Cd)

It should be noted that not all of these sources relate to current practices and materials.

Most agricultural and horticultural soils in technologically advanced countries are regularly amended with fertilisers and many also receive organic manures (usually based on livestock faeces and urine) and lime. The typical heavy metal

Table 3.5 Typical ranges of heavy metal concentrations in fertilisers, farmyard manure, lime and composts (mg/kg). Based on Kabata-Pendias and Pendias [17] with additional data from refs. [18], [22] and this volume

	Phosphate fertilisers	Nitrate fertilisers	Farmyard manure	Lime	Composted refuse*
Ag	—	—	—	—	—
As	2–1200	2.2–120	3–25	0.1–25	2–52
B	5–115	—	0.3–0.6	10	—
Cd	0.1–170	0.05–8.5	0.1–0.8	0.04–0.1	0.01–100
Co	1–12	5.4–12	0.3–24	0.4–3	—
Cr	66–245	3.2–19	1.1–55	10–15	1.8–410
Cu	1–300	—	2–172	2–125	13–3580
Hg	0.01–1.2	0.3–2.9	0.01–0.36	0.05	0.09–21
Mn	40–2000	—	30–969	40–1200	—
Mo	0.1–60	1–7	0.05–3	0.1–15	—
Ni	7–38	7–34	2.1–30	10–20	0.9–279
Pb	7–225	2–27	1.1–27	20–1250	1.3–2240
Sb	<100	—	—	—	—
Se	0.5	—	2.4	0.08–0.1	—
U	30–300	—	—	—	—
V	2–1600	—	—	20	—
Zn	50–1450	1–42	15–566	10–450	82–5894

* Compost data supplied by Warren Springs Laboratory, UK.

concentrations found in these materials are given in Table 3.5 where it can be seen that phosphatic fertilisers and refuse-derived composts can be important sources of heavy metals. Phosphatic fertilisers have been important sources of Cd in soils remote from industrial pollution sources, some such as superphosphate, can have a considerable acidifying effect on soils and hence facilitate the mobilisation and plant uptake of the Cd they introduce to the soil (see Chapter 6).

Some livestock manures can contain high concentrations of certain metals. Pig and poultry manures contain elevated concentrations of Cu and Zn which are fed to improve food conversion efficiency; in the past As was also used for this purpose. In the UK in the mid-1980s the range of Cu in pig manure slurry was 300–2000 mg/kg (mean 870 mg/kg Cu) and the Zn content was 200–1500 mg/kg (mean 600 mg/kg Zn) [12]. The present day concentrations may vary from this but slurries with these values will have been widely applied to soils over many years. In view of the recent concern about the toxicity of Zn to soil microorganisms, livestock manures can be an equally important source as sewage sludge.

3.3.3 Sewage sludges

Sewage sludge is the residue produced from the treatment of domestic and industrial waste waters and large amounts are produced worldwide. In the early 1990s the UK produced 1.1 million tonnes dry sludge solids per year (the USA produced 5.4 million tonnes, West Germany 2.5, France 0.7, the Netherlands 0.28 and Switzerland 0.215 million t) and 6.3 million tonnes in the whole of the European Community.

Table 3.6 The ranges of metal concentrations in sewage sludges reported in the literature and maximum permitted concentrations in the European Union and the USA

Metal	Minimum	Maximum	Maximum permitted concentrations	
			EU	USA
Ag	1	960		
As	3	30		
Cd	<1	3410	20–40	85
Co	1	260		
Cr	8	40600	600*	3000
Cu	50	8000	1000–1750	4300
Hg	0.1	55	16–25	57
Mn	60	3900		
Mo	1	40		
Ni	6	5300	300–400	420
Pb	29	3600	750–1200	840
Sb	3	44		
Se	1	10		
U	<2	5		
Y	20	400		
Zn	91	49000	2500–4000	7500

Values are given as mg/kg dry solids and are taken from refs. [14, 17, 18] and this volume.
* Cr value provisional

Sludge production in the UK is expected to rise to 2.2 million dry tonnes and to 8 million tonnes in the European Community by the year 2006 [13, 14]. In the UK in 1993, 43% (i.e. 465,000 t) was applied to agricultural land and 30% disposed of at sea. This latter disposal route will have to be phased out by 1998 and consequently more sludge will need to be applied to land. At present <1% of arable land receives sludge in any one year but around 10% of all this land has received sludge at some time in the past. In the USA, approximately 22.2% (1.2 million tonnes) of the sludge produced is applied to agricultural land [13].

Sewage sludges are a significant source of plant nutrients and organic matter and some specially treated sludges, such as those containing lime or cement kiln dust, have useful liming properties as well. However, the beneficial properties of sludges are limited by their contents of potentially harmful substances such as heavy metals and organic micropollutants (PAHs, PCBs and pesticides). Although all sludges contain a wide range of metal and other contaminants in varying concentrations, those from industrial catchments generally have higher metal contents than those from mainly suburban domestic areas. Nevertheless, domestic inputs of metals to the sewerage system are still not insignificant, being derived from the corrosion of metal plumbing fittings excretion of metals in the human diet, cosmetics, healthcare products (eg Zn or Se-containing shampoos and Zn-containing baby creams) and other domestic products. It has been estimated that in the UK, 62% of the Cu and 64% of the Zn were from domestic sources [15]. The heavy metals most likely to cause problems for crop production on sludge amended soils are Cd, Cu, Ni and Zn [16]. The ranges of values found in the literature for the concentrations of heavy metals in sewage sludges are given in Table 3.6.

With N and P contents in the dry matter ranging from 0.2–2.2% N and 0.1–3.7% P_2O_5, large amounts of sludges need to be applied in order to supply sufficient amounts of these nutrients for crop requirements. A dressing of 25 t/ha of dewatered sludge is sufficient to maintain an adequate P status for 2 or 3 years [17]. Typical application rates for sludges to arable or pasture land in the UK are 3–5 dry t/ha for liquid sludges and 5–8 dry t/ha for dewatered sludges [18].

Owing to the relatively high concentations of heavy metals in sludges, they are usually the major source of heavy metals in the soils to which they are applied. However, atmospheric and other sources of metals pollution will also contribute to total metal inputs into the soil and crops at sludged sites. Several workers have shown that the transfer of heavy metals from sewage sludge amended soils to crops is significantly lower than those from inorganic sources, such as metal salts or mining wastes [19].

In recent years, the concentrations of most metals in sludges in many technologically advanced countries have shown a pronounced decrease as a result of improved effluent control and waste minimisation; this is illustrated by the comparative data for sludges in the UK in the years 1982/83 and 1990/91 shown in Table 3.7 [20]. These data emphasise the need to use recent analytical data in predictions of soil metal loading from sludges. However, in view of the very long residence periods of heavy metals in most soils (centuries and even millenia), it is still important to be aware of the concentrations of metals in sludges applied to land in the past because they will have contributed to the total concentrations. The metal concentrations in UK soils to which sewage was applied in 1990/91 are given in Table 3.8. This table shows that even the 90th percentile values were well within the regulation limits and also shows that the 28% of the land to which sludge applications were made had pH values of 6 or less. With the exception of

Table 3.7 Comparison of metal concentrations in sludges used on agricultural land in the United Kingdom in 1982/83 and 1990/91 (mg/kg dry solids)

Metal	Ten percentile		Fifty percentile		Ninety percentile	
	Year		Year		Year	
	82/83	90/91	82/83	90/91	82/83	90/91
Cadmium	4.0	1.5	9.0	3.2	33	12
Copper	261	215	625	473	1087	974
Chromium	25	27	124	86	696	489
Lead	164	70	418	217	761	585
Mercury	<2	1.1	3.0	3.2	7.0	6.1
Nickel	21	15	59	37	303	225
Zinc	643	454	1205	889	2058	1471

From DoE (1993) [20]

Table 3.8 Metal concentrations in agricultural soils to which sludge was applied in the United Kingdom in 1990/91 (mg/kg dry soil)

Metal	pH class*	10 percentile	50 percentile	90 percentile	UK regulation limits
Zn	a	23	57	109	200
	b	23	61	121	250
	c	29	67	124	300
	d	40	74	121	450
Cu	a	7	17	37	80
	b	7	19	38	100
	c	8	18	38	135
	d	8	17	34	200
Ni	a	5	14	29	50
	b	5	16	36	60
	c	6	17	33	75
	d	8	20	35	110
Cd	e	0.06	0.55	1.3	3
Pb	e	17	33	78	300
Hg	e	0.06	0.13	0.38	1
Cr	e	11	33	53	400

* pH ranges:
a = pH 5.5–5.0 (land area 3920 ha)
b = pH 5.5–6.0 (land area 7840 ha)
c = pH 6.0–7.0 (land area 27440 ha)
d = pH >7.0 (land area 16800 ha)
e = pH >5.0 (land area 56000 ha)
From MAFF (1993) [24].

Mo, all other metals will be more bioavailable in these acid soils than in those of pH 7 or above.

Sites, such as old sewage farms, which received heavy applications of sludge over many years (100 years in some cases) can be expected to have very high levels of several heavy metals. As a result of the relatively high concentrations of heavy metals in sewage sludges, they are the major source of metals in the soils to which they are applied. However, owing to the comparatively small percentage of agricultural land receiving sludge, they remain a less significant source of some metals, for example Cd, than atmospheric deposition from industrial sources and fertilisers at least in Europe [21].

The limiting metal concentrations for sludged soils in the European Union and the United States are given in Appendices 3 and 4. McGrath *et al.* have reviewed and discussed these values in detail [14].

3.3.4 Fossil fuel combustion

In general, fossil fuel combustion results in the dispersion of a wide range of heavy metals, which can include: Pb, Cd, Cr, Zn, As, Sb, Se, Ba, Cu, Mn, U and V, over a very large area, although not all these elements are present in significant concentrations in all types of coal and petroleum. The metals accumulate in the coal and petroleum deposits as they formed and are either emitted into the environment as airborne particles during combustion, or accumulate in the ash which may itself be transported and contaminate soils or waters, or, may be leached *in situ*. The combustion of petrol (gasoline) containing Pb additives has been the largest source of this metal in the environment and has affected soils over a high proportion of the earth's terrestrial surface. Leaded petrol was first used in 1923 (in the USA) and consumption of Pb in this form rose to more than 375,000 t/yr worldwide [25]. The introduction of Pb-free petrol started in Japan in 1972 and in the USA in 1975. European countries introduced it over the period 1986–1989 and within the European Community, the maximum Pb content of all petrol was fixed as 0.15 g/l [25]. Pb is emitted in the exhausts of vehicles running on Pb-containing petrol as aerosol particles 0.01–0.1 μm in diameter, but these primary particles can cluster to form larger particles (0.3–1 μm). These particles comprise mainly PbBrCl but can react with other air pollutants to form more complex compounds such as a-2PbBrCl.NH_4Cl. The Pb containing particles in motor vehicle exhausts tend to be larger in rural areas and near motorways than in urban areas [26]. In the UK, the Pb content of petrol has varied between 0.14 g/l (level since 1986) and 0.64 g/l (in 1966). The amounts of Pb emitted annually in the UK (c.3000 t/yr) are now lower than at any time since before 1955 but rose to a maximum of 8700 t/yr in 1973 (when the Pb content of petrol was 0.53 g/l) [27]. Using archived samples of herbage from a long-term field experiment at Rothamsted Experimental Station north of London, UK, Jones *et al.* [27] showed that the Pb content of herbage had declined over the period 1966–1988 indicating a decrease in atmospheric deposition of 1.7 g Pb/ha/yr.

3.3.5 *Metallurgical industries*

Metallurgical industries can contribute to soil pollution in several ways: (i) by the emission of aerosols and dusts which are transported in air and eventually deposited on soils or vegetation; (ii) by liquid effluents which may pollute soils at times of flooding; (iii) by the creation of waste dumps in which metals become corroded and leached into the underlying soil. Many heavy metals are used in specialist alloys and steels — V, Mn, Pb, W, Mo, Cr, Co, Ni, Cu, Zn, Sn, Si, Ti, Te, Ir, Ge, Tl, Sb, In, Cd, Be, Bi, Li, As, Ag, Sb, Pr, Os, Nb, Nd and Gd [11]. Hence both the manufacture of these materials, their fabrication into products such as machines and vehicles and their disposal, or recycling in scrap metal can lead to environmental pollution by a wide range of metals. Steel manufacture usually involves a lot of recycling of scrap and so steel works are often discrete point sources of atmospheric aerosols of many different metals.

3.3.6 *Electronics*

A large number of heavy metals are used in the manufacture of semi-conductors, cables, contacts and other electrical components, these include: Cu, Zn, Au, Ag, Pb, Sn, Y, W, Cr, Se, Sm, Ir, In, Ga, Ge, Re, Sn, Tb, Co, Mo, Hg, Sb, As and Gd [11]. Environmental pollution can occur from the manufacture of the components, their accidental contact with soils and their disposal in waste. In addition to metals, old electronic equipment may often also include capacitors and transformers containing polychlorinated biphenyls (PCBs), which are persistent organic pollutants in soils.

3.3.7 Chemical and other industrial sources

Other significant sources of heavy metal pollution of soils and the environment can be either the manufacture and/or use and disposal of the following [11]:

Chlorine manufacture, Hg
Batteries, Pb, Sb, Zn, Cd, Ni, Hg
Pigments and paints, Pb, Cr, As, Sb, Se, Mo, Cd, Ba, Zn, Co
Catalysts, Pt, Sm, Sb, Ru, Co, Rh, Re, Pd, Os, Ni, Mo
Polymer stabilisers, Cd, Zn, Sn, Pb (pollution from incineration of plastics)
Printing and graphics, Se, Pb, Cd, Zn, Cr,Ba
Medical uses, Ag. As, Ba, Cu, Hg, Sb, Se, Sn Pt, Zn
Additives in fuels and lubricants, Se, Te, Pb, Mo, Li

3.3.8 Waste disposal

The disposal of household, municipal and industrial wastes can lead to soil pollution with heavy metals in various ways. The landfilling of municipal solid waste can lead to several metals including Cd, Cu, Pb, Sn, Zn being dispersed into soil, groundwaters and surface waters in leachates if the landfill is not managed properly. Landfill leaches normally have high concentrations of Cl^- and so many of the metals may be present as chloride complexes which are often more mobile and less readily adsorbed than free metal cations. Incineration of wastes can also lead to the emission of metal aerosols (Cd, Pb) if appropriate pollution control equipment is not installed. Temporary waste stockpiles can cause significant soil contamination which may not be discovered until analysis at a later time when the land is no longer used for this purpose. This has been the cause of metal contamination in several urban allotment gardens in the UK. Bonfires and the burial of metal-containing household wastes in domestic gardens can also lead to significant metal accumulations in soils used for growing food crops.

Composts are usually made from the putrescible fraction of municipal solid wastes and once they have become stabilised after a period of aerobic decomposition and associated heating, they can be used as a soil conditioning material, a peat substitute and a growing medium. The amount of compost produced is still relatively small and in many cases the limitation is the concentration of heavy metals and organic pollutants. The most promising type of compost produced so far appears to be that from garden (or 'yard') waste which comprises chipped tree branches, leaves, grass cuttings and other relatively unpolluted organic material which would be much more expensive to dispose of by landfilling or incineration.

3.3.9 *Sports shooting*

Shot pellets composed of Pb with <2% Sb have been used for many years for game bird and clay pigeon shooting. The latter has grown in popularity and high concentrations of these metals have been found at frequently used clay pigeon shooting grounds where total Pb concentrations may rise to levels of several percent. Although most of the metals will be present in the soil as solid pellets and not in a bioavailable form, some exfoliation occurs during the gradual corrosion of the surface of the pellets and the smaller detached fragments become more rapidly dispersed in the soil. Some shot is coated with Ni and so this metal will be an additional contaminant metal in some soils of areas used for shooting. Machines are being develped to reclaim the pellets from heavily contaminated soils but the bioavailability of the residual metals remaining will be important to monitor and, if necessary, control by chemical means. However, since these clay pigeon grounds are not normally used for agriculture, they are of relatively minor toxicological importance, especially if the mechanical removal of the pellets is developed successfully.

From the soil chemistry point of view, shooting results in the input of Pb, Sb and sometimes Ni into a wide range of soil types differing in physico-chemical properties. Although Pb is generally considered to be strongly adsorbed and have a low soil-plant transfer coefficient, the extent of this sorption is likely to vary between soils. Lead has been banned for use as weights on fishing lines owing to the toxicological hazard to water birds and Pb shot is no longer alllowed for wildfowl shooting in wetlands for similar reasons. Alternatives to Pb for shot, including steel, Mo and Bi alloys are currently being evaluated.

3.3.10 *Warfare and military training*

Just as sport shooting has given rise to high concentrations of metals in some soils, so battlefields will have been significantly contaminated by Pb and various alloyed metals from bullets (spent U from anti-tank shells), Cu and Zn from cartridge and shell cases and a wide range of metals from abandoned/destroyed equipment, vehicles, leaking fuels and lubricants, and burnt-out buildings. Consequently, areas anywhere in the world where battles and skirmishes have been fought (especially those in the First World War) will have been significantly contaminated with metals together with various persistent organic micropollutants. These organic pollutants will probably have decomposed over a few decades but in most soils the heavy metals will persist for centuries [11][38][39].

In addition to battlefields, soils in military training areas (especially shooting ranges), camps, air and naval bases are also likely to have been significantly contaminated with heavy metals. This is proving to be an important problem in the former Warsaw Pact countries of Central and Eastern Europe where there are many abandoned military areas and few funds available for surveys and clean-up operations.

3.4 Heavy metals from atmospheric deposition

The atmosphere is an important transport medium for metals from various sources. Soils are often contaminated for up to hundreds of kilometres away from the site of emission. The metals are usually present in air as aerosol particles with a size range of 5 nm–20 μm, but most are between 0.1 and 10 μm in diameter and have an average residence time of 10–30 days [5].

A high proportion of the metals in more recent dust deposits are of anthropogenic origin. Even before the widespread industrial use of metals, the burning of coal contributed considerable amounts of carbon and metals to aerosols in the northern hemisphere [5]. Typical ranges of concentrations of metals found in the air in remote, highly populated and industrialised regions and in the vicinity of volcanoes are shown in Table 3.9. Concentrations are lowest at the South Pole and high in the more industrialised and urbanised locations of Europe and North America (Tables 3.9 and 3.10). Natural inputs of metals into the environment from volcanoes can be seen to be relatively high, and this is the major source of natural atmospheric deposits of metals. Metals in aerosols can be inhaled by people and animals, but their greatest environmental impact in the long term is through deposition under

Table 3.9 Concentrations of selected elements in air at various locations (ng/m^3). From Bowen [5]

Metal	South Pole	Europe median (range)	North America median (range)	Volcanoes (Hawaii/Etna)
Ag	<0.004	1(0.2–7)	1(<0.04–2.4)	30
As	0.007	16(1.5–53)	15(1.7–40)	5.5–850
Au	0.00004	(0.0001–0.006)	(<0.003–0.3)	8
Cd	<0.015	(0.5–620)	(<1–41)	8–92
Co	0.00005	(0.2–37)	3(0.13–23)	4.5–27
Cr	0.005	25(1–140)	60(1–300)	45–67
Cu	0.036	340(8–4900)	280(5–1100)	200–3000
Hg	—	(<0.009–2.8)	(0.007–38)	18–250
Mn	0.01	43(9–210)	150(6–900)	55–1300
Mo	—	(<0.2–3.2)	(<1–10)	—
Ni	—	25(4–120)	90(<1–120)	330
Pb	0.63	120(55–340)	2700(45–13000)	28–1200
Sb	0.0008	8(0.6–32)	12(0.08–55)	45
Se	0.0056	3(0.15–800)	5(0.06–30)	9–21000
Sn	—	(1.5–800)	(<10–70)	—
Tl	—	0.06	0.22	—
U	—	0.02	<0.5	—
W	0.0015	0.7(0.35–1.5	4(0.03–6)	—
Zn	0.03	1200(13–16000)	500(<10–1700)	1000

Table 3.10 Total deposition of metals in various locations (g/ha/yr). From Cawse [29]

Metal	Non-urban locations				Urban locations		
	UK	Tennessee (USA)	Great Lakes (USA)	Texel (Netherlands)	New York (USA)	Swansea (UK)	Gottingen (Germany)
As	8–55	—	—	—	—	61	—
Cu	98–480	280	64	29	—	360	110
Cr	21–88	44	—	—	—	190053	—
Cd	<100	120	—	2.9	9.1	<200	3.9
Ni	35–110	—	37	—	66	220	—
Pb	160–450	230	120	150	790	620	230
Se	2.2–6.5	—	—	—	—	7.3	—
Zn	490–1200	540	530	400	—	1000	470

gravity or wash-out on to vegetation, soil, rivers, lakes and the sea. Some comparative data for the amounts of metals deposited in rural and urban areas are shown in Table 3.10.

Bowen [5] compared the enrichment factors, '*E*' for total deposition calculated by Cawse [30] and Hamilton [31], relative to Sc and Ti, repectively. Cawse [30] gave the following enrichment factors:

E = <1–5 Al, Ce, Cr, En, Fe, K, La, Mn, Rb.
E = 10–50 Ca, Co, Mg, Mo, Ni, Sb, V.
E = 50–500 Ag, As, Au, Cu, I, In, W, Zn.
E = 500–5000 Br, Cd, Pb, Se.

Hamilton's E values differed for Cd (50–500), Se (5–50) and V (50–500), but otherwise agreed with those of Cawse. The high enrichment factors for Cd, Pb and Se indicate the importance of atmospheric deposition as a source of these metals in soils and plants. However, the proximity to sources of contamination has a major effect on atmospheric composition. Enrichment factors for air in rural Northern Nigeria, remote from industrial activity, were lower than those for a site in Nottinghamshire, in the industrialised English Midlands, by factors of 167 for Pb, 117 for Se and 129 for Zn [18].

Yaaqub *et al.* [19] compared aerosols in the UK, with those in Western and Eastern Europe using ratios of constituents with Zn, which was assumed to be a characteristic anthropogenic element. They found that regional aerosols in the UK were enriched in Pb, Cd, soot C and nitrate relative to Zn and the aerosols in the East European sector were deficient in Pb, Cd, soot C and nitrate relative to Zn. These patterns of enrichment were broadly consistent with emission inventories of metals presented by other authors. Yaaqub *et al.* [19] concluded that particulate concentrations would tend to be lower in polluted air transported over the UK in a westerly direction due to greater dilution by mixing in high winds and washout by rainfall, whereas high concentrations of pollutants are likely to occur in air transported from the east due to high pressure systems over Europe, lack of vertical mixing and lower frequency of washout by rainfall.

In Denmark, Hovmand *et al.* [20] found that Cd deposition on to agricultural land from the atmosphere was between 12 and 26 $\mu m/g^3$ during the growing season. This deposition contributed up to 60% of the Cd content of a test crop and was therefore shown to be an important source of the metal. Nearby, in the southernmost part of Norway, Steinnes [21] found that concentrations of Pb, Cd, Zn, As, Sb and Se in moss, soil humus and surface horizons of soil profiles were around 10 times greater than in central Norway. This was due to deposition of metals from air masses moving into southern Norway from more heavily industrialised parts of Europe. Lead deposition in Norway was between 0.6 and 12.9 $mg/m^2/yr$ and that of Cd was 0.02–1.04 $mg/m^2/yr$ [21].

3.5 Heavy metal concentrations in agricultural soils

The total metal content of a soil is the result of inputs of metals from several sources—parent material, atmospheric deposition, fertilisers, agrichemicals, organic wastes and other inorganic pollutants, minus losses in metals removed in cropmaterial, leaching and volatilisation. This can be expressed in the following form:

$$M_{total} = (M_p + M_a + M_f + M_{ac} + M_{ow} + M_{ip}) - (M_{cr} + M_l)$$

where M are heavy metals, p denotes parent material, a atmospheric deposition, f fertilisers, ac agrichemicals, ow organic wastes, ip other inorganic pollutants, cr crop removal, and l losses by leaching, volatilisation, etc.

Although this covers the total metal contents of soils, the bioavailable concentrations will depend upon the soil chemical factors which control the speciation of metals in the soil and plant factors, as outlined in Chapter 2. Although

Table 3.11 Summarised results for the concentrations of selected heavy metals in topsoils (0–15 cm) from England and Wales* (from McGrath and Loveland [35])

Metal	Minimum	Mean	Median	Maximum	CV%
Ba	11	141	121	2973	94
Cd	<0.2	0.8	0.7	40.9	116
Cr	0.2	41.2	39.3	838	106
Co	0.2	10.6	9.8	322	136
Cu	1.2	23.1	18.1	1508	401
Pb	3.0	74.0	40.0	16338	613
Mn	41	3736	3005	62690	277
Ni	0.8	24.5	22.6	440	104
Zn	5.0	97.1	82.0	3648	316

* Based on 5692 samples collected on a 5 km grid comprising 25 subsamples collected from 20 m × 20 m squares at each grid coordinate. CV% = coefficient of variation (SD/Mean %)

Table 3.12 Summarised results for cadmium, copper, nickel, lead and zinc in agricultural soils of the USA (from Holmgren *et al.* [36])

Metal	Minimum	Mean	Median	Maximum	CV%Δ
Cd	<0.01	0.265	0.20	2.0	95
Cu	<0.6	29.6	18.5	495.0	137
Ni	0.7	23.9	18.2	269.0	118
Pb	7.5	12.3	11.0	135.0	61
Zn	<3.0	56.5	53.0	264.0	66

Based on the analysis of 3045 samples of soils from 307 different soil series at sites with healthy crops remote from obvious sources of contamination

these will have been contaminated to a certain extent by normal agricultural practices they will not have been heavily polluted. Comparative values for various degrees of pollution are given in Appendices 5 and 6.

Recent surveys of heavy metals in soils in the USA and England and Wales have been published and these provide a useful indication of 'normal' concentrations of metals which can be encountered in agricultural soils. The summarised data for the survey of soils in England and Wales reported by McGrath and Loveland [35] is shown in Table 3.11 and those for the survey of soils in the USA reported by Holmgren *et al.* [36] in Table 3.12. It should be emphasised that American samples were not collected on an unbiased grid but from sites which were selected because they were remote from sources of overt metal pollution apart from normal agricultural inputs and background levels of deposition from the atmosphere. In contrast, the soils in the survey in England and Wales were collected on a 5 km grid which included sites near to local point sources of pollution, urban/industrial areas and other sources such as metalliferous mining and soils which had received varying amounts of sewage sludge. The two surveys had different objectives, that of Holmgren *et al.* [36] was to determine the extent to which normal agricultural practices caused accumulations of toxicologically important metals in a range of different types of soil, whereas the English and Welsh survey was intended to assess the overall nutrient and metal status of soils and examine regional trends in distribution.

A comparison of Tables 3.11 and 3.12 clearly shows that Cd, Pb and Zn in the American soils from remote sites are much lower than in those from a wide variety of locations in the more densely populated England and Wales. Cadmium, which is of greatest human toxicological importance, has mean and median values around three times higher in the English and Welsh soils, the median for Pb is four times higher and that of Zn is 1.5 times higher than in the American soils.

The maximum values illustrate the differences in sampling policy. As expected, the maximum values for all metals are much lower in the American soils. The greatest difference is found with maximum Pb content, which is 121 times higher in English/Welsh samples and is due to the inclusion of sites in metalliferous mining areas. In general, it was found that most of the samples which had Pb contents above the EC limiting concentrations (for sludge amended soils) of 300 mg/kg were from Pb–Zn mining areas or from urban/industrial regions. It is interesting to note that all the US agricultural soils have Zn contents below the EC limiting concentration for sludged soils of 300 kg. However, recent field experiments in England and Germany have shown this value to be too high for certain soil microorganisms (see Chapters 2 and 13).

The general conclusion from the American survey is that normal agricultural practices, including the use of fertilisers, lime, manures and sprays, have not resulted in dangerously high accumulations in soils of any of the metals monitored. However, just as in England and Wales, there are also many sites in America where soils have been polluted from a wide range of sources but these were avoided in this survey. Values of <49000 mg/kg Pb have been reported in mining contaminated soils in the USA [37] and in the UK values in excess of 12% Pb have been found in some soils which have been heavily contaminated from other sources (Alloway and Merrington, unpublished data).

Bowen [5] estimated the residence time of Cd in the soil to be in the range 75–380 years. Hg persists for 500–1000 years, and more strongly sorbed elements, including As, Cu, Ni, Pb, Se and Zn, have residence times of 1000–3000 years. However, Iimura *et al.* [38, 39] estimated the first half-lives of some of these metals in the soils to be 15–1100 years for Cd, 310–1500 years for Cu and 740–5900 for Pb. Even though the ranges are wide to allow for differing soil conditions, it is very clear that heavy metal pollution of soils is a very long-term matter.

References

1. Mitchell, R.L., in *Chemistry of the Soil*, 2nd edn., ed. Bear, F.E. Reinhold, New York (1964), 320–268.
2. Krauskopf, K.B., *Introduction to Geochemistry*, McGraw-Hill, New York (1967).
3. Rose, A.W., Hawkes, H.E. and Webb, J.S., *Geochemistry in Mineral Exploration*, 2nd edn. Academic Press, London (1979).
4. Kim, K.W. and Thornton I., *Environ. Geochem. and Health* **15**, (1993), 119–133.
5. Bowen, H.J.M., *Environmental Chemistry of the Elements*. Academic Press, London (1979).
6. Duchaufour, P., *Pedology*. George Allen & Unwin, London 1977.
7. Jenny, H., *The factors of Soil Formation*. McGraw-Hill, New York (1941).
8. Fenwick, I.M. and Knapp, B.J., *Soils, Process and Response*, Duckworth, London (1982).
9. Peters, W.C., *Exploration and Mining Geology*, John Wiley & Sons, New York (1978).
10. Pacyna, J.M., in *Lead, Mercury, Cadmium and Arsenic in the Environment*, eds. Hutchinson, T.C., and Meema, K.M SCOPE 31, John Wiley & Sons, Chichester (1987).
11. MAFF, *Advice on Avoiding Pollution from Manures and other Slurry Wastes*, MAFF Booklet 2200, MAFF Publications, London (1986).
12. Alloway, B.J. and Ayres, D.C., *Chemical Principles of Environmental Pollution*, Blackie Academic and Professional, Glasgow (1993).
13. US Environmental Protection Agency, 40 CFR Parts 257, 403 and 503 [FRL-4203-3] *Standards for the Use and Disposal of Sewage Sludge*, US EPA, Washington DC (1992).
14. McGrath, S.P., Chang, A.C., Page, A.L. and Witter, E., *Environmental Reviews* (1994) (in the press).
15. Critchley, R.F. and Agg, A.R., *Sources and Pathways of Trace Metals in the UK*, WRc Report No. ER 822-M, WRc Medmenham (1986), Marlow, UK.
16. Page, A.L., *Fate and Effects of Trace Elements in Sewage Sludge when Applied to Agricultural*

Lands: A Literature Review. US EPA Report No. EPA-670/2-74-005. National Technical Information Service, Springfield VA.
17. Kabata-Pendias, A. and Pendias, H., *Trace Elements in Soils and Plants*. 2nd edition, CRC Press, Boca Raton, Fla (1992).
18. Webber, M.D., Kloke, A. and Tjell, J.C., in *Processing and Use of Sewage Sludge*, eds. L'Hermite, P. and Ott, J.D. Reidel, Dordecht (1984), 371–386.
19. Alloway, B.J. and Jackson, A.P. *Sci Total Environ.* **100** (1991), 151–176.
20. Department of the Environment, UK Sewage Sludge Survey. Final Report. Consultants in Environmental Sciences Ltd, Gateshead (1993).
21. Hutton, M. *Cadmium in the European Community*, MARC Report No. 2, MARC, London (1982).
22. Andersson, A., *Swedish J. Agric. Res.* 7 (1977), 1–5.
23. Ministry of Agriculture, Fisheries and Food, *The Use of Sewage Sludge on Agricultural Land.* Booklet 2409, HMSO, London (1982).
24. Ministry of Agriculture, Fisheries and Food, *Review of the Rules for Sewage Sludge Application to Agricultural Land. Soil Fertility Aspects of Potentially Toxic Elements*, Report No. PB 1561, MAFF, London (1993).
25. Nriagu, J.O., *Sci. Total Environ.* **92** (1990), 13–28.
26. Fergusson, J.E., *The Heavy Elements, Chemistry, Environmental Impact and Health Effects*, Pergamon Press, Oxford (1990).
27. Jones, K.C., Symon, C. Taylor, P.J.L., Walsh, J. and Johnston, A.E., *Atmos Environ.* **25A** (1991), 361–369.
28. Matthews, P.J., Andrews, D.A. and Critchely, R.F., in *Processing and Use of Sewage Sludge.* eds. L'Hermite, P. and Ott, J.D. Reidel, Dordrecht (1984), 244-258.
29. Cawse, P.A., in *Inorganic Pollution and Agriculture* MAFF, HMSO, London (1978).
30. Cawse, P.A., *UK Atomic Energy Report* AERE-R 8398, HMSO, London (1976).
31. Hamilton, E.I., *Sci Total Environ.* **3** (1974), 8.
32. Yaaqub, R.R., Davies, T.D., Jickells, T.D. and Miller, J.M., *Atmos. Environ.* **25A** (1991), 985–996.
33. Hovmand, M.F., Tjell, J.C. and Mossbaek, H., *Environ. Pollut. A* **30** (1983), 27–38.
34. Steinnes, E., in *Lead, Mercury, Cadmium and Arsenic in the Environment* SCOPE 31 eds. Hutchinson, T.C. and Meema, K.M. John Wiley, Chichester (1987), 107–117.
35. McGrath, S.P. and Loveland, P.J. *The Soil Geochemical Atlas of England and Wales*, Blackie Academic and Professional, Glasgow (1992).
36. Holmgren, C.G.S., Meyer, M.W., Chaney, R.L. and R.B. Daniels, *J. Environ. Qual* **22** (1993), 335–348.
37. Levy, D.B., Barabaryk, K.A., Siemer, E.G. and Sommers, L.E., *J. Environ. Qual.* **21** (1992), 185–195.
38. Iimura, K., Ito, H., Chino, M., Marishita, T. and Hiruta, H., *Proc. Int. Sem. SEFMIA*, Tokyo (1977), 357–366.
39. Kabata-Pendias, A., *Trans. XIII Cong. Int. Soc. Soil Sci. Hamburg* I.S.S.S., Hamburg (1987), 570–580.

4 Methods of analysis for heavy metals in soils

A.M. URE

4.1 Introduction: types of analysis

The determination of heavy metals in soils may be carried out for various reasons. These include, firstly, the measurement of the total elementary content. This provides a base-line knowledge of soil components with respect to which changes in soil composition produced by elution, pollution, plant uptake or agricultural manipulation can be assessed. In addition, the analysis may be designed to assess the availability of elements to agricultural crops and hence the likelihood of their entry into the food chain of animals and man. Soil analysis, therefore, plays a vital role in the prediction of plant uptake and in the prediction and diagnosis of deficiency-related disease in crops and farm animals as well as in the assessment of agricultural and environmental toxicity problems.

The extent of the pollution of a soil by a particular element can be gauged by determining its concentration and comparing this with previous analyses or with unpolluted soil contents. It is often more important to be able to estimate the mobile fraction, the readily soluble fraction, the exchangeable fraction, or the plant-available fraction of the heavy metal content of a soil as a more direct indication of the likelihood of deleterious or toxic effects on the environment. This last type of analysis can be regarded as a kind of speciation in which particular phases, chemical forms or oxidation states of an element are identified and their amounts quantified.

Techniques of analysis can be broadly categorised into (i) single-element methods, such as atomic absorption spectrometry, (AAS), or (ii) simultaneous multi-element methods, such as inductively coupled plasma–atomic emission spectrometry (ICP–AES), or X-ray fluorescence spectrometry (XRFS). They can be further categorised into methods such as AAS, which carry out the analysis in solution, i.e. with dissolved samples, or methods, such as XRFS, which analyse solid samples more or less directly. The choice of method for a particular application should take account of those factors as well as their sensitivity, precision and accuracy. The choice of method may well, however, be dominated by the relative cost of the (mainly instrumental) techniques currently in use.

Most of these analytical techniques can be used in manually operated batch systems or with automated sample handling techniques in commercial autoanalysers based on the work of Skeggs [1] or using flow injection procedures following the work of Ruzicka [2]. These automatic methods, as applied to soil analysis, are discussed in some detail by Smith and Scott [3] and will not figure largely in this text.

Detection limits for several different methods of analysis, categorised as solution

Table 4.1 Comparison of detection limits for solid samples in various analytical techniques, viz. flame atomic absorption (FAAS), graphite furnace atomic absorption (GFAAS), inductively coupled with plasma optical emission (ICPAES), inductively coupled plasma mass spectometry (ICPMS), X-ray fluorescence (XRFS) and spark source mass spectrometry (SSMS). Average detection limits for Period 4 elements, i.e. Ca, Ti, V, Cr, Mn, Fe, Co, Ni, Cu, Zn, Ga, Ge, As with K replacing Na and Se, Br and Kr omitted. (Adapted from [4] by permission of The Maculay Land Use Research Institute, Aberdeeen (formerly The Macaulay Institute for Soil Research) and The Royal Society of Chemistry)

	Solution sample techniques				Solid sample techniques	
	FAAS	GFAAS	ICP-AES	ICPMS	XRFS	SSMS
Detection limits in solution (μg/ml)	0.022	1.7×10^{-3}	0.8×10^{-3}	0.01×10^{-3}	—	—
Detection limits in solid (μg/g)					1	0.01
Detection limits in solid (μg/g) analysed in 1% m/v solution	22	0.17	0.08	0.001		
Detection limits in solid (μg/g) analysed in 0.1% m/v solution	22.0	1.7	0.8	0.01		

or solid sample methods, are compared, from a range of heavy metals, in Table 4.1, adapted and expanded from [4]. While comparisons of detection limits are notoriously suspect, these do give some indication of their relative applicability to soil analysis. It also illustrates the fact that for methods requiring solution samples the detection limits quoted will, when applied to solid samples, be poorer by the dilution factor of 100–1000 involved in dissolving them.

4.2 Field sampling

The problem of obtaining representative soil samples [5, 6] is common to all analytical procedures. Standard nomenclature for sampling and sample preparation for analytical atomic spectroscopic and related procedures have been published [7]. Most sampling plans rely on taking a number of sample units from the field or plot and combining them to form a bulk sample representative of that field or plot. In a recommended procedure [6], in wide use, the field is subdivided into smaller sampling areas, perhaps 20 in number, from each of which one sample unit is taken along a zig-zag path through the field. The sample units are taken by a 30–40 mm diameter auger or corer to a uniform depth—usually plough depth, say 15–20 cm, for arable land but less, say 7.5–10 cm, for grassland—and combined to form a bulk sample of about 0.5–1 kg of field-moist soil. For peaty or highly organic soils larger sample may be required. For trace element analysis the sample should be collected in polyethylene bags to minimise sample contamination. (Soils intended for Hg determination should, however, not be stored in polyethylene containers since any elemental Hg, formed for example, by reduction processes in the soil, can escape through polyethylene. This not only invalidates the analysis but can also contaminate other samples stored in the vicinity.)

Sampling tools should be of a material that will not contaminate the sample with elements of interest. In most cases carbon steel, not stainless or high alloy steel, is best. Alumimium may also be used, as well as durable polypropylene or polytetrafluoroethylene (PTFE, Teflon). Tools and apparatus of polyvinylchloride (PVC) should be avoided because of their intrinsic contents of soluble heavy metal (Cd, Pb and Sn) stearates. It is essential that the samples be labelled at this stage with a unique label attached to the outside of the sample bag, not to the inside; the paper label itself may be loaded with fillers and whiteners such as BaO, TiO_2 or ZnO, and marker pen inks may owe their colour to a heavy metal content. Some of these contamination problems are discussed by Scott and Ure [8]. (See also section 4.4).

4.3 Sample drying, storage and subsampling

Soil samples are prepared for drying by breaking down aggregates, spreading the soil on polyethylene sheets in AL trays and drying at 25°C. The dried soil is passed through an Al sieve with a 2 mm mesh. Stones, etc. larger than 2 mm are discarded at this stage. The < 2 mm soil is stored in polyethylene bags or boxes until required.

Generally soils will be analysed on a weighed sample basis but for some purposes, and particularly for rapid routine analysis, a volume basis can be used.

Representative subsampling of the dried soil can be carried out by coning and quartering to the desired weight, or a mechanical riffler may be used. There is a minimum weight of soil, depending on heterogeneity and on grain size, that will represent the whole soil sample. The general criteria expressed by Jackson [9] is that the subsample should contain at least 1000 particles of the sieve opening size, in this case 2 mm. A few examples of minimum sample sizes for different grain sizes are given in Table 4.2 (after Jackson [9]). These values agree well with practical experience, which confirms that for a < 2 mm soil a 10–20 g sample is normally

Table 4.2 Grain size in relation to sample size. Adapted from Jackson [9]

Sieve opening (mm)	Minimum sample wt.* (g)	Optimum sample wt. (g) Approx. 4 × min wt.
4	44	176, say 150
2	5.3	21.2, say 20
1	0.68	2.72, say 2.5
(mm)	(mg)	(mg)
0.16	2.7	10.8, say 10
0.1	0.68	2.72, say 2.5

* Calculated for an assumed soil density of 1.3 g/cm^3

required. An excellent and more rigorous treatment of the sampling of inhomogeneous geochemical materials is given by Ingamells and Pitard [10]. Such samples of the < 2 mm soil are used mainly for the determination of heavy metals in various soil extracts. For the determination of the total contents a finely ground soil is normally required.

Finely ground soil samples are prepared by grinding a representative subsample of about 25 g of < 2 mm soil to a fine powder of < 150 μm in an agate mortar or similar device. Representative subsamples of some 10–50 mg can then be taken for analysis. The required fineness depends to some extent on the analytical method and the weight of sample used, and may range from about 10–20 mg for direct current arc emission spectrometry (DCAES) to 0.1–1 g for XRFS.

4.4 Contamination

Contamination of samples is a constant danger. It can occur at the experimental stage, during sampling and packaging for transport to the laboratory, and in the preparation of the sample for the analysis. Sources of contamination in agricultural laboratories have been discussed in [8], for example, and the general topic has been discussed comprehensively by Zief and Mitchell [11].

It is impossible to avoid all contamination, since some materials have to be used for sampling, handling and treating samples. All that can be done is to choose materials which will not introduce elements important in the particular analysis. For most purposes metal tools should be made of Al or carbon steel. Stainless steel can contain large amounts (up to 30%) of Cr as well as Ni, Mo and Mn, while high-speed steels contain in addition Co, V and W. Grinding equipment employing tungsten carbide should be avoided because of the use of Co in the resin fixative. Commercial equipment such as mills may have to have their metal components replaced by carbon steel or Ti [12]. Plated metals should be avoided as they can be sources of Cr, Ni, Cd, Cu, P and Zn. Galvanised iron is a source of Zn and Cd, while Cu and its alloys are best avoided.

The laboratory itself and the materials used in its construction should be critically considered from the contamination point of view. Paints may contain heavy metal driers as well as metallic (Ba, Ti, Zn, Cd, etc.) pigments. Water-based and emulsion paints may contain Hg compounds as fungicides. The increasing use of plastics for the provision of 'clean' wall and bench surfaces can on occasion be a problem, since PVC and polystyrene can be sources of heavy metals including Cd, Pb and Sn. In general the poly-olefin (polyethylene and polypropylene) plastics are relatively free of these elements, as are the silicone rubbers and polytetrafluoroethylene (Teflon, PTFE). Natural rubber is high in Zn and is a ready source of contamination by this element, as are the paper tissues commonly used in laboratories for wiping pipette tips, etc.

Reagents can themselves introduce contaminants. Together with distilled water, de-ionised water and acids, they must be of high purity and should be checked for blanks. De-ionised water, by its nature, can contain organically complexed heavy metals such as Cu. Acids may require purification by distillation or sub-boiling distillation [13, 14]. Other reagents may need to be purified [15, 16] or selected for low blank levels. It is important that the analyst maintains a record of the batch numbers of reagents, filter paper, etc. so that whenever high blanks are found sources of contamination can be traced back to a particular batch or perhaps to a change in supplier. There are always unsuspected sources of contamination, and continuous monitoring of blank levels and the use of quality control procedures, including the use of reference materials for check analysis, are essential in a trace element laboratory.

4.5 Reference materials

To establish the accuracy of an analytical procedure there are two possible approaches. The first, to compare the results with those from another established procedure, many not be possible because a suitable alternative may not exist or may not be available to the analyst. Furthermore, the check procedure may well have inaccuracies or bias of its own. The second approach is to analyse a reference material whose contents have been established by a variety of analytical procedures and authenticated by the issuing body after careful assessment of the certification analyses. This approach not only serves to validate the analysis but can be used as a routine quality control by regular use.

For verification of analyses of total soil contents the side range of geological Certified Reference Materials (CRMs) available from the US Geological Survey, the US National Bureau of Standards and many others can be used for soil purposes. In addition a number of reference soils and sewage sludges have been issued whose certified contents include many of the heavy metals with which we are here concerned. These soil and sewage sludge reference materials are listed in Appendix 9 with details of the contents of the soils and the issuing authority [17–23]. Reference soils and sludges from the European Community Bureau of Reference (BCR) have also been characterised, although not certified, for a range of aqua regia-soluble heavy metal contents in addition to their certified total element contents. As aqua regia digestion procedures play an increasingly important role in assessing soil pollution, these BCR reference materials are particularly valuable. (See also Section 4.13).

4.6 Methods for total analysis of solid samples

4.6.1 X-ray fluorescence spectrometry (XRF)

X-ray fluorescence spectrometry is a multi-element technique that is capable of detecting most elements (with the exception of those with an atomic number below 8, for which special equipment is necessary). In its wavelength-dispersive form it can determine element concentrations in the range 1μg/g to 100%. Energy-dispersive XRF has been successfully applied to determine the major constituents of soils using computer-aided techniques [24], but its poorer resolution and sensitivity make it less suitable for analysis of minor and trace elements. Wavelength-dispersive XRF is therefore the technique most used in soil analysis. Semiquantitative and qualitative analysis can be carried out directly on the finely ground, dried soil by hydraulically pressing it to form a sample disc whose surface is sealed with boric acid [25].

Since, however, particle size, composition and element form affect the analysis a homogeneous sample is usually prepared for quantitative analysis by fusion with a borate flux. Many of the techniques in common use are derived from that of Norrish and Hutton [26] outlined below.

Fusion procedure for XRFS sample preparation [26]

Fuse 0.28 g ignited soil with 1.5 g flux plus 0.02 g $NaNO_3$.

The flux is made up of 29.6 g $LiCO_3$, 13.2 g lanthanum oxide, fused at 1000°C, quenched and homogenised for use.

This technique has been compared with ICP–AES for soil analysis [39].

A similar method which involves a smaller dilution of the sample has been described [27]. Other fusion procedures include the use of lithimum metaborate [28, 29]. A possible limitation, but not a major one, of these methods and of XRF in general is the relatively large (0.1–1 g) sample size required. For smaller samples recourse to other methods such as the fusion/dissolution/atomic spectrometric method must then be made (see section 4.8). Comprehensive treatments of the theory and practice of XRF as applied to soil analysis has recently been presented by Wilkins [30] and Bain *et al.* [31].

Heavy metal analyses include the determination of Cu, Mn and Zn [32] and Ti and Zr [33–36] at normal soil concentrations. Other heavy metals and metalloids including As, Cr, Ni, Pb and Zn can be determined [32] but the important elements Cd, Hg and Sn cannot normally be determined without preconcentration, and Co and Mo at low soil concentrations are not accessible to XRF. Recent applications have made use of the wavelength-dispersive mode for As [37] and the energy-dispersive mode for B, Cr and V [38].

In conclusion, XRF has a key role in the rapid and very accurate establishment of major element composition and for multi-element surveys of major [37–40] and minor [41–46] elements. As the comparison, in Table 4.3, of detection limits [48] with world mean soil contents [47] show, the sensitivity may be inadequate, however, especially in unpolluted soils. The improved detection limits offered by recently developed total reflection X-ray fluorescence (TXRF) methods have had limited application in soil studies [49] but has considerable potential for soil solution and extract analysis and has been applied successfully to heavy and other metals in sediment digests with detection limits generally < 30 mg/kg [50]. The technique and its applications have recently been reviewed by Klockenkaemper *et al.* [51] and by Reus and Prange [52]. The use of portable EDXRF instruments for the

Table 4.3 Heavy metal determination in soils by X-ray Fluorescence. Data from Jones [48] and Ure and Berrow [47]

Element	Mean soil mg/kg	Detection limit mg/kg	Sample prep.
Ba	568	4	Pressed powder
Cd	0.62	–	Preconc.
Ce	84.2	–	Preconc.
Co	12	<1	Preconc.
Cr	84	<1	Powder/glass
Cs	3	–	Pressed powder
Cu	25.8	<1	Pressed powder
Fe	3.2%	<1	Powder/glass
Ga	21.1	<1	Pressed powder
Ge	3	–	Preconc.
Hg	0.098	–	Preconc.
La	41.2	–	Preconc.
Mn	760	<1	Powder/glass
Mo	1.92	–	Preconc.
Ni	33.7	<1	Pressed powder
Pb	29.2	5	Pressed powder
Rb	120	<1	Pressed powder
Sc	10.1	–	Preconc.
Sn	5.8	–	Preconc.
Sr	278	<1	Pressed powder
Ti	2.18	30	Glass disk
U	5100	1	Pressed powder preconc.
V	108	1	Pressed powder
Y	27.7	<1	Pressed powder
Zn	59.8	<1	Pressed powder
Zr	345	<1	Pressed powder

rapid identification of hot spots on comtaminated land [53] is particularly attractive. Developments in XRF are critically reviewed annually in the Journal of Analytical Atomic Spectrometry (see for example that of Ellis [54]).

4.6.2 Instrumental neutron activation analysis (INAA)

Instrumental neutron activation analysis (INAA) is a multi-element, solid sample technique for the analysis of soils, and plants and biological material, that makes use of the γ-radiation induced by neutron irradiation of the sample. Its considerable potential for heavy metal and other elemental analysis is restricted by the requirement for access to a neutron source, usually a nuclear reactor. While non-reactor sources have been used, generally the lower sensitivity resulting from their lower neutron flux, and their non-isotropic irradiation field, compared to reactor sources, have limited their application. The most common reactor sources in use for INAA are the light-water-moderated 'swimming-pool' reactors.

Sensitive multi-element analysis requires a combination of the appropriate exposure (irradiation) and radiation decay times before analysis of the γ-spectrum,

Table 4.4 Average minimum concentrations of elements detectable in a group of UK soils. Reproduced from Salmon and Cawse [55] by permission

Element	Concentration (μg/g)*	Element	Concentration (μg/g)*
Ag	00.15	K	0.15%
Al	0.15%	La	1.5
As	0.5	Mn	20
Au	0.005	Mo	5
Br	2.0	Na	300
Ca	0.4%	Ni	30
Cd	4	Rb	20
Ce	2	Sb	0.5
Co	0.35	Sc	0.25
Cr	2.5	Se	0.4
Cs	0.3	Sm	0.2
Cu	4	Tb	0.1
Dy	1	Th	0.3
Eu	0.02	Ti	1000
Fe	0.10%	V	10
Hf	0.15	W	0.3
Hg	0.6	Yb	0.3
I	10	Zn	4.0
In	0.05		

* Except where indicated.

and these are dependent on the analyte and on the nature of the matrix components and their interfering radiation. Sophisticated instrumentation and computer facilities as well as experienced staff are thus essential for this technique. These facts have tended to limit the use of INAA to national laboratories or to those with access to a local reactor centre. In a recent review of INAA and its application in soil analysis, Salmon and Cawse describe a protocol for soil analysis making use of two samples of 50–100 mg of finely ground (< 200 μm) soil. One is irradiated for a few seconds in a heavy water reactor with measurements taken after 5 min, 30 min, 3 h and 14 h. A second sample is irradiated for 4–8 h with measurements made 2 days and 20 days after irradiation. Multi-element analysis can therefore be lengthy.

Sensitivity is dependent on the matrix components and their contribution to the background radiation, dominated in the soil situation by Al and Na. Detection limits have been defined for Cu, for example, as 1/100 of the Na content [56], while the induced ^{24}Na radiation may restrict measurements to isotopic species with half-lives > 15 h. Precision is good, often of the order ± 1% RSD, and accuracies of around 10% can be achieved. Minimum detectable concentrations in soils are given in Table 4.4 (from ref [55]). General detection limits for the technique, which, for the majority of elements, are in the range 1 mg/kg to 0.01 mg/kg, are given by Laul [56].

Applications have included multi-element surveys of total soil contents for more than 30 elements. Salmon and Cawse have for example determined, in a range of

UK soils: Ag, Al, As, Au, Br, Ca, Cd, Ce, Cr, Co, Cs, Fe, Hf, Hg, I, In, K, Mn, Mo, Na, Ni, Rb, Sc, Se, Sm Th, Ti, V, W, Zn, and five rare earth elements (REEs). Similar multi-element applications include total contents in soil [57–59], sludge-treated soils [55], sewage sludges [60], soil from a reclaimed refuse dump [61], volcanic ash soils [55, 56]. Pollution studies have examined soils round mining, smelting and refining operations particularly for As, Sb [64, 65], Pb [65] and As, Cu, Ni, Se and Zn [66, 67], with epidemiological studies that included soil/plant analyses for As and a range of heavy metals [68, 69]. INAA has considerable potential for the determination of the halides in soils, and for Br in particular as an indicator of pollution from automobile exhausts [70, 71]. Låg and Steinnes [72] have carried out extensive surveys of Cl, Br, and I in Norwegian organic topsoils. Stable isotope tracer experiments using ^{127}I have also been described [73]. INAA has been important in the elucidation of the extent of soil or plant enrichment in metals deposited on land from the atmospheric aerosol [74, 75].

Other types of INAA analysis include the INAA determination of IR [76] and U [58] in soils, and its use in ^{133}Cs stable isotope studies of Cs transport in sediments [77].

4.6.3 Direct-current arc atomic emission spectrometry (DCAAES)

The development of spectrographic methods of analysis in which samples of non-conducting powders, such as soils, are excited to emit spectra in a DC arc was pioneered in the UK by Mitchell [78, 79] and in other countries by Ivanov [80], Rogers [81], Oertel [82], Pinta [83], Ahrens [84] and Specht *et al.* [85]. It is a simultaneous multi-element technique sensitive enough for the determination of the total soil content for most biologically important elements. Like the other solid-sample methods it requires minimal sample preparation or pretreatment and has, in consequence, fewer problems of sample contamination than solution methods and does not suffer the diminution of sensitivity entailed in dissolving a solid sample. Spectrographic methods, i.e. methods involving the use of photographic plates as detectors, have commonly been employed and in this form offer relatively inexpensive multi-element analysis. More rapid and precise but more expensive direct-reading polychromator methods using photomultiplier detectors have been used [86]. Increasingly AES has been used in conjunction with inductively-coupled plasma (ICP) sources [113] and to a lesser extent with direct-current plasma (DCP) sources for the excitation of spectra. These last two techniques, however require solution samples.

Arc emission spectrometry requires only 10–25 mg of finely ground soil (< 150 μm), and some 70 elements can be detected. In soils the method is sufficiently sensitive for the determination of some 30 elements. Details of the procedures for the semi-quantitive (±30%) analysis of soils are given by Scott *et al.* [87] and Mitchell [78]. The wavelengths used and the limits of detection are presented in Table 4.5 (after Scott *et al.*[87]). From this table it can be seen that for most of the

Table 4.5 Detection limits for direct-current arc atomic emission spectrography using the cathode-layer effect. (From Scott *et al.* [87], p.51, by permission of the Macaulay Land Use Research Institute, formerly The Macaulay Institute for Soil Research, Aberdeen)

Element	Wavelength (nm) μg/g	Detection limit	Element	Wavelength (nm) μg/g	Detection limit
Ag	328.068	1	As	234.984	1000
As	286.045	3000	B	249.773	3
B	249.678	3	Ba	493.409	5
Be	313.042	5	Bi	306.772	30
Bi	289.798	300	Cd	326.106	300
Ce	422.260	500	Co	345.351	2
Cr	425.435	1	Cs	852.110	200
Cs	807.902	300	Cu	324.754	1
Ga	294.363	1	Ge	265.118	10
Ge	303.906	30	Hg	253.652	1000
In	303.936	3	In	451.132	10
La	333.749	10	La	433.373	3
Li	670.784	1	Mn	403.449	10
Mn	280.106	3	Mo	317.035	1
Ni	341.477	2	Pb	283.307	10
Rb	780.023	20	Sb	287.792	300
Sc	391.181	10	Se	241.352	10000
Sn	283.999	5	Sr	460.733	10
Ti	398.976	10	Tl	276.789	50
V	318.540	5	Y	332.788	10
Zn	334.502	300	Zr	339.198	10

heavy metals of biological importance detection limits are in the range 1–10 μg/g. The method is not suitable, however, for volatile elements such as Hg, and the sensitivity for Cd, Cs, Tl, W and the metalloids As, Sb and Se is too poor for practical determination of these important elements.

Matrix interference effects are a major problem which can, however, be overcome by the use of powder standards matched in major components to the sample. Spectral line interferences must be taken into account in choosing lines for analytical use. In addition to the direct analysis of soils the method can readily be applied to the determination of heavy metals in agricultural crop plants by the analysis of the plant ash or in chemical concentrates prepared by an 8-hydroxyquinoline procedure [88]. Soil extracts can be analysed using a similar procedure [89, 90]. By the use of this technique interference from the original matrix components is avoided.

Applications of this method include the determination of some 30 elements in survey analyses of soils [91, 92], the establishment of the typical total contents of 25 elements and extractable contents for 13 elements in sewage sludges from England and Wales [93], and the analysis of the insoluble residues from aqua regia digests of soils [94].

4.6.4 *Total analysis using slurry samples*

While the use of aqueous slurries of powdered materials has been used to improve sample detection limits for analytical methods that use solution samples and at the same time minimise sample preparation time, their use for soils has proved difficult. This technique has been used in FAAS [95], but satisfactory analysis required continuous stirring of the slurry and the use of standards closely matched in matrix composition. In their application for ICP–AES and DCP–AES analysis it has been demonstrated [96] that slurry particles greater than 5–10 μm diameter are not efficiently transported into the plasma. To obtain particle sizes < 10 μm with real soils is generally impracticable, if not impossible. The method has also been investigated with some success for Pb determination in soils and simulated soil matrices with electrothermal AAS, and the need for matrix modifiers demonstrated [97–99]. Slurries of clays (kaolin) [96] and of soils [100] have been analysed by DCP–AES and ICP–MS respectively, while the analysis of homogenised peat samples by GFAAS has been compared, successfully, with analysis by FAAS and by colorimetry of digested samples [101, 102].

4.7 Acid dissolution for total analysis of solution samples

The dilution involved in dissolving soil limits this approach to major soil constituents and a few minor elements. Trace concentrations will generally require preconcentration before analysis and this is often made difficult by the nature of the reagents involved in the dissolution procedure. As atomic spectrometric methods are by far the most commonly employed, dissolution procedures are discussed below in terms of their application to these analytical finishes.

Dissolution with hydrofluoric acid. Since the soil matrix consists to a large extent of silica and silicates, total dissolution requires a procedure capable of attacking these materials. Aqueous solutions of hydrofluoric acid, HF, (approximately 40% by weight) are available commercially. Because of its ability to dissolve silicates it cannot be stored or used in glass vessels. Generally therefore, decompositions using HF are carried out in vessels made of Pt, PTFE, polypropylene or polyethylene. The acid may require purification by distillation [103], or sub-boiling distillation [14], or be prepared in a pure state by bubbling the gas into distilled water [104]. The need to use pure acid can be avoided by decomposition using acid vapour attack [105] in apparatus designed for this purpose [106, 107].

Hydrofluoric acid dissolution without removal of fluoride. These procedures use HF alone, 25–50 ml of 40% HF per 1 g sample, overnight in the cold [108] or at higher temperatures [150–250°C) in a pressure vessel (bomb digestion) [109] using 0.2 g sample plus 5 ml HF for 30–60 min. Following the addition of an excess of boric acid [110], 2.8 g H_3BO_3 to 3 ml HF plus 40 ml water, to complex the excess fluoride, this solution can be handled in glassware for periods long enough (about 1 h) for analysis without significant attack [111]. As can be determined by extraction as AsI_3 after treatment with KI [112].

Dissolution with a combination of HF and an oxidising acid. Because soils contain both organic matter and sulphides it is usually necessary to use an oxidising acid such as HNO_3 and/or perchloric acid ($HClO_4$) in combination with HF. Two typical procedures used for ICP–AES are summarised below (from Lechler *et al.* [113], by permission).

(i) *Procedure for major elements Al, Ca, Fe, K, Mg, Mn, Na, P, Si, Ti.*
100 mg ignited (1050°C) finely ground soil is digested with 2 ml aqua regia plus 3 ml HF in sealed Teflon bombs for 1 h at 140°C. After cooling, 50 ml of 4.4% (m/v) boric acid solution is added and the solution made up to 100 ml with distilled water for direct analysis by ICP–AES. Dilution factor, soil/solution is 100.

(ii) *Trace element procedure*
1.000 g dry soil finely ground, is moistened with distilled water and heated in a 100 ml Teflon beaker with 10 ml conc. HNO_3 and evaporated to small volume. Then 5 ml conc. HNO_3, 5 ml 70% $HClO_4$ and 10 ml conc. HF are added and the whole heated to perchlorate fumes. After 30 min fuming, 10 ml of HCl (1/1, v/v) is added and the mixture boiled for 10 min and then cooled and diluted to 100 ml with distilled water. This is used for the determination of Ba, Cd, Cr, Cu, Mo, Ni, Pb, Se, Th, V and Zn by ICP–AES but is also suitable for AAS. The dilution factor, soil/solution, is 100.

Bomb digestion with HNO_3 or mixed acids in sealed Teflon cups, contained in stainless steel or Al pressure vessels, and heated on a hotplate or in an oven, have been widely recommended for plants and highly organic samples [114, 115], as well as for inorganic materials [103]. This technique can be used for peat digestion but the recommended maximum sample size must not be exceeded because of the danger of explosion [116].

Mircowave dissolution procedures. Dissolution and digestion procedures that make use of microwave ovens are increasingly replacing conventional oven and hot-plate methods and have been recommended by NIST for contaminated environmental materials [117]. Microwave oven dissolution is carried out in sealed, PTFE or similar vessels and exploits the impoved oxidation potential that is achieved at the higher operating temperatures to provide faster dissolution. These intrinsically sealed systems are thus less subject to contamination in the laboratory and provide methods suitable for volatile elements such as As, Cd, Pb, Sb, Se and Tl [118]. They have been applied to the total dissolution of environmental materials such as fly ash, geological materials, coal, sewage sluges [118], peat [119] and sediments [120] using acid mixtures that can include hydrofluoric acid. Conventional aqua regia or nitric acid procedures for the determination of pseudo-total contents (see Section 4.9) of heavy metals in soils can readily be adapted to microwave heating. While domestic microwave ovens can be used successfully for these purposes they require considerable attention to safety in their operation. Commercial systems are now available that provide not only pressure release as a safety measure but are equipped with temperature and pressure monitoring devices that simplify the development of dissolution programmes for different materials and ensure reproducibility. Microwave heating has also benn applied successfully to the drying of geological materials such as clays and limestone [121] but the authors suggest that the long development time required for microwave methods, where a large number of different sample types have to be analysed, minimises its advantages over conventional methods. Compositional, i.e. speciation, changes may be induced by the microwave irradiation that may not be temporally stable. Microwave drying is therefore not recommended for speciation studies or for the preparation of reference materials.

4.8 Fusion methods for total analysis by techniques that use solution samples

For a number of minor minerals, attack by HF alone, or in combination with other acids, is ineffective in attaining total dissolution [122]. For these elements, and for silicate-based materials in general, fusion methods are useful. The most commonly used technique employs fusion with lithium metaborate (with subsequent dissolution of the glass thus formed in HNO_3). This method was developed by Ingamells [122–124]. and has the following advantages:

(i) it is faster than acid attack;
(ii) it dissolves most minor minerals;
(iii) it requires no special pressure vessels;
(iv) it provides a clear solution ready for analysis;
(v) the lithium metaborate acts as a spectrochemical buffer to minimise interference effects [125] in flame spectrometry;
(vi) unlike Na_2CO_3 fusion methods, it does not give rise to severe light-scattering interference effects.

It has the disadvantage, however, of providing solutions with a high salt content and this can degrade detection limits, especially in ICP–AES, because of the increased background emission. In comparison with acid dissolution procedures, the lithium metaborate/nitric acid dissolution method entails a dilution from solid to solution greater by a factor of at least 10, with a further consequent degradation of detection limits.

Lithium metaborate fusion procedures are, therefore, mainly used for major and some minor element determinations. Many trace and ultra-trace analyses can only make use of fusion/dissolution methods in conjunction with preconcentration prior to analysis.

An example of a fusion procedure for the analysis of soils by FAAS [126] and FAES requires only some 50–100 mg of finely ground (< 150 mm) dried soil and provides determination of ten major and minor elements. The fusion/dissolution procedure is outlined below.

Lithium metaborate fusion/nitric acid dissolution procedure. 100 mg < 100 μm dried soil are mixed with 700 mg anhydrous Li metaborate in a 10 ml Pt crucible and heat until sample and flux melt completely. The melt is heated at 900–950°C for 10 min. The melt is allowed to solidify, and when cold the crucible and contents are placed in a 50 ml beaker, covered with distilled water and 4 ml of conc. HNO_3 added. The whole is gently stirred magnetically until the glass is completely dissolved (about 1 h). The contents and rinsings are transferred to a graduated flask and made up to 100 ml with distilled water.

4.9 Pseudo-total analysis: strong acid digestion procedures

Several mineral acids (HCl, HNO_3, $HClO_4$, H_2SO_4) and their mixtures have been used for the dissolution and extraction of elements from soils [127, 128]. They do not dissolve silicates or silica completely but are vigorous enough to dissolve the heavy metals not bound to silicate phases. Most heavy metal pollutants fall into this category.

Bomb digestion methods already discussed [110, 114–116] are particularly valuable for obviating losses of volatile elements. One such procedure uses HCl pressure digestion for the elements As, Bi and Sb in soil [129], the results comparing well with a method which uses dry ashing, in the presence of $MgNO_3$, to prevent losses by volatilisation [130].

Sulphuric acid digestion is not well suited to FAAS analysis because of its complex interference effects. Losses of Pb as insoluble $PbSO_4$ may occur, and other elements may be lost by occlusion in precipitated $CaSO_4$ in soils with high Ca content.

Simple acid digestion methods for both soils and sewage sludges have been used for multi-element analysis [131] and include digestion in open vessels with HCl, HNO_3 and aqua regia [94] as well as HNO_3 digestion in unenclosed, sealed PTFE pots [132].

Digestion by $HClO_4$, usually in combination with HNO_3 acid, has found considerable application in soil analysis [133] despite the danger of explosion with samples high in organic matter. With care $HClO_4$ can be used safely with such materials, and the dangers can be almost entirely eliminated if a preliminary HNO_3 oxidation is carried out and only the final oxidation performed with $HClO_4$.

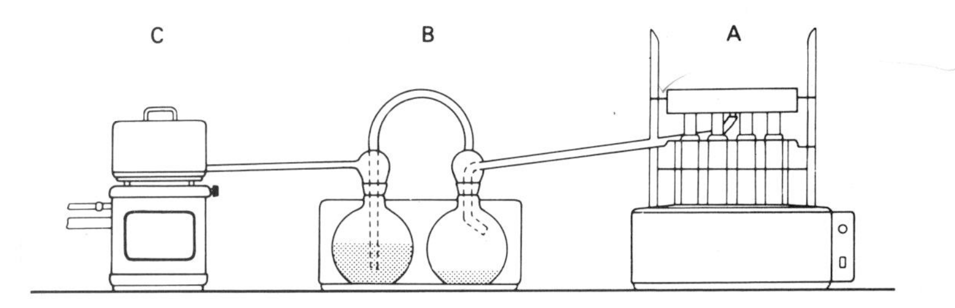

Figure 4.1 Schematic diagram of TURBOSOG acid digestion system, showing the digestion unit A, the pre-separator B, and the centrifugal suction washer C, for safe removal of fumes. Reproduced from the manufacturer's instruction leaflet by permission of C. Gerhardt UK Ltd.

Because perchlorate may accumulate in fume-exhaust systems, there is a danger of explosion where contact with wood or plastic can occur. For this reason, effective fume washing and extraction systems are essential. In addition to well-designed fume cupboards, totally enclosed digestion apparatus is available commercially which incorporates fume gas washing and disposal in the washing water. This is shown schematically in Figure 4.1. With $HClO_4$ digestion, Cr can be lost as volatile chromyl chloride, CrO_2Cl_2, or by retention on acid-insoluble silicate or silica residues [134, 135].

Hydrochloric acid (6 M), HNO_3 (16 M) and aqua regia (HCl + HNO_3, 3 + 1) have been compared in their ability to extract from sludge-treated soils seven of the more important heavy metal pollutants, Cd, Cr, Cu, Mn, Ni, Pb, Zn [94], using both open beakers and beakers covered with watch-glasses. It was concluded that almost identical amounts of six of the elements were extracted by all six procedures. Losses of Cr occurred with all three acids when the beakers were uncovered during digestion.

Decomposition procedures for Se have been considered in detail by Raptis *et al.* [136].

The method most widely adopted for pseudo-total analysis is probably the aqua regia procedure used by the Commission of the European Community Bureau of Reference (BCR) for analysis of their certified reference soils and sludges. This method has been adopted in large part by the UK water industry [137], the principal difference being their use of re-distilled HCl (6 M) in place of concentrated acid in the aqua regia mixture.

Aqua regia digestion for pseudo-total analysis of soils and sludges. To 3 g air-dried, finely ground (< 150 μm) soil or 1 g dried finely ground sludge in a round-bottomed 100 ml flask, 2–3 ml distilled water is added to form a slurry. For each g of dry sample, 7.5 ml of 6 M redistilled HCl and then 2.5 ml conc. HNO_3 are added. The flask is left covered overnight at room termperature. The mixture is boiled gently under reflux using a 40 cm condenser for 2 h and allowed to cool. The condenser is rinsed with 30 ml distilled water into the flask. The solution is filtered through an acid-resistant cellulose filter (hardened ashless, porosity 0.4–1.1 μm) previously washed with 12.5 v/v (HNO_3) and collected in a 100 ml graduated flask. The filter is rinsed with a few ml of warm (50°C) HNO_3. The solution is allowed to cool and made up to 100 ml with 2 M HNO_3.

Aqua regia digestion extracts between 70 and 90% of the total contents of these seven trace elements (Cd, Co, Cr, Cu, Fe, Mn, NI, Pb) while of the major elements, 30–40% of the total Al, 30–60% of Ca, 10–20% of K, 60–70% of Mg but only some 2–5% of Na in uncontaminated soils is extracted [94, 137, 138]. One of the advantages of aqua regia digestion is that BCR reference soils and sludges have indicative values for the aqua regia contents of several elements (see Appendix 9) and these soils and sludges can therefore be used to validate the extraction procedure. Other elements, perhaps including volatile species such as As, are probably substantially recovered from soils with minimal losses. Both FAAS and ICP–AES are capable of determining many of the important heavy metals at the concentrations occurring in digestions of normal, unpolluted soils, although for a few soils low in Cd some difficulties may be encountered. This, and many other, elements are well within the sensitivity capabilites of graphite-furnace atomic absorption spectrometry (GFAAS) but the strong oxidising acid solutions may shorten furnace life.

4.10 Atomic spectrometric methods for the analysis of solutions

Two main techniques, AAS and AES, are currently in wide use for the determination of most metals, the metalloids and some non-metals. Atomic fluorescence spectrometry (AFS) is occasionally applied using apparatus similar to that used in AAS and AES.

4.10.1 Atomic absorption spectrometry (AAS)

Atomic absorption spectrometry makes use of the fact that the free atom of an element absorbs light at wavelengths characteristic of that element and determined by its outer electronic structure. The extent of that absorption is a measure of the number of atoms in the light path. As illustrated in Figure 4.2*a* an atomic absorption spectometer requires: a source of light, A; an energy source such as a flame to decompose the sample into its constituent atoms, i.e. a flame atomiser, B; a monochromator, C, to isolate the required wavelength; a photomultiplier detector, D; and a readout device, E. The light source, A, is usually a hollow cathode discharge lamp (HCL, Figure 4.3*a*), whose cathode is composed of the element to be determined, but, exceptionally, can be an electrodeless discharge lamp (Figure 4,3*c*) excited in a microwave cavity (Figure 4.3*b*) for elements as As, Sb or Se or a metal vapour discharge lamp (Figure 4.3*d*) for mercury or the alkali metals. All

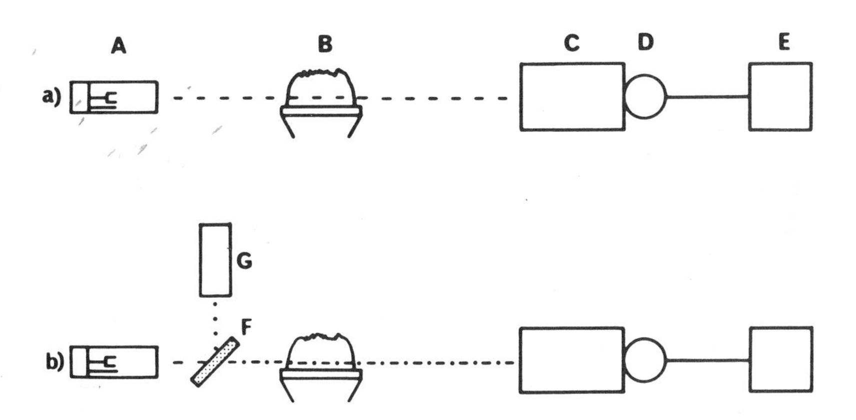

Figure 4.2 Schematic diagram of flame atomic absorption instrumentation, showing (a) the hollow cathode lamp A, the flame atomiser B, the monochromator C, the detector D and the readout device E, and (b) the deuterium background correction lamp G and the beam combiner F.

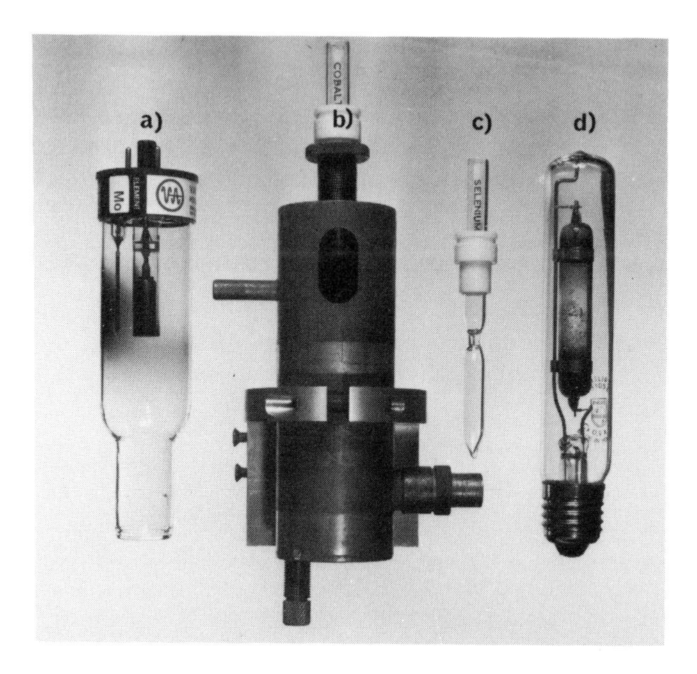

Figure 4.3 Light sources for AAS showing (a) a hollow cathode lamp (HCL), (b) a microwave cavity for (c) an electrodeless discharge lamp (EDL) and (d) a metal vapour discharge lamp.

employ the element to be determined as the source of the spectrum emitted. The limit emitted is thus the characteristic spectrum of that analyte, Cu, for example. The light is passed through the flame, B, into which a fine mist or aerosol of the sample solution is sprayed by a pneumatic nebuliser. The flame desolvates the sample mist and decomposes the resulting clotlets into atoms of the analyte, Cu. The Cu atoms in the flame absorb the Cu spectral radiation at a characteristic wavelength isolated by the monochromator, C, and the decrease in the amount of light reaching the detector when sample atoms are present in the flame is a measure of the concentration of Cu atoms.

Thus

$$\text{absorbance} = \log I_O/I_T = kc$$

where I_O is the intensity of the incident beam, I_T is the intensity of the beam transmitted by the atoms, c is the concentration of atoms in the atomiser and k is a constant.

This linear relationship between the measured absorbance and the concentration of analyte atoms in the flame, which is proportional to the concentration of atoms in the sample solution, holds in practice over some two orders of magnitude. The precise relationship between absorbance and solution concentration of the analyte is established by analysing a series of standard solutions with known analyte

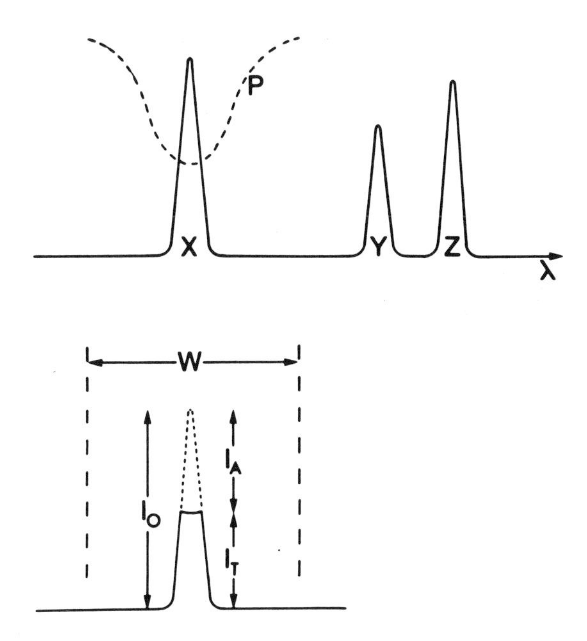

Figure 4.4 The principle of the atomic absorption measurement (see text for details).

concentrations covering the range required or, in some cases, by the method of standard additions (see page 79).

The hollow cathode emits very narrow spectral lines, X, Y, Z, (Figure 4.4) which means that the absorption measurement is made at the maximum of the absorption line, P, and the signal is at a maximum. The monochromation is thus essentially performed by the HCL, and the monochromator is only required to accept the desired line, X, by passing the wavelength interval (bandwidth) W, and to reject the unwanted lines Y and Z. In AAS, as distinct from AES where high-resolution spectrometers are essential, relatively inexpensive, low-resolution monochromators are therefore sufficient. The very narrow widths of lines emitted by the HCl provide excellent monochromation and largely account for the freedom from superimpositional line interference and the high degree of element specificity that are the hallmarks of AAS.

The technique and the apparatus for the analysis of soils by AAS have recently been discussed in some detail [139].

Flame atomic absorption spectrometry (FAAS). With conventional flame atomic absorption and continuous nebulisation of sample solution, detection limits are in the range 0.05–1 mg/l for most metals with a precision of 2–5% RSD. The technique is rapid, and sample-handling, measurement, computation and printout are available in automatic form from all commercial manufacturers. Sample requirements in the 2–10 ml range are usual, and smaller samples of about 50–100 μl can be accommodated by pulse nebulisation techniques.

For most analytical purposes a premixed air/acetylene flame is used with a 10 cm slot burner. A number of elements including Al, Be, Ca, Mg, Mo, V, Ti and Zr form refractory oxides in this flame which are incompletely atomised. Other elements such as Cr, Mn and Fe are subject to depressive matrix interference effects in the air/acetylene flame. While the use of releasing agents such as an excess (5000–1000 mg/l) of Sr or La can ameliorate the problem for Ca and Mg determination, only resorting to the hotter flame of nitrous oxide/acetylene provides a general solution. Used with a reducing flame (excess fuel), this flame is also effective in preventing the formation of these refractory oxides and thus achieving effective atomisation.

The higher the temperature of the flame, however, the greater the ionisation of elements and the corresponding departure from the ideal 100% atom formation. With low-temperature flames, interference problems entailed by incomplete dissociation of analyte compounds, i.e. incomplete atomisation of the analyte, are likely to occur. The usual compromise is to choose the air acetylene flame. For easily ionised elements such as the alkali metals Cs, Rb, K and Na, the alkaline earths and Ba, the ionisation can be suppressed by the addition to sample and standard solutions of an excess, 2000–5000 mg/l, of Cs or another easily ionised element. The use of high concentrations of releasing agents or ionisation suppressors can cause problems because of the impurities thus introduced. The Ar/H flame burning in the entrained air has been used for determination of As [140]. The most important problems in FAAS are chemical interference effects intrinsic to the low energy of flame atomisers.

Electrothermal atomic absorption spectrometry (ETAAS). The limitations of FAAS in sensitivity resulting from the inefficiency of sample introduction by pneumatic nebulisation (< 10% efficient) and the large dilution introduced by the expanding flame gases have been overcome by the use of the electrothermal atomiser [142–145]. This generally takes the form of a cylinder of graphite (graphite furnace atomic absorption spectrometry, GFAAS) which can be heated by the passage of an electric current through it. Microlitre (5–100 μl) samples are introduced into the furnace through a small, central hole (Figure 4.5). The furnace is prevented from oxidation by a surrounding atmosphere of a protective gas, usually Ar or N. In operation, light from the HCL passes through the tube to the monochromator and detector. After injection of the sample solution the atomiser is programmed through a drying stage, a pyrolysis stage, and an atomisation stage by passing progressively higher currents, for selected times, through the atomiser. At the atomisation stage, an atomic absorption measurement is made through the tube. Because the whole sample is atomised, and because the atomic vapour produced is partly confined within the graphite tube, the sensitivity of GFAAS is some 100 to 1000-fold greater than that of FAAS. Detection limits for GFAAS are given in Table 4.6. The technique of GFAAS also has the advantage of requiring small samples (5–50 μl). Despite these unique merits of very high sensitivity and small sample volume requirements, its application in studies of soils and heavy

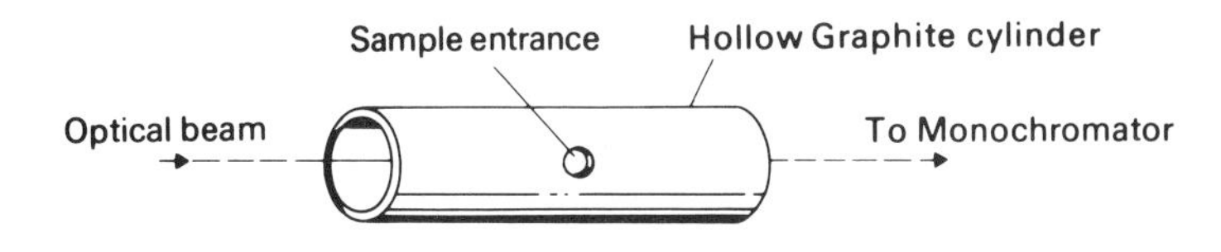

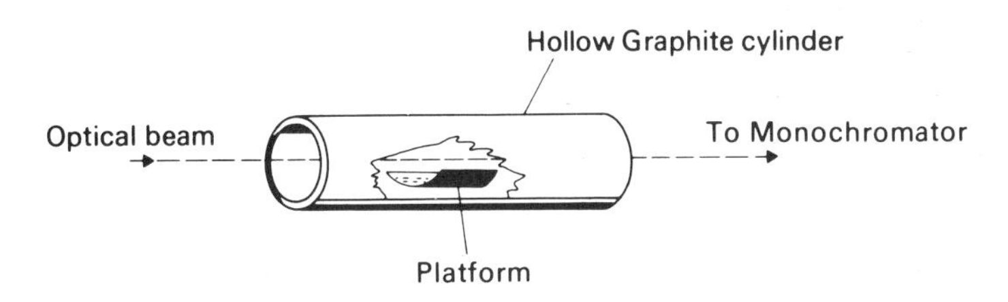

Figure 4.5 Schematic diagram of a graphite furnace atomised. The L'vov platform is mounted in the tube but in poor thermal contact with it. As a result its heating, largely by radiation, is delayed along with the atomisation of the analyte.

metal pollution has been slow to develop because of severe and complex interferences. The nature of many of the interferences is now well understood and effective methods of minimising them are available.

Interferences can be due to solid-phase effects occurring at the tube wall between matrix and analyte, and between the analyte and the graphite of the tube wall itself. Of major importance are the interferences produced by condensation and recombination reactions in the gas phase which reduce the analyte atom population and hence the absorption signal. These are most severe when atomisation takes place into a cooler furnace atmosphere (non-isothermal atomisation) and when volatile elements have to be determined in the presence of halides. Interference

Table 4.6 Detection limits for GFASS (μg/l), 10 μl sample. (Calculated from data in [141])

Element	D/L	Element	D/L
Ag	0.05	Mo	0.08
As	2.5	Ni	3
Au	2.5	Pb	0.5
Bi	3	Sb	1
Cd	0.05	Se	20
Co	1	Sn	20
Cr	2	Te	10
Cu	0.5	Ti	50
Fe	1	V	20
Mn	0.08	Zn	0.02

Table 4.7 Matrix modifiers for different analytes

Modifier	Element	Reference
PO_4	Ag	147
$PO_4 + Mg(NO_3)_2$	Cd, Sn	147
$Mg(NO_3)_2$	Al, Be, Co, Cr, Fe, Ni, Pb, Sb, Zn	147
Ni	As, Au, Bi, Te	147
$Ni + Mg(NO_3)_2$	Se	147
H_2SO_4	Tl	147
$Pd + MG(NO_3)_2$	'Universal'	99, 148–150
Pd + Ascorbic	In, Se	151, 152
PO_4	Pb, Cd	153
Ni, Pt	As, Se	154
Thiourea/EDTA/PO_4	Cd	155

from background effects is much more important in GFAAS than in FAAS and arises from the volatilisation of organic and other matrices, the production of smokes and the presence of graphite particles. These give rise to molecular absorption and non-specific light loss by scattering which result in erroneously high measurements of analyte concentration.

Reliable analysis can usually be carried out if the following principles enunciated by L'vov [146] are observed:

(i) Use of the L'vov Platform (Figure 4.5) to delay atomisation until the furnace atmosphere has reached a high enough temperature to achieve atomisation under isothermal conditions. This is particularly important to minimise recombination and condensation interferences, especially when a volatile matrix such as chloride is present.

(ii) Use of a matrix modifier (Table 4.7) to modify the sample and/or matrix to allow the matrix to be volatilised at the pyrolysis stage without loss of analyte. This not only reduces the impact of background effects but also enables volatile analytes such as Cd and Pb to be retained while potentially interfering matrices such as chloride are removed at the pyrolysis stage.

(iii) Use of effective background correction system (see below) to correct for molecular absorption and scattering effects.

(iv) Use of integrated absorbance (absorbance seconds) instead of peak height (absorbance) measurements. This can compensate for the difference in rates of atomisation that may occur berween samples and standards.

Atomisation in truly isothermal conditions can also be achieved by the use of the probe atomiser [156] for which commercial apparatus is now available. With this technique the furnace tube has a central slot into which a graphite probe can be inserted, sample solution pipetted on to it, and the sample on the probe subjected to a drying and pyrolysis program within the furnace in the usual way. The probe is then withdrawn, the furnace heated up to the atomisation temperature and the probe re-inserted for the sample to be atomised into the already hot furnace.

Standard addition procedures. If these methods are not completely successful then the standard addition procedure can be used instead of direct calibration with external standards. In this procedure the sample solution itself is used as the matrix for the standards. By analysing aliquots of the sample solution to which increments 0, A, 2A, 3A, etc. of the analyte are added the calibration slope is established. This calibration slope can be extrapolated to zero addition to establish the sample content. This method does not, however, compensate for high background.

Background correction systems. In both FAAS and GFAAS the absorption signal can be enhanced over and above the atomic absorption signal by molecular absorption and by non-specific light loss caused by scattering. Light from the hollow cathode lamp can be scattered out of the path to the monochromator by particulate species such as aerosol droplets and clotlets in FAAS and by smokes and carbon particles in GFAAS.

The most common method of background correction makes use of a continuum light source such as a hydrogen or deuterium lamp to measure the background absorption. One instrumental layout shown in Figure 4.3*b* illustrates the use of alternate measurements of atomic absorption using the hollow cathode lamp (A) and background absorption using the deuterium lamp (G). Instrumental subtraction of the two signals provides a background-corrected atomic absorption signal. True double-beam background correction systems are available and have the advantage of providing simultaneous correction which is important in GFAAS where the signal measured is transient. The intensity of deuterium and hydrogen lamps falls off above about 300 nm. At wavelengths above 350 nm they are ineffective, and tungsten filament lamps must be used. Above this wavelength, however, scattering losses are usually small but molecular absorption effects still occur.

For most situations this type of correction system is all that is required but, particularly in GFAAS with organic matrices, a background correction method capable of handling very high background absorption is necessary. Two methods have now become available to meet this demand. The first makes use of the Zeeman effect [157, 158], and is effective up to a background absorbance of 2. The second method employs high-current pulsing of the hollow cathode lamp in order to make use of the wings of the resulting self-absorbed line to make the correction. This method, the Smith–Hieftje method [159], like the Zeeman method, uses the same hollow cathode lamp for atomic absorption and background measurement and thus avoids the need to align the axis of a second lamp—often a difficult procedure. Apart from their ability to handle high backgrounds they can both make effective correction even when the background has a fine structure that is not dealt with by deuterium background correction systems.

Applications of GFAAS. Recent examples of electrothermal atomic absorption spectrometry (ETAAS) using the graphite furnace atomiser include a study of background correction in the determination of Pb and Cd in soil and plant samples with high iron contents [160], the determination of Se in soil extracts [161], and Sn in silicates [162]. Extracts of the iodo-complexes of Pd and Pt [163] have been determined and slurries of soils have been used in the determination of Pb with [164] and without modifiers [165]. ETAAS has also been used for the determination of sub-microgram amounts of Cu and Mn by precipitating their 8-hydroxy quinolates directly into the graphite furnace for analysis [166]. The high sensitivity of GFAAS has been successfully exploited in the determination of Cd in the several steps of a sequential extraction system, the sum of whose contents happily equalled the total Cd content [167]. This was an example of the STPF approach outlined above and other platform methods include a claimed improved performance for a tantalum platform in the determination of Cd in soils [168].

Instrumentation. The instruments now available for atomic absorption are now extremely sophisticated and while few revolutionary advances have been made recently, significant improvements in detail have taken place. These include the incorporation of higher modulation frequencies to make the time between total and background measurements as short as possible to cope with the rapid signal rise times in electrothermal atomic absorption [169]. Automatic sampling facilities are routinely incorporated, usually with the ability to carry out the otherwise tedious standard additions procedure that is often required for highly accurate, interference-free analysis [170]. Most instruments have integral micro-computers by which the instrument parameters are set and the operation of the instrument controlled and monitored. The results of the analyses are computed and printed out automatically together with the basic statistics of the run in what can be a final analytical report. An example of such a modern instrument is shown in Figure 4.6 by courtesy of Varian ltd.

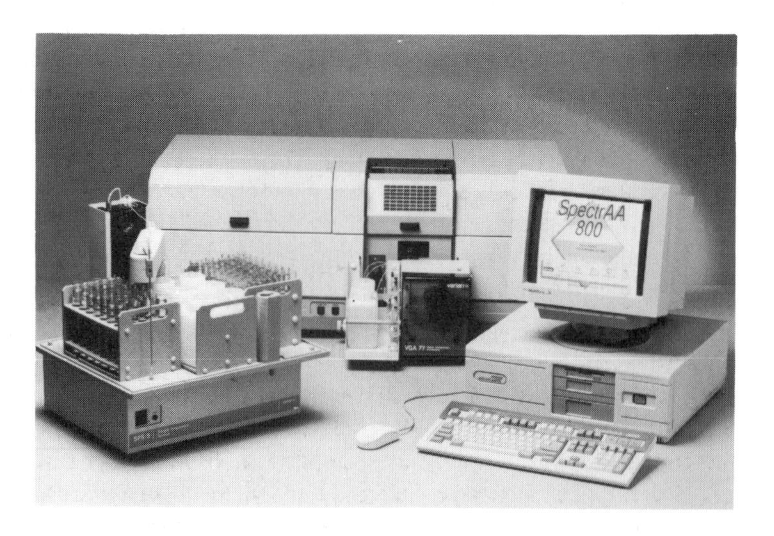

Figure 4.6 A modern instrument for atomic absorption spectrometry (by courtesy of Varian Ltd), showing the controlling computer, automatic sampler, a vapour generation accessory (fitted) and the spectometer.

4.10.2 Atomic emission spectrometry (AES)

In this technique sample solution is nebulised into an energy source such as a flame, an inductively coupled plasma, a DC arc plama or even a graphite furnace. The source acts not only to atomise the sample but also to excite the atoms thus formed to emit their characteristic spectral lines. The intensities of the emitted lines are a function of concentration of the atoms in the exciting source and hence of the solution content. Apparatus for AES consists basically of an exciting source and a monochromator/detector—which may be capable of wavelength scanning for rapid sequential multi-element analysis—or a polychromator with a number of fixed exit slits and detectors for simultaneous multi-element analysis. Atomic emission spectrometry thus differs from AAS in that it can readily provide simultaneous, or very rapid sequential, multi-element analysis of a single sample solution. Atomic absorption, on the other hand, determines only one element at a time in a sample. Commercial AAS instruments are available, however, which automatically alter all the parameters required for determination of one element to another set under the control of a computer program. This approach may work for up to about 10 elements. For up to about six elements it is as fast as sequential ICP–AES, is cheaper in capital and running costs, and suffers less from spectral and background interference effects. As the number of elements to be determined increases, however, emission spectrometry is progressively faster, more economical and convenient.

High-energy sources more readily decompose compounds into their constituent atoms and provide more complete atomisation less subject to chemical interference. For example, the nitrous oxide/acetylene flame does not suffer from the depressive matrix interferences in the determination of Cr, Fe and Mn that occur in the air/acetylene flame. The even higher temperature of the inductively coupled plasma emission source (6500K) is characteristically free from most chemical interference effects. High-temperature sources, however, excite more elements to emit and their spectra are complex and rich in lines and require expensive high-resolution spectrometers to isolate the desired lines.

Flame emission spectrometry (FAES). Historically, this method preceded FAAS; in the form of the Lundegardh flame emission spectrographic technique it was one of the first practical multi-element methods for the analysis of soil extracts [171]. In the 1950s and 1960s, under the name of flame photometry, FAES was in use for the determination of the alkali and alkaline earth metals and a few minor elements (see for example Dean and Rains [172]). With the introduction by Walsh [173] of practical atomic absorption in the 1950s, FAES has declined in importance and now it is used only for the determination of K and Na in soil extracts. The introduction of the nitrous oxide/acetylene flame increased the sensitivity of FAES to the extent that for many elements, especially those with sensitive lines at wavelengths > 350 nm, FAES as sensitive as FAAS or even more so. The use of FAES has largely given way to AAS methods because of the latter's better element specificity and freedom from spectral interference effects, arising from its intrinsically superior monochromation.

Inductively coupled plasma–atomic emission spectrometry (ICP–AES). The analytical potential of the ICP as an emission source was first recognised by Greenfield *et al.* [175] and the early development by this group [176], by Scott *et al.* [177] and Boumans [178. 179] has led to the development of the powerful analytical systems in use today.

The technique uses, in principle, the emission from the flame-like plasma formed on a quartz torch by coupling a radiofrequency (r.f.) electromagnetic field to the electrons in an ionised Ar plasma. The plasma is heated by the passage through the plasma of current, which, because of the high frequency, is concentrated in the outer surface of the plasma. This, in conjunction with the torch parameters, shapes the plasma into a toroidal or 'doughnut' form. Sample aerosol is directed into the

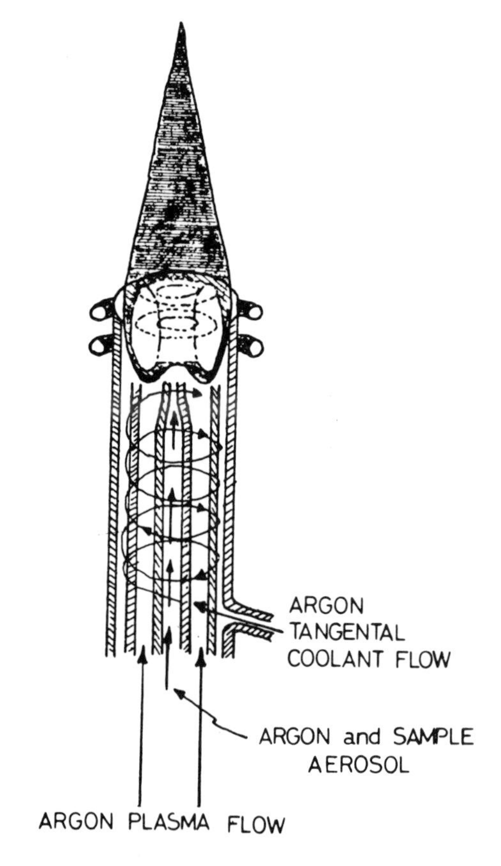

Figure 4.7 Schematic diagram of an ICP torch.

central hole of the plasma doughnut. Temperatures in the plasma are in the range 6000–10000 K and are typically some 6500 K in the analytical measuring zone. The general principles of plasma emission spectrometry are discussed in reference [180] and dealt with in relation to soil analysis by Sharp [181]. At these high temperatures atomisation is virtually complete for most elements, and strong atomic and ionic line emission can occur. Use is made of both atomic and ionic lines. The optical thinness of the atom cell contributes to the fact that calibration graphs are linear over some five or six orders of magnitude.

The ICP source consists of a quartz torch with provision for supplying Ar gas and sample aerosol to it as illustrated schematically in Figure 4.7. Radiofrequency power is supplied by an r.f. generator, most commonly at a frequency of 27.12 MHz, with power outputs of about 2 kW and an Ar consumption of 11–15 l/min.

Samples are usually introduced in the form of aerosols produced by pneumatic nebulisers which are often fed from a peristaltic pump. The design of nebulisers for ICP use is more critical than for AAS since most involve narrow-bore capillaries with lower volumetric flow rates and are prone to blockages. Some of these difficulties have been circumvented by the Babington nebuliser [182], and developments of it, which are capable of handling solutions of high viscosity, high solids content and slurries. The theory, design and construction of nebulisers has been discussed in depth by Sharp [183].

Chemical vapour-generation techniques for the introduction of, for example, the volatile hydrides of elements such as As, Pb, Se, Sb, Sn, etc. have also been used, with order-of-magnitude gains in sensitivity. Such systems have generally made use of continuous-flow methods of hydride generation when combined with ICP–AES [184. 185] while the hydride technique is discussed in depth in references [186] and [187].

Two types of spectrometer are in use: a single-channel scanning monochromator and a multi-channel fixed-wavelength polychromator type. For As, Hg, P and S a vacuum spectrometer is necessary for optimum performance, since their best lines lie below 200 nm. The scanning monochromator has several advantages over the polychromator type; it has higher resolution, it is more flexible in that the number and selection of lines is infinitely variable, background correction can readily be made, and it is less expensive. It is , however, much slower than the polychromator type and as the number of elements increases the speed advantage of the latter increases. For multi-element analysis the sample consumption is greater with the scanning system and for small samples this advantage of the polychromator type may well be the crucial one. Polychromator systems are usually preferred for the routine analysis of a large number of samples and where the range of elements is known in advance. Such systems are expensive, and combined instruments offering both approaches are even more so.

Spectral interferences in ICP–AES are more severe than in FAAS because, in contrast to AAS, the practical resolution attainable will often be insufficient to prevent line overlap. Secondly, the high temperature of the ICP results in complex multi-line spectra and significant background emission. High concentrations of

Table 4.8 Pseudo-total contents of soils in aqua regia digests. Compiled from [125, 181, 31]

Element	World mean soil content (mg/kg)	Expected contents aqua regia digest (mg/l)	Determination limits* FAAS (mg/l)	ICP–AES (mg/l)
Ba	1000	270	0.1	—
Cd	0.62	0.007	0.004	0.015
Co	12	0.2	0.04	0.02
Cu	26	0.5	0.01	0.01
Fe	3.2%	600	0.03	0.01
Mn	760	11	0.01	0.005
Ni	34	0.3	0.04	0.04
Pb	29	0.5	0.1	0.2
Sr	280	—	0.01	—
Ti	0.5%	6	0.25	0.02
Zn	60	1.4	0.005	0.09

* 5 × detection limits.

Al, Ca and Mg, likely to be present in soil samples, can introduce large background errors due to high scattered light levels consequent on the high intensity emitted by these elements. This background interference can often be overcome by choosing an alternative line or by using a correction procedure. Correction can be made by blank subtraction or by making use of the measured contribution of the interfering element at the analyte line wavelength. Blank subtraction is valid only in those few situations in which the interferent concentration is constant over the range of samples. All correction procedures result in a loss of precision and a consequent degradation of the detection limits. Soil analysts should therefore regard published detection limits, perhaps valid in dilute solution, as likely to be optimistic.

Chemical or matrix interferences are generally small but some do occur in the presence of matrix elements which have low ionisation potentials and components of high dissociation energies [188]. Ionisation suppressors are not helpful, since the mechanism is different.

Physical effects due to viscosity, solvent volatility, etc. can affect sample transport and nebulisation efficiency. Generally, matching of sample and standard solutions in solvent or acid concentration will suffice, but in some cases matching of major and/or interfering components will be essential. Internal standardisation is a useful method of improving precision and accuracy and also of reducing matrix interference effects. Two useful approaches have been used, the Myers Tracy [189] and the PRISM [190, 191] methods.

Estimated limits of determination for ICP–AES for a number of heavy metals in aqua regia digests of soils are given in Table 4.8 and compared with those for FAAS and with expected soil average contents.

The sophisticated spectrometer of a modern multi-element ICPAES instrument of the polychromator type is illustrated in Figure 4.8. This shows the controlling and data processing computer, the echelle spectrometer and the charge injection diode array detector system.

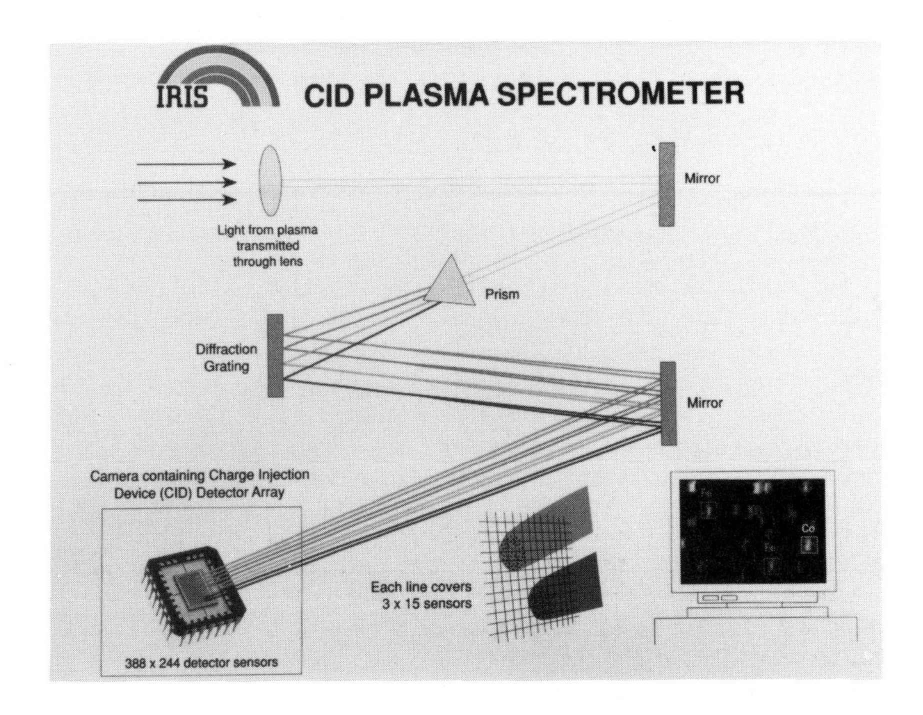

Figure 4.8 The spectrometer and detector system of a sophisticated recent ICPAES instrument of the polychromator type are illustrated with its controlling and data processing computer, the echelle spectrometer and the charge injection diode array detector. (By courtesy of A. Birnie of the Macaulay Land Use Research Institute and Thermo Jarrell Ash Corporation.)

Applications of ICPAES. The multi-element capability of this technique has made it an increasingly powerful competitor to FAAS in the analysis of soil extracts and digests. In general its sensitivity is similar to FAAS and is less subject to chemical and matrix interferences. It is more prone to spectral interferences but chemometric approaches can do much to lessen their effect [192]. Like FAAS its detection limits are in general inferior to those of ETAAS by factors of between 100 and 1000 and for many analyses preconcentration procedures are required. Attractive on-line procedures are being developed for both FAAS [193] and ICPAES [194].

Broekaert [195] has reviewed the use of ICPAES in the analysis of environmental materials and applications include the determination of B [196, 197], and a wide range of elements in soil extracts including ion exchange resin extracts [198]. Garden soils [199] have also been surveyed by this technique. A number of different sample introduction techniques have been used in addition to the conventional nebulisation method. These include slurry methods for clays [200] and sludges [201], electrothermal vaporisation methods [202] and hydride generation procedures [203].

Table 4.9 Detection limits for HGAAS (μg/l) (Calculated from [231])

Element	As	Bi	Ge	Pb	Sb	Se	Sn	Te
Detection limit	8	0.2	4	100	0.5	1.8	0.5	1.5

4.10.3 Atomic mass spectrometry

The principal atomic mass spectrometric technique in recent years has coupled the quadrupole mass spectrometer, and, to a lesser extent the magnetic sector instrument, to the ICP used as an ion source. This has provided comprehensive determination of virtually all the elements in the periodic table with solution detection limits in the ng/1 range. Few soil or agricultural laboratories have a requirement for truly comprehensive analysis and are usually concerned with less than a dozen elements. Furthermore different extracting reagents are required for different elements, reducing still further the need for the analysis of solutions for a large number of elements. These facts and the very high cost of these instruments have meant that the impact of this powerful new technique on soil and soil extract analysis has been minimal and largely confined to specialised topics such as the determination of the radionuclides [204, 205].

4.11 Special methods in atomic spectrometry

4.11.1 Hydride generation methods

In hydride generation atomic absorption spectrometry (HGAAS) or atomic emission spectrometry (HGAES) the sample is presented to the atomiser, not as an aerosol, but as a gaseous species, the volatile hydride, formed from it. The technique was introduced by Holak [206], who generated arsine, AsH_3, by a Zn/HCl reduction reaction and collected it in a liquid nitrogen trap. The arsine was then released and borne by a stream of N_2 into an air/acetylene flame for determination by AAS. The method has been successfully applied to eight elements, and, with the exception of Pb, provides markedly superior detection limits (Table 4.9) to FAAS. Early methods used a variety of reduction reactions, but the use of sodium borohydride, introduced by Braman [208], is now almost universal and was first applied in AAS by Schmidt and Royer [209] for the determination of As, Bi, Sb and Se. Its use has been extended to Ge [210], Sn and Te [211] and Pb [212]. Since the conditions for hydride generation are not too dissimilar for the different elements, simultaneous determination by ICP–AES is possible [213]. Atomisation sources have included the air/acetylene flame [206] and the $Ar/H_2/$ entrained-air flame developed by Kahn and Schallis [214]. The similar $N_2/H_2/$ entrained-air flame has also been used [215, 216]. These cool flames suffer from molecular absorption effects and have largely been superseded by the silica tube furnace heated electrically [217,

218] or heated by an air/acetylene flame [212, 219, 220]. A flame-in-tube atomiser has been described in which the hydride is decomposed by a small flame supported by the accompanying H_2 with the addition of a little O_2 [221–224]. In this context it should be noted that O_2 takes part in the atomisation process and its presence in small amounts can enhance the sensitivity [224, 225].

The graphite furnace has also been used to atomise the hydrides of As, Bi, Sb and Se [226, 227]. It has also been employed as a combined trap and atomising device with exceptionally good absolute detection limits, 0.2 pg for Sb [228], 70 pg for Se [228] and 40 pg for As [230]. An automated continuous-flow system for the determination of antimony by AAS using a flame-heated atomiser has been applied to the determination of Sb in soils and sewage sludges following an aqua regia digestion [231]. Aqua regia digestion has been found to be one of the best methods in a study of 10 procedures for As [232]. Automated flow injection has been applied in hydride generation for ICP–AES analysis [184].

4.11.2 Atom trapping atomic spectrometry (ATAAS)

Atom trapping was first introduced by Lau and Stephens [233] in 1976 and has been developed subsequently at the Macaulay Institute for Soil Research in Aberdeen as an *in situ* preconcentration technique [234, 235]. It has been further developed by Thomson [236]. It has been applied successfully to the analysis of

Table 4.10 Atom-trapping AAS detection limits, μg/ml (2 min. collection on alumina-coated tubes)

Ag	0.0008	Mn	0.010
As	0.15	Pb	0.001
Au	0.035	Sb	0.048
Bi	0.030	Se	0.050
Cd	0.0003	Tl	0.011
Cu	0.019	Zn	0.0017

Table 4.11 Contents of Cd and Pb in aqua regia digests of some top soil samples by conventional FAAS, (A) and the atom-trapping technique, (B). Results in μg/g in dry soil. Reproduced from Lau *et al.* [237], by permission of The Macaulay Land Use Research Institute, formerly The Macaulay Institute for Soil Research, Aberdeen

Sample	Cd content		Pb content		Sample	Cd content		Pb content	
	(A)	(B)	(A)	(B)		(A)	(B)	(A)	(B)
1	<0.25	0.21	16.0	15.8	7	<0.25	0.14	14.0	13.8
2	<0.25	0.10	9.3	8.5	8	0.25	0.28	26.5	27.3
3	<0.25	0.28	14.0	14.5	9	<0.25	0.16	22.0	23.0
4	<0.25	0.18	55.0	52.0	10	0.25	0.25	22.0	23.0
5	<0.25	0.32	29.0	29.0	11	<0.25	0.21	24.5	24.8
6	<0.21	0.21	23.3	25.5	12	<0.25	0.25	21.0	21.8

Table 4.12 Cadmium extractable by 0.05 M $CaCl_2$ in 16 Scottish tip soils, as μg/g in dry soil. Reproduced from Fraser *et al.* [238], by permission of The Macaulay Land Use Research Institute, formerly The Macaulay Institute for Soil Research, Aberdeen

Soil No.	Extractable Cd (μg/g)	Soil No.	Extractable Cd (μg/g)
1	0.007	9	<0.033
2	<0.004	10	<0.004
3	0.031	11	0.004
4	0.044	12	0.007
5	0.072	13	0.004
6	0.044	14	0.041
7	0.051	15	0.073
8	0.035	16	0.024

soil digests and soil extracts, particularly Pb and Cd [237, 238]. Interference effects are small and gains in senstivity of up to 40-fold compared with conventional FAAS are obtained. The detection limits with this technique are shown in Table 4.10. The results of analysing aqua regia digests of soils by this method are compared, in Table 4.11, with conventional FAAS for Cd and Pb [237]. The method is applicable to some 12 elements and is well suited to the determination of Cd and Pb in $CaCl_2$ extracts of soils where the concentrations in the extract are too low for conventional FAAS and where the high chloride content makes GFAAS methods difficult. The results of such analysis [238] for Cd are presented in Table 4.12.

4.11.3 Pre-concentration and separation techniques for atomic spectrometry

A useful general treatment of pre-concentration methods is given by Zolotov [239]. By far the most common has been solvent extraction of the metal combined or complexed with an organic reagent. These techniques are discussed in depth in references [240, 241]. The most usual procedure makes use of ammonium pyrrolidene dithiocarbamate (APDC) [242] as the complexing agent and methyl isobutyl ketone (MIBK) as the extracting solvent. The elements Ag, As, Bi, Cd, Co, Cr, Cu, Fe, Hg, In, Mn, Mo, Ni, Pb, Pd, Se, V and Zn have all been treated by this method at various pHs and the conditions for the extraction of many metals discussed in depth [243]. Data on APDC, 8-HQ, and ion-pair methods of solvent extraction are tabulated by Kirkbright *et al.* [244]. Problems with APDC methods in strong acid digests of soils or geological materials can be obviated by the use of metal-iodide ion-association systems for the determination of, for example, Ag, Cd, Se, Te and Tl [245]. The use of dithizone [246], generally with chlorinated hydrocarbon extraction, has largely been neglected in atomic spectrometry because of the dangers associated with its use in flames where the poisonous gas phosgene can be formed. It can, however, be used safely in electrothermal atomisers and it has thus been applied successfully to the determination of Cd and Pb in soil extracts [247, 248].

4.11.4 Cold vapour atomic absorption spectrometry (CVAAS)

Normal flame methods are not sensitive enough for soil Hg determination. The cold vapour atomic absorption method, with detection limits in the 1–10 μg/l range, is well suited for the determination of Hg in unpolluted or polluted soils and sewage sludges. An oxidative acid digestion procedure is required to destroy organic matter and this is followed by reduction of Hg compounds to elemental Hg for analysis in the vapour phase. Since Hg vapour is monatomic, an atomic absorption measurement can be made in the cold Hg vapour released from the reduced solution. Most methods for soils and peats follow that of Hatch and Ott [249] or developments of it [250–252]. The technique has been reviewed and discussed in detail by Ure [253] and Chilov [254].

Important interferences occur from I, Se and the noble metals Au, Pd and Pt, and are especially severe from Pd [251]. Commercial hydride generation kits generally include facilities for cold vapour determination of Hg (see Figure 4.6).

4.12 The analysis of soil extracts

4.12.1 Soil extraction procedures

A range of empirically derived extraction procedures has been developed over the years to simulate the availability of essential elements, and some toxic elements, to plants. These have been developed locally for a variety of diagnostic purposes including the prediction of trace element deficiency in farm crops and animals. The factors affecting the availability of trace elements in soils have been discussed by Mitchell [255]. The different extraction methods cannot be discussed here in detail and reference should be made to detailed texts such as references [9, 87, 128, 256–263]. The techniques most widely used for soils employ a single extractant whose content for one element correlates with plant-available content and can be used to predict plant uptake or the likelihood of deficiency or toxicity symptoms occurring in plant or animal. Such an approach is well established in soil science for predicting the uptake, at natural concentrations, of essential heavy metals such as Cu, Co, Mn and Zn and of potentially toxic elements such as Mo and Ni. For the heavy metals associated with pollution from sewage sludges or industrial wastes and effluents such as Cd, Cr, Hg, Pb and Sn, or even for the essential elements at elevated concentrations, extractant methodology is not well established and few predictions of uptake from soil extraction analysis can reliably be made. Some examples of soil extractants are listed in Table 4.13.

In general, it is apparent that although a number of well-established extraction procedures with some useful predictive power exist, many of them are specific to one element, are relevant only to specific crops and may be restricted in use to particular soil types. Perhaps the most generally useful for heavy metal analysis are 0.01 M or 0.05 M EDTA and 0.005 M DTPA. For the important heavy metal pollutants Cd and Pb, 0.05 M $CaCl_2$ [264] and 0.1 M $NaNO_3$ [265] are equally

Table 4.13 Some soil extractants for heavy metals

Extract	Milieu	Elements	Correlated plant content	References
Water	Soil	Zn	Wheat	333
Water	Sludged soil	Cd	Lettuce	334
Water	Sludged soil	Cd, Cu, Zn	Lettuce, wheat	335
EDTA	Soil	Cd,Cu, Ni, Pb, Zn	General	336–338
EDTA	Greenhouse soil	Se, Mo		339
DTPA	Calcareous soil	Zn		338, 340
DTPA	Soil	Fe, Mn, Zn	Sorghum corn	341
DTPA	Soil	Cd, Cu, Zn	Beans	342
DTPA	Soil	Zn	Lettuce	335
DTPA	Soil	Zn	Wheat	333, 335
DTPA	Sludged soil	Cd	Maize	343
DTPA	Sludged soil	Ni		339
DTPA	Soil	Mn, Zn	Wheat	344
HOAc	Soil	Cd, Cu, Ni, Pb, Zn	General	337, 345
HOAc	Soil	Cu, Co		346
HOAc	Soil	Cd, Ni		347, 348
HOAc	Soil	Ni		349
HOAc	Soil	Cr	Grass	350
Amm. OAc	Soil	Zn	Swiss chard	347
Amm. OAc	Soil	Zn	Rice	351
Amm. OAc	Soil	Ni	Sorghum	349
Amm. OAc	Soil	Pb	Oats	174
Amm. OAc	Soil	Mo	Herbage	352, 353
Amm. OAc + EDTA	Soil	Cu, Zn		354
Amm. OAc + EDTA	Soil	Cu, Zn		344
Amm. OAc + EDTA	Soil	Zn	Wheat yield	333
$CaCl_2$	Soil	Cd, Pb	Vegetables	264
$NaNO_3$	Soil	Cd, Pb	Vegetables	265

good predictors of plant uptake although only for a limited range of plants. Analytically 0.1 M $NaNO_3$ is the more difficult of the two because the amounts extracted by it are much lower than those extracted by 0.05 M $CaCl_2$, too low for FAAS, but accessible by GFAAS. The 0.05 M $CaCl_2$ extractable Pb and Cd contents are, in some soils, near the limit of detection by FAAS, are difficult to determine by GFAAS because of the chloride matrix, but are readily determined by atom-trapping AAS [238]. For elements such as Cd and Pb whose behaviour in soils is pH-dependent these weaker extractants are to be preferred. This argues in favour of the soil solution itself as a diagnostic of plant uptake of heavy metals, but the contents of the soil solution are so low that only the msot sensitive analytical methods such as GFAAS are likely to be practicable. This approach has been recommended [266] and demonstrated for Pb [267], and for Cd, Cu and Pb in radish, spinach and grass [268]. Soil solution contents have been shown to correlate well with plant uptake of Cu, Ni and Zn [269]. Different methods of obtaining the soil solution have also been described and include displacement [270, 271] and centrifugation [272].

4.13 Speciation

Speciation of heavy metals in soils can be defined in many ways. First, it can be defined as the extraction and quantification of a soil phase which is functionally designated in that its element content is, for example, the plant-available content. Most of the single extractants discussed above for diagnostic purposes fall into this category, i.e. the species extracted is defined by its role or function which in these cases would be availability to plants.

Secondly, speciation can be defined operationally, i.e. by the extraction technique itself. The extraction will usually be designed to extract the element associated or bound to a particular soil phase but, in practice, the soil phase may be ill-defined and the extractant not specific to the phase. The use of selective extractants to quantify the element content in a particular phase is illuminated by the concept (due to Viets [273]) of 'pools' in soils, of elements of different solubilities and mobilities, that can be selectively sampled by extractants of different strengths. This concept has been further clarified by Jones *et al.* [274] and is illustrated in Figure 4.9. From this figure it is obvious that most of the extractants are not specific

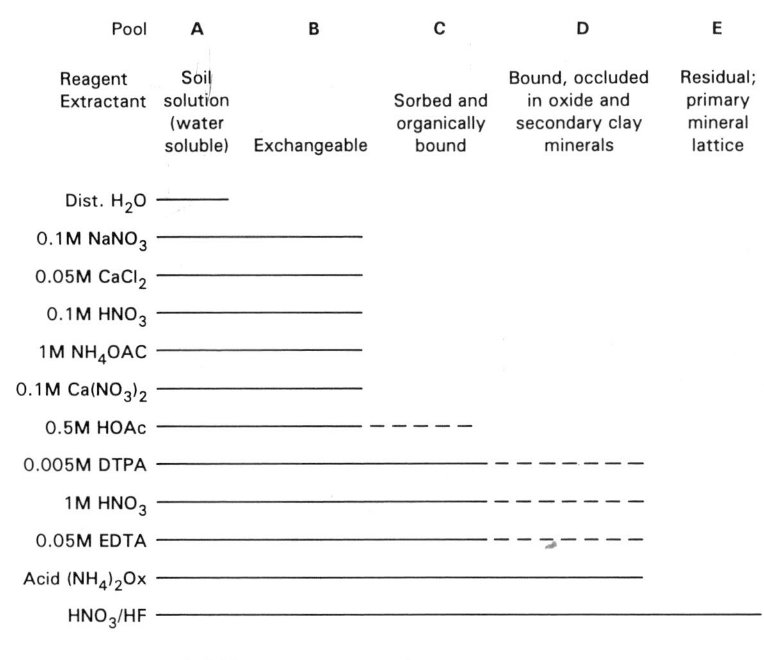

Figure 4.9 Diagrammatic illustration of the ability of different extractant to extract element from particular pools or phases. Adapted from Jones *et al.* [274] by permission, and supplemented by data from references [273, 275].

to a single phase or a particular element form. By making use of a series of sequential extractions with reagents of increasing power, however, the individual phases can be more precisely isolated, at least in principle. The subject has been reviewed by Pickering [276] and much work has been done on sequential extraction schemes for sediments [273–284]. Many of these are improvements and modifications to the most widely used sequential extraction scheme, that of Tessier [285]. Other studies include references to soils [286–294], sludge-treated soils [295, 296] and sludges [297].

Because of the very many different single and sequential extraction schemes in use, results from different laboratories cannot easily be compared. This makes adherence to international guidelines on heavy or toxic metal levels in the environment difficult. In pursuit of this quality control aspect of metal speciation the Community Bureau of Reference (BCR) of the Commission of the European Communities has been carrying out a programme of harmonisation of methods of extraction and analysis of soils and sediments with a group of some 35 European laboratories. One of the aims was the eventual preparation of soil (and sediment) reference materials certified for contents of species extracted by agreed and validated procedures. Two such reference soils should be available from BCR, by the end of 1994, with certified heavy metal contents extractable by 0.05 mol/l EDTA (pH 7) and by 0.043 mol/l acetic acid under closely prescribed protocols [298]. Preparation

Table 4.14 Sequential extraction scheme for BCR harmonisation study

SAMPLE

Extract overnight (16 h) with 0.11 mol/1 acetic acid (40 ml extractant/g soil) at 20°C; centrifuge, analyse supernatent (Solution 1), retain Residue 1.

RESIDUE 1

Extract overnight (16 h) with 0.1 mol/NH_2OH. HCl, pH 2 with nitric acid (40 ml extractant/g soil) at 20°C; centrifuge, analyse supernatent (Solution 2), retain Residue 2.

RESIDUE 2

Treat with 10 ml of 8.8 mol/l H_2O_2 per g of soil (pH 2 with nitric acid) for 1 h at room temperature and 1 h at 85°C. Reduce to a few ml, add fresh H_2O_2 and treat for 1 h at 85°C. Again reduce volume to a few ml and extract this with 50 ml of 1 mol/l ammonium acetate solution (pH 2 with nitric acid) per g of original soil overnight at 20°C. Centrifuge, analyse supernatent (Solution 3).

Solution analysed	Phase(s) extracted
Solution 1	Water and acid soluble, exchangeable
Solution 2	Reducible, Fe/Mn oxide
Solution 3	Organic + sulphide

of a similar sediment reference material certified for heavy metals extracted in a short three-stage sequential procedure [299], Table 4.14, is also in progress. Chemical extraction methods for speciation have been compared with electro-ultrafiltration methods [300] and with the use of progressive acidification [301] for assessing the mobility and uptake of heavy metals.

A last and more precise definition of speciation is the determination of precise chemical forms or oxidation states of an element. This can take the form, for example of the isolation and determination of the content of methyl mercury or the measurement of the amount of Cr^{III} in a soil or soil extract. Speciation in this narrowly defined sense is a very difficult task, especially at trace concentrations, because most treatments or extraction procedures will themselves change the speciation. Direct, non-destructive methods for establishing the combinational form of an element i.e. the speciation, such as ESR, IR, XRD, NMR and Mössbauer spectroscopies are sufficiently sensitive only for major element constituents of soil. In a few cases, however these methods are feasible for heavy metals. Mössbauer [302] and FTIR [303] spectroscopies have, for example, been used for the assessment of iron complexes. The use of ESR for the study of fulvic acid metal complexes has been compared with fluorescence methods [304]. Most studies have therefore been confined to the soil solution or soil pore waters. Haswell *et al.* [305] have characterised soil pore waters in terms of their content of arsenate, arsenite and monomethyl-arsonic acid using anion-exchange HPLC coupled to a continuous-flow HGAAS system. Studies in the speciation of the metals Cd and Pb in soil solutions from polluted soils have been made by Tills and Alloway [306].

4.14 Recent applications of atomic spectrometry

The technique of ICP–AES for geological materials has recently been reviewed [306] and it has been compared with AAS for plant and soil analysis [307].

Solvent extraction methods for both ICP and AAS methods have included procedures for Mo [308], Mo and Sn [309], As [310] and Ag, Bi, Cu, Cd, In, and Sb [311].

Extensive development of hydride methods has occurred and automated procedures for Se [312, 313], As [313, 314] and Sb [314], including flow injection methods for As and Se using the DCP [315], have been described. Methods for Sn in sludges [316] and Bi [317] and the use of a slotted tube flame atomiser have been presented [318]. Hydride preparative methods have been reviewed [319] and methods for the reduction of interferences on As and Se have been given [313, 320].

Speciation studies of As [321, 322] and Sb [323] have been made and alkyl Pb in soils and run-off have been discussed [324].

Atomic fluorescence has only rarely been applied in this field, mainly in the non-dispersive mode [325] for the determination of Hg [326, 327]. It has, however been used for Bi [328], and laser-induced AFS has recently been applied to soil analysis [329].

Recent XRF applications include the use of the wavelength dispersive mode for As [330] in geological materials and energy dispersive XRF for Cr and V [331]. It has been compared with ICP–AES for soil analysis [332].

4.15 Concluding comments

For reasons of brevity, many of the analytical techniques which have played, and continue to play, a role in soil analysis have not been considered here. The emphasis in this treatment has been largely on atomic spectrochemical approaches, reflecting the recent dominance of these methods and particularly of AAS. The emphasis is rapidly changing, and future years are likely to see the rapid spread of ICP–AES with its advantages of multi-element analysis. The related technique of ICP–MS, despite its comprehensive element coverage and high sensitivity, is likely to be more limited in its application to soil trace analysis, because of cost. Its capability for isotopic analysis and especially its role in stable isotope tracer experiments should, however, make a major contribution to soil research in the future. Developments in this field and in XRF and the other atomic, spectrometric techniques of analysis will, as in the past, be critically reviewed annually in *Atomic Spectrometry Updates*, published in the *Journal of Analytical Atomic Spectrometry* by the Royal Society of Chemistry.

References

1. Skeggs, L.T., *Am J. Clin. Pathol.* **28** (1957), 311.
2. Ruzicka, J. and Hansen, E.H., *Flow Injection Analysis*, Wiley, New York (1981).
3. Smith, K.A. and Scott, A., in *Soil Analysis*, ed. K.A. Smith, Marcel Dekker, New York (1983) Chapter 3.
4. Ure, A.M., *Anal Proc.* **17** (1980) 409.
5. Kubota, J., *Ann. New York Acad. Sci.* **199** (1972), 104.
6. Scott, R.O., Mitchell, R.L., Purves, D. and Voss, R.C., *Spectrochemical Methods for the Analyses of Soils, Plants and Other Agricultural Materials*, Consultative Committee for Development of Spectrochemical Work, Bulletin No. 2 (1971) The Macaulay Inst. for Soil Research, Aberdeen, 8.
7. Ure, A.M., Butler, L.R.P., Scott, R.O and Jenkins, R., *Pure and Appl. Chem.* **60** (1988), 1461.
8. Scott, R.O and Ure, A.M., *Anal. Proc.* **9** (1972), 288.
9. Jackson, M.L., *Soil Chemical Analysis*. Prentice-Hall, Englewood Cliffs, NJ (1958), 30.
10. Ingamells, C.O. and Pitard, F.F., in *Applied Geochemical Analysis*, John Wiley, New York (1986), Chapter 1 and 593–606.
11. Zeif, M. and Mitchell, J.W., *Contamination Control in the Laboratory*, Wiley, New York (1976).
12. Shand, C.A., Aggett, P.J. and Ure, A.M. in *Proceedings of Conference on Trace Element Analytical Chemistry in Medicine and Biology*, Munich, eds. Bratter, P. and Schramel, P. (1983) Walter Gruyter, Berlin, 1025.
13. *NBS Tech. Bull.* May (1972), 104.
14. Moody, J.R. and Beary, E.S. *Talanta* **29** (1982), 1003.
15. Mitchell, J.W. *Talanta* **29** (1982), 993.
16. Tschöpel, B., *Pure and Appl. Chem.* 54 (1982), 913.
17. Colinet, E., Gonska, H., Griepink, B. and Mantau, H. Commission of the European Communities, Brussels Report EHR8833EN (1983).
18. Colinet, E., Gonska, H., Griepink, B. and Mantau, H. Commission of the European Communities, Brussels Report EHR8834EN (1983).

19. Colinet, E., Gonska, H., Griepink, B. and Mantau, H. Commission of the European Communities, Brussels Report EHR8835EN (1983).
20. Colinet, E., Griepink, B. and Mantau, H. Commission of the European Communities, Brussels Report EHR8836EN (1983).
21. Colinet, E., Griepink, B. and Mantau, H. Commission of the European Communities, Brussels Report EHR8836EN (1983).
22. Colinet, E., Griepink, B. and Mantau, H. Commission of the European Communities, Brussels Report EHR8837EN (1983).
23. Colinet, E., Griepink, B. and Mantau, H. Commission of the European Communities, Brussels Report EHR8838EN (1983).
24. van Greiken, R., van't Dack L., Costa Dantas, C. and da Silviera Dantas, H., *Anal Chim Acta* **108** (1979), 93.
25. Childs, C.W. and Furkert, R.J. *Geoderma* **11** (1974), 67.
26. Norrish, K. and Hutton, J.T., *Geochim. Cosmochim. Acta* **33** (1969), 431.
27. Hutton, J.T. and Elliott, S.M., *Chem. Geol.* **29** (1980), 1.
28. Hauka, M.T. and Thomas, I.L., *X-ray Spectr.* **6** (1977), 204.
29. Thomas, I.L. and Hauka, M.T., *Chem Geol.* **21** (1978), 39.
30. Wilkins, C., in *Soil Analysis*, ed. K.A. Smith, Marcel Dekker, New York (1983), 195.
31. Bain, D.C., Berrow, M.L., McHardy, W.J., Paterson, E., Russell, J.D., Sharp, B.L., Ure, A.M. and West T.S., *Anal Chim. Acta* **180** (1986), 163.
32. Wilkins, C., J. *Agric. Sci.* **92** (1979), 61.
33. Fanning, D.S. and Jackson, M.L., *Soil Sci.* **103** (1967), 253.
34. Bain, D.C., *J. Soil Sci.* **27** (1976), 68.
35. Hubert, A.E. and Chao, T.T., *Anal. Chim. Acta* **92** (1977), 197.
66. Murad, E., *J. Soil Sci.* **29** (1978), 219.
37. Watts, S.H., *Geochim, Cosmochim. Acta* **41** (1977), 1164.
38. Wakatsuki, T., Furukama, H. and Kyama, K., *Geochim. Cosmochim. Acta* **41** (1977), 891.
39. Williams, C. and Rayner, J.H. *J. Soil Sci.* **28** (1977), 180.
40. Drees, L.R. and Wilding, L.P., *Soil Sci. Amer. Proc.* **37** (1973), 523.
41. Evans, L.J. and Adams, W.A., *J. Soil Sci.* **26** (1975), 319.
42. Oertel, A.C., *J. Soil Sci.* **12** (1961), 119.
43. Gilkes, R.J., Scholz, G. and Dimmock, G.M., *J. Soil Sci.* **24** (1973), 523.
44. Childs, C.W., *Geoderma* **13** (1975), 141.
45. Cowgill, U.M., *Dev. Appl. Spectrosc.* **5** (1966), 3.
46. Bradley, I., Rudeforth C.C. and Wilkins, C., *J. Soil Sci.*, **29** (1978), 258.
47. Ure, A.M., and Berrow, M.L., The Elemental Constituents of Soils' in *Environmental Chemistry, Vol. 2*, ed. Bowen, H.J.M. The Royal Society of Chemistry, London (1982) 94.
48. Jones, A.A. 'X-ray Fluorescence Analysis', in *Soil analysis* 2nd edn., ed. Smith, K.A. Marcel Dekker, New York, 1991, p 287.
49. Mukhtar, S., Haswell, S.J., Ellis, A.T., and Hawkes, D.T., *Analyst*, **116** (1991), 333.
50. Prange, A., Knoth, J., Stoessel, R.P., Boeddeker, H. and Kramer, K., *Anal. Chim. Acta*, **195** (1987), 275.
51. Klockenkaemper, R., Knoth, J., Prange, A. and Schwenke, H., *Anal. Chem.* **64** (1992), 1115A.
52. Reus, U. and Prange, A., *Spectrosc. Europe* **5** (1993), 26.
53. Report ONRL/TM-11385, Order No. AD-A216871.
54. Bacon, J.R., Ellis, A.T., McMahon, A.W., Potts, P.J. and Williams, J.G. *J. Anal. Atom. Spectrom.*, **8** (1993), 261R.
55. Salmon, L. and Cawse, P.A., in *Soil Analysis*, ed. K.A. Smith, Marcel Dekker, New York (1983), 299.
56. Laul, J.C., *Atomic Energy Review* **173** (1979), 611
57. Khera, A.K. and Steinnes, E., *Agrochimica* **13** (1969), 524.
58. Murmann, R.P., Winters, R.W. and Martin T.G., *Soil Sci. Amer. Proc.* **35** (1971), 647.
59. Buenafama, H.D., *J. Radioanal. Chem.* **18** (1973), 111.
60. Weaver, J.N., Hanson, A., McGaughey J. and Steinkruger, F.J., *Water Air Soil Pollution* **3** (1974), 327.
61. van der Klugt, N., Poelstra, P. and Zwemmer, E., *J. Radioanal. Chem.* **35** (1977), 109.
62. Borchard, G.A. and Harward, M.E., *Soil Sci. Soc. Amer. Proc.* **35** (1971), 626.
63. Borchard, G.A., Harward, M.E. and Schmitt, R.A., *Quatern. Res.* **1** (1971), 247.

64. O'Toole, J.J., Clark, R.G., Malaby, K.L. and Trauger, D.L., in *Nuclear Methods in Environmental Reserch*, eds Vogt, J.R., Parkinson, T.F. and Carter, R.L., University of Missouri, Columbia (1971), 172.
65. Crecelius, A.E., Johnson, C.J. and Hofer, C.C., *Water Air Soil Pollution* **3** (1974), 337.
66. Pattenden, N.J., AERE Harwell Report No. R 7729, HMSO, London (1974).
67. Bowen, H.J.M., *Trace Elements in Biochemistry*. Academic Press, New York (1966).
68. Jervis, R.E., Paciga, J.J. and Chattopadhyay, A.A., *Trans. Amer. Nucl. Soc.* **21** (1975), 95.
69. Gounchev, H. and Dimchev, T., *Pochvozn. Agrokhim.* **8** (1973), 77.
70. Farmer, J.G. and Cross, J.D., *Water Air Soil Pollution* **9** (1978), 193.
71. Oakes, T.W., Furr, A.K., Adair, D.J. and Parkinson, T.F., *J. Radioanal. Chem.* **37** (1977), 881.
72. Låg, J. and Steinnes, E., *Isotopes and Radiation in Soil Plant Relationships Including Forestry*. IAEA, Vienna, (1972).
73. Benes, J., Frana, J. and Mastalka, A., *Collect. Czech. Chem. Commun.* **39** (1974), 2783.
74. Cause, P.A., in *Inorganic Pollution and Agriculture*, Proc. ADAS Conf. London (1977), HMSO, London 180.
75. Cause, P.A. AERE Harwell Report No R7669, HMSO London (1974).
76. Stefanov, G. and Daieva, L., *Isotopenpraxis* **4** (1972), 146.
77. Hakonson, T.E. and Whicker, F.W., *Health Phys.* **28** (1975), 699.
78. Mitchell, R.L., Tech. Commun. No 44A, Commonwealth Bur. Soils, Commonwealth Agric. Bureaux, Farnham Royal, Bucks, UK (1964).
79. Mitchell, R.L., *Soil Sci.* **83** (1971), 1.
80. Ivanov, D.N., *Pedology* **34** (11) (1939), 4.
81. Rogers, L.H., *J. Opt. Soc. Am.* **31** (1941), 260.
82. Oertel, A.C., *J. Counc. Scient. Ind. Res. Aust.* **17** (1944), 225.
83. Pinta, M., *Am. Agron.* **2** (1955), 189.
84. Ahrens, L.H., and Taylor, S.R., *Spectrochemical Analysis*, 2nd edn. Pergamon, London (1961).
85. Specht, A.W., Myers, A.T., and Oda, U., in *Methods of Soil Analysis*, eds. Black, C.A., Evans, D.D., White, J.L., Ensminger, L.E. and Clark, F.E., Amer. Soc. Agronomy Inc. Madison (1965), 822.
86. Scott, R.O., Burridge, J.C., and Mitchell, R.L., *Proc. XV Coll. Spectros. Internat.*, Madrid, 1 (1970), 56.
87. Scott, R.O., Mitchell, R.L., Purves, D. and Voss, R.C., Bull. Consult. Comm. Devel. Spectogr. Work. Macaulay Inst., Aberdeen No.2 (1971).
88. Scott, R.O. and Mitchell, R.L., *J. Soc. Chem. Ind. London* **62** (1943), 4.
89. Mitchell, R.L. and Scott, R.O., *Appl. Spectrosc.* **11** (1957), 6.
90. Mitchell, R.L. and Scott, R.O., *J. Soc. Chem. Ind. London* **66** (1947), 330.
91. Mitchell, R.L., in *Trace Elements in Soils and Crops*, Min. Agric. Fish. Food, Tech. Bull. 21 (1971), 8.
92. Mitchell, R.L., in *Chemistry of the Soil*, ed. F.E. Bear, Reinhold, New York (1964), 428.
93. Berrow, M.L. and Webber, J., *J. Sci. Fd. Agric.* **2** (1972), 93.
94. Berrow, M.L. and Stein, W.M., *Analyst* **108** (1983), 277.
95. Stupar, J. and Ajlec, R., *Analyst* **107** (1982), 144.
96. Ebdon, L. and Sparks, S.T., *European Winter Conf. on Plasma Spectrochemistry*, (1985) Leysin, Switzerland.
97. Hinds, W.W. and Jackson, K.W., *J Anal. Atom. Spectrom.* **3** (1988), 997.
98. Hinds, W.W. and Jackson, K.W., *J Anal. Atom. Spectrom.* **2** (1987), 441.
99. Hinds, W.W., Kutyal, M. and Jackson, K.W., *J. Anal. Atom. Spectrom.* **3** (1988) 83.
100. Williams, J.G., Gray, A.L., Norman, P. and Ebdon, L., *J. Anal. Atom. Spectrom.* **2** (1987), 469.
101. Carrondo, M.J.T., Perry, R. and Lester, J.N., *Sci. Total Envirn.* **12** (1979), 1.
102. Stoveland, S., Astruc, M., Perry, R. and Lester, J.N., *Sci. Total Environ.* **13** (1979), 33.
103. Tölg, G., *Pure and Appl. Chem.* **44** (1975), 645.
104. Tatsumoto, M., *Anal. Chem.* **41** (1969), 2088.
105. Zilbershtein, K.I., Piryutko, M.M., Nikitina, O.N., Fedorov, Y.F. and Nenarokov, A.N., *Zavodsk. Lab.* **29** (1963), 1266.
106. Mitchell, J.W. and Nash, D.L., *Anal. Chem.* **46** (1974), 327.
107. Woolley, J.F., *Analyst* **100** (1975), 896.

108. Block, R., *Decomposition Methods in Analytical Chemistry*, trans. Marr, I.L. Int. Textbook Co., London (1979), 57.
109. Langmyhr, F.J. and Paus, P.E., *Anal. Chem.* **43** (1968), 397.
110. Bernas, B., *Anal. Chem.* **40** (1968), 1682.
111. Pakalus, P., *Anal. Chim. Acta* **69** (1974), 211.
112. Kreshkov, A.P., Myshlyaeva, L.V. and Sibirtseva, A.B., *Zh. Analit. Kim.* **24** (1969), 1194.
113. Lechler, P.J. Roy, W.R. and Leininger, R.K., *Soil Sci.* **130** (1980), 238.
114. Holak, W., Kritiz, B. and Willaims, J.C., *J. Assoc. Offic. Anal. Chem.* **55** (1972), 741.
115. Holak, W., *J. Assoc. Offic. Anal. Chem.* **58** (1975), 777.
116. Borggard, O.K. and Willems, M., *J. Assoc. Offic. Anal. Chem.* **65** (1982), 762.
117. ASTM Spec, Tech. Publ., **1062** (1990), 259.
118. Bettinelli, M., Baroni, U. and Pastorelli, N., *Anal. Chim. Acta*, **225** (1989), 159.
119. Papp, C.S.E. and Fischer, L.B., *Analyst* **112** (1987), 337.
120. Nakashima, S., Sturgeon, R.E., Willie, S.N. and Berman, S.S., *Analyst* **113** (1988), 159.
121. Beary, S., *Anal. Chem.* **60** (1988), 742.
122. Ingamells, C.O., *Talanta* **11** (1964), 665.
123. Ingamells, C.O., *Anal. Chem.* **38** (1966), 1228.
124. Suhr, N.H. and Ingamells, C.O., *Anal. Chem.* **38** (1966), 730.
125. Ure, A.M., in *Soil Analysis*, ed. Smith, K.A., Marcel Dekker, New York (1983), 30.
126. Verbeek, A.A., Mitchell, M.C. and Ure, A.M., *Anal. Chim. Acta* **135** (1982), 215.
127. Brotzen, O., Kvalheim, A. and Marmor, V. in *Geochemical Prospecting, in Fennoscandia*, Interscience, New York (1967), 89.
128. Hesse, P.R., *A Textbook of Soil Chemical Analysis*, Murray, London (1971).
129. Pahlavanpour, B., Thompson, M. and Thorne, L., *Analyst* **105** 91980), 756.
130. Pahlavanpour, B., Weatley, M. and Thompson, M., *Int. Conf. on Environmental Pollution*, London, CEP Consultants, Edinburgh (1984), 816.
131. McGrath, S.P. and Cunliffe, C.H., *J. Sci. Fd. Agric.* **87** (1985), 163.
132. Hawke, D.T. and Lloyd, A., *Analyst* **113** (1988), 413.
133. Archer, F.C., in *Inorganic Pollution and Agriculture*, MAFF Reference Book 326, HMSO. London (1980), 184.
134. Scott, K., *Analyst* **103** (1978), 754.
135. Chao, S.C. and Pickett, E.E., *Anal. Chem.* **52** (1980), 335.
136. Raptis, S.E., Kaiser, G. and Tölg, G., *Fresenius' Z. Anal. Chem.* **316** (1983), 105.
137. *Methods for the Determination of Metals in Soils, Sediments and Sewage Sludge and Plants by Hydrochloric-citric Acid Digestion*...HMSO, London (1987).
138. Berrow, M.L. and Ure, A.M., *Environ. Technol. Letts.* **2** (1981), 485.
139. Ure, A.M., in *Soil Analysis*, ed. Smith, K.A., Marcel Dekker, New York (1983), Chapter 1.
140. Nakamura, Y., Nagai, H., Kubota, D. and Himeno, S., *Bunseki Kagaku* **22** (1973), 1543.
141. Van Loon, J.C., *Selected Methods of Trace Metal Analysis*, John Wiley, New York (1985), 30.
142. L'vov, B.V., *Ing. Fiz. Zhur.* **11** (1959), 44.
143. L'vov, B.V., *Spectrochim. Acta* **17** (1961), 761.
144. Massmann, H., *Z. Anal. Chem.* **225** (1967), 203.
145. Massmann, H., *Spectrochim. Acta* **23B** (1968), 215.
146. L'vov, B.V., *Spectrochim Acta* **36B** (1978), 153.
147. Slavin, W., Carnrick, G.R., Manning, D.C. and Prusekowska. E., *At. Spectrosc.* **3** (1983), 69.
148. Shan, X.-Q. and Ni, Z.-M., *Can. J. Spectrosc.* **27** (1982), 75.
149. Shan, X.-Q. and Wang, D.-X., *Anal. Chim. Acta* **173** (1985), 315.
150. Schlemmer, G. and Welz, B., *Spectrochim. Acta* **41B** (1986), 1157.
151. Shan, K.-Q and Ni, Z.-M., *Anal. Chim. Acta* **171** (1985), 269.
152. Knowles, M.B. and Brodie, K.G., *J. Anal. Atom. Spectrom.* **3** (1988), 511.
153. Jin, F. and Liu, F., *Guangpuxue Yu Guangpu Fenxi* **6** (1986), 45.
154. Bauslaugh, J., Radziuk, B., Saeed, K. and Thomassen, Y., *Anal. Chim. Acta* **165** (1984), 149.
155. Lin, F. and Liu, F., *Guangpuxue Yu Guangpu Fenxi* **6** (1986), 45.
156. Littlejohn, D., *Lab. Practice* **36** (1987), 121.
157. Koizumi, H. and Yasuda, K., *Spectrochim. Acta* **31B** (1976), 523.
158. Grassam, E., Dawson, J.B. and Ellis, J.D., *Analyst* **102** (1977), 804.
159. Sotera, J. and Kahn, H.L., *Am. Lab.* **14** (1982), 100.

160. van der Lee, J.J., Temminghoff, E., Houba, V.G., and Novozamsky, I., *Appl. Spectrosc.* **41** (1987), 388.
161. Poole, S., *Commun. Soil Sci. Plant Anal.* **19** (1988), 1681.
162. Elsheimer, H.N. and Fries, T.L., *Anal. Chim. Acta*, **239** (1990), 145.
163. Brooks, R.R., and Lee, B.-S., *Anal. Chim. Acta*, **204** (1988), 333.
164. Hinds, M.W. and Jackson, K.W., *J. Anal. Atom. Spectrom.* **3** (1988), 997.
165. Hinds, M.W., Latimer, K.E. and Jackson, K.W., *J. Anal. Atom. Spectrom.* **6** (1991), 473.
166. Akatsuka, K. and Atsuya, I., *Anal. Chim. Acta*, **202** (1987), 223.
167. Dubois, J.P., *J. Trace Microprobe Tech.* **9** (1991), 149.
168. Ma, Y., Bai, J., Wang, J., Li, Z., Zhu, L., Li Zheng, H. and Li, B., *J. Anal. Atom. Spectrom.* **7** (1992), 425.
169. Littlejohn, D., Egila, J.N., Gosland, S.M., Kunwar, U.K., Smith, C. and Shan, Xiao-Quan, *Anal. Chim. Acta*, **250** (1991), 71.
170. Ure, A.M., Thomas, R. and Littlejohn, D., *Internat. J. Environ. Anal. Chem.* **51** (1993), 65.
171. Lundegardh, H., *Die Quantitative Spektralanalyse der Elemente*, Vol. a Gustav Fischer, Jena (1929).
172. Dean, J.A. and Rains, T.C., *Flame Emission and Atomic Absorption Spectrometry*, Vol. 3, Marcel Dekker, New York (1975).
173. Walsh, A., *Spectrochim. Acta* **7** (1955), 108.
174. John, M.K., *J. Environ. Qual.* **1** (1972), 295.
175. Greenfield, S., Jones, I.L. and Berry, C.T., *Analyst* **89** (1964), 713.
176. Greenfield, S., McGeachin, H. McD., *Chem. Brit.* **16** (1980), 653.
177. Scott, R.H., Fassel, V.A., Kniseley, R.N and Nixon, D.E., *Anal. Chem.* **46** (1974), 76.
178. Boumans, P.W.J.M. and de Boer, F.J., *Spectrochim. Acta* **27B** (1972), 391.
179. Boumans, P.W.J.M. and de Boer, F.J., *Spectrochim. Acta* **30B** (1975), 309.
180. Anderson, T.A., Burns, D.W. and Parsons, M.L., *Spectrochim. Acta.* **39B** (1984), 559.
181. Sharp, B.L., in *Soil Analysis*, 2nd edn, ed. Smith, K.A., Marcel Dekker, New York (1989).
182. Babington, R.S., US Patents 3421 692; 3421 699; 3425 058; 3425 059; 3504 859.
183. Sharp, B.L., *J. Anal. Atom. Spectrom.* **3** (1988) Part I, 652; Part II, 939.
184. Liversage, R., Van Loon, J.C. and Andrade, J.C., *Anal. Chim. Acta.* **161** (1984) 275.
185. Wang, X. and Barnes, R.M., *Spectrochim. Acta* **41** (1986), 967.
186. Welz, B. and Schubert-Jacobs, M., *J. Anal. Atom. Spectrom.* **1** (1986), 23.
187. Agterdenbos, J., van Noort, J.P.M., Peters, F.F. and Box, D., *Spectrochim. Acta.* **41B** (1986), 283.
188. Ramsey, M.H. and Thompson, M., *J. Anal. Atom. Spectrom.* **1** (1986), 185.
189. Myers, S.A. and Tracy, D.H., *Spectrochim. Acta.* **39B** (1983), 1227.
190. Ramsey, M.H. and Thompson, M., *Analyst* **110** (1985), 519.
191. Ramsey, M.H. and Thompson, M., *J. Anal. Atom. Spectrom.* **2** (1987), 497.
192. Zhang, Peixun, Littlejohn, D. and Neal, P., *Spectrochim. Acta.* **48B** (1993), 1517.
193. Mohammad, B., Ure, A.M. and Littlejohn, D., *J. Anal. Atom. Spectrom.* **8** (1993), 325.
194. Alexandrova, A. and Arpadjan, S., *Analyst*, **118** (1993), 1309.
195. Broekaert, NATO ASI Ser. G (Metal Speciation in the Environment) **23** (1990), 213.
196. Zarcinas, B.A. and Cartwright, B., *Analyst*, **112** (1987), 1107.
197. Jeffrey, A.J. and McCallum, L.E., *Commun. Soil Sci. Plant Anal.* **19** (1988), 663.
198. Somasiri, L.L.W., Birnie, A. and Edwards, A.C., *Analyst*, **116** (1991), 601.
199. Boon, D.Y. and Soltanpour, P.N., *Commun. Soil Sci. Plant Anal.* **22** (1991), 369.
200. Ebdon, L. and Collier, A.R., *J. Anal. Atom. Spectrom.* **3** (188), 557.
201. Hawke, D.J. and Lloyd, A., *Analyst*, **113** (1988), 413.
202. Sturgeon, R.E., Willie, S.N., Luong, Van T. and Berman, S.S., *J. Anal. Atom. Spectrom.* **5** (1990), 635.
203. Hwang, J.D., Huxley, H.P., Diomiguardi, J.P. and Vaughn, W.J., *Appl. Spectrosc.* **44** (1990), 491.
204. Morita, S., Kim, C.K., Takatu, Y., Seki, R. and Ikeda, N., *Applied Rad. Isot.* **42** (1991), 531.
205. Kim, C.K., Otsuji, M. Takatu, Y., Kawamura, H., Shiraishi, K., Igarashi, Y., Igarashi, S. and Ikeda, N., *Radioisotopes* **38** (1989), 151.
206. Holak, W., *Anal. Chem.* **41** (1969), 1712.
207. Lakanen, E., and Ervio, R., *Acta Agr. Fenn.* **123** (1971), 223.
208. Braman, R.S., Justin, L.L. and Foreback, C.C., *Anal. Chem.* **44** (1972), 2195.

209. Schmidt, F.J. and Royer, J.L., *Anal. Letts.* **6** (1973), 17.
210. Pollock, E.N. and West, S.J., *At. Absorpt. Newslett.* **12** (1973), 6.
211. Fernandez, F.J., *At. Absorpt. Newslett.* **12** (1973), 93.
212. Thompson, K.C. and Thomerson, D.R., *Analyst*, **99** (1974), 595.
213. Thompson, M., Pahlavanpour, B. and Walton, S.J., *Analyst* **103** (1978), 568.
214. Kahn, H.L. and Schallis, J.E., *At. Absorpt. Newslett.* **7** (1968), 5.
215. Fiorino, J.A., Jones, T.W. and Capar, S.G., *Anal. Chem.* **48** (1976), 120.
216. May, I. and Greenland, P.L., *Anal. Chem.* **49** (1977), 2376.
217. Chu, R.C., Barman, P.G. and Baumgarner, P.A.W., *Anal. Chem.* **44** (1972), 1476.
218. Van Cleuvenbergen, J.A.R., Van Mol, W.E. and Adams, F.C., *J. Anal. Atom. Spectrom.* **3** (1988), 169.
219. Collett, D.L., Fleming, D.E. and Taylor, G.A., *Analyst* **103**, (1978), 1074.
220. Anderson, R.K., Thompson, M. and Culbard, E., *Analyst* **111** (1986), 1143.
221. Siemer, D.D. and Hagemann, L., *Anal. Lett.* **8** (1975), 323.
222. Siemer, D.D., Koteel, P. and Jarawala, V., *Anal. Chem.* **48** (1976), 836.
223. Siemer, D.D., Vitek, R.K., Koteel, P. and Houser, W.L., *Anal. Lett.* **10** (1977), 357.
224. Dedina, J. and Rubeska, I., *Spectrochim. Acta.* **35B** (1980), 119.
225. Goulden, P.D. and Brooksbank, P.A., *Anal. Chem.* **46** (1974), 1431.
226. Knudson, E.J. and Christian, G.D., *Anal Letts.* **6** (1973), 1039.
227. McDaniel, M., Shendrikar, A.D., Reiszner, K.D. and West, P.W., *Anal. Chem.* **14** (1976), 2240.
228. Sturgeon, R.E., Willie, S.N and Berman, S.S., *J. Anal. Atom. Spectrom.* **57** (1985), 2311.
229. Willie, S.N., Sturgeon, R.E. and Berman, S.S., *Anal. Chem.* **58** (1986), 1140.
230. Sturgeon, R.E., Willie, S.N. and Berman, S.S., *J. Anal. Atom. Spectrom.* **1** (1986), 115.
231. Mohammad, B., The Development of a Continuous Flow Hydride Generation Atomic Absorption Spectrochemical Method for Determination of Antimony in Environmental Materials (1989). MSc Thesis, Strathclyde University, Glasgow.
232. Van Der Veen, N.G., Keukens, H.J. and Vos. G., *Anal. Chim. Acta.* **171** (1985), 285.
233. Lau, C., Held, A. and Stephens, R., *Can. J. Spectrosc.* **21** (1976), 100.
234. Khaligie, J., Ure, A.M. and West, T.S., *Anal. Chim. Acta.* **107** (1979), 191.
235. Lau, C.M., Ure, A.M. and West, T.S., *Anal. Chim. Acta.* **141** (1982), 213.
236. Hallam, C. and Thompson, K.C., *Analyst* **110** (1985), 497.
237. Lau, C.M., Ure, A.M. and West, T.S., *Anal. Chim. Acta.* **146** (1983), 171.
238. Fraser, S.M., Ure, A.M., Mitchell, M.C. and West, T.S., *J. Anal. Atom. Spectrom.* **1** (1986), 19.
239. Zolotov, Yu.A., *Pure and Appl. Chem.* **50** (1978), 129.
240. Morrison, G.H. and Freiser, H., *Solvent Extraction in Analytical Chemistry*, Wiley, New York (1957).
241. Cresser, M.S., *Solvent Extraction in Flame Spectroscopic Analysis*, Butterworths, London (1978).
242. Malissa, H. and Schoffman, E., *Microchim. Acta.* **1** (1955), 187.
243. Lakanen, E., *At. Absorpt. Newslett.* **5** (1966), 17.
244. Kirkbright, G.F. and Sargent, M., *Atomic Absorption and Fluorescence Spectrometry*, Academic Press, London (1974), 491.
245. Ebervia, B., Macalalad, E. Rogue, N. and Rubeska, I., *J. Anal. Atom. Spectrom.* **3** (1988), 199.
246. De, A.K., Khopkar, S.M. and Chalmers, R.A., *Solvent Extration of Metals*, Van Nostrand Reinhold, London (1970), 128.
247. Ure, A.M. and Mitchell, M.C., *Anal. Chim. Acta.* **87** (1976), 283.
248. Ure, A.M. and Hernandez, M.P., *Anal. Chim. Acta.* **94** (1977), 195.
249. Hatch, W.R. and Ott, W.L., *Anal. Chem.* **40** (1968), 2085.
250. Iskandar, I.K., Syers, J.K., Jacobs, L.W., Keeney, D.R and Gilmour, J.T., *Analyst* **97** (1972), 388.
251. Ure, A.M. and Shand, C.A., *Anal. Chim. Acta.* **72** (1974), 63.
252. Hoover, W.L., Melton, J.R. and Howard, P.A., *J. Assoc. Offic. Anal. Chem.* **54** (1971), 860.
253. Ure, A.M., *Anal. Chim. Acta.* **76** (1975), 1.
254. Chilov, S., *Talanta* **22** (1975), 205.
255. Mitchell, R.L., *Geol. Soc. Am. Bull.* **83** (1972), 1069.

256. Cox, F.R. and Kamprath, E.J., in *Micronutrients in Agriculture*, eds. Morvedt, J.J., Giordano, P.M. and Lindsay, W.L., Soil Sci. Soc. Amer., Madison, Wisconsin (1972), 289.
257. Chapman, H.D., ed. *Diagnostic Criteria for Plants and Soils*, University of California Press (1966).
258. Dolar, S.G. and Keeney, D.R., *J. Sci. Fd. Agric.* **22** (1971) 273 and 279.
259. Borggard, O.K., *Acta. Agric. Scand.* **26** (1976), 144.
260. Lindsay, W.L. and Norvell. W.A., *Soil Sci. Soc. Amer. J.* **41** (1978), 421.
261. Ministry of Agriculture, Fisheries and Food, *The Analysis of Agricultural Materials*, Tech,. Bull. 27 HMSO, London (1973).
262. Sillanpaa, M., *Micronutrients and the Nutrient Status of Soils*, FAO SOil Bulletin 48, FAO Rome (1982).
263. Leschber, R., Davis, R.D. and L'Hermite, P. eds., *Chemical Methods for Assessing Bioavailable Metals in Sludges and Soils*, Elsevier, Amsterdam (1985).
264. Sauerbeck, D.R. and Styperek, P., in *Chemical Methods for Assessing Bioavailable Metals in Sludges and Soils*, eds. Leschber, R., Davis, R.D. and L'Hermite, P. Elsevier, Amsterdam (1985), 49.
265. Häni, H. and Gupta, S., in *Environmental Effects of Organic and Inorganic Contaminants in Sewage Sludge*, eds. Davis, R.D., Hucker, G. and L'Hermite, P.D., Reidel, D., Dordrecht (1982), 121.
266. Davis, R.D., *J. Sci. Fd. Agric.* **30** (1979), 937.
267. Ellis, R.H. and Alloway, B.J., *Proc. Int. Conf. on Heavy Metals in the Environment*, Heidelberg, CEP Consultants, Edinburgh (1983), 358.
268. Christensen, T.H. and Tjell, J.C., *Proc. Int. Symp. Processing and Use of Sewage Sludge* (1983), Brighton (1984), 358.
269. Sanders, J.R., McGrath, S.P. and Adams, T.McM., *J. Sci. Fd. Agric.* **37** (1986), 961.
270. Adams, T.McM, and Sanders, J.R., *Int Conf. on Environmental Contamination*, London, CEP Consultants, Edinburgh (1984), 400.
271. Sanders, T.R., *J. Soil Sci.* **34** (1983), 315.
272. Linehan, D.J., Sinclair, A.H. and Mitchell, M.C., *Plant and Soil* **86** (1985), 147.
273. Viets, F.G., *J. Agric. Fd. Chem.* **10** (1962), 174.
274. Jones, K.C., Peterson, P.J. and Davies, B.E., *Geoderma* **33** (1984), 157.
275. Tiller, K.G., Honeysett, J.L. and DeVries, M.P.G., *Soil Sci.* **10** (1972), 165.
276. Pickering, W.F., *Ore Geology Reviews* **1** (1986), 100.
277. Chao, T.T., *Soil Sci. Soc. Amer. Proc.* **36** (1972), 764.
278. Salomans, W. and Förstner, U., *Environ. Lett.* **1** (1980), 506.
279. Heath, G.R. and Dymond, J., *Geol. Soc. Am. Bull.* **88** (1977), 723.
280. McKeague, J.A., *Can. J. Soil Sci.* **5** (1967), 167.
281. De Groot, A.J., Zschuppel, K.H and Saloman, W., *Hydrobiologia* **92** (1982), 689.
282. Kersten, M. and Förstner, U., *Water Sci. Technol.* **18** (1986), 121.
283. Meguellati, M., Robbe, D., Marchandise, P. and Astruc. M., *Proc. Int. Conf., Heavy Metals in the Environment*, Heidelberg, CEP Consultants, Edinburgh (1983), 1090.
284. Lum, K.R. and Edgar, D.C., *Analyst* **108** (1983), 918.
285. Tessier, A., Campbell, P.G.C. and Bisson, M., *Anal. Chem.* **51** (1979), 844.
286. Fischer, W.L. and Fechter, H., Z. *Pfanzenernahr. Bodenk.* **145** (1982), 151.
287. Shuman, L.M., *J. Soil Sci. Amer.* **46** (1982), 1099.
288. Miller, W.P., McFee, W.W. and Kelly, J.M., *J. Environ. Qual.* **12** (1983), 579.
289. Shuman, L.M., *Soil Sci.* **140** (1985), 11.
290. McLaren, R.G. and Crawford, D.V., *J. Soil Sci.* **24** (1973), 172.
291. Shuman, L.M., *Soil Sci.* **127** (1979), 10.
292. Sposito, G., Lund, L.J. and Chang, A.C., *J. Soil Sci. Soc. Amer.* **46** (1982) 260.
293. Stoner, R.G., Sommers, L.E. and Silviera, D.J., *J. Water Pollution Control Fed.* **48** (1976), 2165.
294. Gatehouse, S., Russell, .W. and Van Moort, J.C., *J. Geochem. Explor.* **8** (1977), 483.
295. Soon, Y.K., Bates, T.E. and Moyer, J.R., *J. Environ. Qual.* **9** (1980), 497.
296. Petrozelli, G. Giudi, G. and Lubrano, L., *Proc. Int. Conf. Heavy Metals in the Environment*, Heidelberg, CEP Consultants, Edinburgh (1983), 475.
297. Lakanen, E. and Ervio, R., *Acta. Agr. Fenn.* **123** (1971), 233.
298. Ure, A.M., Quevauviller, Ph., Muntau, H. and Griepink, B., *Internat. J. Environ. Anal. Chem.*,

S51 (1993), 135.
299. Eadem, Report EUR 14763 EN, Commission of the European Communities, Community Bureau of Reference, Brussels 1993, 85 pp.
300. Murthy, A.S.P. and Schoen, H.G., *Plant and Soil* **102** (1987), 207.
301. Rudd, T., Lake, L., Mehotra, I., Sterritt, R.M., Campbell, J.A. and Lester, J.N., *Sci Total Environ.* **74** (1988), 149.
302. Kodama, H., Schnitzer, M. and Murad, E., *Soil Sci. Soc. Amer. J.* **52** (1988), 994.
303. Hareland, W.A., Grant, S.E., Ward, S.P. and Anderson, D.R., *Appl. Spectrosc.*, **41** (1987), 1428.
304. Senesi, N., *Anal. Chim. Acta.* **232** (1990), 51 and 77.
305. Haswell, S.J., O'Neill, P. and Bancroft, K.C.C., *Talanta* **32** (1985), 69.
306. Tills, A.R. and Alloway, B.J., *Int. Conf. Heavy Metals in trhe Environment*, Heidelberg Vol. 2, CEP Consultants, Edinburgh (1983), 1211.
307. Walsh, J. and Howie, R.A., *Appl. Geochem.* **1** (1986), 161.
308. Thompson, M. and Zao, L., *Analyst* **110** (1985), 229.
309. Welsch, E.P., *Talanta* **32** (1985), 996.
310. Huang, Y.Q. and Wai, C.M., *Commun. Soil Sci. Plant Anal.* **17** (1986), 125.
311. Donaldson, E.M., *Talanta* **33** (1986), 233.
312. Chan, C.Y., *Anal. Chem.* **57** (1985), 1482.
313. Terashima, S., *Geostandard Newslett.* **10** (1986), 127.
314. Terashima, S., *Bunseki Kagaku* **33** (1984), 561.
315. Ek, P. and Hulden, S.G., *Talanta* **34** (1987), 495.
316. Legret, M. and Divet, L., *Anal. Chim. Acta.* **189** (1986), 313.
317. Crock. J.G., *Anal. Lett.* **19** (1986), 1367.
318. Yang, M. and Guo, X., *Guangpuxui Yu Gecangpu Fenxi* **6** (1986), 53.
319. van der Veen, N.G., Keubens, H.J. and Vos, G., *Anal. Chim. Acta.* **171** (1985), 285.
320. Hon, P.K., Lau, O.W. and Tsui, S.K., *J. Anal. Atom Spectrom.* **1** (1986), 125.
321. Sarx, B. and Baechmann, K., *Atom. Spektrom. Spurenanal.* **2** (1986), 619.
322. Pohl, B. and Baechmann, K., *Fres. Z. Anal. Chem.* **323** (1986), 859.
323. Cutter, G.A., *Anal. Chem.* **57** (1985), 2951.
324. Blais, J.S. and Marshall, W.D., *J. Environ. Qual.* **15** (1986), 255.
325. Larkins, P.L., *Anal. Chim. Acta.* **173** (1985), 77.
326. Yang, Z. and Mao, Z., *Fenxi Huaxue* **14** (1986), 457.
327. Zhang, J. nd Zhang, Q., *Yankuang Ceshi* **5** (1986), 37.
328. Yang, M. and Guo, X., *Fexi Huaxue* **14** (1986), 333.
329. Bol'shov, M.A., Zybin, A.V., Kolomiiskii, Yu, R. Koloshnikov, V.G. Loginov, Yu. M. and Smirenkiva, I.I., *Zh. Anal. Khim.* **41** (1986), 402.
330. Hemens, C.M. and Elson, C.M., *Anal. Chim. Acta.* **188** (1986), 311.
331. Potts, P.J., Webb, P.C., Watson, J.S and Wright, D.W., *J. Anal. Atom. Spectrom.* **2** (1987), 67.
332. Coetzee, P.P., Hoffmann, P., Speer, R. and Lieser, K.H., *Fres. Z. Anal. Chem.* **363** (1986), 254.
333. Bansal, R.L., Taklar, P.N., Sanota, N.S. and Mann, M.S., *Field Crops Research* **3** (1980), 43–51.
334. Mahler, R.J., Bingham, F.T., Sposito, G. and Page, A.L., *J. Environ. Qual.* **9** (1980), 359–364.
335. Mitchell, G.A., Bingham, F.T. and Page, A.L., *J. Environ. Qual.* **7** (1974), 165.
336. Davis, R.D., *J. Sci. Fd. Agric.* **30** (1979), 937–47.
337. Clayton, P.M. and Tiller, K.G., *CSIRO Div. of Soils. Tech. Papers* **41** (1971), 1–17.
338. Haq, A.U. and Miller, M.H., *Agron. J.* **64** (1972), 779.
339. Williams, C. and Thornton, I., *Plant and Soil* **39** (1973), 149–159.
340. Soltanpour, P.N. and Schwab, A.P., *Commun. Soil Sci. Plant Anal.* **8** (1977), 195.
341. Lindsay, W.L. and Norvell, W.A., *Water Pollution Control* **82** (1978), 421–8.
342. Latterell, J.J., Dowdy, R.H. and Larson, W.E., *J. Environ. Qual.* **7** (1978), 435–440.
343. Street, J.J., Lindsay, W.L. and Sabey, B.R., *J. Environ. Qual.* **6** (1977), 72–77.
344. Sillanpaa, M., FAO Soil Bulletin 48, *Micronutrients and the Nutrient Status of Soils*, FAO, Rome, (1982).
345. Ellis, R.J. and Alloway, B.J., *Proc. Int. Conf. on Heavy Metals in the Environment*, Heidelberg, CEP Consultants, Edinburgh (1983), 358.

346. Mitchell, R.L., Reith, JW.S. and Johnston, I.M., *J. Sci. Fd Agric.* **8** (1957), 51–59.
347. Haq, A.U., Bares, T.E. and Soon, K., *J. Soil Sci. Soc. Amer.* **44** (1980), 772–777.
348. Burridge, J.C. and Berrow, M.L., *Proc. Int. Conf. on Environmental Contamination*, London, CEP Consultants, Edinburgh (1984), 215.
349. Misra, S.G. and Pande, P., *Plant and Soil* **41** (1974), 697–700.
350. Fleming, G.A. and Murphy, W.E., Internal report on An Investigation of a Nitrogenous Waste Water as a Source of Nitrogen for Grassland. Johnstown Castle Research Centre, Wexford, Ireland (1985).
351. Sedberry, J.E., Miller, B.J. and Said, M.B., *Commun. Soil Plant Anual.* **10** (1979), 689–701.
352. Berrow, M.L., Burridge, J.C. and Reith, J.W.S., *J. Sci. Fd. Agric.* **34** (1983), 53–54.
353. Pierzynski, G.M., Crouch, S.R. and Jacobs, L.W., *Commun. Soil Sci. Plant Anal.* **17** (1986), 419.
354. *Macaulay Institute Annual Report*, No. 52 (1981/2), 58.

Section 2 — Individual elements

5 Arsenic

P. O'NEILL

5.1 Introduction

Arsenic has achieved great notoriety because of the toxic properties of a number of its compounds. Fortunately there are great differences in the toxicity of different compounds, and the species that are most commonly found in soils are not the most toxic. The uptake of As by many terrestrial plants is not very great so that, even on relatively high As soils, plants do not usually contain dangerous levels of As.

Arsenic compounds appear to have been used by humans for several thousand years. Kipling [1] provides a general account of the history of As usage (and abusage) and reviews [2, 3, 4] of health effects and metabolic changes in animals, plants and humans give more details than can be provided in this short chapter.

The major present-day uses of As compounds are as pesticides, wood preservatives, and as growth promoters for poultry and pigs.

A review [5] of the global recycling of As suggested that natural inputs to the atmosphere were 45000 t As/yr whereas anthropogenic sources added 28000 t As/yr. In soils the natural levels are dependent on the source rock type, and the normal range is 1–40 mg As/kg with most soils being in the lower half of this range [6, 7, 8, 9]. Levels may be elevated due to mineralisation, contamination from industrial activity (especially Cu smelters) and the use of As-based pesticides.

Arsenic differs from many of the common heavy metals in that the majority of the organo As compounds are less toxic than inorganic As compounds. Whilst having many chemical similarities to P, the soil chemistry of As is much more diverse because it can exist in more than one oxidation state under the normal range of soil conditions and As can form bonds with S and C much more readily than does P.

5.2 Geochemical occurrence

Over 200 As-containing minerals have been identified, with approximately 60% being arsenates, 20% sulphides and sulphosalts and the remaining 20% including arsenides, arsenites, oxides and elemental As [6]. The most common of the As minerals is arsenopyrite, FeAsS, and As is found associated with many types of mineral deposits especially those including sulphide mineralisation [7]. The concentration of associated As can range from a few parts per million up to percentage quantities.

There is relatively little difference in the concentration of As in rocks unless the levels have been raised by associated mineralisation. Mean values of the order of 2 mg/kg appear to be those most commonly found for igneous and sedimentary rocks, but the finer grained argillaceous rocks and phosphorites average 10–15 mg As/kg. A detailed study [10] of the Peterborough Member of the Oxford Clay Formation (an organic-rich, argillaceous rock) indicated that the As concentration (5–11 mg/kg) appears to be related to the reduction of sulphate to sulphide as sea water percolated through the sediment. Higher levels of As are often associated with the presence of sulphide minerals such as pyrites. The As contents of metamorphic rocks reflect those of the original igneous or sedimentary rocks.

Arsenic has often been used as a 'pathfinding' or indicator element when geochemical prospecting methods have been utilised to identify mineral deposits. It is a particularly good indicator because it is associated with a wide variety of mineral deposits; the general background levels of As in rocks are low; it often forms more volatile and soluble phases than the major elements with which it is associated and therefore the As halo is more widely dispersed; sensitive methods of analysis have been developed. It is especially useful for Au and Ag deposits but is also associated with Bi, Cd, Co, Cu, Fe, Hg, Mo, Ni, Pb, Pt metals, Sb, Se, Sn, U, W and Zn deposits.

Whilst the major simple As minerals such as arsenopyrite, orpiment, As_2S_3, realgar, AsS, and enargite, Cu_3AsS_4, have been mined in their own right as sources of As, nowadays most As is produced as a by-product of the extraction of Cu, Pb, Au and Ag from their ores.

5.3 Origin of arsenic in soils

5.3.1 Soil parent materials

There is little difference between the various types of igneous rocks with concentrations of < 1 –15 mg As/kg. The argillaceous sedimentary rocks (shales, mudstones, slates) have significantly higher As levels (< 1–900 mg/kg), than sandstones and limestones (< 1–20 mg/kg). Phosphate rocks have range of < 1–200 m/kg.

The ability of As to bind to S ligands means that it tends to be found associated with sulphide-bearing mineral deposits either as separate As minerals or as a trace or minor constituent of other sulphide minerals. This leads to elevated levels in soils in many mineralised areas.

An overall mean As value for 2691 uncontaminated soils was 10 mg/kg [8]. In Poland 127 soils [11] had a geometric mean (*GM*) As concentration of 2.63 mg/kg (range 0.5–15 mg/kg) with sandy soils having the lower mean concentration (*GM* 1.99 mg/kg) and silty soils having higher concentrations (*GM* 4.62 mg/kg). Similar low values (*GM* 1.1 mg As/kg) have been found in sandy soils in Germany [12]. Background levels in the USA (*GM* 5.2 mg As/kg) [13] are similar to those in

Alaska (*GM* 6.7 mg As/kg) [14]. Results from 2982 soils in mainland China, including Tibet, gave higher values (*GM* 9.2 mg As/kg) with lithosols (terra rossa and purplish) and cold-highland soils (felic inceptisol) having the highest geometric mean concentrations 16–17 mg As/kg and alfisols the lowest concentrations (*GM* 4.95 mg As/kg) [15]. These values can be compared to those for a mineralised area and its surroundings where the 0–5 cm surface soils had mean concentrations of 424 mg/kg and 29–51 mg/kg for apparently non-mineralised areas also in SW England [16].

5.3.2 Agricultural materials

Arsenic compounds have been widely used as pesticides for over a hundred years, but their use is now declining, having probably halved in the decade from 1970 to 1980. The phytotoxic effects of As compound made them attractive as herbicides and as desiccants to allow cotton to be easily harvested after defoliation. However, there has been concern about the build-up of As residues in soils and lake sediments which has occurred after the use of large quantities of inorganic As compounds. As a consequence other pesticides have replaced As compounds such as lead arsenate, which was commonly used in orchards to control insect pests, and sodium arsenite, which was extensively employed as a herbicide to clear aquatic weeds and defoliate seed potatoes.

Worldwide usage has been recently estimated to be 8000 t As/yr as herbicide, 12000 t As/yr as cotton desiccant and 16000 t As/yr in wood preservatives [5]. The rate of application of the pesticides is generally in the range of 2–4 kg As/ha, but larger quantities of dimethylarsinic acid may be used with application rates being up to three times greater [2]. In addition, small quantities of organo As compounds are used as animal feed additives, at the rate of 10–50 mg As/kg feed, to promote growth in chickens, turkeys and pigs. The compounds are rapidly excreted, often with little chemical change having apparently taken place.

Soils have generally been less badly affected by As build-up than aquatic sediments unless lead arsenate has been used, when residuals of 100–200 mg As/kg soil with highs of over 2500 mg As/kg have been reported [17]. The lack of long-term build up of As in soils has been explained by referring to the production of volatile As compounds by microorganisms and leaching. The mobility of the As in aquatic sediments appears to have been reduced by the presence of abundant hydrated iron oxides and/or sulphide [18].

Phosphate fertilisers are a potential source of As. The concentration of As in the fertiliser will vary with the source of the phosphate rock used to produce the fertiliser, but estimates for the UK indicate an average burden of 7.7 mg As/kg rock phosphate [19]. This suggests that, using a weighted average fertiliser application rate of 54.5 kg P_2O_5/ha, the annual addition of As to arable land in the UK is about 0.12 mg As/m^2. This is equivalent to an increase in As concentration of less than 0.005% (if a mixing depth of 20 cm is assumed).

The levels of As in N and K fertilisers will be insignificant and the quantities of As added via liming will also be small, reflecting the small concentrations of As generally found in limestones. The concentrations of As in manure will reflect the dietary intake of the animals concerned. There is a general low level in feedstuffs except when As compounds are added in growth promoters. In these cases the concentration of As in the manure may be up to 30–40 mg/kg dry weight, but there does not seem to be a significant increase in As concentration in the crops grown on amended soils and the added As was rapidly removed [17].

5.3.3 *Atmospheric deposition*

The relatively high volatility of a number of As compounds means that the geochemical cycle of As contains significant fluxes passing through the atmosphere. However, the vapour phase has been estimated to make up only about 7% of the atmospheric burden, with the remaining As associated with the particulate phase.

An estimate of the atmospheric flux suggested a value of 73 540 t/yr with a 60:40 split between natural and anthropogenic sources [5]. This compares with previous estimates of 31 400 t/yr and a natural:anthropogenic ratio of 25:75 or of 296 470 t/yr and a ratio of 70:30. Nriagu has estimated anthropogenic atmospheric As fluxes as 18 800 t/yr. [20].

Measurements of As deposition at various rural areas in the UK have given values of 0.8–5.5 mg/m^2/yr [21]. If the average rate of deposition is taken as 1 mg/m^2/yr then the average increase in mass of As in the top 5 cm of soil would be 0.15% (this assumes average concentration in soil is 10 mg/kg and density is 1.4 g/cm^3). However, as about 35% of the atmospheric flux has been estimated to be due to low-temperature volatilisation [5] of As from soil, the net increase would be 0.1% per annum. On a global scale the average rate of increase would be about 0.05% for the Northern Hemisphere and 0.02% for the Southern Hemisphere, if the average deposition rates of 0.44–0.50 mg As/m^2/yr for the Northern Hemisphere and 0.16–0.21 mg As/m^2/yr for the Southern Hemisphere are accepted.

Volcanic action is the next most important natural source of As after low-temperature volatilisation, and on a local scale it will be the dominant atmospheric source.

Smelting of Cu is the largest single anthropogenic input, representing about 40% of the anthropogenic total, with coal combustion being the next most significant at about 20% of the total. There can be widely different quantities of anthropogenic atmospheric emissions, depending upon the industrialisation of a country and the degree of pollution control that is utilised. An estimate for emissions in Europe in 1979 [22] gave a total emission of 6500 t As from 28 countries, but 65% of the total came from only three countries (i.e., USSR, the German Federal Republic and Poland) with three other countries (Belgium, Spain and France) being responsible for another 14%.

5.3.4 *Sewage sludges*

The levels of As in sewage sludge reflect the degree of industrialisation of the area served by the sewage system. The As is mainly derived from surface run-off, bringing in atmospherically deposited As plus residues from pesticide usage. Phosphate detergents add small quantities, and industrial effluents, particularly from the metal processing industry, can add significant quantities. Ranges of 0–188 mg As/kg dry weigh have been quoted [17] and for the UK an average of 8 mg As/kg dry weight is suggested for sewage sludge disposed of on agricultural land, compared to 10 mg As/kg dry weight for all UK sludges [19]. In the UK the more contaminated sludges are usually incinerated, and they average 29 mg As/kg dry weight.

The disposal of sewage sludge on land is not thought to cause a significant increase in crops. At a typical addition rate of 5 t sludge/ha the rate of As addition is 4 mg/m^3. This is over 30 times greater than for additions from phosphate fertiliser, and represents an increase of about 0.15 % in the upper 20 cm of the soil. Because the area used for the disposal of sewage sludge is much smaller than that to which phosphate fertiliser is applied, about 2.5 t As/yr is added with sludge and 6.1 t As/yr with phosphate fertiliser in the UK as a whole. An estimate [23] of the total of global input of As to soils from sewage sludge gives a range of $0.01–0.25 \times 10^6$ kg/a. The maxiumum acceptable concentration in sludge applied to land varies from 10 mg/kg DW (Norway, Belgium) to 75 mg/kg (Canada).

5.3.5 Other sources of arsenic

The elevated As level within the gangue minerals associated with an economic ore mineral leads to the risk of wind dispersal and leaching of As from the finely ground spoil-tip and tailings-dam materials. This may lead to very high As concentrations close to old dumps, e.g. over 40 000 mg As/kg in Virginia, USA and over 25 000 mg/kg in SW England [24], though levels drop off rapidly with distance away from the dump. The rate of decline in As concentration is influenced by the degree of stabilisation of the spoil material, which affects wind dispersal, and the drainage pathways. It should be noted that in both the sites mentioned above vegetation of a limited type had colonised the sites despite the high As levels.

The large increase in the use of As compounds (either as copper chromium arsenate or ammonium copper arsenate) does not appear to be causing much direct contamination of soils. Some elevated As levels have been found in soils adjacent to treated posts (10–220 mg/kg) but these have declined to back-ground levels within a few centimetres [17]. Some of these posts had been in the ground for 30 years. The disposal of treated wood in the future may cause local problems if the wood is burnt rather than being disposed of at controlled landfill sites. With usage of As for wood treatment forecast to rise above the present 16000 t/yr, the potential for undesirable disposal of As is likely to increase.

As well as releasing As to the atmosphere, coal combustion also produces As containing ash. Levels of As are generally in the range 7–60 mg/kg, but values of over 200 mg As/kg have been reported [25]. In the UK pulverised fuel ash is commonly used to reclaim land previously used for sand and gravel extraction.

Oil shales which could be used as a source of petroleum compounds contain about 50 mg As/kg, and almost all of this As would be retained in the spent shale that needed to be disposed of after oil extraction. The large quantities of materials that would be utilised could cause problems of control if the extraction process ever became economic.

Irrigation with waters containing As might be expected to increase the concentration in As in the soil in arid areas as the water evaporated. Levels of As in surface and ground waters, whilst being generally low (2–3 μg/l), can be very high (up to 35 mg/l) if associated with hydrothermal activity [1, 2]. In Montana [26], however, irrigation of sandy soils by water with a mean concentration of 50 μg As/l appeared to be causing elevation of groundwater As concentrations from 26 μg/l to 150 μg/l by the recharging of a shallow aquifer with As-enriched water. Geothermal power production, where solutions containing high levels of As are used to raise steam, provides potential soil pollution problems.

Dredged material from rivers and estuaries may contain high As concentrations due to anthropogenic contamination of the rivers, e.g. the Rhine. The dredging of Rotterdam harbour produces sludge (23 mg As/kg) that is being used to reclaim land [27]. Whilst the initial porewater As concentration was 8 μg/l it is expected to rise to 100–160 μg/l and remain at that level, with little migration of As, for hundreds of years.

5.4 Chemical behaviour in the soil

Arsenic has an outer electronic configuration of $4s^2 4p^3$ and is in the nitrogen group of elements (N, P, As, Sb, Bi). The reduction in electronegativity that is found on descending this group is not sufficient to give As much metallic character or to produce simple cations. Arsenic is often described as a metalloid element, but for the purposes of describing its chemical behaviour in soil it can be thought of as a non-metal forming covalent compounds or being found in anionic species.

There is an apparent similarity between the chemistry of As and P in that they both commonly form oxyanions (arsenate and phosphate) in the +5 oxidation state in soils. However, phosphate is stable over a much wider range of E_h and pH conditions than is arsenate. Arsenic is also found in soils in the +3 oxidation state, and ligands other than O form stable species that are not found with P.

The natural sources of As in soils are mainly oxysalts and S containing minerals. The normal oxidising conditions at the Earth's surface under which weathering takes place lead to the formation of oxyanions based on the +5 oxidation state. The ranges of E_h and pH in soils can lead to either As(V) or As(III) with microbial activity causing methylation, demethylation and/or change in oxidation state and the presence of S species may, if the redox potential is low enough, favour the formation of arsenic sulphide minerals [28]. A further complicating factor may be the presence of clay minerals, Fe and Al oxides and organic matter which can influence solubility and rate of oxidation.

The equilibria for arsenous acid (As(III)) and arsenic acid (As(V)) in aqueous solutions are given below. The pK_a values indicate that the species that should be thermodynamically most stable over the normal soil pH ranges of 4–8 will be (i) H_3AsO_3 (up to about pH 9); (ii) $H_2AsO_4^-$ (approximately pH 2–7); (iii) $HAsO_4^{2-}$ (above pH 7);

Arsenic acid

$$H_3AsO_4 + H_2O \rightleftarrows H_2AsO_4^- + H_3O^+ \quad pK_a\ 2.20$$
$$H_2AsO_4^- + H_2O \rightleftarrows HAsO_4^{2-} + H_3O^+ \quad pK_a\ 6.97$$
$$HAsO_4^{2-} + H_2O \rightleftarrows AsO_4^{3-} + H_3O^+ \quad pK_a\ 11.53$$

Arsenous acid

$$H_3AsO_4 + H_2O \rightleftarrows H_2AsO_3^- + H_3O^+ \quad pK_a\ 9.22$$
$$H_2AsO_3^- + H_2O \rightleftarrows HAsO_3^{2-} + H_3O^+ \quad pK_a\ 12.13$$
$$HAsO_3^{2-} + H_2O \rightleftarrows AsO_3^{3-} + H_3O^+ \quad pK_a\ 13.4$$

When the E_h value drops below about +300 m V at pH 4 and −100 m V at pH 8, H_3AsO_3 becomes the thermodynamically more stable As species [3] in the absence of complexing species and methylating organisms. The rate of change in oxidation state with change in E_h/pH conditions does not always appear to be very rapid in aqueous systems. Therefore, the proportion of the various As species present in soil porewaters may not correspond to the expected distribution.

A change in the ratio of As(V) to As(III) can be brought about purely by an inorganic mechanism with Eh/pH changes, but the presence of microorganisms can also influence the reaction pathway. In particular, methylation of the oxyanions may occur forming monomethylarsonic acid, $CH_3AsO(OH)_2$; dimethylarsinic acid (cacodylic acid), $(CH_3)_2AsO(OH)$; trimethylarsenic oxide, $(CH_3)_3AsO$; trimethylarsine, $(CH_3)_3As$; dimethylarsine, $(CH_3)_2AsH$.

The biomethylation reactions that actually occur depend upon both the microorganisms and the As compounds present [3]. Some microorganism can fully methylate As compounds over a wide range of pH conditions, whereas many other microorganisms appear much more limited in the substrates that they can methylate and the degree of methylation they can produce.

$(CH_3)AsO(OH)_2$ has pK_a values of 4.19 and 8.77 at 25°C, therefore the monohydrogen ion, $(CH_3)AsO_2(OH)^-$, will generally be its main species found at normal soil pHs. However, as the pKa of dimethylarsinic acid is 6.27 the change from the neutral form to the anionic form, $(CH_3)_2AsO_2^-$, will occur close to pH 6.

The analytical difficulties in determining the As species in soils is reflected by the sparsity of the data. Problems in isolation and preservation of minor solid phases and any soluble species in soil porewaters have meant that much of the information about the soil chemistry of As is inferred from simplified systems containing only a limited number of components.

Soil porewaters collected in the Tamar Valley district of SW England, where there are naturally elevated As levels associated with hydrothermal Sn/Cu/As vein deposits, contained arsenate, arsenite and occasionally mono-methylarsonic acid

[29]. Arsenate made up about 90% of the dissolved As species in aerobic soils from both mineralised and unmineralised areas, but only 15–40% in the anaerobic waterlogged soils. No dimethylarsinic acid was detected in any of the porewaters. Monomethylarsonic acid was found in most of the samples from the mineralised areas where it provided 3–11% of the soluble As, but was only occasionally present in the waters from unmineralised areas. In the waterlogged soils the major As species was As(III), with only traces of monomethylarsonic acid. In general, the sum of the percentage of As(III) and monomethylarsonic acid was relatively constant with an inverse relationship between the proportion of each. The reduction of the arsenate to arsenite can be brought about either by the action of micro-organisms or by a change in the physicochemical conditions irrespective of microbial activity. The formation of monomethylarsonic acid indicates that microbial reactions are occurring in some soils. Soils treated with arsenical pesticides have been found to contain dimethylarsinic acid as the major methylated species [17].

In the Tamar estuary, which receives sediment from the mineralised area discussed above, the closer to the mouth of the estuary samples were collected, the the higher the levels of methylated As species [30]. In the upper reaches of the estuary close to the mineralised area there was very little methylated As. The results appeared to be related to the activity of both microorganisms and macrophytes. The presence of monomethylarsonic acid was related to microbial activity in the sediment and the levels of dimethylarsinic acid were attributed to the output of macrophytes.

It is not clear to what extent volatile arsenic compounds are produced in soils as opposed to aquatic sediments. A number of studies have suggested that volatile arsines are produced from lawns and moist soils, and that arsine and methylarsines are produced from soils treated with arsenate, arsenite, monomethylarsonate or dimethylarsinate. However, it has been suggested that organic compounds have not been detected in the atmosphere away from obvious point sources [31].

Despite the apparent stability of soluble ionic species, the concentration of As in soil porewaters is usually low (less than 10 μg/l) unless the area is mineralised. The leaching of As from soils is inhibited by the presence of hydrated oxides of Fe and Al, clays and organic matter.

Pierce and Moore's [32] investigation of the sorption of arsenate and arsenite on amorphous Fe oxide indicated that two processes were operating. The initial sorption of the As species could be described by means of a Langmuir-type adsorption isotherm where the plot of c/s against c is linear, c is the solution phase concentration of As species, s is the sorbed phase concentration of As species.

The Langmuir isotherm describes the situation where the surface of the solid consists of an array of adsorption sites of equal energy with each site being capable of adsorbing one species. This is the situation one might expect to find on the surface of a pure crystalline solid with few defects. Langmuir-type isotherms have been found to describe the sorption of arsenate [28] and arsenite [34] on aluminium hydroxide, and arsenate on soils [35]. However, after 0.5 mmol As(III)/g Fe $(OH)_3$

had been adsorbed, a linear isotherm gave the best description of the data. A plot of *s* against *c* gives a straight line for the higher sorption densities. The range of surface coverage where the isotherm type changed was well below that at which Pierce and Moore expected site limitation to be a problem. They proposed a heterogeneous site model and explained the high sorptive capacity (> 50 mmol As(V)/g Fe $(OH)_3$) by visualising a loose, highly hydrate structure, permeable to hydrated ions that were free to diffuse through the structure. This meant that sorption was not confined to the surface as would be expected with a crystalline solid. The pH of the suspension was found to be critical, with maximum sorption of As(V) at pH 4 (where $H_2AsO_4^-$ was the major species) and at pH 7 for As(III) as H_3AsO_3. In general it has been found that the sorption of As(V) is greater for both Fe and Al hydroxides than is the sorption of As(III).

As well as the homogeneous adsorption mechanism, indicated by Langmuir-type isotherms and the heterogeneous sorption mechanism indicated by linear isotherms, there is a third type of sorption process indicated by Freundlich-type isotherms. With Freundlich-type isotherms plots of log *s* against log *c* are linear, and these are thought to indicate surface heterogeneity with groups of different homogeneous adsorptive sites of differing energy each following Langmuir-type interactions. This set of features can be envisaged in a soil consisting of a range of minerals each with its own individual surface properties and with possible associated coatings of various hydrated oxides of Fe, Mn and Al.

Elkhatib *et al.* [33] found that the sorption of arsenite on A and B horizons of five West Virginian (USA) soils was best described by a Freundlich-type isotherm. Differences in the rates of sorption by the various soils was explained by differences in pH, Fe_2O_3 content and organic C content. The amount of desorption of As(III) was very low, but could be related in part to pH and Fe_2O_3%.

A linear isotherm gave the best description of the arsenite sorption characteristics of 15 Japanese soils [36]. The dithionite-extractable Fe content of the soils (thought to be equivalent to the amorphous Fe oxides and hydroxides) correlated well (r = 0.90) with the variations in As(III) sorption on the different soils.

Studies [37] carried out at a range of redox potentials (+500 mV to –200 mV) and pH (4.0 to 7.5) on a suspension of lake sediment indicated that As solubility increased about 25 times on moving from 500 mV, pH 4.0 to –200 mV, pH 6.9. About 50% of the total As was solubilised under the more reducing conditions. However, even at –200 mV 15% of the soluble As was As(V) indicating the relatively slow rate of conversion to As(III). No methylated species was found at any of the pH or Eh values studied. The changes in total As concentration in solution were highly correlated ($p < 0.01$) with total Fe in solution suggesting that, as iron oxyhydroxides were dissolved, sorbed As species were released.

Laboratory based studies [38] of the structure of freshly precipitated ferrihydrite and its behaviour on ageing in the presence of As(V) solutions indicate that the

> As(V) absorbs on ferrihydrite by the formation of binuclear, inner-sphere complexes with no detectable change in As–Fe bonding after ageing. As–O–As bonding was not observed and the As–O–Fe bond distances and coordination numbers were

significantly different from those of ferric arsenate (scorodite), indicating that neither arsenic oxides nor ferric arsenate formed at the surface.

Follow-up studies [39] on the kinetics of arsenate absorption on ferrihydrite indicate that diffusional processes may be extremely important in controlling rates of absorption and desorption from surface sites. However, interactions due to the great variety of ions present in natural solutions make it difficult to apply laboratory data to natural systems.

The dissociation constants for orthophosphoric acid, H_3PO_4, are very similar to those of arsenic acid (in parentheses) being $pK_1 = 2.124(2.20)$, $pK_2 = 7.20(6.97)$, $pK_3 = 12.40(11.53)$. As a consequence, similarly charged As and P species will be competing for the sorption sites on the soil components. The smaller size of the phosphate species, compared to the corresponding arsenate species, and their higher charges than arsenic species, leads to the expectation that phosphate will bind more strongly than arsenate or arsenite. This expectation has been shown to be correct [35, 40], but the relative concentrations of the various species must be taken into account because like all equilibrium reactions the position of the equilibrium can be displaced. This is illustrated by the use of higher concentration arsenate solutions to displace phosphate from soils. Phosphate also interferes with the sorption of arsenate and arsenite by humic acids [41]. The maximum sorption of As(V) by the humics occurred at about pH 5.5, with the As(III) maxima being found at higher pHs and showing a more varied relationship between pH and type of humic material. In general the humics sorbed 20% less As(III) than As(V), and the sorption behaviour could be described by a Langmuir-type isotherm with some deviation at higher concentrations. The uptake mechanism appeared to be largely explainable by the assumption that humics were acting as anion-exchangers. As the pHs became more alkaline the humics themselves became soluble and their ability to remove arsenic from solution was decreased.

Monomethylarsonates are sorbed by soil components in a similar manner to arsenates, but dimethylarsinates are less strongly sorbed, presumably reflecting their higher pK values of 6.27 and larger size [42].

The level of sulphide is generally low in soils and because of the relatively high redox potentials any sulphide introduced from the bedrocks will be oxidised to sulphate and leached out. As a consequence it is unusual to find As sulphides in soils even though they would be the stable solid phase under reducing conditions such as would be found in waterlogged soils. In lakes, rivers, estuaries and marine sediments, the presence of sulphide and reducing conditions can lead to the precipitation of As_2S_3 [43]. A study of Lake Ohakuri provided no evidence for the presence of As_2S_3, in the sediments, though Fe sulphides were present [44]. An increase in concentrations of Fe and As (up to 6000 mg/kg) in the surficial sediments was probably due to reduction of Fe(III) to Fe(II) at depth causing increased solubility of Fe and co-precipitated As, followed by re-precipitation at the higher E_h at the top of the sediments. The interstitial porewaters were found to contain As(III) (up to 90% of total As) and As(V), but no monomethylarsonic acid or dimethylarsinic acid were detected even though the sediments contained about 10% organic material.

The distribution of As in some Canadian acid sulphate soils collected from upper well-drained sites and from downslope poorly drained sites has been described [45, 46]. The concentration of As in the well-drained soils was in the 5–10 mg/kg range for the top of the A horizons and increased to 30–50 mg/kg for the B and C horizons. This profile was similar to other forested soils of the Gray Luvisol type, except for the acid sulphate soils containing about 10 times as much As. The poorly drained, Gleysolic soils had similar As concentrations to their upslope neighbours, indicating a lack of downslope transportation. The organic-rich surface horizons of the Gleysolic soils contained very little As. This was taken as evidence that microbial activity, producing volatile As compounds was responsible for the majority of the loss of As from the surface horizons. The increase in As levels in the B horizon compared to the C horizons was small, suggesting that little translocation from the A horizon had occurred and providing further support for the hypothesis of loss by volatilisation. These soils also fitted in well with As sorption mechanisms in that, the finer the material, the higher the As concentration with the concentration in the clay being about 4 times higher than in the silt fraction.

5.5 Soil–plant relationships

The level of As in edible plants is generally low, often being close to the limit of detection, even when the crops are grown on contaminated land [2, 47]. The data indicate that when each soil type contains similar As concentrations, lower levels are found in plants grown on clays and silts, with their higher clay mineral and Fe/Al oxide content, than in plants grown on lighter soils, e.g. sands or sandy loams. This reflects the sorption characteristics of soils.

A summary of the phytotoxic levels of soil arsenic [48] show the narrow margin between background and toxic concentrations of As in soils. For inorganic sources monocotyledonous and dicotyledonous plants do not differ significantly in response to soil As, but the inorganic As in sandy soils is about five times more available with regard to its toxic effects than in clay soils (toxicity threshold for sandy soils is about 40 mg/kg compared to 200 mg/kg for clays). Residues from the use of organic-based As pesticides are more toxic than inorganic sources.

In common with most trace elements, the degree of uptake varies widely from species to species. Unlike some marine and freshwater organisms where very high concentrations (over 1000 mg As/kg fresh weight, similar to those in the sediments, have been found in some macrophytes), the levels in terrestrial plants remain well below the level in the soil. In general roots contain higher levels than stems, leaves or fruit. Whilst it is rare to obtain levels of As approaching 1 mg/kg fresh weight in food crops, grasses growing on spoil tips from old As mines in SW England [49] contained up to 3460 mg/kg dry weight with spoil material levels up to 26530 mg/kg. Similar grasses growing in urban areas on soil containing 20 mg As/kg were found to have a maximum of 3 mg As/kg dry weight.

When a number of crops were grown on dredged spoil (35–108 mg As/kg) from the rivers Rhine and Meuse [50], the highest As levels were shown by radishes with 0.8–21 mg As/kg dry weight. Whilst the levels of As in the plants did not directly reflect the As concentrations in the soils, the relative order and magnitude of plant As accumulation remained constant. If the dry matter As concentration for each species is compared with the dry matter As concentration in radishes grown on the same soil type a relative concentration factor is obtained (given below in parentheses) that is constant for the spoil samples and on uncontaminated fluvial loam (8 mg As/kg). The order of decreasing As accumulation is: raddish (1) > grass (0.33) > lettuce (0.26) > carrot (0.17) > potato tuber (0.07) > spring wheat grain (0.04).

The availability of the As in soils is affected by changes in pH. In general there is an increase in As toxicity effects on plants as the soils become more acid, particularly at pHs below 5 when As-binding species such as Fe and Al oxycompounds become more soluble. However, the uptake of As by plants may be increased on higher pH soils [51]. Pot and field trials on the effect of applying sewage sludge, containing arsenate, to a sandy loam (pH 6) and to a calcareous loam (pH 8) indicated enrichment factors that were 3–4 times greater in the lettuces and rye grass grown on the calcareous loam. The highest level of As reached in the crops was 2 mg/kg dry weight in lettuce. The increases in plant tissue As concentration were significantly correlated ($p < 0.01$) with soil As for both types of soil.

The phytotoxic effects of As are indicative of a sudden decrease in water mobility, as suggested by root plasmolysis and discolouration followed by necrosis of leaf tips and margins. Seed germination is also arrested. The sensitivity of a plant to As appears to be determined by the plant's ability to either not absorb or not translocate the As to sensitive sites. Beans and other legumes are amongst the most sensitive plants to As toxicity and the NRCC report [3] provides a table of the variability of crop tolerance to As.

The uptake and toxicity of As species also vary. The degree of uptake and concentration of As species absorbed from nutrient solutions by bean roots was arsenate > arsenite > monomethylarsonate > dimethylarsinate, with toxicity being directly proportional to the root concentration.

The 'available' As content of the soil is a better indicator of phytotoxicity than is the total As concentration. The quantity of soluble or potentially soluble As in a soil varies widely with pH, E_h and the presence of other soil components such as Fe, Al and clay minerals and organic matter. The value obtained for the 'available' As level in a soil is a function of the extracting agent used, and because different workers have used different extracting agents detailed comparison of results is difficult.

The addition of phosphate may or may not reduce As toxicity effects [31]. With silt- and clay-rich soils the addition of phosphate has improved yields of crops. With more sandy soils either no improvement was found or there was an increase in As toxicity (there was also an increase in 'available' As in the soil). The different effects depend upon the relative magnitudes of the competition between phosphate and arsenate.In the clay- and silt-rich soils, where there are large numbers of sorption sites on the soil minerals, the competition for sites will be less important and the competition for uptake by the plant roots will be more important. However, in the sandy soils where there are fewer sorption sites the phosphate may displace some of the bound arsenate ions which move into the solution phase and become available for uptake by the plant roots.

Deep ploughing can be used to reduce the effects of surface applied As by (i) dilution; (ii) provision of extra binding sites, especially if Fe/Al/clay-rich B horizons are exposed; and (iii) moving the As below the root zone. Variable results have been obtained in the reduction of As toxicity in soils by the addition of Al, Fe and Zn compounds, S, lime and organic matter [17].

The uptake of As compounds used as pesticides with their application to growing plants results in much of the uptake being foliar rather than from the soil. As well as inorganic arsenate and arsenite compounds, monomethylarsonic acid and dimethylarsinic acid and their salts have been extensively utilised. Residues from the application of these compounds can lead to high levels of As in the soil and continued phytotoxic effects long after application has ceased. Particularly long-lived have been lead arsenate residues in orchards [17] where application rates were much higher in the past than rates now used for arsenical pesticides.

Because As concentrations in terrestrial plants are generally low, the uptake of As by animals from this source is also low. However, direct ingestion of As from soil can be a major source of dietary As for grazing livestock [52]. The actual amount of soil ingested varied from season to season, being lowest in early summer when ground cover by grass is at its maximum. The average proportion of the total intake of As that was estimated to come from the soil was 60–75% but the range was 2–90%. It was also estimated that about 1% of the As in the soil was actually ingested by the cattle, with the remainder being excreted directly.

5.6 Polluted soils

Contamination of soils by mining operations tends to be localised and, because of the phytotoxic effects of As, not easily overlooked. However, because As often exhibits an extensive halo effect, the area around the mine will also have soils with a naturally high level of As. In SW England 722 km^2 (7.9% of the land area surveyed) has been described as contaminated on the basis of correlations between soil As concentrations and stream sediment As concentrations [53]. The majority of this contamination is a natural consequence of mineralisation rather than due to anthropogenic activities. Crops grown in the area do not generally have unusually high levels of As, though grasses, etc, growing on the spoil tips can have elevated As concentration [49].

Smelting of metals, expecially Cu, and fossil fuel combustion, particularly if there are low chimneys, can lead to contamination of soils and crops in surrounding areas. In Poland the As concentration in vegetables varied in relationship to their proximity to various industrial sources [54]. The average As concentration for 16 types of vegetables in a non-polluted area was 0.05 mg/kg fresh weight. For the same group of vegetables the average became:

0.09 mg/kg fresh weight, near fossil-fuelled power stations;
0.15 mg/kg fresh weight, near a superphosphate plant;
0.18 mg/kg fresh weight, near smelters.

These values compared to the Polish Ministry of Health's regulations that As concentrations should be below 0.2 mg/kg fresh weight in foods containing up to 20% of dry mass. Whether the increased vegetable concentrations were due to foliar uptake or via the soil was not discussed.

As well as the atmospheric distribution of As from coal-fired power stations there is also the As contained in the fly-ash and furnace-bottom ash to be considered. In the UK about 60% of the ash is disposed of to landfill [55]. The proportion of the As that is water soluble varies, and the value obtained depends upon the method of estimation. 1% of the As in the coal fly-ash from 3 UK power plants was water soluble using a high solid/liquid ratio, whereas fly-ash in the USA was reported to contain 4% of the As in a soluble form using a high liquid/solid ratio. The As in the water extract was determined to be all As(V). The use of pulverised fuel ash on reclaimed land in the UK has not been reported to be leading to any As toxicity problem in the crops growing on this land.

Contamination of soils from the smelting of non-ferrous metals can be very significant, with estimated average atmospheric emission factors of 1.5 kg As/t Cu produced, 0.4 kg As/t Pb produced and 0.65 kg As/t Zn produced [5]. The actual emission factors vary widely from smelter to smelter and from time to time as the As concentration in the ore varies and as the pollution control equipment varies. The Tacoma Cu smelter (Washington State, USA) had emission factors that varied during the period 1970–80 from 1.8–16.8 kg As/t Cu produced. The degree of contamination of the soils around this smelter, which is now closed, were a function of wind direction and distance from the stack. On islands to the NNE of the smelter (upwind) the soils contained 90–340 mg As/kg, whereas, to the SSW (downwind) the corresponding levels were 1–90 mg As/kg. Much higher levels have been found associated with gold smelters at Yellowknife, Canada [56]. Surface soil horizons contained over 20000 mg As/kg at 0.28 km from a smelter, 10000 mg As/kg at 0.8 km and 600 mg As/kg at 8 km. The tailings deposits also led to contamination of surrounding soils. The vegetation that grew in these contaminated areas generally contained low concentrations of As except where soil levels were above 1000 mg/kg which either produced phytotoxic effects or growth of a few tolerant genotypes.

Arsenic-contaminated soils in Japan were found to contain arsenate, arsenite, monomethylarsonic acid and dimethylarsinic acid in the extractable As fraction, which was always less than 50% of the total As [57]. The proportion of dimethylarsinic acid increased as the proportion of monomethylarsonic acid decreased. This was explained as being due to the fact that formation/degradation rate for monomethylarsonic acid from/to arsenite was relatively constant under both aerobic and anaerobic conditions, but under anaerobic conditions the dimethylarsinic acid was produced more rapidly from monomethylarsonic acid than it was degraded.

The European Community has proposed that the maximum acceptable concentration of As in agricultural soils treated with sewage sludge should be 20 mg/kg. In the UK the Department of the Environment has suggested 'trigger' concentrations of 10 mg As/kg air-dried soil for domestic gardens and allotments and 40 mg As/kg air-dried soil for parks, playing fields and open spaces. The trigger level is thought to define a threshold at which expert assessment and judgement should be applied to determine whether an undesirable effect will occur and, hence, whether remedial action may be required.

5.7 Concluding comments

Whilst the main features of the chemistry of As in soil have been investigated, the details of many of the processes involved have not yet been determined precisely enough to allow us to be confident that we can correctly forecast the fate of any As added to a particular soil. The universality of microbial methylation reactions and the degree to which As compounds are mobilised by conversion to gas-phase or solution phase species still requires to be determined.

The difficulties involved in identifying and quantifying the various As species in soils and associated solutions and gases are still a major problem. The varied chemistry and toxicity of the different As species means that it is essential that changes in species rather than total As are monitored. The requirement of preventing As toxicity problems in humans, other animals and plants, especially if As-containing wastes are being disposed of on land, requires guidelines based on actual or potential As species concentrations. Fortunately, in terrestrial systems, as opposed to aquatic systems, contamination of food crops via uptake from soil is comparatively rare. However, the relationship between As in soil and in water still needs to be fully determined.

References

1. Kipling, M.D., in *The Chemical Environment*, ed. Lenihan, J. and Fletcher, W.W. Blackie, Glasgow (1977), Chapter 4.
2. National Academy of Science, *Medical and Biologic Effects of Environmental Pollutants: Arsenic*, Washing D.C. (1977).
3. National Research Council of Canada, *Effects of Arsenic In the Canadian Environment*, NRCC No. 15391, Ottawa, Canda (1978).
4. Fowler, B.A., Vahler, M., Pershagen, G. and Squibb, K.S., in *Biological and Environmental Effects of Arsenic*, ed. Fowler, B.A., Elsevier, Amsterdam (1983), Chapters 4–7.
5. Chilvers, D.C. and Peterson, P.J., in *Lead, Mercury, Cadmium and Arsenic in the Environment*, eds. Hutchinson, T.C. and Meema, K.M. John Wiley, New York (1987), Chapter 17.
6. Onishi, H., in *Handbook of Geochemistry*, ed. Wedepohl, K.H. Springer-Verlag, New York (1969).
7. Boyle, R.W. and Jonasson, I.R., *J. Geochem. Explor.* **2** (1973), 251.

8. Berrow, M.L. and Reaves, G.A. in *Proc. Int. Symp. On Environmental Contamination*, London (1984), 333.
9. Tanaka, T., *Appl. Organometall. Chem.* **2** (1988), 283.
10. Norry, M.J., Dunham, A.C. and Hudson, J.D., *J. Geol. Soc.* **151** (1994), 195.
11. Dudka, S. and Markert B., *Sci. Total Environ.* **122** (1992), 279.
12. Severson, R.C., Gough, P.L. and Van Den Boom, G., *Water, Air and Soil Pollution*, **61** (1992), 169.
13. Schacklette, H.T. and Boerngen, J.G., Element Concentrations in Soils and other Surface Materials of the Continuous United States. *US Geol. Survey Professional Paper* 574-B (1984).
14. Gough, P.L., Severson, R.C. and Schacklette, H.T., Element Concentrations in Soils and Other Surficial Materials of Alaska. *US Geol. Survey Professional Paper* 1458 (1988).
15. Chen, J., Wei, F., Zhey, C., Wu, Y. ad Adriano, D.C., *Water, Air and Soil Pollut.* **57/58** (1991), 699.
16. Culbard, E.B. and Johnson, L.R., in *Proc. Int. Symp. on Environment Contamination*, London (1984), 276.
17. Woolson, E.A., in *Biological and Environmental Effects of Arsenic*, ed. Fowler, B.A. Elsevier, Amsterdam (1983), Chapter 2.
18. Siami, M., McNabb, C.D., Batterson, T.R. and Glandon, R.P., *Env. Toxicol. Chem.* **6** (1987), 595.
19. Hutton, M. and Symon, C., *Sci. Total Environ.* **57** (1986), 129.
20. Nriagu, J.O., *Environ. Pollut.* **50** (1988), 139.
21. Cawse, P.A., in *Inorganic Pollution and Agriculture*, MAFF Reference Book 326, HMSO, London (1980), Chapter 2.
22. Pacyna, J.M., in *Proc. Int. Symp. on Heavy Metals in the Environment*, Heidelberg, Vol 1. (1983), 178.
23. Alloway, B.J. and Jackson, A.P., *Sci. Total Environ.* **100** (1991), 151.
24. Porter, E.K. and Peterson, P.J., *Environ. Pollut.* **14** (1977), 255.
25. Wodge, A., Hutton, M. and Peterson, P.J., *Sci. Total Environ.* **54** (1986), 13.
26. Sonderegger, J.L. and Ohguchi, T., *Environ. Geol. Water Sci.* **11** (1988), 153.
27. Nijssen, J.P.J. and Wijnen, E.J.E., in *Proc. Int. Symp. on Heavy Metals in the Environment*, Athens, Vol. 2 (1985), 204.
28. Moore, J.N., Ficklin, W.H. and Johns, C., *Environ. Sci. Technol.* **22** (1988), 432.
29. Haswell, S.J., O'Neill, P. and Bancroft, K.C.C., *Talanta* **32** (1985), 69.
30. Walton, A.P., Ebdon, L. and Millward, G.E., *Anal. Proc.* **23** (1986), 22.
31. Andrea, M.O., in *Organometallic Compounds In The Environment*, ed. Graig, P.J. Longman, London (1986), Chapter 5.
32. Pierce, M.L. and Moore, C.B., *Environ. Sci. Technol.* **14** (1980), 214.
33. Elkhatib, E.A., Bennett, O.L. and Wright, R.J., *Soil Soc. Am. J.* **48** (1984), 1025.
34. Gupta, S.K. and Chen, K.Y., *J. Water Pollut. Control Fed.* **50** (1978), 493.
35. Livesey, N.T. and Huang, P.M., *Soil Sci.* **131** (1981), 88.
36. Sakato, M., *Environ. Sci. Technol.* **21** (1987), 1126.
37. Masscheloyn, P.J., Delaune, R.D. and Patrick, W.H., *J. Environ. Qual.* **20** (1991), 522.
38. Waychunas, G.A., Rea, B.A., Fuller, C.C. and Davis, J.A., *Geochim. Cosmochim. Acta.* **57** (1993), 2251.
39. Fuller, C.C., Davis, J.A. and Wagchunas, G.A., *Geochim. Cosmochim. Acta.* **57** (1993), 2271.
40. Roy, W.R., Hassett, J.J. and Giffin, R.A., *Soil Sci. Soc. Am. J.* **50** (1986), 1176.
41. Thanabalasingam, P. and Pickering, W.F., *Environ. Pollut. Ser. B.* **12** (1986), 233.
42. Gosh, M.M. and Yuan, J.R., *Environ. Progress* **6** (1987), 150.
43. Moore, J.N., Ficklin, W.H. and Johns, *C., Environ. Sci. Technol.* **22** (1988), 432.
44. Aggett, J. and O'Brien, G.A., *Environ. Sci. Technol.* **19** (1985), 231.
45. Dudas, M.J., *Can. J. Soil Sci.* **67** (1987), 317.
46. Dudas, M.J., Warren, C.J. and Spiers, G.A., *Commun. in Soil Sci. Plant Anal.* **19** (1988), 887.
47. Ministry of Agriculture, Fisheries and Food, *Survey of Arsenic In Food*, HMSO, London (1982).
48. Sheppard, S.C., *Water, Air and Soil Pollution*, **64** (1992), 539.
49. Porter, E.K. and Peterson, P.J., *Environ. Pollut.* **14** (1977), 255.
50. Smilde, K.W., Van Driel, W. and Van Luit, B., *Sci. Total Environ.* **25** (1982), 225.
51. Campbell, J.A. Stark, J.H. and Carlton-Smith, C.H., in *Proc. Int. Symp. on Heavy Metals in the Environment*, Athens, Vol. 1 (1985), 478.

52. Thornton, I. and Abrahams, P.W., *Sci. Total Environ.* **28** (1983), 287.
53. Abrahams, P.W. and Thornton, I., *Trans. Instn. Min. Metall. (sec. B: Appl. Earth Sci.)* **96** (1987), B1.
54. Grajeta, H., *Rocz. Pantw. Zakl. Hig.* **38** (1987), 340.
55. Wodge, A. and Hutton, M., *Environ. Pollut.* **48** (1987), 85.
56. Hocking, D., Kucher, P., Plambech, J.A. and Smith, R.A., *J. Air Pollut. Control Assoc.* **28** (1978), 133.
57. Takamatsu, T., Aoki, H. and Yoshida, Y., *Soil Sci.* **133** (1982), 239.

6 Cadmium

B.J. ALLOWAY

6.1 Introduction

Cadmium belongs to the group IIB of the Periodic Table, and is a relatively rare metal, being 67th in order of elemental abundance. It has no essential biological function, and is highly toxic to plants and animals. However, the concentrations of Cd normally encountered in the environment do not cause acute toxicity. The major hazard to human health from Cd is its chronic accumulation in the kidneys where it can cause dysfunction if the concentration in the kidney cortex exceeds 200 mg/kg fresh weight [1]. Food is the main route by which Cd enters the body, but tobacco smoking and occupational exposures to CdO fumes are also important sources of the metal. The FAO/WHO recommended maximum tolerable intake of Cd is 400 to 500 μg/week, which is equivalent to about 70 μg/day [1]. Average dietary intakes of Cd around the world range between 25 and 75 μg/day [2] and there is clearly a problem where the intake is near the top of the range. People who smoke can add an extra 20 to 35 μg Cd/day to their intake.

In view of the dangers of the chronic accumulation of Cd in the human body, the factors influencing its concentration in the components of the diet are of great importance. Since concentrations of the metal in uncontaminated soils are usually low, sources of contamination and the behaviour of Cd in the contaminated soils will be the main concern. With the estimated half-life for Cd in soils varying between 15 and 1100 years [3], this is obviously a long-erm problem and pollution needs to the prevented or minimised wherever possible. Several countries have restricted the use of Cd, or are planning to, but nearly all have a legacy of pollution from its many sources.

Cadmium pollution of the environment has been rapidly increasing in recent decades as a result of rising consumption of Cd by industry. Environmental pollution is an inevitable consequence of metal mining, manufacture and disposal. Unlike Pb, Cu, and Hg, which have been utilised for centuries, Cd has only been widely used this century. More than half of the Cd ever used in industry was produced in the last 25 years [4]. It is obtained as a by-product of the smelting of Zn and other base metals, and no ores are used primarily as a source of Cd. World production of Cd increased from 11 000 t in 1960 to 20 200 t in 1990 [5, 6]. Its principal uses are (i) as protective plating on steel, (ii) in various alloys, (iii) in pigments (for plastics, enamels and glazes), (iv) as a stabiliser for plastic, (v) in Ni–Cd dry-cell batteries and (v) other miscellaneous uses, including photovoltaic cells and control rods for nuclear reactors [7].

Sources of soil contamination by Cd are the mining and smelting of Pb and Zn;

atmospheric pollution from metallurgical industries; the disposal of wastes containing Cd, such as the incineration of plastic containers and batteries; sewage sludge application to land; and the burning of fossil fuels [8]. Even before Cd was used commercially, contamination was occurring from a wide range of materials containing Cd as an impurity. Phosphatic fertilisers are an important example of this; their Cd contents vary, but their continual use has led to significant increases in the Cd contents of many agricultural soils. The deposition of aerosol particles from urban/industrial air pollution also affects the soils in most industrial countries, and Cd from this source can also be absorbed directly into plants through the foliage.

6.2 Geochemical occurrence

The average concentration of Cd in the Earth's crust is estimated to be in the region of 0.1 mg/kg [9, 10]. Cadmium is closely associated with Zn in its geochemistry; both elements have similar ionic structures and electronegativities (a property related to the ionisation potential) and both are strongly chalcophile (see Chapter 3) although Cd has a higher affinity for S than Zn. The average Zn:Cd ratio for all rocks is around 500:1, but ranges from 27:1 to 7000:1 [11]. Cadmium is obtained as a by-product from smelting of sulphide ore minerals in which it has substituted for some of the Zn. The most abundant sources of Cd are the ZnS minerals sphalerite and wurtzite and secondary minerals, such as $ZnCO_3$ (smithsonite) which typically contain 0.2–0.4% Cd although concentrations of up to 5% Cd can be found [7, 12].

Sedimentary rocks show a greater range of Cd concentrations than other rock types, with phosphorites (sedimentary Ca phosphates) and marine black shales having the highest contents (Table 6.1). Phosphorites and black shales also contain anomalously high concentrations of several other heavy metals in addition to Cd (see Chapter 3). Both types of rock are formed from organic-rich sediments under anaerobic conditions, and the heavy metals accumulated as sulphides and organic complexes.

6.3 Origin of cadmium in soils

6.3.1 *Soil parent materials*

Page and Bingham [17] suggest that soils derived from igneous rocks would have Cd contents of 0.1–0.3 mg/kg, those on metamorphic rocks would contain 0.1–1.0 mg/kg Cd and those derived from sedimentary rocks 0.3–11 mg/kg Cd. In general, most soils can be expected to contain < 1 mg/kg, except those contaminated from discrete sources or developed on parent materials with anomalously high Cd contents, such as black shales.

Cadmium concentrations in soil. From a review of the literature, Kabata-Pendias and Pendias [3] conclude that mean Cd contents reported from most analytical

Table 6.1 Cadmium concentrations in rocks (mg/kg or % where indicated). Data from [11–16]

	Range	Mean
Igneous rocks		
Rhyolites	0.03–0.57	0.23
Granites	0.01–1.60	0.20
Basalts	0.01–1.60	0.13
Metamorphic rocks		
Gneisses	0.007–0.26	0.04
Schists	0.005–0.87	0.02
Sedimentary rocks		
Shales and clays	0.017–11	—
Black shales	0.30–219.0	—
Sandstones and conglomerates	0.019–0.4	—
Carbonates	0.007–12	0.065
Phosphorites	<10–980	—
Coal	0.01–300	
Crude oil	0.01–10000	
Sulphide ore minerals		
Sphalerite (ZnS)	0.2–0.4 (< 5%)	
Galena (PbS)	<0.5%	
Tetrahedrite-tennartite (Cu.Zn) (Sb.As)S	<0.24%	
Metacinnabar (HgS)	11.7%	

surveys of background levels of metals in soils lie in the range 0.06–1.1 mg/kg with a calculated worldwide mean of 0.53 mg/kg. A survey of agricultural soils in the USA comprising 3045 samples representing 307 different soil series collected from sites remote from obvious sources of metal contamination gave a mean topsoil Cd concentration of 0.265 mg/kg, a median of 0.2 mg/kg and a range of < 0.01–2.0 mg/kg [18]. The highest Cd concentrations were mostly found in the areas of California underlain by the relatively Cd-rich Monterey Shales (see below) and in Colorado alluvial soils contaminated by Ag mining activity. In the Great Lakes states, Oregon and Florida, organic soils used for vegetable production showed accumulations of Cd which may have been due to the heavy use of P fertilisers or sewage sludges [18]. Other slightly higher concentrations of Cd tended to be found in soils which received high levels of P fertiliser especially those manufactured from Western USA rock phosphates which tend to contain more Cd than those from Florida [18]. A survey of 2746 Japanese paddy (rice growing) soils from apparently uncontaminated sites showed mean Cd concentration of 0.4 mg/kg [19].

A recently published survey of 5692 soil samples collected on a regular grid over the whole of England and Wales showed the mean Cd concentration in topsoils to be 0.8 mg/kg with a median of 0.7 mg/kg and a range of < 0.2–40.9 mg/kg [20]. Unlike the biased sampling in the USA, this study included both 'normal' agricultural land and obviously contaminated sites. Some of the soils included in this analytical survey had received sewage sludge and others had been contaminated by metalliferous mining activities. (See Chapter 3). An earlier survey of UK soils by Archer [21] based on the analysis of 659 samples gave the median Cd content as 1.0 mg/kg and the range as 0.08–10 mg/kg. Ure and Berrow [22] gave the geometric mean concentration of Cd in world soils as 0.62 with a range of < 0.005–8.1.

A regional geochemical reconnaissance survey of England and Wales based on the analysis of approximately 50 000 samples of stream sediments from tributary drainage showed that around 1200 km^2, or 0.33% of the area surveyed, contained significantly elevated concentrations of Cd due to both geochemically anamalous soil parent materials and to various sources of environmental contamination [23]. This is probably a conservative estimate because the data were subjected to a statistical 'smoothing' procedure which tended to remove isolated anomalous concentrations.

In a study of the soils in 94 urban vegetable gardens in England, Moir and Thornton [24] reported a geometric mean Cd concentration of 0.53 mg/kg with a range of < 0.2–5.9 mg/kg. Culbard *et al.* [25] showed that 579 garden soils in London boroughs had a geometric mean Cd concentration of 1.3 mg/kg and a range of < 1–4.0 mg/kg.

Soils developed on parent materials with anomalously high Cd contents, especially black shales, can have significantly elevated total Cd concentrations even in the absence of marked contamination from anthropogenic sources. Cadmium concentrations reported for soils on black shales include < 22 mg/kg on outcrops on the Monterey shale in California, USA [14], < 24 mg/kg in soils on Carboniferous black shales in Derbyshire, UK [23] and < 11 mg/kg in alluvial soils in an area of black shale and black slate outcrops in South Korea [26]. (See Chapter 3).

Within soil profiles, Cd is normally found concentrated in the surface horizon which is due to a combination of factors: it is the zone with the highest organic matter content and metals may be retained in this strongly adsorptive horizon after reaching it as a result of cycling through vegetation, or from applications of Cd-containing fertilisers and manures, or from the wet and dry deposition from the atmosphere. However, unlike Cu and Pb, Cd (together with Zn and Ni) does have a tendency to move down the profile although the extent and rate will depend on various soil and site factors [3, 27]. Holmgren *et al.* [18] reported data for topsoils and subsoils (layer immediately below tilled layer) for 26 sites where the topsoil Cd content was > 1 mg/kg. In three out of the ten sites in mineral soils, the Cd concentration was higher in the subsoil but the ratio of surface to subsoil Cd concentrations for the ten sites ranged from 11.0 (1.8 mg/kg in subsoil). For the 16 sites in organic soils, 15 had higher Cd concentrations in the surface soil and the ratios for all 16 ranged from 5.9 (1.3 mg/kg surface and 0.22 mg/kg subsoil) to 0.61 (1.1 mg.kg surface and 1.8 mg/kg subsoil). In laboratory leaching experiments, Tyler [28] found that it took 6 years to reduce the Cd content of the O (Mor humus) horizon when the pH was maintained at 4.2, but only 3 years to reduce it by the same amount at pH 3.2.

In soils treated with sewage sludge it is generally considered that little downward movement of heavy metals occurs in the short term (5–10 years) [29]. However, many factors need to be considered, including the climate (relative balance of precipitation and evaporation), pH, and soil permeability (including presence of macopores and cracks). In semi-arid soils in California, Williams *et al.* [30] found that even after applications of 1800 t/ha of sewage sludge, Cd and other metals remained within the zone of incorporation after 9 years. In the more humid conditions of the UK, Davis *et al.* [31] reported that 60–100% of Cd, Cr, Cu, Mo, Ni, Pb and Zn remained in the surface 10 cm of sewage-treated grassland soils after several years. Likewise, several authors have found that little downward movement occurs in sewage sludge-treated soils under forestry.

However, there are also reports in the literature of a greater degree of movement of Cd down the profiles of some sewage sludge treated soils. Legret *et al.* [32] found that in a long-term field experiment on a coarse textured soil in South Western France which had received heavy applications of sludge, Cd had moved down the profile to depths of 60–80 cm. In an old sewage farm site on a chalky boulder clay soil in England it was found that Cd had moved down to 100 cm or deeper in the profile [33, 34]. These latter findings are important because they relate to a much longer time period (over 100 years) compared with many other published studies.

6.3.2 Agricultural materials

Phosphatic fertilisers are widely regarded as being the most ubiquitous source of Cd contamination of agricultural soils. Relatively high concentrations of Cd (< 500 mg/kg) can be found in phosphorites (rock phosphates) used for the manufacture of fertilisers (Table 6.2). However, it can be seen from Table 6.2 that the Cd concentrations found in P fertilisers can vary widely depending on the origin of the phosphorite raw material. Phosphate fertiliser manufacture is also regarded as being a key source of Cd contamination in the environment (solid wastes and effluents). In the USA, P fertilisers made from Florida phosphorite containing < 10 mg/kg Cd contributed 0.3 to 1.2 g Cd/ha/yr to soils in long-term soil fertility experiments [39]. In contrast, P fertilisers manufactured from western phosphorite deposits containing an average of 174 mg/kg Cd contributed 100 g Cd/ha/yr to the soils in a 36 year-field trial in California, USA. The concentration of Cd in the soil was raised from 0.07 mg/kg in the controls to 1.0 mg/kg Cd in the fertilised plots [40]. Significant increases in Cd concentrations in crops on these soils were noted in California, but not where fertilisers made from Florida phosphorite were used [39, 40]. Around 70% of the P fertiliser used in the USA is produced from the relatively low-Cd Florida phosphorite. Australian fertilisers generally contain between 25 and 50 mg/kg Cd [36]. Phosphatic fertilisers with contents of 3–8 mg/kg Cd used in long-term field experiments at Rothamsted, UK, contributed 2 g Cd/ha/yr to an arable soil and 7.2 g Cd/ha/yr to the soil under permanent grass [35]. The higher accumulation of Cd in the grassland soil could be partly due to lack of mixing by cultivations, to adsorption by the larger quantity

Table 6.2 Cd concentrations in phosphatic fertilisers

Origin of the phosphorite	Range (mg Cd/kg fertiliser)	Ref.
Various	0.1–170	3
Various	3.3–40	35
Pacific islands	18–91	36
Western USA	<200	37
Florida, USA	<20	37
	Range (mg Cd/kg P)	
Morocco	137	8
USA	80	8
Togo	367	8
Senegal	584	8
Russia	1.8	8
Tunisia/Algeria	137	8
Israel/Jordan	82	8
P Fertiliser used in various countries	Range (mg Cd/kg fertiliser)	
Canada	2.1–9.3	38
Australia	18–91	
USA	7.4–156	
Netherlands	9–30	
Sweden	2–30	

of organic matter and to greater interception of atmospheric pollutants by the grass sward [35].

Over the last 10 years or so, estimates of Cd inputs from P fertilisers have included 4.3 g/ha/y in the UK [41] and 3.5 g/ha/y in the former West Germany [42]. Nriagu and Pacyna [43] used a range of Cd concentrations in fertilisers of 0.2–15 mg/kg to estimate the annual worldwide input of Cd to soils from all types of fertilisers. The total input to soils from the use of fertilisers was given as 30–250 t Cd/y.

With the increasing awareness in recent years of this source of Cd contamination of soils, some major manufacturers have changed to sources of phosphorite raw material with lower Cd concentrations and this has resulted in a reduction of the concentration in P fertilisers in some countries or regions. A recent survey of 66 fertiliser samples in the UK showed a mean Cd content in the material of 36.7 mg Cd/kg P_2O_5 (or 84 mg Cd/kg P) (MAFF pers comm.). In the EU, fertiliser inputs are estimated to be around 300 t Cd/yr and are expected to rise to 346 t Cd/yr by the year 2000 [8]. Almost all soils used for commercial agriculture will have had their Cd content raised, at least to a slight extent, by phosphatic fertilisers.

Although Australian diets tend to be amongst the lowest in the world for Cd content (15 μg/day) there has been growing concern in recent years about the accumulation of Cd derived from P fertilisers in potatoes and other food crops. Potatoes are considered to be the source of 55% of Cd in the Australian diet compared with 24% in the USA. The Australian Government has introduced a maximum permissible Cd concentration for Cd in some foods of 50 μg/kg (or 0.05 mg/kg) FW [44]. Phosphatic fertilisers are considered to be the main source of Cd in agricultural soils and some of these fertilisers have tended to be manufactured from phosphorites with relatively high Cd contents. It is not uncommon for P fertilisers in Australia to contain < 300 mg Cd/kg P and these can add 30–60 g Cd/ha to potato growing soils [44]. This is due to some of the Australian soils being very deficient in P and to the high P requirements of the crop and therefore it is necessary to apply relatively large amounts of the fertiliser to each potato crop. In experiments in South Australia, Cd inputs in P fertilisers for a potato crop varied from 2.8 g Cd/ha with mono-ammonium phosphate (MAP) to 40 g Cd/ha with single superphosphate (SSP). Cadmium removal in potato tubers varied from 0.8 g/ha (MAP) to 9.6 g/ha (SSR) [44]. Some authors have reported differences in the bioavailability of Cd in different forms of P fertiliser in greenhouse pot experiments (Cd in di-ammonium phosphate less plant available than triple superphosphate) [45]. However, McLaughlin *et al.* [44] did not find significant differences between P fertiliser types in field trials with potatoes in South Australia and they considered that it was the residual Cd from earlier fertiliser applications which was the most important source to the crop.

Farmyard manure (FYM). Values of 0.3–1.8 mg/kg Cd in the dry matter have been quoted for farmyard manure [3, 36]. Large annual applications (35 t fresh weight/ha) of manure were found to be a more significant source of Cd in a long-term field experiment at Rothamsted than the combined inputs from phosphatic fertilisers and atmospheric deposition [35].

6.3.3 Atmospheric deposition of cadmium

Cadmium concentrations in air normally range from 1 to 50 ng/m^3, depending on the distance from emitting sources [37]. Typical ranges of atmospheric Cd concentrations in Europe are 1–6 ng/m^3 for rural areas, 3.6–20 ng/m^3 for urban regions and 16.5 to 54 ng/m^3 for industrial locations, rising to 11000 ng/m^3 near a metal recovery plant [8]. The major sources of atmospheric emissions are non-ferrous metal production, fossil fuel combustion refuse incineration and iron and steel production, as shown in Table 6.4 [38]. The relatively high volatility of Cd when heated above 400°C accounts for the significance of atmospheric emissions from these sources [10]. The total annual atmospheric emission of Cd throughout the world was estimated by Nriagu [49] to be 8100 t (800 t from natural sources and 7300 t from anthropogenic sources). Average annual emission rates from global anthropogenic sources have been increasing from 3400 t (during 1951–60), to 5400 t (during 1961–70) and to 7400 t (during 1971–80) [50]. Total inputs of Cd from both dry deposition and rainfall range from 2.6 to 19 g/ha/yr in rural areas, with 3 g/ha/yr considered as the representative value for Cd deposition on to

Table 6.3 Estimated Cd atmospheric emissions from major sources in Europe in 1979 (t/yr). From Pacyna [48]

Primary non-ferrous metal production	1631.4
Coal combustion	143.7
Oil combustion	108.2
Refuse incineration	83.6
Iron and steel manufacture	59
Industrial applications of metals	19.7

agricultural land in the European Community [8]. Atmospheric deposition has been found to vary from 0.06 g/ha/yr in Greenland, to 44.4 g/ha/yr in New York City and 135.6 g/ha near the major Pb–Zn smelter at Avonmouth in England [51]. Nriagu and Pacyna [43] estimated the worldwide atmospheric deposition of Cd onto soils in 1988 to be within the range 2200–8400 t Cd/yr. Sposito and Page [52] estimated deposition of Cd on to soils in Europe and the USA to range from < 0.2 to 978 g/ha/yr, with heavier levels of deposition in Europe. They estimated increases of Cd concentrations in the topsoil (0–15 cm) due to atmospheric deposition in the USA to be < 0.089 μg/kg/yr in rural areas, 29 μg/kg/yr in industrial regions and 0.98 μg/kg/yr in metropolitan regions [52]. Nriagu [49] estimated the global deposition of Cd from the atmosphere to be 5700 t/yr over land and 2400 t/yr over the oceans.

6.3.4 Sewage sludges

Sewages vary in composition and contain Cd from various souces, including human excretion, domestic products containing Zn, storm waters containing particles of rubber tyres and various industrial effluents. Almost all the Cd accumulates in the insoluble sludge produced during sewage treatment and a wide range of Cd concentrations in sewage sludges has been reported from around the world (< 3650 mg/kg DM). In the past, median values for British sludges have ranged between 17 and 23 mg/kg DM [53, 54] and median values of 16 mg/kg have been quoted for US sludges [55]. Nriagu and Pacyna [43] used a range of 1.0–20.0 mg/kg for Cd in municipal sewage sludges as range 20–340 t Cd/yr. However, over the last decade or more, the Cd contents of sewage sludges have been lowering through the introduction of various waste minimisation technologies and effluent controls. As shown in Chapter 3, the median (50 percentile) values for sewage sludges used on agricultural land in the UK have decreased from 9.0 mg/kg DM in 1980/81 to 3.2 mg/kg DM in 1990/91 and for the upper levels, the 90 percentile values have decreased from 33 mg Cd/kg in 1980/81 to 12.0 mg Cd/kg in 1990/91 [56]. The median Cd content of soils to which sludge has been applied in 1990/91 was 0.55 mg/kg (cf UK/EU maximum permissible value of 3.0 mg/kg Cd) [56].

In the USA a national sewage sludge survey was carried out in 1990 and this included a questionnaire survey of all 11 407 publicly owned treatment works in the country from which 208 representative works were chosen and for an intensive analysis of samples of sewage sludge which included 412 analytes in each sample

Table 6.4 Maximum allowable Cd concentrations in sewage sludges used on agricultural land and in sludge-treated soils and annual loading limits (After McGrath *et al.* [58])

Country	Year	Maximum sludge	Concentration in sludge treated soil (mg/kg DM)	Annual loading limits (kg Cd/ha/y)
European Union	1986	20–40	1–3	0.15
France	1988	20	2	0.15
Germany	1992	10*	1.5†	0.15
Spain	1990	20	1.0	0.15
Denmark	1990	1.2	0.5	0.008
	1995	0.8	—	—
Finland	1995	1.5	0.5	0.0015
Sweden	1995	2.0	0.5	0.002
USA	1993	8.5	2.0	1.9

* Sludge concentration limit in Germany reduced to 5 mg Cd/kg where receiving soil < 5% clay or pH 5–6; and † maximum soil Cd is for soils > pH 5, at pH 5–6 the Cd limit is 1.0 mg/kg

[57]. The results for Cd from this survey were a median value of 7.0 mg/kg, a 95th percentile value of 21 mg/kg and a 98th percentile of 25 mg/kg. The data from this survey were then used in conjunction with exposure assessments to determine the limiting concentrations for sludges and the loading limits for soils.

The maximum allowable Cd concentrations of sewage sludge used on agricultural land in European countries and in the USA have been summarised and discussed by McGrath *et al.* [58]. The limiting values are given in Table 6.4 together with the maximum limits for Cd in sewage sludge-treated soils. Within the European Union there are marked differences in the allowable maxima but all are very much lower than those now used in the USA under the new Environmental Protection Agency (EPA) Part 503 Rule (1992). There is obviously a major difference in policy between the US EPA and officials in both the European Union and the ministries concerned with environmental quality in the individual countries of the European Union. Basically, the US EPA approach is based on the analysis of 14 different exposure pathways together with the data from the National Sewage Sludge Survey of 1990 from which cumulative pollutant loading limits were determined. The regulations set concentration limits for high quality 'clean' sludge which could be applied at a rate of 10 t/ha for 100 years before exceeding the loading limits. The cumulative loading principle adopted in the USA relies on the capacity of the soil to assimilate and reduce the bioavailability of the pollutants. In contrast, the European countries tend to use the 'metal balance approach'. Since the inputs of metals in the soils of industrialised countries always exceed the losses through plant offtake and leaching, several countries, such as the Netherlands, Sweden and Denmark, aim to minimise metal inputs to soil from sludges as far as possible. A more pragmatic approach is taken in the UK but in November 1993 a special committee set up by the UK Ministry of Agriculture, Fisheries and Food recommended a reduction in the maximum permissible Zn concentration in sludged soils from 300 mg Zn/kg to 250 mg Zn/kg as a result of findings that soil microorganisms are significantly affected at levels above this lower value. This is the 'lowest observed adverse effect concentration' approach which is based on actual observations of toxic effects in various types of key organisms [58].

The Soil Protection Policy developed in the Netherlands was originally designed for the assessment and remediation of contaminated land and comprised 'A', 'B' and 'C' values. The 'A' values are reference values based on soils in nature reserves which are only exposed to background atmospheric contamination. The 'B' values are no longer used but were originally trigger values indicating a need for further investigation. The 'C' values are the 'intervention values' derived from ecotoxicological and human toxicological risk assessments. These values, like those of the US EPA, were based on exposure pathways and are the concentrations above which there is a danger of the soil functional properties being seriously affected. The Netherlands 'A' and 'C' values for Cd are 0.8 and 12 mg Cd/kg, respectively [58, 59]. Soils which have received abnormally large applications of sewage sludge are discussed in section 6.6.

Although an expedient form of waste disposal, and a source of N and P macronutrients, the application of sewage sludges to land does result in soils being significantly contaminated with Cd and other non-essential metals, which will inevitably lead to their increased uptake by crops.

6.3.5 Other sources of cadmium

The other major sources of Cd which can cause contamination of soils are the mining, ore-dressing and smelting of Cd-containing sulphide ores which can contain up to 5% Cd. Dispersion of particulates from these sources can be by gravity from spoil tips, by wind, and by water through the erosion and fluvial transport of tailings from old mines and mineral dressing floors. Soils severely polluted by Pb–Zn mining and smelting have been found to contain up to 750 mg/kd Cd [60]. Cadmium-polluted soils are discussed in more detail in section 6.6.

6.3.6 Summary of cadmium inputs into soils

For individual western countries the relative contributions of Cd from the major anthropogenic sources have been estimated to be: phosphatic fertilisers 54–58%, atmospheric deposition 39–41%, and sewage sludge 2–5% [61, 62]. These sources give rise to an average annual increase in the Cd content of agricultural soils in Denmark of 0.6% [62]. However, much higher inputs occur at sites near metallurgical works emitting Cd or where sewage sludge is applied to land.

6.4 The chemical behaviour of cadmium in soils

6.4.1 Speciation of cadmium in the soil solution

It is important to be able to identify the forms of metals in the soil, especially in the soil solution, in order to more fully understand the dynamics of the metal in agricultural and natural ecosystems. The toxic effect of a metal is determined more by its form than by its concentration. The free ion Cd^{2+} is more likely to be adsorbed on the surfaces of soil solids than other species, such as neutral or anionic species. The principal species of Cd in the soil solution is Cd^{2+} but the metal can also form the following complex ions: $CdCl^+$, $CdOH^+$, $CdHCO_3^+$, $CdCl_3^-$, $CdCl_4^{2-}$, $Cd(OH)_3^-$ and $Cd(OH)_4^{2-}$ together with organic complexes [3].

The concentrations and speciation of a metal such as Cd in the soil solution will depend partly on the concentration of ligands in the soil solution and the stability constants of the ligand–metal complexes. There are several models available for the prediction of the species present in aquatic systems; but GEOCHEM was developed by Mattigod and Sposito [63] specifically for soils. This model contains a database with typical values of stability constants for individual metal–ligand pairs. The insertion of parameters, such as pH, organic C, cation and anion concentrations enables a prediction of the predominant species to be made. Using this model it has been predicted that the principal chemical species of Cd in oxic soil solutions are (in decreasing order): Cd(II), $CdSO_4(0)$ and $CdCl_4^+$ in acid soils, and Cd(II), $CdCl^+$, $CdSO_4(0)$ and $CdHCO^+$ in alkaline soils [52]. However, some authors have found that GEOCHEM predictions for the free-ion contents of some metals in soil solutions were not in close agreement with those determined experimentally. McGrath *et al.* [64, 65] found that the percentages of Cu, Ni and Pb in the solutions of sewage sludge-amended soils were overestimated to a considerable extent by GEOCHEM, although the model predictions were in close agreement for well-defined solutions. A new version of this model called SOILCHEM has been developed and many of the problems encountered with GEOCHEM have been overcome.

The predominant soluble Cd species in soils polluted from several different sources were fractionated by Tills and Alloway [66] using a combination of ion-exchange and reverse-phase chromatography. They found that the free ion Cd^{2+} predominated, with neutral species, such as $CdSO_4$ or $CdCl_2$, present in increasing amounts where the pH was greater than 6.5. The proportion of organically bound Cd in the soil solutions was relatively small; a soil which had been heavily amended with sewage sludge had only 13.2% organically complexed Cd. The method may have underestimated the proportion of organically bound Cd, but the results were comparable to those predicted for sludged soils by Mahler *et al.* [67] using the GEOCHEM model.

6.4.2 The adsorption of cadmium in soils

The dynamic equilibrium between Cd in the soil solution and that adsorbed on the solid phase of the soil depends on the pH, the chemical nature of the metal species, the stability of Cd complexes, the binding power of the functional groups and the ionic strength of solutions and competing ions [68, 69]. The adsorption of any heavy metal in the soil is rarely a simple relationship between its distribution coefficient and soil properties because more than one species of each metal usually occurs in the soil solution as a result of the formation of inorganic or organic complexes. Some aspects of the chemical behaviour of Cd in soils can be explained by the hard–soft Lewis acid–base principle (HSAB). Cadmium is a soft Lewis acid and will therefore react and complex most readily with soft Lewis bases, such as chloride and hydroxyl groups [70].

Cadmium tends to be more mobile in soils and therefore more available to plants than many other heavy metals, including Pb and Cu. In Chapter 2 it was shown that the selectivity of several soil adsorbents for Cd was lower than for Cu or Pb. Nevertheless, adsorption mechanisms are still very important in the dynamics of Cd in soils and the following brief review is intended to show the major factors influencing adsorption.

One of the problems in reviewing the work on Cd adsorption in soils is that unrealistically high concentrations of the metal have been used in some cases and therefore the findings cannot be easily extrapolated to field situations [71]. Christensen [71] maintains that levels should be < 50 μg/l Cd (50 ppb) in solution and < 20 mg/kg total Cd, but some workers have used up to 500 mg/l Cd in solution [72]. Adsorption processes rather than precipitation appear to control the distribution of Cd between soluble and soil-bound forms at the concentrations normally encountered even in the majority of polluted soils. However, at extremely high Cd concentrations, precipitates of Cd phosphates and carbonates could be expected to form [72].

Adsorption isotherms. In many cases it has been found that Cd adsorption by suspensions of soils or their constituents fits either the Langmuir or Freundlich isotherm equations [73, 74]. The choice of the model is not very important since Cd is usually well below saturation when realistic concentration ranges are used [75]. Gerritse and Van Driel [69] found the Langmuir isotherms most appropriate for Cd adsorption by 33 polluted soils from the Netherlands, Britain and France. The adsorption/desorption of Cd and Zn showed a greater sensitivity to pH than that of Pb and Cu. Between 10 and 50% of the adsorbed Cd, Zn and Cu were exchangeable, compared with 1–5% of the adsorbed Pb.

In two separate studies on the specific adsorption of Cd by a wide range of soils, it was shown that the data fitted the Freundlich adsorption isotherm in two distinct sections [76, 77]. This suggested two types of adsorption sites: (i) low-capacity, high-affinity sites at low solution concentrations of Cd, and (ii) lower-affinity but higher-capacity sites at higher solution concentrations. This type of two-section curve has also been found by other workers for Zn, Ni, Cu and Pb. Adsorption of Cd is normally rapid with more than 90% occurring within the first 10 min [71].

Effects of pH. Christensen [71] showed that Cd adsorption by sandy and loamy soils increased by a factor of 3 for every pH unit increase between pH 4 and 7.7. Farrah and Pickering [78] showed that Cd adsorption increased markedly with pH up to pH 8. Naidu *et al.* [79] found that Cd adsorption by two Australian oxisol soils, an andept and a fragiaqulf from New Zealand increased with increasing pH as a result of increasing surface negative charge. Increases in ionic strength caused a reduction in Cd adsorption. Their results indicated that depending on pH and PZC, there was evidence of both specific and non-specific adsorption of cadmium. Garcia-Miraga and Page [80] found that between pH 6 and 7, soils with high contents of either organic matter or hydrous Fe oxides adsorbed more Cd than those with large amounts of 2:1 clays, even though the latter had higher CECs. Pickering [68] found that goethite removed all the Cd from solution at approximately three pH units below the theoretical pK_i value. With increasing pH, solution Cd concentrations decreased due to increases in: (i) hydrolysis, (ii) adsorption density, and (iii) pH-dependent negative charges. Using multiple regression, Alloway *et al.* [77] showed that pH was one of the key factors, together with organic matter and hydrous oxide contents, controlling the specific adsorption of Cd by 22 different soils. The amount of Cd adsorbed by hydrous Mn oxides has been found to increase almost linearly with pH up to a maximum, and this pH-dependent adsorption process is largely reversible [68].

Effects of competition from other metal ions. Competition with other metal ions, such as Ca, Co, Cr, Cu, Ni and Pb can inhibit the adsorption of Cd. Christensen [71] found that increasing the solution Ca concentration by a factor of 10 (10^{-3} to 10^{-2} M) reduced the adsorption capacity of a sandy loam soil by 67%. Cowan *et al.* [81] observed a strong competitive effect between Cd and Ca for adsorption on hydrous Fe oxide and they considered this to occur via mass action on mutually accessible sites. Likewise, Naidu *et al.* [79] found that Ca competed with Cd for adsorption on oxisols, and andept and a fragiaqulf. Christensen [82] found that Zn had the greatest inhibitory effect on Cd adsorption and this could be explained by a competitive Langmuir model. Even though Cd adsorption was reduced, the shape of the adsorption isotherms was the same [82].

Adsorption on calcite. Alloway *et al.* [83] have shown that soils containing free $CaCO_3$ can sorb Cd and reduce its bioavailability. The adsorption of Cd by calcite has been investigated in detail by McBride [84] and Papadopoulos and Rowell [85]. It was found that calcite had a high affinity for Cd and gave a linear adsorption isotherm at low Cd concentrations (< 1 μmol/g). However, with high Cd concentrations $CdCO_3$ precipitation predominated. The chemisorption of Cd at low concentrations was considered to involve the replacement of Ca by Cd in the surface crystals of calcite [85].

Effects of organic ligands. Complexation of Cd with certain organic ligands in the soil solution can have major effects on the amounts sorbed. Farrah and Pickering [86] found that EDTA prevented the adsorption of Cd over the pH range 3–11. The ligand NTA forms anionic complexes but these dissociate at low pH, which results in some Cd being adsorbed. An excess of the amino acid glycine caused precipitation to be displaced to regions of higher pH, but tartaric acid had no effect on adsorption. It was concluded that clays do not adsorb anionic metal complexes to any significant extent and that adsorption of cationic species can be reduced significantly through competition from protonated ligands. Elliot and Denneny

[87] also found that NTA and EDTA inhibited adsorption of Cd by soils through the formation of non-adsorbing complexes, but oxlate and acetate had no effect. All sorption isotherms showed a typical pH dependency with a maximum at around pH 7. With increasing acidification of the solution, less metal was complexed by the ligands because H^+ was preferentially bound at low pH. Although Cd forms anionic complexes with humic and fulvic acids [88], they are less stable than those formed with Cu and Pb [75, 89].

Neal and Sposito [90] found S-shaped adsorption isotherms for Cd with sludged soils, indicating that ligands in the aqueous solution had a stronger affinity for low concentrations of Cd than the surface charges on the soil. However, after washing the soils to remove the soluble organic ligands, the adsorption isotherms were the more usual L-curve. Organic ligands not only enhance the solubilities of trace metals, but also reduce their toxic effects on plants, because the free (hydrated) ion appears to be more toxic than very stable inorganic complexes, such as $CdCl^-$ and organic complexes [91]. Baham *et al.* [92] found that the Cd in fulvic acid solutions extracted from sewage sludges was only weakly bound (less strongly than Zn and Ni), but variations could occur in different sludge-derived fulvic acids. Stevenson [93] determined the stability constants of Cd complexes with humic acids using potentiometric titrations. Cadmium was more weakly bound than Pb and Cu, especially at lower pHs. It was concluded that carboxyl and phenoxyl groups were involved in the binding of all the metals. Fletcher and Beckett [94] found that the soluble OM from digested sludge has two groups of exchange sites; one binds Ca, Mg, Zn, Ni, Co, Mn, Cd, Pb and Fe(III) and the other binds only Cu, Pb and H^+.

Effects of chloride ions. Cadmium forms very stable soluble complexes with Cl^- ligands and several authors have reported a decrease in adsorption/increase in mobilisation in soils with high concentrations of water-soluble Cl^- ions, such as in saline soils, irrigation with saline water, and contamination by landfill leachate. Evans *et al.* [95] found that the reduction in metal adsorption in the presence of Cl^- was: Zn < Pb < Cd < Hg and this was directly related to the ability of the metals to form complexes with Cl^-. In South Australia, McLaughlin *et al.* [96] found that the uptake of Cd by potatoes was higher where saline irrigation waters were used. They found that Cd concentrations in potato tubers in that region were positively correlated with water-extractable Cl^- in soil.

6.5 Soil–plant relationships

The soil chemical processes affecting the availability of Cd for uptake by plants are particularly important in considerations of the impact of soil contamination on human health. Although Cd toxicity can occur in plants on severely polluted soils, its accumulation in food crops at sub-phytotoxic levels is a greater cause for concern due to the risk of increased dietary exposure in consumers. Even slightly elevated Cd concentrations in foods can have a significant effect in the long term.

Mitchell *et al.* [97] found the order or toxicity to wheat and lettuce plants on acid soils to be Cd > Ni > Zn. Chaney *et al.* [98] stress that it is not possible to rely on the onset of visible symptoms of Cd toxicity to act as a warning when food crops have accumulated excessive amounts of metals, such as Cd, which could be hazardous to health. Relatively large concentrations of Cd can accumulate in edible portions without the plant showing symptoms of stress. Acute Cd toxicity is manifested by leaf chlorosis, wilting and stunted growth, but is rarely found. Many cases of toxicity on heavy metal-polluted soils are due to excesses of other elements present in far higher concentrations.

The amount of Cd taken up by plants depends on a combination of soil and plant factors, which are discussed below.

6.5.1 Soil factors affecting the uptake of cadmium by plants

The soil cadmium content. Although various soil parameters can affect the availability of Cd, the total amount of the element present in the soil is one of the major factors affecting the Cd contents of plant. Kabata-Pendias and Pendias [3] refer to data which show that Cd in potato stalks and barley grain have steep linear relationships with Cd in soil, whereas spinach leaves showed a log-normal relationship with soil Cd. Alloway [34] found total Cd to be closely correlated with the Cd contents in the edible portions of cabbage, carrot, lettuce and radish grown on 50 different soils polluted from various sources. Chumbley and Unwin [94] reported highly significant correlations between the total Cd content of sewage sludge-amended soils and the Cd contents of lettuces and cabbages. Lund *et al.* [100] also found significant correlations between Cd in the soil and Cd concentrations in the leaves of several crop species. Hornberg and Bummer [101] found that Cd concentrations in wheat grain were linearly correlated with the total Cd content of the soil.

The origin of the Cd in the soil can also affect its bioavailability. Alloway *et al.* [77] found that the Cd in soils contaminated from inorganic sources, such as metalliferous mining and smelting, tended to be more readily accumulated in the edible portions of vegetables than that from soils amended with sewage sludge. However, the lowest accumulation ratios (Cd in crop/Cd in soil) of all were found in crops grown on the calcareous, mining-polluted soils from Shipham (section 6.6) [83]. Sewage sludge is the most common source of relatively high concentrations of Cd in soils. The organic matter applied in the sewage sludge increases the metal adsorbing capacity of the amended soil.

Several workers have reported that crops grown on soils spiked with Cd salts take up more Cd than those grown on soils containing the equivalent amount of Cd from sewage sludge. Korcak and Fanning [102] showed that uptake by maize was 5 to 18 times greater from $CdSO_4$-spiked soils compared with equivalent amounts of Cd in sludge. Experiments by Mahler *et al.* [103] with $CdSO_4$-spiked sludge-amended soils and soils without sludge showed that Cd was less available from spiked sludged soils. Liming the soils resulted in reduced Cd uptake, and this effect was greatest on sludged soil. Alloway [34] found that Cd was more available from soils spiked with metal salts, than from soils collected from polluted field sites, even though they had been left to equilibrate for many months after treatment.

Bingham *et al.* [104] found that Cd^{2+} activity in soil solutions correlated better with plant uptake by Swiss chard than total soluble Cd or free-ion and ion-pair concentrations. They suggested that the GEOCHEM model was the most convenient way of assessing the major soluble Cd species.

Soil pH. Soil pH is the major factor determining the availability of Cd in the soil, because it affects all adsorption mechanisms and the speciation of metals in the soil solution. Cadmium uptake is inversely related to soil pH. Page *et al.* [2] reported that the Cd content of Swiss chard leaves increased by factors of between 2 and 3.9 when the soil pH was reduced from 7.4 to 4.5. The uptake of Cd by rice decreased when the pH was increased from 5.5 to 7.5, and wheat showed a similar response [105]. Jackson and Alloway [106] reported that the addition of lime to raise the pH of a wide range of different sludged soils to pH 7.0 had the effect of reducing Cd concentrations in cabbages by an average of 43% and the Cd contents of lettuces by 41%; however, no significant reductions in Cd concentrations in potato tubers were observed.

Andersson and Nilsson [107] found that the addition of CaO to soils reduced the uptake of Cd by fodder rape due to both an increase in pH and to competition between Ca^{2+} and Cd^{2+} ions. With increasing acidity, the increased activity of Cd^{2+} is partly due to the dissolution of hydrous oxides and their co-precipitated metals and reduced adsorption on colloids due to a decreased pH-dependent negative charge. An exception to the relationship between pH and uptake was reported by Pepper *et al.* [108] who did not observe a reduction in the Cd content of maize after liming a soil amended with anaerobically digested sludge to pH 6.5.

Alloway *et al.* [109] found that pH (measured in 0.01 M $CaCl_2$) was the second most influential factor (after total Cd) in the multiple regression equations derived to describe the accumulation of Cd by four crops on 50 different polluted and control soils. The highest Cd accumulation ratios tended to occur in plants grown on acid soils [83].

Sorptive capacity of soils. Several workers have demonstrated that the Cd content of plants is inversely proportional to the CEC of the soils on which they were grown [110–112]. Alloway *et al.* [77] found an inverse relationship between the distribution coefficients (K_d) for Cd determined from the isotherms obtained from specific adsorption experiments and the Cd content of cabbages in several soils (Table 6.5).

Organic matter contributes part of the soil's CEC but also adsorbs heavy metals by forming complexes. Hinesly *et al.* [110] reported that Cd uptake by maize was inversely related to the CEC of soils spiked with $CdCl_2$, but was not correlated with CEC in soils amended with Cd-containing sewage sludge. Mahler *et al.* [103] found no consistent relationship between soil CEC and the Cd content of either lettuce or Swiss chard leaves. Likewise, Alloway and Jackson [109] did not find

Table 6.5 The variation in Cd uptake and soil parameters for soils of two different textures and redox conditions from a field experiment with identical applications of sewage sludge. From Alloway *et al* [77].

	pH	% Organic matter	Hydrous Fe (%)	Oxides Mn (mg/l)	Soil total Cd (mg/kg)	K_d	Cabbage Cd (mg/kg dry matter)
Clay soil	5.4	13.6	2.07	98	7.8	297	9.5
Sandy loam	5.5	8.5	4.18	535	6.1	582	4.9

CEC to be an important variable in models derived for the uptake of Cd by four crops on a range of soils.

The relationship between CEC and plant uptake remains unclear because cation exchange is only one of several adsorption mechanisms affecting the solubility of Cd in soils. Although hydrous oxides do not contribute much to the CEC of a soil below pH 8 (Chapter 2), they specifically adsorb considerable amounts of Cd. It could therefore be concluded that CEC itself is not the most appropriate parameter to indicate the ability of a soil to adsorb metals such as Cd.

Redox conditions. Rice is a unique food crop in that it can be grown both under reducing conditions in flooded paddy fields, and also in oxidising conditions, when the paddies are drained and allowed to dry. Some varieties of rice are also grown as a field crop without fooding (upland rice). It has been shown that rice grown under flooded (reducing) conditions accumulates much less Cd and shows lower yield losses than when grown in oxidising conditions [113]. This is due to the formation of solid state CdS in the anoxic paddy soils. When the sulphide is oxidised there will also be some acidification which will contribute to Cd availability. In the Jintsu Valley of Japan where 'itai-itai' disease was first reported in the 1950s, the Cd content of the paddy rice was found to be correlated with the number of days the paddies were drained and allowed to aerate before harvest [2] (section 6.6).

Although many other crop species cannot tolerate prolonged reducing conditions, many species can be affected by the indirect effects of reducing conditions (gleying). Most cultivated clay soils are gleys which have oxic conditions in the topsoil for much of the year as a result of field drainage and cultivations. Cadmium can be more available in these soils than in ungleyed soils due to their lower adsorptive capacities for Cd which are caused by lower contents of hydrous Fe and Mn oxides. This is shown in Table 6.6. Cabbages grown in pots of a clay (drained gley) soil and a sandy loam (freely drained) from field trails with identical sewage sludge treatments showed a greater uptake on the clay soil. This clay soil had lower contents of hydrous Fe and Mn oxides and a lower K_d value for Cd than the sandy loam soil [77].

Effects of other elements in the soil. Relative excesses of Cu, Ni, Se, Mn and P can reduce the uptake of Cd by plants [2]. The situation with Zn is less clear and appears to depend on the Cd content of the soil. Zinc has been found to have an antagonistic effect on Cd uptake in soils with low Cd concentrations, and either a synergistic or a nil effect with relatively high Cd contents [2]. Smilde *et al.* [114] observed antagonistic effects of soil-applied Zn on the uptake of soil-applied Cd by plants over five consecutive years of pot trials with lettuce, spinach, spring wheat, endive and maize in sandy and a loam soil. Cd also had an antagonistic effect on Zn uptake but not so pronounced and in the loam soil there was evidence of a synergistic effect where Zn uptake increased with applied Cd. Oliver *et al.* [115] found that applications of low rates of Zn (5 kg/ha) at sowing markedly reduced Cd concentrations in wheat grain in areas of marginal to severe Zn deficiency in South Australia. They suggest that this effect could be partly due to the rectification of the damage to root tissue caused by Zn deficiency as well as to competition between Cd and Zn for uptake. Lead is considered to have a synergistic effect on Cd uptake due to it being preferentially adsorbed, thus leaving more Cd in solution [15]. The translocation of Cd to plant shoots can also be inhibited by relative excesses of other elements. [98]. Villarroel *et al.* [116] showed that the application of N fertiliser to Swiss Chard growing in a sludge treated soil had the effect of increasing the uptake of Cd. With the increased yield the uptake of Cd increased by 50% but that of Zn remained unchanged. It was concluded that the uptake of Cd and Zn was limited by the rate of desorption of the metals. The increased root growth resulting from the N fertilisation appears to have facilitated the desorption of Cd in the rhizosphere.

6.5.2 Plant factors affecting the uptake of cadmium from soil

Plant genotype. Plant species and varieties (cultivars) differ widely in their ability to absorb, accumulate and tolerate heavy metals. Davis and Calton-Smith [117] showed that lettuce, spinach, celery and cabbage tended to accumulate relatively high concentrations of Cd, while potato tubers, maize, french beans and peas accumulated only small amounts of Cd. Many researchers have found lettuce to be greatest accumulator of Cd among food crops [15]. Tomato leaves were found to accumulate 70 times more Cd than carrot leaves from the same culture solution [118]. Bingham *et al.* [119] gave the following order of decreasing sensitivity to Cd toxicity, based on the Cd concentration in the soil causing a 25% decrease in yield: spinach > soyabean > curly cress > lettuce > maize > carrot > turnip > field bean > wheat > radish > tomato > squash > cabbage > Swiss chard > upland rice. However, this order is only for one cultivar of each crop; cultivars within a species can differ widely in their tolerance of trace elements. In South Australia, McLaughlin *et al.* [120] compared the uptake of Cd by 14 commonly grown potato cultivars at 12 sites and found significant differences between cultivars at most sites and an average range of concentration of 30–50 μg/kg FW. At some of the sites, individual cultivars exceeded the maximum permissible concentration of 50 μg/kg FW.

Sposito and Page [52] estimated the Cd removal from soils by harvested crops (in g/ha/yr) to be: potatoes 0.79, tomatoes 0.22, spinach 0.57, wheat 0.06. When considered in relation to the estimates for Cd additions to soil (section 6.3), it can be seen that a positive balance exists, with deposition exceeding removal.

Distribution of cadmium in plants. Cadmium, together with Mn, Zn, B, Mo and Se are recognised as being the trace elements which are readily translocated to plant tops after absorption through the roots [98]. MacLean [123] showed that Cd was present in higher concentrations in the roots than in other organs of oats, soya bean, timothy grass, alfalfa, maize and tomato, but none of these is grown for the consumption of its roots. However, in lettuce, carrot, tobacco and potato, Cd contents were highest in the leaves [123]. In soya bean plants, 2% of the accumulated Cd occurred in the leaves and 8% in the seeds [124].

The speciation of the Cd in tissue of food plants is an important factor in determining its accumulation in the human body. Cadmium has been found to be bound to cytoplasmic proteins that usually contain cysteine and are collectively called phytochelatins. These proteins have been identified in mushroom, bean, soya bean, cabbage, wheat and other plants [125]. It is not yet known whether any soil factors can influence the speciation of Cd in food crops, apart from determining the amount of metal taken up. Elevated concentrations of Cd in plant tissues may trigger the formation of phytochelatins.

In addition to uptake through the roots, Cd can be effectively absorbed into foliage and translocated around plants and this is a significant route for Cd into the food chain in areas affected by atmospheric pollution [126]. (Chapter 3.)

6.6 Cadmium-polluted soils

Apart from the small, but significant, inputs of Cd from atmospheric deposition and P fertilisers, which are relatively ubiquitous, the most common sources of high concentrations of pollutant Cd in soils are: (i) Pb–Zn mining and smelting, and (ii) heavy applications of sewage sludge over many years.

6.6.1 Contamination from metalliferous mining and smelting

Wherever ZnS, $ZnCO_3$ or other sulphide ores are mined or smelted, there exists the possibility of Cd pollution. The major source of metals in soils surrounding old mines is the heaps of finely ground tailings which usually contain 1–10 mg Cd/kg but can have < 500 mg/kg. In a study around two historic Pb–Zn mines in the UK, Merrington and Alloway [127] found that wind-blown tailings particles contributed up to 3.3 kg Cd/yr in the soils within 300 m of tailings heaps. Up to 4.8 kg Cd/yr were transported from the heaps by adjacent streams and 88% of this was in solution as a result of the weathering of residual sphalerite. Davis and Roberts [128] reported total Cd concentrations of up to 540 mg/kg in soils polluted by Pb–Zn mining in North Wales, UK. In Montana, USA, Buchauer [129] found up to 750 mg/kg Cd in soils in the vicinity of a Zn smelter. Anomalous Cd concentrations in soils were detected as far as 40 km from a smelting complex in South Wales, UK [13] and up to 10 or 15 km from the Avonmouth smelter, near Bristol, UK [130].

The only clearly established case of soil and water pollution causing Cd poisoning in people was among rice farmers in the Jintsu Valley, in the Toyama Prefecture of Japan. A Pb–Zn mine had been causing extensive pollution of the river water and paddy soils in the flood plain of the Jintsu Valley for many years. During and after the Second World War, more than 200 elderly women who had borne several children developed skeletal deformations and kidney damage, and 65 died. This condition ('itai-itai' disease) was primarily caused by Cd toxicity, exacerbated by dietary deficiencies of Ca, vitamin D and protein, plus the effects of pregnancies and ageing. Both the locally grown rice and the drinking water had been markedly contaminated with Cd [1]. Average Cd contents in rice were 10 times higher than local controls (0.7 mg/kg and 0.07 mg/kg fresh weight, respectively), with a maximum of 3.4 mg/kg Cd [1, 15]. A recent survey of rice in 22 countries showed an average concentration of 0.029 mg/kg Cd, but the mean for Japan was 0.065 mg/kg Cd [15]. The Cd intake by the inhabitants of the Jintsu Valley was estimated to have been 600 μg/day, which is about ten times greater than the maximum tolerable intake [1]. In Japan, 9.5% of paddy soils have been found to be polluted with Cd, determined on the basis of producing rice with a Cd content of 1 mg/kg Cd or above. A further 3.2% of upland soils and 7.5% of orchard soils are also Cd polluted [131]. Rice is the source of more than 60% of the Cd consumed in areas considered unpolluted in Japan [131]. The yield of rice is not reduced by Cd toxicity until concentrations of Cd in the rice are much higher than 1 mg/kg Cd maximum permissible concentration for human consumption [121]. Kjellstrom [132] found that background exposure levels of Cd in Japan were three times higher than those in Sweden or the USA, which suggests that the staple diet of rice contains more Cd than the components of the diet in the other countries.

Another case of large-scale environmental pollution by Cd, Pb and Zn in a populated area occurred in the village of Shipham in Somerset, UK, where Zn was mined during the eighteenth and nineteenth centuries. Large-scale expansion of the village occurred between 1951 and 1981, when most of the new housing was built on the sites of the old mines. Geochemical and soil surveys revealed that there were very high concentrations of Zn, Pb and Cd in the soils of the village [133]. In view of the possible health effects of this pollution, a comprehensive survey was carried out in 1979 to monitor heavy metals in soils, house dusts and crops, and to assess the health of the population in Shipham and a nearby control village.

The ranges and median values (mg/kg), respectively, for 329 soil samples from Shipham were: Cd 2–360 (91), Zn 250–37200 (7600) and Pb 108–6540 (2340). The mean Cd concentration (fresh weight) of almost 1000 samples of vegetables was 0.25 mg/kg, which was nearly 17 times higher than the national average of 0.015 mg/kg Cd. The highest Cd concentrations were found in leafy vegetables, such as spinach, lettuce and brassicas. The most contaminated vegetables contained 15 to 60 times more Cd than those grown in ordinary soils. Health studies on 500 people (about 50% of the population) revealed small but significant differences in some biochemical parameters, but there was no evidence of adverse health effects in the participating members of the population [134]. Marked differences exist between the situation in Shipham and that in the Jintsu Valley. The people of Shipham had a nutritionally superior diet, with only a small proportion of their diet produced on the polluted soils, and many people had lived in the area for only a few years.

In laboratory experiments with female mice, Bhattacharyya [135] found that excess dietary Cd could cause a dose-dependent loss of Ca from the bones of multiparous mice dams, but not from single pregnacy dams. She concluded that Cd contributed to the pathogenesis of Cd and probably acts directly on the bones rather than indirectly as a result of kidney dysfunction.

6.6.2 *Soil contamination from heavy applications of sewage sludge*

Recommendations exist in many countries for maximum soil loadings of Cd from sewage sludge. The EU mandatory limit for the rate of addition will be 0.15 kg Cd/ha/yr with a maximum permissible total soil Cd content of 3 mg/kg [58]. However, many sites used for the large-scale disposal of sewage sludge in the past, such as sewage farms, will have much higher soil Cd concentrations.

Pike *et al.* [33] reported Cd contents of up to 61 mg/kg Cd, 2470 mg/kg Pb and 2020 mg/kg Cr in the soil on a disused sewage farm in Leicester, UK. Alloway *et al.* [83] investigated more than 20 sites which had received heavy applications of sewage sludge and found soil total Cd contents of up to 64.2 mg/kg Cd (with up to 938 mg/kg Pb, 1748 mg/kg Zn, 770 mg/kg Cu, 333 mg/kg Ni and 6000 mg/kg Cr). Chumbley and Unwin [99] found Cd concentrations of up to 16.8 mg/kg dry matter in lettuce and 8.0 mg/kg dry matter in spinach grown on soils which had received heavy applications of sewage sludge for several years with total Cd contents of up to 26.2 mg/kg. Mahler *et al.* [103] found Cd contents of up to 96.3 mg/kg dry matter Cd in maize and up to 53.2 mg/kg in Swiss chard on previously sludged soils.

McGrath [136] found that anomalously high contents of Zn, Cu, Ni, Cd and Cr from sewage sludge had persisted in the topsoil of a field trail for more than 40 years after application of the sludge. Some decrease had occurred in both total and extractable metal concentrations of treated plots, due to the movement of soil out of the plots during cultivations. In a review of other work, McGrath [136] found that metal uptake by crops in the years after sludge application remained constant in 9 out of the 11 studies considered. Of the two studies which reported a decrease in plant uptake during the residual period, one was complicated by large pH changes but the other, by Hinesly *et al.* [137], showed a decrease in crop uptake during the four years after sludge application. A rapid decrease in Cd uptake (g/ha) by maize has been subsequently reported by Bidwell and Dowdy [138]. They showed that sludge treatments increased the Cd and Zn concentrations in maize stover and grain, but after the termination of applications, these concentrations decreased rapidly in the first two or three years after sludge applications ceased. Cadmium contents had decreased by 80% over six years after sludging. Neither the DTPA nor the HNO_3 soil tests predicted this decrease in plant uptake. Hinesly *et al.* [137] considered that the current years' application of Cd in sewage sludge was more important than the cumulative application in terms of bioavailabilty. Heavy applications of sewage sludge add both metals and metal-adsorbing materials to soils. Corey *et al.* [139] pointed out that at high rates of application the sludge itself becomes the major factor determining the bioavailability of sludge-borne metals rather than the properties of the soil. Jing and Logan [140] showed that plant uptake of Cd from a sludged soil was strongly correlated with the Cd content of the sludge and with Cd:P ratios in the sludges. They considered that this supported the Corey hypothesis and suggest that 'clean' sludges pose less of a risk to the food chain than more contaminated sludges. Hooda and Alloway [141] found that Cd uptake by ryegrass from sludge-soil mixtures with 50 t/ha and 150 t/ha of sludge were higher from the lower sludge treatment. This implied that although the higher rate contributed more Cd and other metals to the growing medium, the greater quantity of organic matter and other adsorbents rendered the metals less available in this treatment.

The bioavailability of metals in sludged soils during the residual period is subject to variations in pH. As the organic matter is mineralised, the soil is likely to become more acid. Liming residual sludged soils to pH 7 brings about a reduction in Cd uptake by crops [103, 106]. Jackson and Alloway found that liming 18 different heavily sludged soils to pH 7 reduced the Cd content of cabbage by an average of 43% and that of lettuce by 41% [106]. Ryan *et al.* [142] calculated that maintaining sludged soils at pH 7 would enable almost three times more Cd/ha to be applied to soils used for food crop production than could be applied to acid soils (pH 5.6).

As discussed in Chapter 2 (Section 2.4.9), it has been found that the relatively high concentrations of metals present in soils which have received heavy applications of sludge have had an inhibitory effect on the soil microbial biomass. This has been found to be most pronounced in the case of *Rhizobium leguminosum* bv *trifolii* which has been shown to be very susceptible to Cd toxicity [143]. In many sludged soils, the high Zn content is considered to pose a greater microbial toxicity problem than Cd (which is often present at much lower concentrations) but the Cd is more toxic to microorganisms. The order of toxicity was found to be Cu > Cd > Ni > Zn [143].

In addition to reducing pH, the mineralisation of organic matter could also decrease the Cd adsorption capacity of some soils, especially sandy soils which have low contents of inorganic adsorbents. Alloway [34] found evidence of this in a sludge-amended sandy soil from a field trail in SW France, which had an organic matter content of only 1.8% and a high proportion of soluble Cd (Table 6.6). However, McGrath [136] did not find a trend towards increasing Cd bioavailability on a sandy loam soil nearly 30 years after the cessation of sludge applications; but the climate was not as conducive to the oxidation of organic matter as that in SW France. Hooda and Alloway [141] showed that both the pH and organic matter content of sludge–soil mixtures decreased over a two year period after mixing.

Table 6.6 The total and soluble Cd contents of soils contaminated from several different sources. %OM is the percentage loss on ignition, pH in water, soluble Cd extracted by centrifugation at field capacity, S/T% = soluble Cd as % of total Cd. From Alloway [34].

Soil	Source of Cd	pH	%OM	Total Cd	Soluble Cd	S/T%
Jintsu Valley (Japan)	Pb–Zn mine	5.1	7.5	3.0	0.119	3.97
Shipham						
Garden	Zn mine	7.5	10.2	134	0.053	0.04
Field	Zn mine	7.8	8.6	365	0.158	0.04
W Wales	Pb–Zn mine	4.1	12.4	1.4	0.227	16.2
France	Sludge	6.4	1.8	80.2	2.652	3.31
Sewage farm 1	Sludge	5.1	28.4	20.0	0.236	1.18
Sewage farm 2	Sludge	6.5	26.9	64.24	0.099	0.15
Sewage farm 3	Sludge	5.5	19.6	59.8	0.250	0.43

This decline was relatively rapid during the first two months and subsequently more gradual. The pH of the mixture with the lower rate of sludge application (50 t/ha) declined to a level significantly below that of the untreated control and that of the higher rate reached the same pH as that of the control. These findings indicate that sludge application can have an overall acidifying effect on soils with a low buffering capacity. The predicted changes in global climate over the next 40 years due to the 'greenhouse effect' are likely to have significant effects on the organic matter content and extent of leaching in soils. Consequently, the bioavailability of Cd and other heavy metals in polluted soils in many areas is likely to change during this time.

Table 6.6 gives some results for mining-contaminated and sewage sludge-amended soils, related to the cases discussed above. It can be seen that the Shipham soils have the lowest proportion of Cd in the soil solution due to their high pH and carbonate contents. The mining-polluted soil from W. Wales has the highest relative solubility but the lowest total Cd content. This high solubility is probably due to its very low pH. The sewage sludge-amended soil from France has a low organic matter content and a relatively high proportion of soluble Cd, compared with the sewage farm soils which have higher organic matter contents and similar or lower pHs. The Jintsu Valley soil was in an aerated state rather than its more characteristic waterlogged anoxic condition. Its low adsorptive capacity for Cd was probably due to its small content of hydrous oxides and relatively low pH and organic matter content [34].

6.6.3 Amelioration and rehabilitation of cadmium-contaminated soils

The hazard to plant, animal and human health from Cd-contaminated soils can be reduced or removed by the following approaches:

(i) Complete removal of the contaminated soil and its safe disposal (such as in a licensed landfill) and replacement with clean (uncontaminated) soil. This is very disruptive of the site and very expensive.

(ii) Covering the contaminated soil with a layer (usually < 1 m thick) of uncontaminated soil ('coverloam'). Frequently, a protective membrane is placed over the underlying contaminated soil to minimise upward capillary movement of solutes. Cadmium-polluted paddy soils are rehabilitated by spreading a 30 cm layer of clean soil over the contaminated soil. However, the reducing conditions at depth facilitate this remediation because much of the Cd will remain as insoluble CdS. As with the removal of contaminated soil, the addition of a layer of coverloam is also expensive and can create engineering problems in some cases.

(iii) Liming to pH 7 to reduce bioavailability—this is the most widely used remedial treatment and requires regular monitoring of the soil pH to maintain the elevated pH.

(iv) Mixing additional organic matter (clean sewage sludge, farmyard manure, slurry and other types of organic wastes) into the topsoil to increase the sorptive capacity of the soil. This is a possible remedial strategy but is rarely used on its own.

(v) Mixing adsorptive industrial minerals such as zeolites, kieselguhr (diatomaceous earth) and possibly hydrous Mn oxides into the topsoil to increase the adsorptive capacity of soils for Cd and other potentially toxic elements. Gworek [144] showed that the addition of pellets of synthetic zeolites to contaminated soils significantly reduced (by < 86%) the uptake of Cd in oats and ryegrass in pot experiments. Although Mench *et al.* [145] found hydrous Mn oxides to be the most effective of five adsorptive mineral compounds in reducing the plant uptake of Cd and Pb from a smelter-contaminated soil, they expressed concern about the long-term stability of this oxide in low pH or low E_h conditions.

(vi) Reducing the Cd content of the contaminated soil by leaching with either acids or chelates, either *in situ* or *ex situ*. Care must be taken to avoid groundwater pollution. Where acid leaching is used there is a risk that the bioavailability of the metal may increase even though the total Cd content is significantly reduced.

(vii) Flooding to create reducing conditions and convert the metal to insoluble CdS—this is only really appropriate for rice-growing regions or areas where ponds or swamps can be created.

(viii) Growing hyperaccumulator crops to remove large amounts of bioavailable Cd and other metals. This technique is very much in its infancy but shows considerable potential at least for Zn [146].

(ix) Growing non-food crops, or, in cases of slight contamination, growing species or cultivars of crops with a low potential to accumulate the metal. Alternatively, forage crops for ruminants could be grown and the Cd-accumulating organs in the animals, such as kidneys, rejected for human consumption.

6.7 Concluding comments

The highly labile behaviour of Cd in soils, especially those contaminated with relatively high concentrations of the metal, is an important factor in the accumulation of Cd in the human diet. Emphasis needs to be placed on the identification of all soils with anomalously high Cd concentrations in order that they can be managed appropriately. It must be remembered that, in most cases, these soils will remain contaminated for hundreds of years (possibly more than 1000 years) and that the bioavailability of Cd will vary during this time as a result of changes in soil properties. Although 'clean-up' procedures may be possible in some circumstances, they are unlikely to be practicable for the majority of agricultural soils.

Holmgren *et al.* [18] who carried out the survey of 'uncontaminated' agricultural soils in the USA, concluded that the accumulation of Cd in soils did not constitute a major problem. However, in many countries, relatively large areas of soils, may have been, or will be, contaminated from various sources such as atmospheric emissions, sewage sludges or high Cd P fertilisers.

It would be logical to avoid or minimise further contamination of soils by Cd-containing materials but P fertilisers and sewage sludge pose problems. Phosphatic fertilisers are essential for modern intensive methods of farming and so the only possible way of reducing inputs from this source is to use raw materials with low Cd contents or extract the metal during their manufacture. Although the Cd contents of phosphorite deposits vary considerably, it is unlikely to be economically feasible to use only those with low concentrations of the metal. However, the technology may become available to remove Cd during the fertiliser manufacturing process. In the case of sewage sludge, there is likely to be increasing disposal onto land, at least in maritime European countries as a result of the prohibition of the disposal of sewage into the sea. Fortunately, Cd concentrations in sewage sludges are decreasing, but even though levels of Cd in sludge-treated soils will be kept below the limiting concentrations, the metal will remain in a potentially bioavailable form for many years. From the research point of view, there is a need to monitor the behaviour of Cd and other metals in sludged soils for an indefinite period. Most of the work published on this subject only relates to periods of up to 10 years. In their review of long-term sludge experiments, Juste and Mench [147] found that out of the 40 field experiments for which they could find details, the oldest was started in 1942 (Woburn, UK) and the second oldest was started in 1958 (Bonn, Germany). Fifty two years is a very short time period in relation to the persistence of the metals in the soil and therefore these long-term experimental sites need to be preserved and monitored. Sites with longer histories of sludging exist and need to be identified in order that they can also be studied. The most important criterion is the time interval since the last sludge was applied.

Overt contamination around old mines and derelict industrial complexes is more straightforward to deal with, and is less likely than sludge disposal to affect agricultural crops. With derelict land the options are to isolate and confine the contaminated material and soil or to clean up the site. In most cases sites are used for industrial, commerical or amenity purposes and are not returned to agriculture. However, where derelict industrial land is used for housing, care must be taken to ensure that the garden soils are safe for growing vegetables.

Although the Shipham study in the UK [134] did not reveal any serious effects of Cd contaminated soil on human health, much lower levels of Cd contamination (<2.5 mg/kg) from industrial emissions in a village (Luykgestel) in Kempenland, in the Netherlands, have been found to cause increased body burdens and altered kidney function [148]. There is an obvious requirement for more research to be carried out on the risk assessment of Cd contaminated soils but the safest policy would appear to be to minimise inputs of Cd to soils wherever possible and restrict its bioavailability in the soil–plant–animal pathway. Apart from the obvious roles played by pollution control and soil chemistry, plant breeding can make a vital contribution through the selection and utilisation of crop genotypes which accumulate the least Cd (and other potentially toxic elements). On the other hand, major heavy metal accumulators may prove to be of use in the *in situ* cleaning-up of contaminated soil.

References

1. Fassett, D.W., in *Metals in the Environment* ed. Waldron, H.A. Academic Press, London (1980, 61–110).
2. Page, A.L., Bingham, F.T. and Chang, A.C., in *Effect of Heavy Metal Pollution on Plants*, Vol. 1 ed. Lepp, N.W. Applied Science, London (1981), 72–109.
3. Kabata Pendias, A. and Pendias, H., *Trace Elements in Soils and Plants* 2nd edition, CRC Press, Baton Rouge, Fa. (1992).
4. Hutton, M., in *Lead, Mercury, Cadmium and Arsenic in the Environment*, SCOPE 31, eds. Hutchinson, T.C. and Meema, K.M. John Wiley, Chichester (1987), 35–41.
5. Nriagu, J.D., *Environ. Pollut.* **50** (1988), 139–161.
6. Aylett, B.J., in *The Chemistry, Biochemistry and Biology of Cadmium*, ed. Webb, M. Elsevier, Amsterdam (1979), 1–43.
7. World Resources Institute, *World Resources 1992/93*, Oxford University Press, New York (1992).
8. Hutton, M., *Cadmium in the European Community*, MARC Rep. No. 2, MAARC, London (1982).
9. Heinrichs, H., Schultz-Dobrick, B. and Wedepohl, K.J., *Geochim. Cosmochim. Acta.* **44** (1980), 1519–1532.
10. Bowen, H.J.M., *Environmental Chemistry of the Elements*, Academic Press, London (1979).
11. Fleischer, M., Sarofim, A.F., Fassett, D.W., Hammond P., Shacklette, H.T., Nisbet, I.C.T. and Epstein, S., *Environ. Health Perspect.* 7 (1974), 253–323.
12. Rose, A.W., Hawkes, H.E. and Webb, J.S., *Geochemistry in Mineral Exploration*, 2nd edn. Academic Press, London (1979).
13. Holmes, R., *The Regional Distribution of Cadmium in England and Wales.* Unpublished PhD Thesis, University of London (1976).
14. Page, A.L., Chang, A.C. and Mohamed El-Amamy, in *Lead, Mercury, Cadmium and Arsenic in the Environment* SCOPE 31, eds. Hutchinson, T.C. and Meema, K.M. John Wiley, Chichester (1987), 119–146.
15. Adriano, D.C., *Trace Elements in the Terrestrial Environment*, Springer-Verlag, New York (1986).
16. Thornton, I., Sources and pathways of cadmium in the environment, in Nordberg, G.F., Alessio, L. and Herber, R.F.M., *Cadmium in the Human Environment*, International Agency for Research on Cancer (IARC), Lyon (1992).

17. Page, A.L. and Bingham, F.T., *Residue Rev.* **48** (1973), 1–43.
18. Holmgren, C.G.S., Meyer, M.W., Chaney R.L. and R.B. Daniels, *J. Environ Qual.* **22** (1993) 335–348.
19. Yamagata, N., Cadmium in the environment and humans. In Tsuchiya, K. (ed) *Cadmium Studies in Japan: A Review.* Elsevier/North Holland Biomedical Press, Amsterdam (1978) 19–43.
20. McGrath, S.P. and Loveland, P.J., *The Soil Geochemical Atlas of England and Wales*, Blackie Academic and Professional, Glasgow (1992).
21. Archer, F.C., Trace elements in England and Wales, in Anon (ed) *Inorganic Pollution and Agriculture* (Reference Book 326) Ministry of Agriculture, Fisheries and Food, London (1980) 184–190.
22. Ure, A.M. and Berrow, M.L, in *Environmental Chemistry*, ed. Bowen, H.J.m. Roy. Soc. Chem., London (1982).
23. Marples, A.E., and Thornton, I., The distribution of cadminum derived from geochemical and insustrial sources in agricultural and pasture herbage in parts of Britain. In *Cadmium '79*, Proc. of Second International Cadmium Conference, Cannes, 1979, Metal Bulletin, London (1980) 74–79.
24. Moir, A.M. and Thornton, I., *Environ. Geochem. Health* **11** (190), 113–120.
25. Culbard, E.B., Thornton, I., Watt, J., Wheatley, M., Moorcroft, S. and Thompson, M., *J. Environ. Qual.* **17** (1988) 226–234.
26. Kim, K.W. and Thornton, I., *Environ. Geochem. and Health* **15** (1993) 119–133.
27. Merington, G. and Alloway, B.J., *Applied Geochemistry* (1994) (in the press).
28. Tyler, G., *Water, Air and Soil Pollut.* **15** (1981) 353.
29. Alloway, B.J. and Jackson, A.P., *Sci. Total Environ.* **100** (1991) 151–176.
30. Williams, D.E., Vlamis, J., Pukite, A.H. and Corey, J.E., *Soil Sci.* **143** (1987) 124–131.
31. Davis, R.D., Carloton-Smith, C.H., Stark, J.H. and Campbell, J.A., *Environ. Pollut.* **49** (1988) 99–115.
32. Legret, M.L., Divet, L. and Juste, C., *Water Res.* **22** (1988) 953–959.
33. Pike, E.R., Graham, L.C. and Fogden, M.W., *J. Ass. Publ. Analysts* **13** (1975), 48–63.
34. Alloway, B.J., Unpubl. Res. Report, DOE PECD 7/8/05 (1986).
35. Jones, K.C., Symon, K.C. and Johnston, A.E., *Sci. Total Environ.* **67** (1987), 75–90.
66. Williams, C.H. and David, D.J., *Aust. J. Soil. Res.* **11** (1973), 43–56.
37. Jones, K.C., Symon, K.C. and Johnston, A.E., *Sci. Total Environ.* **67** (1987), 75–90.
38. Tiller, K.G., *Advances in Soil Sci.* **9** (1989) 113–142.
39. Mortvedt, J.J., *J. Environ.Qual.* **16** (1987), 137–142.
40. Mulla, D.J., Page, A.L. and Ganje, T.J., *J. Environ. Qual.* **9** (1980), 408–412.
41. Hutton, M. and Symon, C., *Sci. Total Environ.* **57** (1986) 129–150.
42. Kloke, A., Sauerbeck, D.R. and Vetter, H., The contamination of plants and soils with heavy metals and transport of metals in terrestrial food chains. In Nriagu, J.O. (ed) *Changing Metal Cycles in Human Health*, Springer Veriag, Berline (1984) 113–141.
43. Nriagu, J.O. and Pacyna, J.M., *Nature (London)*, **333** (1988) 134–139.
44. McLaughline, M.J., Maier, N.A., Freeman, K., Tiller, K.G., Willaims, C.M.J. and Smart, M.K., *Fert. Res.* (1994) (in the press).
45. Reuss, J.O., Dooley, H.L. Griffiss, W., *J. Environ. Qual.* **7** (1978) 128–133.
46. McGrath, S.P., *J. Agric. Sci. Camb.* **103** (1984), 25–35.
47. Bennett, B.G., *Exposure Committment Assessments of Environmental Pollutants* Vol. 1., No. 1. MARC, London (1981).
48. Pacyna, J.M., in *Lead, Mercury, Cadmium and Arsenic in the Environment* SCOPE 31, eds. Hutchinson, T.C. and Meema, K.M., John Wiley, Chichester (1987), 69–87.
49. Nriagu, J.E. (ed.), *Cadmium in the Environment, 1: Ecological Cycling*, John Wiley, New York (1980).
50. Nriagu, J.O., *Nature* **279** (1979), 409–411.
51. Willaims, C.R. and Harrison, R.M., *Experientia* **40** (1984) 29–36.
52. Sposito, G. and Page, A.L., in *Metal Ions in Biological Systems*, ed. Sigel, H. Marcel Dekker, New York (1984).
53. Davis, R.D., in *Heavy Metals in the Environment*, CEP Consultants, Edinburgh (1983), 330–337.
54. Williams, J.H., *Water Pollut. Control* (1975), 635–642.

55. Sommers, L.W., *J. Environ. Qual.* **3** (1977), 225–232.
56. Department of the Environment, *UK Sewage Sludge Survey. Final Report.* Consultants in Environmental Sciences Ltd, Gateshead (1993).
57. Chaney, R.L. *Biocycle* **31** (10), (1990) 68–73.
58. McGrath, S.P., Chang, A.C., Page, A.L. and Witter, E., *Environmental Reviews* (1994) (in press).
59. Ministry of Housing, Physical Planning and Environment, Director General for Environmental Protection (Netherlnds), *Environmental Standards for Soil and Water*, Leidschendam (1991).
60. Fergusssion, J.E., *The Heavy Elements: Chemistry, Environmental Impact and Health Effects*, Pergammon Press, Oxford (1990).
61. Yost, K.J and Miles, L.J., *J. Environ. Sci. Health. A* **14** (1979), 285–311.
62. Tjell, J.C., Hansen, J.A., Christensen, T.H. and Hovmand, M.F., in *Characterisation, Treatment and Use of Sewage Sludge*, eds. L'Hermite, P. and Ott, H.D., D. Reidel, Dordrecht (181), 1493–1498.
63. Mattigod, S.V. and Sposito, G., in *Chemical Modelling in Aqueous Systems*, ed. Jenne, A., American Chemical Soc., Washington D.C. (197(), 837–856.
64. McGrath, S.P., Sanders, J.R., Tancock, N.P. and Laurie, S.H., in *Soil Contamination*, CEP Consultants, Edinburgh (1984), 707–712.
65. McGrath, S.P., Sanders, J.R., Laurie, S.H. and Tancock, N.P., *Analyst* **111** (1986), 559–565.
66. Tills, A.R. and Alloway, B.J., *J. Soil Sci.* **34** (1983), 769–781.
67. Mahler, R.J., Bingham, F.T., Sposito, G. and Page, A.L., *J. Environ. Qual.* **9** (180), 359–364.
68. Pickering, W., in *Cadmium in the Environment. Part 1 Ecological Cycling*, ed. Nriagu, J.O., John Wiley, New York (1980), 365–397.
69. Gerritse, R.G. and Van Driel, W., *J. Environ. Qual.* **13** (1984), 197–204.
70. Pulls, R.W. and Bohn, H.L., *Soil Soc. Am. J.* **52** (1988), 1289–1292.
71. Christensen, T.H., *Water, Air & Soil Pollution* **21** (1984), 105–114.
72. Street, J., Lindsay, W.L. and Sabey, B.R., *J. Environ. Qual.* **6** (1977), 72–77.
73. Levi-Minzi, R., Soldatini, G.FD. and Riffaldi, R., *J. Soil Sci.* **27** (1976), 10–15.
74. Cavallaro, N. and McBride, M.B., *Soil Sci. Soc. Am. J.* **42** (1978), 550–556.
75. Tjell, J.C., Christensen, T.H. and Bro-Rasmussen, B., *Ecotoxicology & Environ. Safety* **7** (1983), 122–140.
76. Jarvis, S.C. and Jones, L.H.P., *J. Soil Sci.* **31** (1980), 469–479.
77. Alloway, B.J., Tills, A.R. and Morgan, H., in *Trace Substancers in Environmental Health* **18** (1985), 187–201.
78. Farrar, H. and Pickering, W.F., *Water, Air & Soil Pollution* **8** (1977), 189–197.
79. Naidu, R., Bolan, N.S., Kookana, R.S. and Tiller, K.G., *European J. Soil Sci.* (1994) (in press).
80. Garcia-Miragaya, J. and Page, A.L., *Water, Air & Soil Pollution* **8** (1978), 289–299.
81. Cowan, C.E., Zachara, J.M. and Resch, C.T., *Environ. Sci. Technol.* **25** (1991) 437–446.
82. Christensen, T.J., *Water, Air & Soil Pollution* **34** (1987), 305–314.
83. Alloway, B.J., Thornton, I., Smart, G.A., Sherlock, J. and Quinn, M.J., *Sci. Total Environ.* **75** (1988), 41–69.
84. McBride, M.B., *Soil Sci. Soc. Am. J.* **43** (1980), 26–28.
85. Papadopoulos, P. and Rowell, D.L., *J. Soil Sci.* **39** (1988), 23–36.
86. Farrar, H. and Pickering, B., *Aust. J. Chem.* **30** (1977), 1417–1422.
87. Elliot, H.A. and Denneny, C.M., *J. Environ. Qual.* **11** (1982), 658–662.
88. Duffy, S.J., Hay, G.W., Micklethwaite, R.K. and Van Loon, G.W., *Sci Total Environ.* **76** (1988), 203–215.
89. Livens, F.R., *Environ. Pollut.* **70** (1991) 183–208.
90. Neal, R.H. and Sposito, G., *Soil Sci.* **142** (1986), 164–172.
91. Sposito, G., in *Applied Environmental Geochemistry*, ed. Thornton, I., Academic Press, New York (1983), 123–170.
92. Baham, J., Ball, N.B. and Sposito, G., *J. Environ. Qual.* **7** (1978), 181–188.
93. Stevenson, F.J., *Soil Sci. Soc. Am. J.* **40** (1976), 665–672.
94. Fletcher, P. and Beckett, P.H.T., *Water Res.* **21** (1987), 1163–1172.
95. Evans, L.J. Lumsdon, D.G. and Bolton, K.A., The influence of pH and chloride on the retention of zinc, lead, cadmium and mercury by soil, in *Proc. Technology Transfer Conf. The Multi-Media Approach: Integrated Environmental Protection* Vol 1, Environment, Canada, Toronto (1991) 123–130.

96. McLaughlin, M.J. Palmer, L.T., Tiller, K.G., Breech, T.A. and Smart, M.K., *J. Environ. Qual.* **23** (1994) (in press).
97. Mitchell, G.A., Bingham, F.T., Page, A.L. and Nash, P., *J. Environ. Qual.* **7** (1978), 165–171.
98. Chaney, R. and Giordano. P.M., in *Soils for the Management of Organic Wastes and Waste Waters*, eds. Elliot, L.F. and Stevenson, F.J. Soil Sci. Soc. Am., Am. Soc. Agron & Crop Sci. Soc. Am., Madison (1977), 235–279.
99. Chumbley, C.G. and Unwin, R.J., *Environ. Pollut.* **B 4** (1982), 231–237.
100. Lund, L.J., Betty, E.E., Page, A.L. and Elliott, R.A., *J. Environ. Qual.* **10** (1981), 551–556.
101. Hornberg, V and Brummer, G.W. Cadmium availability in soils and content in wheat. In Anke, M., Braunlcih, H., Bruckner, C. adn Groppel, B (eds) *Fifth Symposium on Iodine and other Elements*, Schiller Univ., Jena (1986) 916. Cited in Kabata-Pendias and Pendias [3].
102. Korcak, R.F. and Fanning, D.S., *Soil Sci.* (1985), 23–34.
103. Mahler, R.J., Bingham, F.T. and Page, A.L., *J. Environ. Qual.* **7** (1978), 274–281.
104. Bingham, F.T., Strong, J.E. and Sposito, G., *Soil Sci.* **135** (1983), 160–165.
105. Bingham, F.T., Peryea, F.J. and Jarrell, W., in *Metal Ions in Biological Systems* Vol. 20, ed. Sigel, H. Marcel Dekker, New York (1986), 119–156.
106. Jackson, A.P. and Alloway, B.J., Transfer of cadmium from soils to the human food chain, in Adriano, D.C., *Biogeochemistry of Trace Metals*, Lewis Publisher, Baton Rouge, Fla. (1992) 109–158.
107. Andersson, A. and Nilsson, K.O., *Ambio* **3** (1974), 198–200.
108. Pepper, I.L., Bedizeck, D.F., Baker, A.S. and Sims, J.M., *J. Environ. Qual.* **12** (1983) 270–275.
109. Alloway, B.J., Jackson, A. and Morgan, H., *Sci. Total Environ.* (1989) in press.
110. Hinesly, T.D., Redborg, K.E., Ziegler, E.L. and Alexander, J.D., *Soil Sci. Soc. Am. J.* **46** (1982), 490–497.
111. John, M.K., Van Laerhoven, C.J and Church, H.H., *Environ. Sci. Technol.* **6** (1972), 1005-1009.
112. Miller, J.E., Hassett, J.J. and Koeppe, D.E., *J. Environ. Qual.* **6** (1976), 18–20.
113. Bingham, F.T., Page, A.L., Mahler, R.J. and Ganje, T.J., *Soil Sci. Soc. Am. J.* **40** (1976), 715–719.
114. Smilde, K.W., van Luit, B. and van Driel, W., *Plant and Soil* **143** (1992) 233–238.
115. Oliver, D.P., Hannam, R., Tiller, K.G., Wilhelm, N.S., Merry, R.H. and Cozens, G.D., *J. Environ. Qual.* **23** (1994) (in press).
116. de Villarroel, J.R., Chang, A.C. and Amrhein, C., *Soil Sci.* **155** (1993) 197–205.
117. Davis, R.D. and Calton-Smith, C., *Crops as Indicators of the Significance of Contamination of Soil by Heavy Metals*, WRC, Stevenage TR140 (1980).
118. Turner, M.A., *J. Environ. Qual.* **2** (1977), 118–119.
119. Bingham, F.T., Page, A.L., Mahler, R.J. and Ganje, T.J., *J. Environ. Qual.* **2** (1975), 207–211.
120. McLaughlin, M.J., Williams, C.M.J., McKay, A., Kirkham, R., Gunton, J., Jackson, K.J., Thompson, R., Dowling, B., Partington, D., Smart, M.K. and Tiller, K.G., *Austr. J. Agric. Res.* (1994) (in press).
121. Mench, M. and Martin, E., *Plant and Soil* **132** (1991) 187–196.
122. Kuboi, T., Noguchi, A. and Yazaki, I., *Plant & Soil* **104** (1987), 275–280.
123. Maclean, A.J., *Can. J. Soil Sci.* **56** (1976), 129–138.
124. Cataldo, D.A., Garland, T.R. and Wildung, R.E.C., *Plant Phys.* **68** (1981), 835–839.
125. Spivey Fox, M.R., *J. Environ. Quad.* **17** (1988), 175–180.
126. Tjell, J.C., Christensen, T.H. and Bro-Rasmussen, B., *Ecotoxicology & Environ. Safety* **7** (1983), 122–140.
127. Merrington, G. and Alloway, B.J., *Water, Air and Soil Pollut.* **73** (1994) 333–334.
128. Davies, B.E. and Roberts, L.J., *Environ. Pollut.* **6** (1975), 49–57.
129. Little, P. and Martin, M.H., *Environ. Pollut.* **3** (1972) 159–172.
130. Asami, T., in *Changing Metal Cycles and Human Health*, ed. Nriagu, J.O, Springer-Verlag, Berlin (1984) 95–111.
131. Kjellstrom, T., *Environ. Health Persp.* **28** (1979), 169–197.
132. Sims, D.L. and Morgan, H., *Sci. Total Environ.* **75** (1988), 1–10.
133. Morgan, H. and Sims, D.L., *Sci. Total Environ.* **75** (1988), 135–143.
134. Bhattacharyya, M.J., *Water, Air and Soil Pollut.* **57–58** (1991) 665–673.
135. McGrath, S.P., in *Pollutant Transport and Fate in Ecosystems*, eds. Coughtrey, P.J., Martin, M.H. and Unsworth, M.J. Blackwell, Oxford (1987), 301–307.

136. Hinesly, T.D., Ziegler, E.L. and Barrett, G.L., *J. Environ. Qual.* **8** (1979), 35–38.
137. Bidwell, A.M. and Dowdy, R.H., *J. Environ. Qual.* **16** (1987), 438–442.
138. Corey, R.B., King, L.D., Lue-Hing, C., Fanning, D.S., Street, J.J. and Walker, M., Effects of sludge properties on accumulation of trace elements by crops, in Page, A.L., Logan, T.J. and
139. Ryan, J.A. (eds) *Land Application of Sludge: Food Chain Implications*, Lewis Publishers, Chelsea, MI (1987) 25–27.
140. Jing, J. and Logan, T.J., *J. Environ. Qual.* **21** (192) 73–81.
141. Hooda, P.S. and Alloway, B.V., *J. Soil Sci.* **44** (1993) 97
142. Ryan, J.A., Pahren, H.R. and Lucas, J.B., *Environ. Res.* **27** (1982), 251–302.
143. Chaudri, A.M., McGrath, S.P. and Giller, K.E., *Soil Biol. Biochem.* **24** (1992) 625–632.
144. Gworek, B., *Environ. Pollut.* **75** (1992) 269–271.
145. Mench, M., Didier, V., Gomez, A. and Loffler, M., *Remediation of metal contaminated soils in Eijsackers.*
146. McGrath, S.P., Sidoli, C.M.D., Baker, A.J.M. and Reeves, R.D., The potential for the use of metal-accumulating plants for the *in situ* decontamination of metal-polluted soils, in Eiijackers, H.J.P. and Hamers, T., eds. *Integrated Soil and Sediment Research, A Basis for Proper Protection*, Kluwer Academic Publishers, Dordrecht (1993).
147. Juste, C. and Mench, M., Long-term application of sewage sludge and its effect on metal uptake by crops, in Adriano, D.C., eds. *Biogeochemistry of Trace Metals*, Lewis Publishers, Boca Raton, Fla (1992) 159–193.
148. Kreis, I.A., Wijga, A. and van Wijnen, J.H., *Sci Total Environ.* **127** (1992) 281–292.

7 Chromium and Nickel

S.P. McGRATH

7.1 Introduction

Chromium is a *d*-block transition metal of Group VIB of the Periodic Table. It has an atomic number of 24 and an atomic weight of 51.996; of the five known radioisotopes, ^{51}Cr (half-life 27.8 days) is the most commonly used in experimental work.

Chromium has been used in alloy steels since about 1877, and chrome plating dates from about 1926. Chromium metal is grey and brittle and can be highly polished. It is resistant to attack by oxidation, which leads to its use in alloys that are resistant to corrosion. The presence of Cr in alloys also increases hardness and resistance to mechanical wear. It occurs in +3 and +6 oxidation states in the environment, though Cr(III) (Cr^{3+}) is the most stable; ionic radii are 0.052–0.053 nm for Cr^{6+} and 0.064 nm for Cr^{3+}.

Nickel is a transition metal of Group VIII of the Periodic Table. Its atomic number is 28, and its atomic weight 58.71. Of the seven known radioisotopes, 63Ni (half-life 92 years) is the most useful in soil-plant studies. Nickel can occur in a number of oxidation states, but only Ni(II) is stable over the wide range of pH and redox conditions found in the soil environment. The ionic radius of Ni(II) is 0.0065 nm (close to those of Fe, Mg, Cu and Zn). Nickel can replace essential metals in metallo-enzymes and cause disruption of metabolic pathways. Various Ni-steel alloys were developed in the nineteenth century, and their resistance to corrosion resulted in their use for manufacture of motor vehicles, armaments, aircraft and tableware. More recent uses include batteries and electronic components. Nickel is a whitish silver metal which is hard but brittle, polishable, and conducts both heat and electricity.

7.1.1 *Usage*

Chromium is produced from the ore chromite, which is a mixed oxide with the general formula $FeO.Cr_2O_3$ but also contains variable amounts of Mg and Al. Of the 10^7 t of Cr produced annually [1], about 60–70% is used in alloys [2], including stainless steel which contains Fe, Cr and Ni in varying proportions according to the properties required in the final product. Alloy steels contain 10–26% Cr. The refractory properties of Cr are exploited in the production of refractory bricks for lining furnaces and kilns, accounting for approximately 15% of the chromate ore

S. Smith was a co-author of this chapter in the first edition of this book.

Table 7.1 Some uses of chemicals containing Cr. After Stern [2]

Antifouling pigments	Metal finishing
Antiknock compounds	Metal primers
Catalysts	Mordants
Ceramics	Phosphate coatings
Corrosion inhibitors	Photosensitisation
Drilling mud	Pyrotechnics
Electronics	Refractories
Emulsion hardeners	Tanning
Flexible printing	Textile preservatives
Fungicides	Textile printing and dyeing
Gas absorbers	Wash primers
High-temperature batteries	Wood preservatives
Magnetic tapes	

used. About 15% is also used in the general chemical industries e.g. chrome alum (Cr(III)) for tanning leather, pigments and wood preservatives (sodium dichromate). Major uses of Cr chemicals are shown in Table 7.1. About 4% is converted into chromic acid and used for electroplating or as an oxidant.

Nickel is extracted from sulphide and oxide ores, and there are two economically exploitable Ni ores: lateritic oxides and pentlandite, a sulphide. The latter is the most commercially important, and like chromite, is found in association with mafic and ultramafic rock formations [3]. World production of Ni is approximately 0.8 $\times 10^6$ t/yr [1] and the single largest use, as with Cr, is the manufacture of stainless steels. Other major uses are electroplating, alloys, Ni–Cd batteries, electronic components, catalysts for hydrogenation of fats, methanation and in petroleum products.

7.1.2 Biological requirements

The essentiality of Cr was first shown in rats by Mertz and co-workers in 1955 [4]. Rats fed on a diet based on *Torula* yeast developed an intolerance to glucose, a decreased glycogen reserve and a syndrome similar to classic diabetes. When these rats were fed small amounts of brewer's yeast they recovered and showed normal glucose tolerance. Later, a so-called glucose tolerance factor (GTF) was isolated from brewer's yeast but was absent from *Torula* yeast. GTF has been identified as an active dinicotinato Cr(III)-glutathione complex, but it has never been crystallised. Although simple Cr salts can be effective, GTF isolated from brewer's yeast or pigs' kidney is much more biologically efficient. Studies with humans suffering from impaired glucose tolerance (e.g. diabetics, old or malnourished people) have in some cases shown an improvement after Cr supplementation, but it appears that other factors are also involved. Small-scale epidemiological studies have indicated a correlation between a lack of dietary Cr and maturity-onset diabetes and cardiovascular problems [5], and it has been reported that countries with high levels of soil Cr have low death rates due to cardiovascular diseases [6]. In fact, an objective of several soil–plant studies has been to find ways of increasing the Cr content of crop plants in order to supplement the diet [7, 8]. Normal dietary intakes are likely to be below 200 μg per day [9]. whilst several studies have reported stimulatory effects of Cr on plants, the essentiality of the element for plants has not been demonstrated.

In contrast to Cr, it has been suggested for many years that Ni may play an essential role in metabolic processes of higher plants [10]. In 1975 it was found that Ni is a component of the enzyme urease in beans [11]. Claims were made that Ni was essential for all legumes and possibly all other higher plants [12], but tests for essentiality in non-legumes were not reported until 1987 [10]. The latter authors point out that the following criteria must be satisfied to establish essentiality of a nutrient; (i) that the organism cannot complete its life cycle in the absence of the element, and (ii) no other element can substitute for the essential element. Brown *et al.* [10] tested the essentiality of Ni for barley (*Hordeum vulgare* L.) and found reduced germination and seedling vigour, followed by disruption of grain filling of those plants which grew to maturity. Only 'low-Ni' seed from plants grown for several generations in nutrient solutions containing less than 30 ng Ni/l showed these responses. Grains containing < 30 ng/kg dry weight were not viable; others with low Ni content had very low rates of germination, unless incubated with 1.0 μM $NiSO_4$, which restored germination to near 100%. The authors state that Ni deficiency is unlikely in conventional experiments with culture solutions because the levels of Ni present as contaminants in nutrient salts and water will be enough to satisfy the plant's requirements. By implication, it is unlikely that deficiency would occur in plants grown in soil, because the small amounts necessary should be available. Evidence for lack of substitution of the elements Al, Cd, Sn and V for Ni in soya bean growth has also been reported [12].

Nickel has been shown to be essential for the growth of some microorganisms. The growth of a strain of blue–green algae *Oscillatoria* spp. and of the bacterium *Alcaligenes eutrophus* has been shown to depend on the presence of Ni in specially purified media [13, 14]. However, in neither case was the biochemical role of Ni identified. Nickel appears to be necessary for the growth of marine microalgae grown on urea as the sole source of nitrogen [15]. Presumably, Ni is required for synthesis and activity of urease in these organisms.

It has been demonstrated in a number of experiments that Ni is an essential element for animals [16] and that by implication it may play an essential role in human metabolism, but the precise funcion of Ni is unclear. The effects of diets with very small concentrations of Ni (e.g. 40 μg/kg) include impaired liver metabolism, decreased iron absorption and lower activity of many enzymes [16, 17]. Deficiency is unlikely to occur in animals or humans because intakes are much higher than in experiments on Ni deficiency. Mean Ni intakes in western countries are likely to be between 200 and 300 μg/day [18].

As with other trace elements, Ni and Cr are also toxic to plants and animals under conditions of above average exposure. In cases of intense human exposure, such as the occupational route, Ni and Cr can be both toxic and carcinogenic [19, 20].

Table 7.2 Concentrations of chromium and nickel in various types of rock (mg/kg). After Cannon [21]

	Cr		Ni	
	Average	Range	Average	Range
Ultramafic igneous	1800	1000–3400	2000	270–3600
Basaltic igneous	200	40–600	140	45–410
Granitic igneous	20	2–90	8	2–20
Shales and clays	120	30–590	68	20–250
Black shales	100	26–1000	50	10–500
Limestone	10	—	20	—
Sandstone	35	—	2	—

7.2 Geochemical occurrence

Chromium is the seventh most abundant element on Earth, but 21st in abundance in the crustal rocks, with an average concentration of 100 mg/kg rock. Chromium is found in igneous rocks (Table 7.2) where it readily substitutes for Fe which has an ionic radius of 0.067 nm, close to that of Cr(III). Mafic and ultramafic rocks are richest in Cr, containing up to 3400 mg/kg of Cr along with 3600 mg/kg Ni (Table 7.2). The only commercially exploitable ore is chromite which is a spinel also known as 'chrome iron ore'. Chromite is associated with mafic and ultramafic rocks. The composition of the ores varies between 42–56% Cr_2O_3 and 10–26% FeO, together with varying amounts of MgO, Al_2O_3 and SiO_2. Three types of ores are recognised, according to end use: metallurgical, which has a minimum of 48% Cr_2O_3 and a Cr:Fe ratio of 3:1; refractory, which must have high Cr_2O_3 and Al_2O_3 but low Fe; and chemical, with high Cr_2O_3 but low SiO_2 and Al_2O_3. Chromium(III) also replaces Fe^{3+} and Al^{3+} in many other minerals, resulting in Cr tourmalines, garnets, micas and chlorites. Trace amounts of Cr give colour to other minerals, e.g. the green of emerald and the red of ruby (due to chromic oxide).

The Earth's crust has an average concentration of 75 mg Ni/kg rock. Nickel, the 24th most abundant element, is twice as abundant as Cu. Sulphide ore bodies are formed after the separation of Ni, Cu and Fe as sulphides from molten mafic magma. The mineral pentlandite (Ni, $Fe)_9S_8$) is a Ni-Fe sulphide and this is the most important commercial ore. Nickel also substitutes for Fe in other sulphide formations such as pyrite. Siderophile (iron-loving) elements such as Ni, Cr and Co tend to be enriched in ultramafic rocks. Chromium tends to have localised occurrences of extremely high concentrations, because of discrete bodies of chromite, but Ni^{2+} substitutes for Mg^{2+} in magnesium minerals, and so is more uniformly distributed in ultramafites than Cr [22]. Nickel-silicates, particularly the hydrous magnesium silicate, garnierite $(Ni, Mg)_6Si_4O_{10}(OH)_8$, are important sources of Ni and were first recognised in the deposits of New Caledonia. Another Ni-containing group of minerals are the laterites that form after prolonged weathering of ultrabasic peridotite. Weathering of the original rock leaves a concentrated residue of Ni and silica, from which silicate minerals form.

Most of the important ore-bearing rocks are Ni sulphides and are mined in deep mines. The ore bodies are usually found in areas with mafic and ultramafic igneous rocks. By contrast, Ni oxides are lateritic and mined by open-cast techniques.

The occurrence of Ni varies between different rock types, as shown in Table 7.2. Ultramafic rocks such as peridotite, dunite and pyroxenite have the most Ni, followed by mafic (gabbro and basalt) and intermediate rocks. Igneous rocks rich in ferromagnesian and sulphide minerals are rich in Ni (e.g. pyroxene, olivine, biotite and chlorite). In these minerals, Ni substitutes for Fe and Mg because of the similarity of the ionic radii of these elements (see Section 7.1). Acid igneous rocks have less Ni than the above, and alkaline rocks and sedimentary rocks are particularly low in Ni.

7.3 Origin of chromium and nickel in soils

7.3.1 Nickel and chromium in soils and parent materials

The average concentration of Ni in world soils is probably around 20 mg/kg, which obscures much variation between soil types. The content of Ni in a soil depends very much on the nature of the parent material. For example, soils formed on serpentine can contain 100–7000 mg/kg [22] and also contain unusually large Cr, Mg and Fe concentrations but small concentrations of Ca and Si. The exact cause of poor plant growth on serpentine soils remains controversial, but the consensus is that Ni is more likely to be toxic than the relatively large concentrations of Cr and Co present [22]. Toxicity of Ni and high Mg:Ca ratios are likely to be the main limiting factors for crop growth on these soils. The averge concentration of Ni and its range in various rock types and soils are shown in Tables 7.2 and 7.3. Vinogradov [23] gave an average value for Ni in soils of 40 mg/kg. Ure and Berrow [24] in a more recent survey of 13 000 published results from the world literature quoted averages of 84 mg Cr/kg and 34 mg Ni/kg. In a survey of 2944 and 4122 Scottish soil samples, geometric mean concentrations of Cr and Ni were 62 and 27 mg/kg

Table 7.3 Reported concentrations of Cr and Ni in soils (mg/kg)

Cr			Ni			
Mean	Median	Range	Mean	Median	Range	Reference
200	—	—	40	—	—	23
—	—	5–3000	—	—	10–800	27
—	70	5–1500	—	50	2–750	28
—	6.3	—	—	17	—	29
—	54	1–2000	—	19	5–700	30
84	—	0.9–1500	34	—	0.1–1523	24
150	62*	0.5–10000	53	27*	0.5–5000	25
41	39	0.3–837	25	20	0.8–440	26
—	—	—	24	18	1–269	31

* Geometric mean

Table 7.4 Statistical summary of the concentrations of chromium and nickel in topsoils of England and Wales, classified according to soil texture class (after McGrath and Loveland 1992 [26])

Soil texture	No. of samples	Minimum	Percentile			Maximum
			25th	*50th*	*75th*	
Chromium						
Clayey	479	18.7	50.3	59.0	69.6	837.8
Fine loamy	202	5.0	35.3	43.5	54.0	692.9
Fine silty	1063	4.9	38.4	48.0	57.6	285.4
Coarse silty	184	6.3	29.5	39.3	47.4	143.5
Coarse loamy	1143	0.2	21.1	27.4	36.0	356.3
Sandy	229	0.2	9.4	13.2	18.0	91.5
Peaty	557	0.2	6.2	12.2	24.8	153.7
ALL SOILS	5692	0.2	26.5	39.3	52.6	837.8
Nickel						
Clayey	479	10.5	31.4	38.2	44.5	194.6
Fine loamy	2002	1.6	19.0	25.3	32.8	439.5
Fine silty	1063	2.3	20.7	28.2	36.2	298.8
Coarse silty	184	4.5	17.7	22.4	31.0	89.7
Coarse loamy	1143	0.8	10.8	15.8	21.9	436.4
Sandy	229	0.8	4.5	7.5	12.1	74.3
Peaty	557	0.8	4.5	6.6	12.7	123.9
ALL SOILS	5692	0.8	14.0	22.6	32.4	439.5

respectively [25], whilst McGrath and Loveland in a recently completed survey of almost 6000 topsoils in England and Wales, reported geometric mean concentrations of 34 mg Cr and 20 mg Ni/kg soil [26].

The influences of geology and soil forming processes on the concentrations of Cr and Ni in soil are clearly demonstrated in Table 7.4. This shows that for both Cr and Ni, coarse loamy, sandy and peaty soils contain less than the median amounts of these metals, whilst clay-rich soils contain above median concentrations.

Many studies have been made of the distribution of Ni in soil profiles with conflicting results. Depending on the origin of the soil and pedogenic processes, the surface, or the subsoil may be relatively enriched, or have the same Ni concentrations [3]. In some soils Ni, like Fe and Mn, may accumulate in the B horizon as mixed oxides, whilst in others Ni may build up in surface litter and humus. Berrow and Reeves [25] found that, on average, the organic debris accumulated in the surface 0–6 cm of Scottish soils had smaller concentrations of Cr and Ni than the mineral soil, but that below 6 cm there were only small changes in Cr and Ni concentrations with depth.

7.3.2 Agricultural materials

The total annual input of Cr and Ni into soils worldwide has been estimated to be between 480–1300 and 106–544 $\times 10^3$ respectively [32].

Table 7.5 Concentrations of Cr and Ni in fertilisers, limestone and animal manures (mg/kg)

	Cr	Ni	Reference
Fertilisers			35
Nitrogen	tr–50	tr80	
Phosphorus	tr–1000	tr–300	
Potassium	tr–1000	tr–80	
Mixed compounds	tr–900	tr–800	
Limestone	tr–300	tr–130	36
Animal wastes			
Cattle manure	20–31	—	36
Poultry waste	6	—	36
Pig waste	14	—	37
Farmyard manure	12	17	37, 38
Cow manure	56	29	39

Fertilisers contain more Cr than Ni, with phosphates being the richest in both elements (Table 7.5). The National Research Council of Canada [33] reported a range of 30–3000 mg Cr/kg in phosphate fertilisers. Phosphate deposits normally contain 30 mg Ni/kg but some are enriched with 1000 mg/kg or more [34]. However, most phosphate fertilisers contain only small amounts of Ni, and phosphates with average Ni contents used at standard rates should not cause a large enough build-up in the soil to cause concern about possible food chain effects. Although the amount of Cr entering soil via use of phosphate fertilisers is uncertain, it is likely to exist as Cr(III) in soil and, as such, is not likely to be toxic.

Unlike the situation for Zn and Cu, animal manures contain little Cr or Ni (Table 7.5). Chromium and Ni are not used in agricultural pesticides but, like cement, limestone used to correct soil acidity contains Cr. Published values vary: <1–120 mg Cr/kg limestone rock, average 10 mg/kg [33]; 10–60 mg Cr/kg [32]. Limestone contains smaller amounts of Ni (Table 7.5).

Comparison of Table 7.5 with Table 7.3 shows that the concentrations of Cr and Ni in very few phosphates, limestones and manures are greater than the concentrations already in soil. Therefore, it is unlikely that there will be a large build-up of these metals in the soil as a result of application of most fertilisers and agricultural wastes.

7.3.3 Atmospheric deposition

The largest total amounts of Cr released to the atmosphere by human activity are from metallurgical industries in the form of particles, e.g. from electric arc furnaces. In an air emissions inventory of the USA [40], ferrochrome production was the most important of these industries, with estimated emissions, even after air pollution control, of 12360 t/yr. The next most important source of atmospheric chromium

Table 7.6 Estimated global emissions of Cr and Ni to the atmosphere from natural and anthropogenic sources in 1983

	Cr	Ni
	(emissions 10^3 t/yr)	
*Anthropogenic sources**		
Coal combustion	2.92–19.63	3.38–24.15
Oil combustion	0.45–2.37	11.00–43.14
Non-ferrous pyrometallurgical industry		
Mining		0.80
Pb production		0.33
Cu–Ni production		7.65
Steel and iron	2.84–28.40	0.04–7.10
Refuse incineration		
Municipal	0.098–0.98	0.098–0.42
Sewage sludge	0.15–0.45	0.03–0.18
Phosphate fertilisers	—	0.14–0.069
Cement production	0.89–1.78	0.09–0.89
Wood combustion	—	0.60–1.80
Motor transport	—	0.9†
Total	7.34–53.61	25.05–88.05
Average	30.48 (36%)	55.65 (87%)
Natural sources		
Soil suspension	50†	4.8
Volcanoes	3.9†	2.5
Vegetation	—	0.82
Forest fires	—	0.19
Meteoric dust	—	0.18
Sea salt	—	0.009
Total	53.9 (64%)	8.5 (13%)
Grand total	84.38	64.15

* Derived from Nriagu and Pacyna [32]
† Derived from Schmidt and Andren [41]

was refractory brick production, releasing 1630 t/yr, followed closely by combustion of coal which released 1564 t/yr, steel production released 520 t/yr. Nriagu and Pacyna concluded that the iron and steel industry is the largest anthropogenic source of Cr emissions globally [32].

The largest anthropogenic source of Ni is the burning of fuel and residual oils, emitting 26 700 t Ni/yr globally [41]. Nickel concentrations in diesel exhaust are 500–10 000 mg/l [42]. Oil contains more Ni than does coal and, like Pb, Zn and Cu, there is evidence of a gradient of Ni concentrations in soil and grass with distance away from major roads [43]. Combustion of coal is the next most important emission, followed by Ni mining and smelting which is likely to have locally severe impacts close to mining and smelting sites, e.g. Sudbury in Ontario, Canada, and Clydach in South Wales, UK.

There are also large natural resources of Cr and Ni present in the atmosphere, e.g. windblown soil, volcanic activity, forest fires, meteoric dust and sea salt spray or particles, although only the first two are important for Cr. Biogenic emissions from vegetation are small, and known for Ni [41]. Another difference between the metals is that natural sources are more important for Cr, but more than 80% of emissions of Ni are of anthropogenic origin (Table 7.6). It should be pointed out, however, that data for emissions are imperfect and many assumptions are made in calculating total emissions [32, 41]. In particular, accurate data are lacking for Cr from the metallurgical industry and from the chemical industry. Also, values for emissions from the secondary non-ferrous industry and recycling of scrap metal are absent for both elements.

Nriagu and Payna [32] estimated the worldwide fall out of Cr and Ni from the atmosphere to soils to be between 5.1 and 38×10^3 t/yr for Cr and $11–37 \times 10^3$ t/yr for Ni. This can be viewed in context against their estimates of median *total* global inputs from all sources to soil of 896 and 325×10^3 t/yr respectively. The majority of the total inputs come from disposal of wastes of various types on land. However, the distribution of the metals derived from wastes is likely to be less uniform and more concentrated than those from atmospheric fallout. Obviously, however, the severity of atmospheric inputs will also be intense at locations with strong local sources. Deposition of metals from the atmosphere results from either wet or dry deposition. Large particles settle out quickly and are deposited near the source, having in some cases a devastating impact on nearby soils, plants and animals (see Section 7.6).

Measured total deposition of Ni varies from 2 $kg/km^2/yr$ in rural areas to 88 $kg/km^2/yr$ near industry and in urban areas [41]. The species of Ni present in deposition probably include soil minerals, oxides and sulphates [41]. Bulk deposition of Cr is less than 0.2 $kg/km^2/yr$ in remote regions, whilst in rural areas it is 0.5–5 $kg/km^2/yr$ and in urban locations deposition is variable but generally more than 10 $kg/km^2/yr$ [44]. Cawse [45] expressed the importance of long-term deposition of elements from the atmosphere by reference to the total concentrations present in the top 5 cm of soil at a monitoring station in Chiltern, UK. In 30 years, deposition at average rates would increase soil Cr by 1% and Ni by 13%, compared with the starting values of 1973. There is little information on the forms of Cr present in deposition. The amount of Cr that is soluble in rain water from different parts of the world seems to be < 2% [44]. Although there is no general information on the valency of Cr in deposition, specific industrial uses can generate aerosols that contain Cr(III) or Cr(VI) [44], e.g. 4% of dust emitted from ferrochrome smelters contains Cr(VI).

7.3.4 Sewage sludges

Metals from natural, domestic and industrial sources tend to concentrate in the organic residue at sewage treatment works. Removal of Cr and Ni from wastewater by sludge depends on the process used and the 'age' of the sludge. In experimental

Table 7.7 Concentrations of chromium and nickel in sewage sludges (mg/kg dry weight)

Location	Cr	Ni	No. of samples	Reference
	(median and range)			
UK*	86 24–1255	37 15–900	—	56
Ontario, Canada	530 100–9740	120 23–410	17	57
England and Wales	250 40–8800	80 20–5300	42	58
England and Wales	335	94	193	59
USA	1290 169–14000	190 36–562	16	39
USA	890 10–99000	82 2–3520	165	60
Sweden	— 20–40615	— 16–2120	93	61

* This information for the UK consists of medians for sludge used in agriculture and the range is the 10 and 90 percentile of all sludges

continuous flow systems, using activated sludge, nearly 100% of Cr(III) and 92–100% of Ni was removed, the former mainly by precipitation. However, only 26–48% of Cr(VI) was removed [46]. Industrial sources contributed 68% of Cr and 83% of Ni in the influent to sewers in New York City, USA [47]. As much as 70% of the total Ni input was from electroplating industries. Nickel is also present in many domestic cleaning products, e.g. soap, 100–700 mg/kg; powdered detergents, 400–700 mg/kg; and powdered bleach, 800 mg/kg [48]. Chromium-containing effluents are released by the following activities: metal plating, anodising, ink manufacture, dyes, pigments, glass, ceramics, glues, tanning, wood preserving, textiles and corrosion inhibitors in cooling water. Both Cr(III) and Cr(VI) can be present in these wastewaters. An example in which Cr(VI) predominates is in raw effluent from plating works [49]. However, many effluents are either treated on site to decrease the potentially toxic load in waste water, or at the sewage treatment works, where Cr(VI) is reduced by organic matter and consequently the Cr in sludge exhibits the chemistry of Cr(III) compounds [50].

Nriagu and Pacyna [32] estimated that between 1.4 and 11×10^3 t of Cr and 5–22×10^3 t of Ni are added to soils each year as a result of disposal of sewage sludge globally. Many metals accumulate to high concentrations in the surface layer of soils treated with sewage sludge. The depth of the contaminated soil depends on the depth to which sewage sludge is physically incorporated by ploughing and other cultivations or upon the presence of cracks or channels in the soil at the time of application of liquid sludge. After addition, however, there is little evidence of significant downward movements of metals in sludged soils, including Ni and Cr, even after long periods of time [51, 52, 53]. The residence time of metals derived from sludge is probably of the order of 10^3–10^4 years [51]. It should be noted, however, that there is emerging evidence from laboratory experiments that

Table 7.8 Concentrations of Cr and Ni in surface soils of sewage sludge farms (mg/kg soils)

	Cr	Ni	Reference
Braunschweig (Germany)			62
Treated	202	32	
Untreated	38	10	
Beaumont Leys (UK)	2020	385	63, 64
Stoke Bardolf (UK)	1500	387	65

negatively charged complexes of Cd with soluble organic matter can move deep within the soil after fresh sludge is applied [54]. Nevertheless, at least 80% of the Cr and Ni applied to a long-term field experiment could be accounted for in the cultivated layer after 45 years [55].

Many countries now regulate the amounts of metals which can be applied to soils in sewage sludge, or maximum metal concentrations in the soil itself (see Section 7.7). In the past, such rules did not exist and the concentrations which built up at particular sites depended on the amount of sludge applied and the concentrations of metals within the sludge. The presence of industries like tanneries and plating works explains the upper concentrations of Cr and Ni reported in sewage sludge from many countries (Table 7.7). In general, sludge from rural areas will have the lowest concentrations of Cr and Ni. As a result, fields receiving sludge from urban industrial areas may have larger contents of Cr and Ni than similar ones in rural areas. Sludge farms, i.e. dedicated land which routinely receives sludge from a treatment works for many years, may have extremely large concentrations of Cr and Ni and many other metals if they receive sludge from urban areas. Concentrations in soil at some of these farms are shown in Table 7.8. One example is Stoke Bardolph sewage farm near Nottingham, UK where 600 kg Ni/yr has been disposed of for over 100 years in sewage sludge containing an average of 550 mg Ni and 2600 mg Cr/kg dry solids.

An experiment on sandy loam soil with starting concentrations of 26 mg Cr and 11 mg Ni/kg soil that received sewage sludge from West London, UK containing average concentrations of 919 mg Cr/kg and 188 mg Ni/kg between 1942 and 1961 [51] contained 126 mg Cr ad 27 mg Ni/kg treated soil in 1985 [55]. The availability of Cr from sludged soils appears to be low [51, 66]. Uptake of Cr by plants and transport to the aerial tissues is minimal at near-neutral soil pH, characteristic of Cr(III) [67]. In comparison with Cr, Ni in sludge treated soils is relatively available, less so than Zn and Cd, but more than Cu which is more strongly complexed by organic matter [68].

7.3.5 Other sources of chromium and nickel

Disposal of fly-ash on land is the largest single input of both Cr and Ni to soils [32]. Table 7.9 shows the average concentrations of Cr and Ni in coal and fly-ash.

Table 7.9 Concentrations of Cr and Ni in coal and fly-ashes (mg/kg). (From Adriano *et al.* [72])

	Cr	Ni
Coal	15	15
Fly-ash		
Bituminous	172	11
Sub-bituminous	50	1.8
Lignite	43	13

Fly-ash is enriched in Cr and, as a result, soils around coal-fired electricity generators may be slightly enriched with Cr [69]. The dumping of large amounts of pulverised fuel ash on soil leads to large increases in concentrations of Cr compared to background soils. Concentrations of Ni in some coals can reach 70 mg/kg and at these high levels in rare cases fly-ashes can contain up to 900 mg Ni/kg [70]. In general, however, the burden of Ni in fly-ash is smaller than Cr. Although Cr in fly-ash can greatly increase soil Cr, little appears to be taken up by crops grown on fly-ash-amended soil [71]. Similarly, the Ni in fly-ash does not appear to be taken up by plants in substantial amounts, unlike the Ni in sewage sludge [72]. More soluble elements such as B, Mo and Se are of greater concern following disposal of fly-ash than Cr and Ni.

Emissions of Ni from smelters and Cr in the waste from chromate smelters bring about large increases in soil Ni and Cr [33, 73]. It has been estimated that, worldwide, 22–64 $\times$ 10^3 t/yr of Ni are emitted to soils by release of mine tailings [32]. Chromium(VI) in heaps of chromate waste in the Croal Valley in northern England was intensely phytotoxic [74]. Chemical reduction of Cr(VI) to Cr(III) has been suggested as a solution which would decrease the toxicity and permit re-vegetation of the site.

Other smaller sources of Cr include wear of Cr-containing asbestos brake linings in vehicles and aerosols produced from Cr catalysts used in emission reduction systems for treating exhaust fumes. Both of these may have the greatest impact on roadside soils.

To conclude this section on inputs of Cr and Ni to soils, it is useful to compare these with those to air and water. It is clear that soils are gaining much larger

Table 7.10 Estimated emissions of Cr and Ni to air, water and soil on a global basis (10^3 t/yr). (From Nriagu and Pacyna [32])

	Cr	Ni
Air	30	56
Water	142	113
Soil	896	325

amounts of these metals than the other environmental media (Table 7.10). Three other points have important implications for future environmental protection policies: (i) much of the Cr and Ni emitted to air and some released to waters will impact on soils, (ii) air and water are subject to a much greater mixing and therefore dilution of metal load than soil; and as a result of the above (iii) the accumulation of metals in surface soils is often more intense on both local and global scales than in the other components of the biosphere.

7.4 Chemical behaviour in the soil

Chromium may exist in a number of oxidation states, but the most stable and common forms are Cr(III) and Cr(VI). These have sharply contrasting chemical properties: Cr(IV) exists as an anion, it is more readily extracted from soil and sediment particles and is considered the more toxic form. Chromate is in pH-dependent equilibrium with other forms of Cr(VI) such as $HCrO_4^-$ and dichromate ($Cr_2O_7^{2-}$) with CrO_4^{2-} the predominant form at pH > 6. Chromium(III) on the other hand, is much less mobile and adsorbs to particulates more strongly. The solubility of Cr(III) decreases above pH 4 and above pH 5.5 complete precipitation occurs.

Chromium (VI) is the more stable form in equilibrium with atmospheric oxygen. However, Cr(VI), with its high positive reduction potential, is a strongly oxidising species, and in the presence of soil organic matter Cr(VI) is reduced to Cr(III) [8, 50, 75, 76]. Reduction is more rapid in acid than alkaline soils [8, 77]. Thus in the majority of soils, the relatively insoluble and less mobile Cr(III) form predominates and it generally occurs as insoluble hydroxides and oxides [8, 50, 78]. For example, Cary *et al.* [8] observed that additions of soluble Cr as either Cr(VI) or Cr(III) to soils reverted to forms which were insoluble and unavailable to plants. Following reduction of Cr(VI), it was concluded that insoluble Cr in the soils was present as hydrated oxides of Cr(III) complexing with soluble organic acids (e.g. citric acid, DTPA, fulvic acids and soil extracts of water-soluble organic matter) maintains small amounts of Cr(III) in solution above the pH at which uncomplexed Cr precipitates and is therefore a means of enhancing its mobility [79, 80]. Bartlett and Kimble [75] and Bartlett and James [76] drew attention to the similarity of adsorption and solubility behaviour between Cr(III) and Al(III) in response to changes in pH and phosphorus status of soils. Oxidation of a proportion of Cr(III) is considered a likely occurrence in soils with a pH greater than 5 and rich in oxidised Mn, although there are difficulties in detecting the transition because drying and storage seemingly destroy the capacity of soil to oxidise the element [81]; in essence, the soil pH is lowered by drying [76]. Bartlett and James [81] found that the amount of Cr(III) oxidised to Cr(VI) was proportional to the Mn reduced and also to the amount reducible by hydroquinone. In situations where the Cr(VI) form remains in soil, Bartlett and Kimble [75] contend that its adsorption capability is similar to that of orthophosphate and the anion will remain mobile only if its concentration exceeds both the adsorbing and the reducing capacities of the soil. In fact, adsorption of Cr(VI) in certain soils can offset its reduction [82].

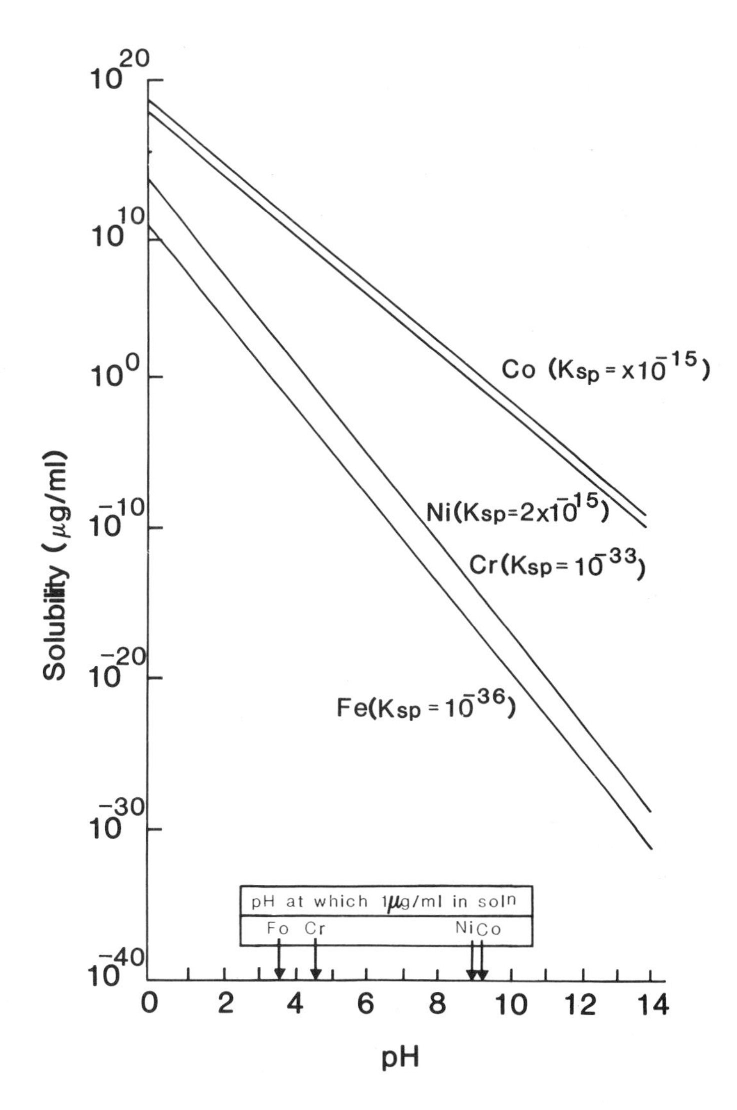

Figure 7.1 The solubility of hydroxides of Cr and Ni together with those of Co and Fe (μg/ml) expressed as a function of pH (after Brooks [22]).

By comparison, the soil chemistry of Ni is much simpler and is based on the divalent metal ion (Ni^{2+}). The solubility of the hydroxides of Cr (Cr(III)) and Ni, together with other siderophilic elements, at different pH values gives some indication of the relative mobility of these entities in soils (Figure 7.1). Both become increasingly soluble at lower pH values, but clearly Ni is the more soluble.

On the basis of thermodynamic stability models, Ni ferrite ($NiFe_2O_4$) is the most probable solid phase that can precipitate in soils [83, 84]. Where the soil environment is acid and reducing, the sulphides of Ni are likely to control the concentration of Ni in the soil solution. The hydroxy-complex $Ni(OH)^+$ and Ni^{2+} ions are the most likely major forms in the soil solution above pH 8, whilst in acid soils Ni^{2+}, $NiSO_4^0$ and $NiHPO_4$ are important although the relative proportions would depend on the levels of SO_4^{2-} and PO_4^{3-}. Sposito and Page [85] are in broad agreement but emphasise more the involvement of bicarbonates and carbonates of Ni; under oxic conditions in acid soils the principal chemical species are predicted to be Ni^{2+}, $NiSO_4^0$, $NiHCO_3^+$ and organic complexes, and in alkaline soils $NiCO_3^0$, $NiHCO_3^+$, Ni^{2+} and $NiB(OH)_4^+$.

Sequential extractions of soil using chemical reagents have been used to identify operationally defined 'fractions' of metals in soil (see Chapter 4). Over 50% of Ni in soils may be associated with the residual fraction (HF and $HClO_4$ soluble) and in the region of 20% in the Fe–Mn oxide fraction, with much of the remainder bound up with the carbonate fraction, and only a relatively small proportion in the exchangeable and organic fractions [86]. In soils amended with sewage sludge the organic fraction assumes a greater importance [87]. Selectivity sequences with respect to clay minerals and Fe oxides indicate that Ni is one of the least tenaciously sorbed of the transition elements, e.g. with kaolinite and montmorillonite in a solution without competing ligands, sorption followed the sequence Cd > Zn > Ni [88] and the binding strength on external goethite surfaces decreased in the order Zn > Cd > Ni [89].

The most important factor determining the distribution of Ni between the solid and solution phases would seem to be pH, while factors such as the clay content and the amount of hydrous Fe and Mn oxides in the soil are of secondary importance [90]. The mobility of Ni in soils increases as the pH and CEC decrease [91–93] and a marked increase in soluble Ni has been recorded in sewage amended soils as pH was decreased, especially below about pH 6 [94]. Furthermore, in such soils, the amount of extractable Ni has been found to be a function of metal loading, pH and CEC [95]. Increased loading of Ni to a soil tends to enhance the proportion adsorbed and the distribution coefficient, K_d (i.e. the ratio of the amount of metal adsorbed by soil to that in the soil solution) of Ni also increases in the presence of clays with a high CEC [96]. However, Ni added to soils as a complex with EDTA results in a greater proportion of Ni in the soluble phases of the soil [91]. In sludge-amended soils, soluble Ni is composed largely of organic and inorganic complexes [97]. Acid sulphate soils from Finland and the Ni-contaminated soils from the Sudbury region of Canada affected by industrial sulphurous emissions both show increased amounts of extractable Ni due to the low pH of the soils [73, 98–100].

The amounts of Cr soluble in acetic acid are commonly below limits of detection and less than 0.1% of the total value, and this extraction method is regarded as unsatisfactory for measuring bioavailable Cr. Shewry and Peterson [101] compared the efficiency of a whole range of extractants in serpentine soils from Scotland and found that soluble Cr (soluble in dilute acetic acid) was less than 0.05% of the total, and that only oxalate treatment removed a substantial amount of Cr. Brooks [22] compared total and extractable element concentrations in serpentine soils from a number of studies covering a wide geographic range and concluded that there is little uniformity in the values, although one trend worthy of note is the low solubility of Cr compared to Ni.

It has been suggested that pyrophophate ($Na_4P_2O_7$) removes organically bound Cr(III), while HCl probably extracts mostly inorganic forms of Cr(III), such as hydroxides and phosphates. The minute quantitites of Cr(III) extracted using ammonium acetate and sodium fluoride represent the mobile forms in equilibrium with the organic fraction [75]. Potassium dihydrogen phosphate (KH_2PO_4) has been used to extract sorbed Cr(VI) [75], but the efficiency of this has been questioned and alternatives include aqueous ammonia or aqueous tris(hydroxymethyl)methylamine [77].

Application of sewage sludge to soils may result in changes in the proportion of physicochemical forms of metals in both components. For example, Ni in aqueous extracts of sludge-amended soils has been found to be less than in the sludge itself, whereas the water solubility of Cr has been reported to increase on mixing sludge with soil [102, 103]. Yet a reduction in both 0.05 M EDTA-extractable Ni and 0.5 M acetic acid-extractable Cr has been reported in sludge/soil mixtures relative to sludge [102]. Raising the pH by liming affects extractability of metals in sludge-amended soils, but Bloomfield and Pruden [103] found no consistent trends; whilst the amounts of water- and acetic acid-soluble Ni decreased after liming, it did not cause a particularly large decrease in solubility and that of Cr actually increased; this is consistent with the observation that Cr toxicity increases with soil pH, which is presumably related to an increase in the Cr(VI) form in the soil.

Beckett *et al.* [104] recorded a greater extractability of Ni in sludge-treated soils relative to the indigenous soil. In fact Lake *et al.* [102] not that there is an overall shift in the solid-phase metal forms in sludge-amended soils such that the sulphide/residual fraction decreases whilst that extracted with milder reagents increases. For example, in three unamended soils 80–94% of Ni resided in the sulphide/residual fraction but this was decreased to 61–69% following sludge application and a weathering period of 21 months. Other studies have noted an increase in DTPA-extractable Ni in soil–sludge mixtures with time and this has been ascribed to dissolution of metal precipitates such as carbonates, hydroxides and phosphates through changes in pH or gas composition of the soil resulting from microbial activity. Sanders *et al.* [95] manufactured a series of soil–sludge mixtures in which about 10% of the added Ni was recovered in $CaCl_2$ extracts; over a 21 month incubation period as soil pH values decreased, the average concentrations of Ni increased in $CaCl_2$-extracts and displaced solutions; in these solutions Ni was present predominantly as the free ion (Ni^{2+}).

Evaluation of vertical metal movement in soils has received a lot of attention, and particularly in sewage-treated soils. Whilst the evidence appears contradictory, the overwhelming majority of reports have found a general lack of leaching below the zone of sludge incorporation [105, 106]. In one experiment Cr and Ni analyses of soils which had received metal contaminated sludge treatment from 1942 to

1961 showed no evidence 25 years later of significant movement below the depth to which the soil was cultivated [51]. Where significant vertical movement has been reported [107–110] soil acidification and texture (including cracking) are probably important factors, and also the degree of saturation may be significant [109].

7.5 Soil–plant relationships

Concentrations of Cr in plant-available form are extremely small in the majority of soils, and this lack of solubility is reflected in the low concentrations of the element in plants. Concentrations in the foliar parts of plants show little relationship with the overall content of Cr in the soil; non-contaminated or background concentrations in plants are of the order of 0.23 mg/kg, and in general concentrations are less than 1 mg/kg even across a wide range of Cr soil values [28]. The subject of the bioavailability of Cr to plants has been reviewed by Cary [111] and by the National Research Council of the NAS [112]. The situation is aptly demonstrated in serpentine soils. Many of these soils with an average pH of 6.8 have a greater total Cr concentration than Ni but, because the amount of available Cr is extremely small, it is widely believed that Ni is much more significant as a 'serpentine factor' than Cr [22]. In fact, plants grown on serpentine soils rarely contain Cr concentrations larger than 100 mg/kg, and Brooks [22] states that a Cr concentration in excess of this value is due to soil contamination. For example, serpentine plants of the Great Dyke, Zimbabwe, had a maximum Cr concentration of 77 mg/kg [113] and in species of *Geissois* from New Caledonia concentrations ranged up to 45 mg/kg [114]. Marked differences in concentrations of Cr between species growing on some Scottish serpentine soils have been noted by Shewry and Peterson [101] and between-site variation was clearly not a simple function of either HNO_3 soluble or exchangeable Cr.

Concentrations of Cr in plants growing on mine spoil and various types of Cr waste are commonly in the range 10–190 mg/kg, but toxic concentrations may accumulate in plants growing on chromate waste in which the more soluble r(VI) form predominates [74]. Bartlett and James [81] induced Cr toxicity in mustard, barley and alfalfa by adding Cr(III) to fresh moist soils that were rich in Mn oxides, but noticeably less injury was recorded after additions of Cr(III) to air-dried soils; the difference was attributed to the formation of Cr(VI) from Cr(III) in fresh soils. However, concentrations of Cr in crop plants grown on soils treated with sewage sludge containing elevated levels of Cr are barely above background [115, 116], which reflects the formation of very stable organic complexes or precipitates with Cr(III).

Concentrations of Ni in plants growing in non-contaminated and non-serpentinic soils are generally in the range 0.1–5 mg/kg [22, 73, 117–119]. Slightly higher values have been found in crops grown on highly organic soils [120] and a number of forest tree species [118]. Concentrations of Ni in plants from serpentine areas are commonly in the range 20–100 mg/kg. However, the serpentine flora contain a group of plants with Ni concentrations an order of magnitude higher (> 1000 mg/kg) and these have been termed as 'hyperaccumulators' of Ni [22] e.g. the New Caledonian tree *Sebertia acuminata* has a blue sap that contains 11% Ni on a fresh basis [121].

Nickel hyperaccumulating plants have been shown to drive a 'Ni cycle' in which Ni uptake and transport followed by leaf fall continuously enriches the topsoil with readily available Ni [122]. As a result, marked Ni tolerance was found in soil bacteria growing underneath hyperaccumulator trees, but not under non-accumulator trees or in deforested areas on similar serpentine soils.

There has been recent interest in the concept of cropping and removing the above ground biomass of hyperaccumulators of nickel and other metals for decontamination of polluted surface soils [123]. However, nickel-polluted soils are rather rare, compared to those impacted by Zn, Cd and Pb.

Elevated concentrations of Ni also occur in a variety of plants growing in the highly contaminated soils in the vicinity of the large Ni–Cu smelting complex at Sudbury, Ontario, Canada; e.g. foliar concentrations of over 900 mg/kg have been found in *Deschampsia flexuosa* and *Vaccinium angustifolium* a few kilometres from the source [73]. In fact, the grass *Deschampsia cespitosa* has evolved multiple metal tolerance on soils contaminated by Cu, Ni and Co in the Sudbury mining and smelting area [124]. Other Ni-tolerant genotypes include *Agrostis gigantea* and *Phragmites communis* [124–126].

The concentration of Ni in plants generally reflects the concentration of the element in the soil, although clearly the relationship is more directly related to the concentration of soluble ions of Ni and the rate of replenishment of this mobile pool [73, 127]. Factors which increase the solubility and exchangeability of Ni in soils also lead to an increased concentration of the element in plants, and Ni absorption in a variety of crops has been related to water-soluble or exchangeable Ni in sludge-amended soils [128]. Mycorrhizal fungi can decrease the concentrations of metals in trees, however, seemingly by restricting translocation of metals, including Ni, to the leaves [129]. It is well established that exchangeable Ni in soils increases roughly in proportion to an increase in acidity, and the absorption of Ni by plants increases as the exchangeable fraction in soils increases [98, 118, 130]. In a comparison of the relationship between various extraction agents and plant uptake, it has been shown that $CaCl_2$-extractable Ni (compared with buffered EDTA and DTPA extracts) produced a more pronounced increase with decreasing pH of soil–sludge mixtures and correlated more significantly with concentrations of Ni in ryegrass [131]. Liming and additions of organic matter to soils result in a decrease of both extractable Ni (ammonium acetate) and the amount taken up by plants [132]. There are reports that exchangeable and plant-available Ni increase in poorly drained soils [118]. In addition to pH, a reduction in CEC may enhance Ni mobility in soils and hence increase plant uptake [91]. Also, Ni supplied as an EDTA complex enhances water solubility and plant accumulation of the element. Within the plant, Ni is considered to be a highly mobile element, and Cataldo *et al.* [133] suggest that it behaves in a similar fashion to Cu and Zn.

In soya bean, during vegetative growth, Ni was found to accumulate mainly in leaves, but during senescence a considerable fraction was mobilised to seeds. Sauerbeck and Hein also found that the concentrations of Ni were larger in cereal grain than straw, and postulated a physiological relationship between transport of photosynthates and Ni [134].

The prevailing view is that Cr(VI) is more toxic than Cr(III); for example Skeffington *et al.* [135] observed that Cr(VI) inhibited the growth of both roots and shoots of barley seedlings to a greater extent than the Cr(III) form, although it was noted that uptake of Cr as Cr(III) was greater than that of Cr(VI) in the roots. However, an alternative explanation put forward by McGrath [67] maintains that the perceived difference in toxicity between the two forms is due to the much reduced bioavailability of Cr(III) at $pH > 5$. By adjusting conditions to achieve equal concentrations of Cr(III) and Cr(VI) in continuous-flow cultures it was shown that both forms were toxic to oat seedlings and, at the concentrations tested, inhibition of root growth was greater with Cr(III) than with Cr(VI). Thus, toxicity of Cr(III) is unlikely, except in extremely acid soils [67] and is therefore regarded as relatively nontoxic, but Cr(VI) is invariably toxic to plants. Chromate anions, however, are more available at high pH [50] and amelioration of soils of low organic matter status or waste heaps polluted with Cr(VI) depends largely on increasing the potential for reduction of Cr(VI) to the less toxic Cr(III) by adding organic matter; a process that occurs naturally in most soils [74, 77].

Regardless of the form of Cr that the plant is exposed to, most of the Cr taken up remains in the root tissue; for example Skeffington *et al.* [135] observed a hundredfold drop in concentration of Cr across the hypocotyl in barley seedlings exposed to Cr(III) or Cr(VI) and in nine different crop plants 98% of either Cr(III) or Cr(VI) absorbed by the plant was retained in the roots [136]; similar trends have been reported by Ramachandran *et al.* [137]. Even exposing plants to organically complexed Cr has failed to increase sorption of the element relative to inorganic Cr(III) or Cr(VI). Moreover, translocation within the plant is not increased by supplying Cr in the form of organic acids [7]. Uptake of Cr from soils treated with K_2CrO_4 by different plant species indicates that the leaves of cereal crops are less effective in accumulating Cr than leafy vegetables. It has been suggested that plants that tend to accumulate Fe also accumulate Cr.

In *Leptospermum scoparium*, a trioxalate–Cr(III) compound has been identified as the chemical form of the element in the plant tissues [138] and a Cr-containing compound with similar properties has been found to occur in ethanol extracts of barley and cauliflower plants [135, 136]. Chromium in lucerne and alfalfa appears to occur as anionic complexes of molecular weight about 2900 Da which is distinguishable from the glucose tolerance factor isolated from brewer's yeast [123, 139–141].

It has been reported that over 90% of Ni in root and leaf tissues of soya bean is associated with the soluble fraction [133]. Soluble and translocated forms of Ni are thought to be citrate complexes, especially in the Ni hyperaccumulators, such as *Sebertia acuminata* [22, 121, 123]. When the Ni–EDTA complex was added to soils, it appeared as Ni–EDTA in spinach plants; even when taken up as a positive ion it was promptly complexed and appeared in plant extracts as neutral and negative complexes [91]. In xylem exudates from soya bean plants, a number of anionic and cationic Ni complexes have been found and the major Ni-containing component was identified as a tripeptide [142]. The nickel hyperaccumulator *Dichapetalum gelonoides* from the Philippines contained more than 2% Ni in the leaf dry matter, most of which was present in the form of anionic citrate or malate complexes [143].

7.6 Polluted soils

Examples of soils contaminated with Cr as a result of sewage sludge disposal or dumping of chromate wastes have been given above. Two examples of Ni-contaminated soils will be discussed here, one from Canada and one in Wales, UK.

Probably the best-documented case of Ni pollution is that in the area of the Sudbury Basin in Ontario, Canada [144, 145]. The defoliation and lack of vegetation in parts of this area are due to the emission of Ni, Cu, Fe and acidic SO_2 from large Ni smelters. The ore pentlandite is mined in deep mines in the district, and its smelting in the early days involved roasting wood. This caused the S in the mineral to burn, releasing sulphurous smoke for months after each firing. In 1972 a 300 m stack was installed at Copper Cliff. Freedman and Hutchinson [100] estimated that, despite the high stack, more than 40% of the Ni and Cu emitted is deposited within 60 km of the smelter.

The concentrations of Ni and Cu in the soils (and vegetation) decrease with increasing distance from the smelters (Table 7.11). Furthermore, the largest build-up of Ni is in the surface soil and litter layers. This accumulation in the surface

Table 7.11 Concentrations of Ni and Cu in soils near nickel smelters in Sudbury, Ontario, Canada (mg/kg). (From Hutchinson and Whitby [144])

Distance from smelter (km)	Ni	Cu
1.1	5104	2892
1.6	1851	2416
2.2	2337	2418
2.9	1202	1657
7.4	1771	1371
10.4	282	287
13.5	271	233
19.3	306	184
24.1	101	45
32.1	35	46
38.6	39	2
49.8	35	26

horizons has had deleterious effects on soil microbial activity, seed germination and plant growth in the area [73].

In the area around Clydach in Wales where Ni has been smelted since 1900 the concentrations of Ni in soils were above background over an area of 6 km^2 [146]. These loadings may be due to tipping of solid wastes in heaps, and older methods of smelting Ni which produced dust and fumes. However, the electrolytic Mond process is now used. This involves the extremely toxic gas nickel carbonyl, and takes place in a closed system to prevent toxicity to smelter employees. Whilst emissions have been reduced, nevertheless atmospheric deposition of Ni (and Co and Cu) has been identified as a major source of soil contamination in the area [147, 148]. As is the case for other metallic cations, increasing soil pH by adding lime is an effective and practical means of ameliorating the toxicity of Ni^{2+} in soil [149]. This method is likely to work on soils receiving sewage sludge and emissions from smelters. Increasing the organic matter content can also reduce the availability of Ni because of the binding of the metal in organic complexes [150].

7.7 Soil protection against chromium and nickel

Regulations and guidelines have evolved as part of environmental protection policies in many western countries. One type of guideline seeks to limit the rate of addition of metals to soils whilst another imposes upper maxima of metal concentrations in soils receiving wastes such as sewage sludge. However, such

Table 7.12 Guidelines for upper permitted concentrations of Cr and Ni in near neutral soils receiving sewage sludge and comparison with assumed background concentrations (mg/kg soil) (Limits from McGrath *et al.* [151])

	Cr	Ni
Limits		
European Community	—	30–75
Denmark	30	15
Federal Republic of Germany	100	30
Finland	200	60
France	150	50
Italy	150	50
Norway	100	30
Spain	100	30
Sweden	30*	15*
UK	400*	75
US	1500	210
Background concentrations		
Federal Republic of Germany	30	30
England and Wales	50	25

* Provisional

preventive measures can only be applied before soil becomes contaminated. A second type of regulation is the classification of soils that are already polluted, for example by industrial activities, according to how 'safe' they are for various uses. In some countries soils above certain limits are designated as requiring expensive 'cleaning-up' processes. Examples of upper limits to metal concentrations in soil receiving sewage sludge are given in Table 7.12. The limits for Cr vary widely and this, in part, is due to the lack of a sufficient number of experiments with enough different concentrations of Cr in soil in which yield effects have been measured. Because of this, the Commission of the European Communities set no limit for Cr in 1986 [152]. Subsequently, available evidence from field and pot trails was assessed, with rightful emphasis on the former, and a suggested EC limit of 150–250 mg Cr/kg recommended for soils treated with sewage sludge [153]. This seems conservative for the concentration of Cr(III) in soil, but included a margin for safety. However, this suggested limit has now been dropped. The US EPA [154] has recently set regulations for the loadings of metals applied in sewage sludge to soil, based on a risk assessment of fourteen different potential pathways for metal impacts. The resulting concentrations of Ni and Cr in soil, when the loading limits are reached, assuming mixing to 20 cm depth, are very high compared to other limits, as shown in Table 7.12.

In the UK, 'trigger' values for concentrations of metals in soils have been adopted [155]. These are threshold values that are used when an area is scheduled for development. If the concentration of metals in soil is less than the threshold trigger value, the land is regarded as uncontaminated. The threshold varies according to the proposed use of the land: 70 mg Ni or Cr/kg soil wherever crops are to be grown; 600 Cr mg/kg soil for domestic gardens and allotment, increasing to 1000 mg Cr/kg for less intensive uses, such as parks, playing fields and open space [155]. Recognising that Cr(VI) is more toxic, the threshold for this was set lower, at 25 mg Cr(VI)/kg soil for all end uses. The Netherlands has a soil assessment system in which the 'A' or reference value is smilar to the lowest UK threshold trigger concentration, but if the concentration in soil exceeds the 'intervention' value further investigation is required, in the form of a full risk assessment of the contaminated land, and depending on the outcome of this, a soil cleansing operation may be required (Table 7.13).

Finally, the reader is referred to four reviews which contain additional

Table 7.13 Values for concentrations of Cr and Ni in the soil used for assessment in the Netherlands (mg/kg) (from van den Berg *et al.* [156])

	Reference	Intervention
Cr	100	380
Ni	35	210

information and references on the geochemistry of Cr and Ni, their chemistry in soils, their biological availability and effects on organisms [157–160].

References

1. Papp, J.F., in US Bureau of *Mines Minerals Yearbook 1986*, Volume 1: *Metals and Minerals*. US Dept of Interior, Washington, DC (1988), 225–244.
2. Stern, R.M., in *Biological and Environmental Aspects of Chromium*, ed. Langård, S. Elsevier, Amsterdam (1982), Chapter 2.
3. Adriano, D.C., *Trace Elements in the Terrestrial Environment*. Springer-Verlag, New York (1986).
4. Mertz, W. and Schwarz, K., *Arch Biochem. Biophys.* **58** (1955), 504–508.
5. Anderson, R.A., *Sci. Total Environ.* **17** (1981), 13–29.
6. Cannon, H.L. and Hopps, H.C., *Geol. Soc. Am. Spec.* Paper No. 140. Boulder, Col. (1970).
7. Cary, E.E., Allaway, W.H. and Olsen, O.E., *J. Agric. Food Chem.* **25** (1977), 300–304.
8. Cary, E.E., Allaway, W.H. and Olsen, O.E., *J. Agric. Food. Chem.* **25** (1977), 305–309.
9. Guthrie, B.E., in *Biological and Environmental Aspects of Chromium*, ed. Langård, S. Elsevier, Amsterdam (1982), Chapter 6.
10. Brown, P.H., Welch, R.M. and Cary, E.E., *Plant Physiol.* **85** (1987), 801–803.
11. Dixon, N.E., Gazzola, C., Blakely, R.L. and Zerner, B., *J. Am. Chem. Soc.* **97** (1975), 4131–4133.
12. Eskew, D.L., Welch, R.M. and Norvell, W.A., *Plant Physiol.* **76** (1984), 691–693.
13. Repaske, R. and Repaske, A.C., *Appl. Environ. Microbiol.* **32** (1976), 585–591.
14. van Baalen, C. and O'Donnell, R., *J. Gen. Microbiol.* **105** (1978), 351–353.
15. Oliverira, L. and Antia, N.J., *Canadian J. Fisheries & Aquatic Sciences* **43** (186), 2427.
16. Welch, R.M., *J. Plant Nutr.* **3** (1981), 345–356.
17. Kirchgessner, M. and Schnegg, A., in *Nickel in the Environment*, ed. Nriagu, J.O. John Wiley, New York (1980), Chapter 27.
18. Clemente, G.F., Rossi, L.C. and Santaroni, G.P., in *Nickel in the Environment*. ed. Nriagu, J.O. John Wiley, New York (1980), Chapter 19.
19. Levis, A.G. and Bianchi, V., in *Biological and Environmental Aspects of Chromium*, ed. Langard, S. Elsevier, Amsterdam (1982), Chapter 8.
20. Furst, A. and Radding, S.B., in *Nickel in the Environment*, ed. Nriagu, J.O. John Wiley, New York (1980), Chapter 24.
21. Cannon, H.L., *Geochem Environ.* **3** (1978), 17–31.
22. Brooks, R.R., *Serpentine and Its Vegetation*. Croom Helm, London (1987).
23. Vinogradov, A.P., *The Geochemistry of Rare and Dispersed Chemical Elements in Soils*. Consultants Bureau Inc., New York (1959).
24. Ure, A.M. and Berrow, M.L., in *Environmental Chemistry*, Vol. 2, Royal Society of Chemistry, London (1982), Chapter 3.
25. Berrow, M.L. and Reaves, G.A., *Geoderma* **37** (1986), 15–27.
26. McGrath, S.P. and Loveland, P.J., *The Soil Geochemical Atlas of England and Wales*. Blackie Academic and Professional (1992).
27. Mitchell, R.L., *Chemistry of the Soil*. ed. Bear, F.E., Reinhold, New York (1964), 320–368.
28. Bowen, H.J.M., *Environmental Chemistry of the Elements*. Academic Press, London (1979).
29. Rose, A.W., Hawkes, H.E. and Webb, J.S., *Geochemistry in Mineral Exploration*. 2nd ed. Academic Press, London (1979).
30. Shacklete, H.T. and Boerngen, J.G., *Element Concentration in Soils and other Surficial Materials of the Conterminous United States*. US Geol. Surv. Prof. Paper 1270. Govt. Printing Office, Washington (1984).
31. Holmgren, G.G.S., Meyer, M.W., Chaney, R.L. and Daniels, R.B., *J. Environ. Qual.* **22** (1993), 335–348.
32. Nriagu, J.O. and Pacyna, J.M., *Nature (Lond).* **333** (1988), 134–139.
33. National Research Council of Canada, Effects of Chromium in the Canadian Environment. NRCC/CNRC, Ottawa (1976).
34. Boyle, R.W. and Robinson, H.A., in *Nickel and Its Role in Biology. Metal Ions in Biological Systems*, Volume 23. eds. Sigel, H. and Sigel, A. Marcel Dekker, New York (1988).

35. Mattigod, S.V. and Page, A.L., in *Applied Envrionmental Geochemistry*, ed. Thornton, I. Academic Press, London (1983), Chapter 12.
36. Caper, S.G., Tanner, J.T., Friedman, M.H. and Boyer, K.W., *Environ. Sci. Technol.* **12** (1978), 785–790.
37. Arora, C.L., Nayyar, V.K. and Randhawa, N.S., *Indian J. Agric. Sci.* **45**(1975), 80–85.
38. McGrath, S.P., *J. Agric. Sci.* **103** (1984), 25–35.
39. Furr, A.K., Lawrence, A.W., Tong, S.S.C., Grandolfo, M.C., Hofstader, R.A., Bache, C.A., Guttenmann, W.H. and Lisk, D.J., *Environ. Sci. Technol.* **10** (1976), 683–687.
40. GCA Corporation, *National Emissions Inventory of Sources and Emissions of Chromium*. US National Technical Information Service No. PB 230–034 (1973).
41. Schmidt, J.A. and Andren, A.W., in *Nickel in the Environment*, ed. Nriagu, J.O. John Wiley, New York (1980), Chapter 4.
42. Frey, J.W. and Corn, M., *Am. Ind. Hyg. Assoc. J.* **28** (1967), 468.
43. Lagerwerff, J.V. and Specht, A.W., *Environ. Sci. Technol.* **4** (1970), 583–586.
44. Nriagu, J.O., Pacyna, J.M., Milford, J.B. and Davidson, C.I., in *Chromium in the Natural and Human Environment*, eds. Nriagu, J.O. and Nieboer, E. John Wiley, New York (1988), Chapter 5.
45. Cawse, P.A., in *Pollutant Transport and the Fate in Ecosystems.* Special Publication No. 6 of the British Ecological Society. Blackwell Scientific, Oxford (1987), 89–112.
46. Sterritt, R.M., Brown, M.J. and Lester, J.N., *Environ. Pollut. Series A* **24** (1981),, 313–323.
47. Yapijakis, C., and Papamichael, F., *Water Science and Technology* **19** (1987), 133–144.
48. Gurnham, C.F., Ritchie, R.R., Smith, A.W. and Rose, B.A., in *Source and Control of Heavy Metals in Municipal Sludge*. Peter F. Loftus Crop., Chicago (1979).
49. Beszedits, S., in *Chromium in the Natural and Human Environment*, ed. Nriagu, J.O. and Nieboer, E. John Wiley, New York (1988), Chapter 9.
50. Grove, J.H. and Ellis , B.G., *Soil Sci. Soc. Am. J.* **44** (1980), 238–242.
51. McGrath, S.P. in *Pollutant Transport and Fate in Ecosystem*. Special Publication No. 6 of the British Ecology Society. Blackwell Scientific, Oxford (1987), 301–317.
52. Davis, R.D., Carlton-Smith, C.H., Stark, J.H. and Cambell, J.A., *Environ. Pollut.* **49** (1988), 99–115.
53. Baxter, J.C., Aguilar, M. and Brown, K., *J. Environ. Qual.* **12** (1983), 311–316.
54. Christensen, T.H., *Water, Air and Soil Pollut.* **44** (1984), 43–56.
55. McGrath, S.P. and Lane, P.W., *Environ. Pollut.* **60** (1989), 236–256.
56. UK D.o.E. *UK Sewage Sludge Survey. Final Report.* CES, Beckenham, Kent (1993).
57. Ontario Ministry of Environment, *Plant Operating Summary*. Water Pollution Control Projects, Toronto, Canada. (1977).
58. Berrow, M.L. and Webber, J., *J. Sci. Fd. Agric.* **23** (1972), 93–100.
59. Department of the Environment. *Standing Technical Committee Reports No. 20.* National Water Council, London (1981).
60. Sommers L.E., *J. Environ. Qual.* **6** (1977), 225–232.
61. Berggren, B. and Oden, S., *Analyresultat Rorande Fung Metaller Och Klorerade Kolvaten* I *Rötslam Fran Svenska Reningsverk*, 1968–1971. Institutioren für Markvetenskap Lantbrukshogskolan, 750 07 Uppsalan, Sweden (1972).
62. El-Bassam, N., Tietjen, C. and Esser, J., in *Management and Control of Heavy Metals in the Environment.* CEP Consultants, Edingburgh (1979), 521 –524.
63. Pike, E.R., Graham, L.C. and Fogden, M.W., *J. Assoc. Publ. Analysts*, **13** (1975), 19–33.
64. Pike, E.R., Graham, L.C. and Fogden, M.W., *J. Assoc. Publ. Analysts*, **13** (1975), 48–63.
65. Rundle, H.J. and Holt, C., in *Heavy Metals in the Environment*, Vol. 1. Proceedings of an International Conference, Heidelberg, 1983. CEP Consultants, Edinburgh (1983), 353–357.
66. Dowdy, R.J. and Larson, W.E., *J. Environ. Qual.* **4** (1975), 229–233.
67. McGrath, S.P., *New Phytol.* **92** (1982), 381–390.
68. Logan, T.J. and Chaney, R.L., in *Utilization of Municipal Wastewater Sludge on Land*, eds. Page, A.L., Gleson, T.L., Smith, J.E., Iskandar, I.K. and Sommers, L.E., US Environmental Protection Agency, Washington (1984), 235–326.
69. Klein, D.H. and Russel, P., *Environ. Sci. Technol.* **7** (1973), 357.
70. Swaine, D.J., in *Nickel in the Environmental*, ed. Nriagu, J.O., Chapter 4. John Wiley, New York (1980).
71. Furr, A.K., Kelly, W.C., Bache, C.A., Guttenmann, W.H. and Lisk, D.J., *J. Agric. Fd. Chem.* **24** (1976), 885–888.

72. Adriano, D.C., Page, A.L., Elseewi, A.A., Chang, A.C. and Straughan, I., *J. Environ. Qual.* **9** (1980), 333–344.
73. Hutchinson, T.C., in *Effect of Heavy Metal Pollution on Plants, Vol. 1*, ed. N.W. Lepp. Applied Science Publishers, London (1981), Chapter 6.
74. Breeze, V.G., *J. Appl. Ecol.* **10**, (1973), 513.
75. Bartlett, R.J. and Kimble, J.M., *J. Environ. Qual.* **5** (1976), 379–383.
76. Bartlett, R.J. and James, B.R., in *Chromium in the Natural and Human Environment*, ed. Nriagu, J.O. and Nieboer, E., John Wiley, New York (1988), Chapter 10.
77. Bloomfield, C. and Pruden, G., *Environ. Pollut.* Series A 23 (1980), 103–114.
78. Smith, S., Peterson, P.J. and Kwan, K.H.M., *Toxicol. Environ. Chem.* **24** (1989), 241–251.
79. James, B.R. and Bartlett, R.J., *J. Environ. Qual.* **12** (1983), 169–172.
80. James, B.R. and Bartlett, R.J., *J. Environ. Qual.* **12** (1983), 173–176.
81. Bartlett, B.R. and James, B.R., *J. Environ. Qual.* **8** (1983), 31–35.
82. James, B.R. and Bartlett, R.J., *J. Environ. Qual.* **12** (1983), 177–181.
83. Sadiq, M. and Enfield, C.G., *Soil Science* **138** (1984), 262–270.
84. Sadiq, M. and Enfield, C.G., *Soil Science* **138** (1984), 335–340.
85. Sposito, G. and Page, A.L., in *Circulation of Metal Ions in the Environment. Metal Ions in Biological Systems*. Vol. 18, ed. Sigel, H. Marcel Dekker, New York (1984).
86. Hickey, M.F. and Kittrick, J.A., *J. Environ. Qual.* **13** (1984), 372–376.
87. Dudley, L.M., McNeal, B.L. and Baham, J.E., *J. Environ. Qual.* **15** (1986), 188–192.
88. Puls, R.W. and Bohn, H.L., *Soil Sci. Soc. Am. J.* **52** (1988), 1289–1292.
89. Bruemmer, G.W., Gerth, J. and Tiller, K.G., *J. Soil Science* **39** (1988), 37–52.
90. Anderson, P.R. and Christensen, T.H., *J. Soil Science* **39** (1988), 15–22.
91. Willaert, G. and Verloo, M., *Plant and Soil* **107** (1988), 285–292.
92. Kiekens, L. in *Utilisation of Sewage Sludge on Land: Rates of Application and Long-term Effects of Metals. Proc.* Summary of Commission of the European Communities, Uppsala (1983).
93. Verloo, M., Kiekens, L. and Cottenie, A., *Pedologie* **30** (1980), 163–175.
94. Sanders, J.R. and Adams, T. McM., *Environ. Pollut.* **43** (1987), 219–228.
95. Sanders, J.R., McGrath, S.P. and Adams, T. McM., *Environ. Pollut.* **44** (1987), 193–220.
96. Reddy, M.R. and Dunn, S.J., *Environ. Pollut. Series B* **11** (1986), 303–313.
97. Dudley, L.M., McNeal, B.L., Baham, J.E., Coray, C.S. and Cheng, H.H., *J. Environ. Qual.* **16** (1987), 341–348.
98. Palko, J. and Yli-Halla, M., *Acta. Agric. Scand.* **38** (1988), 153–158.
99. Whitby, L.M. and Hutchinson, T.C., *Environ. Conserv.* **1** (1974), 191–200.
100. Freedman, B. and Hutchinson, T.C., *Can. J. Biol.* **58** (1980), 108–132.
101. Shewry, P.R. and Peterson, P.J., *J. Ecol.* **64** (1976), 195–212.
102. Lake, D.L., Kirk, P.W.W. and Lester, J.N., *J. Environ. Qual.* **13** (1984), 175–183.
103. Bloomfield, C. and Pruden, G., *Environ. Pollut.* **8** (1975), 217–232.
104. Beckett, P.H.T., Warr, E. and Brindley, P., *Water Pollution Control* **82** (1983), 107–113.
105. Williams, D.E., Vlamis, J., Pukite, A.H. and Corey, J.E., *Soil Science* **140** (1985), 120–125.
106. Chang, A.C., Warnedke, J.E., Page, A.L. and Lund, L.J., *J. Environ. Qual.* **13** (1984), 87–91.
107. Kirkham, M.B., *Environ. Sci. Technol.* **9** (1975), 765–768.
108. Schirado, T., Vergara, I., Schalscha, E.B. and Pratt, P.F., *J. Environ. Qual.* **15** (1986), 9–12.
109. Welch, J.E. and Lund, L.J., *J. Environ. Qual.* **16** (1987), 403–410.
110. Legret, M., Divet, L. and Juste, C., *Water Research* **22** (188), 953–959.
111. Cary, E.E., in *Biological and Environmental Aspects of Chromium*, ed. Langard, S. Elsevier, Amsterdam (1982), 49–64.
112. National Research Council, *Committee on Biological Effects of Atmospheric Pollutants, Chromium*. National Academy of Science, Washington D.C. (1974).
113. Brooks, R.R. and Yang, X.H., *Taxon* **33** (1984), 392–399.
114. Jaffré, T., Brooks, R.R. and Trow, J.M., *Plant Soil* **51** (1979), 157–162.
115. Mortvedt, J.J. and Giordano, P.M., *J. Environ. Qual.* **4** (1975), 17–174.
116. Chang, A.C., Granato, T.C. and Page, A.L., *J. Environ. Qual.* **21** (1992), 521–536.
117. Vanselow, A.P., in *Diagnostic Criteria for Plants and Soil*, ed. Chapman, H.D. Quality Printing Co., Abilene, Texas (1966), 302–309.
118. Farago, M.E. and Cole, M.M., in *Nickel and Its Role in Biology. Metal Ions in Biological Systems*. Vol. 23, eds. Sigel, H. and Sigel, A. Marcel Dekker, New York (1988), Chapter 3.

119. Hutchinson, T.C., Freedman, B. and Whitby, L., in *Effects of Nickel in the Canadian Environment*. National Research Council, Canada (1981), Chapter 5.
120. Hutchinson, T.C., Czuba, M. and Cunningham, L.M., in *Trace Elements in Environmental Health Symposium 8*, ed. D.D. Hemphill. University of Missouri, Columbia (1974), 81–93.
121. Jaffré, T., Brooks, R.R., Lee, J., and Reeves, R.D., *Science* **193** (1976), 579–580.
122. Schlegel, H.G., Cosson, J.-P. and Baker, A.J.M., *Bot. Acta.* **104** (1991), 18–25.
123. Baker, A.J.M., Brooks, R.R. and Reeves, R., *New Scientist* (10 March 1988), 44–48.
124. Cox, R.M. and Hutchinson, T.C., *New Phytol.* **84** (9180), 631–647.
125. Hogan, G.D. and Rauser, W.E., *New Phytol.* **83** (1979), 665–670.
126. Cox, R.M. and Hutchinson, T.C., *Nature* **279** (1979), 231–233.
127. Duneman, L., Von Wiren, N., Schulz, R. and Marschener, H., *Plant and Soil.* **133** (1991), 263–269.
128. Keefer, R.F., Singh, R.N. and Horvath, D.J., *J. Environ. Qual.* **15** (1986), 146–152.
129. Wilkins, D.A., *Agric. Ecos. Environ.* **35** (1991), 245–260.
130. Mizuno, N., *Nature* **219** (1968), 1271–1272.
131. Sanders, J.R., McGrath, S.P. and Adams, T.McM., *J. Sci. Food Agric.* **37** (1986), 961–968.
132. Halstead, R.L., Finn, B.J. and MacLean, A.J., *Can. J. Soil Sci.* **49** (1969), 335–342.
133. Cataldo, D.A., Garland, T.R., Wildung R.E. and Drucker, H., *Plant Physiol.* **62** (1978), 566–570.
134. Sauerbeck, D.R. and Hein, A., *Water, Air Soil Pollut.* **57/58** (1991), 861–871.
135. Skeffington, R.A., Shewry, P.R. and Peterson, P.J., *Planta* **132** (1976), 209–214.
136. Lahouti, M. and Peterson, P.J., *J. Sci. Food Agric.* **30** (1979), 136–142.
137. Ramachandran, V., D'Souza, T.J. and Mistry, K.B. *J. Nuclear Agric. Biol. 9* (1980), 126–128.
138. Lyon, G.L., Peterson, P.J. and Brooks, R.R., *Planta* **88** (1969), 282–287.
139. Blincoe, C., *J. Sci. Food Agric.* **25** (1974), 973–979.
140. Starich, G.H. and Blincoe, C., *Sci. Total Environ.* **28** (1983), 443–454.
141. Starich, G.H. and Blincoe, C., *J. Agric. Food Chem.* **28** (1983), 458–462.
142. Cataldo, D.A., Wildung, R.E. and Garland, T.R., *J. Environ. Qual.* **16** (1987), 289–295.
143. Homer, F.A., Reeves, R.D., Brooks, R.R. and Baker, A.J.M., *Phytochem.* **30** (1991), 2141–2145.
144. Hutchinson, T.C. and Whitby, L.M., *Environ. Conserv.* **1** (1974), 123–132.
145. Hutchinson, T.C. and Whitby, L.M., *Water Air & Soil Pollut.* **7** (1977), 421–438.
146. Davies, B.E., in *Applied Soil Trace Elements*. ed. Davis, B.E., John Wiley, Chichester (1980), Chapter 9.
147. Goodman, G.T. and Roberts, T.M., *Nature* **231** (1971), 287–292.
148. Goodman, G.T. and Smith, S., in *Report of a Collaborative Study on Certain Elements in Air, Soil, Plants, Animals and Humans in the Swansea-Neath-Port Talbot Area together with a report on a Moss-Bag Study of Atmospheric Pollution across South Wales*. Welsh Office (1975).
149. Bingham, F.T., Page, A.L., Mitchell, G.A. and Strong, J.E., *J. Environ. Qual.* **8** (1979), 202–207.
150. Leeper, G.W., *Managing Heavy Metals on the Land*. Marcel Dekker, New York (1978).
151. McGrath, S.P., Chang, A.C., Page, A.L. and Witter, E., in *Environmental Reviews*. 2 in press.
152. Commission of the European Communities. Council directive on the protection of the environment, and in particular of the soil, when sewage sludge is used in agriculture. *Official Journal of the European Communities* No. L 181, Annex 1A, p. 10 (1986).
153. Williams, J.H., *Chromium in Sewage Sludge Applied to Agricultural Land*. Commission of the European Communities SL/124/88, Brussels (1988).
154. US EPA. *Standards for the use or disposal of sewage sludge*. Federal Register 58:9248–9415, (1993).
155. Inter-departmental Committee on the Redevelopment of Contaminated Land. *Guidance on the Assessment and Redevelopment of Contaminated Land.* 59/83, 2nd edn. Department of the Environment, London (1987).
156. Berg. R. van den, Denneman, C.A.J., and Roels, J.M., in *Contaminated Soil '93*. eds. F. Arendt *et al.*, Kluwer Academic Publisher, The Netherlands (1993).
157. Nriagu, J.O., *Nickel in the Environment*, John Wiley, New York (1980).
158. Langård, S., in *Biological and Environmental Aspects of Chromium*. Elsevier, Amsterdam (1982).

159. Nriagu, J.O. and Nieboer, E., *Chromium in the Natural and Human Environment.* John Wiley, New York (1988).
160. Sigel, H. and Sigel, A., *Nickel and Its Role in Biology. Metal Ions in Biological Systems*, Vol. 23. Marcel Dekker, New York (1988).

8 Copper

D.E. BAKER AND J.P. SENFT

8.1 Introduction

Copper is one of the most important, essential elements for plants and animals. In the metal state, Cu is reddish coloured, takes on a bright metallic luster, and is malleable, ductile, and a good conductor of heat and electricity. The principal uses of Cu are in the production of wire and of brass and bronze alloys. In nature, Cu forms sulphides, sulphates, sulphosalts, carbonates and other compounds, and also occurs under reducing environments as the native metal. Cu ranks 26th behind Zn, in abundance in the lithosphere. An average for abundance of Cu in the lithosphere is considered to be 70 mg/kg, while values reported for the earth's crust range from 24 to 55 mg/kg. For soils of the world, the older literature value of 20 mg/kg Cu has been changed recently and reported as 30 mg/kg. Copper is associated with soil organic matter, oxides of iron and manganese oxides, soil silicate clays and other minerals.

Extractable Cu, sometimes called 'available', refers to an amount of this element in soil that correlates statistically with concentrations absorbed and assimilated by plants. The 'availability' of Cu to plants refers to the readiness with which the available ion $[Cu(H_2O)_6]^{2+}$ is absorbed by plants in acid soils and $Cu(OH)_2^0$ in neutral and alkaline soils. The availability of Cu is related to the chemical potentials (analogous to pH) of the respective species in the soil solution. The level and distribution of total and extractable Cu in the soil profile varies with soil type and parent material. Copper is specifically adsorbed or 'fixed' in soils, making it one of the trace metals which moves the least. Higher concentrations of Cu in the surface horizon of a soil is an indication of soil additions from smelter, fertilisers, sewage sludges and other wastes, fungicides or bactericides or manures from swine and poultry fed selected Cu containing compounds for increased feed efficiency and greater growth rates.

The plant abundance of essential micronutrients generally rank in the order Fe > Mn > B > Zn > Cu > Mo > Cl. Copper was confirmed as an essential element for plants, in the 1930s. Copper was the element of special reference in the manuscript by Arnon and Stout in 1939 [1] establishing criteria of essentiality for elements involved in plant nutrition. Typically, Cu concentrations in plants range from 5–20 mg/kg Cu which may extend from 1–30 mg/kg. Cu accumulation differs among plant species and cultivar differences. Therefore, it is not possible to give single values for Cu deficiency or toxicity concentrations. Copper in plants functions as part of the prosthetic groups of enzyme systems, and as a facultative activator of enzyme systems.

In animal nutrition, Cu deficiency is almost entirely confined to grazing cattle and sheep due to very low levels of Cu in the forage or to the normal to low–normal levels of Cu accompanied by elevated intakes of Mo, S, Fe, or soil materials that are sufficient to limit Cu absorption and retention. Under normal conditions Cu is a benign agent to humans. The total body Cu of an adult is about 100 to 150 mg, normal diet provides 1 to 5 mg Cu/day and it is difficult to construct a diet with less than 1 mg Cu/day.

Hereditary Cu toxicosis in humans known as Wilson's disease was first described in 1912. For this condition, the Cu-binding ligands involved in Cu homeostasis are deficient and toxic levels accumulate in several tissues.

8.2 Geochemical occurrence of copper

While much is known about Cu in soils and soil–water systems, the quantitatively reliable accounts are limited to well defined solid phases and complexed ions in relatively concentrated systems. For example, using electron spin resonance it was shown that when a mono-layer of water occupies the interlamellar region of Cu-hectorite and the silicate layers are parallel, Cu(II) is coordinated to four water molecules in the *xy*-plane and two silicate oxygens along a symmetry axis, *z*, perpendicular to the silicate layers. If several layers of water molecules occupy the interlamellar region of Cu-hectorite, $Cu(H_2O)_6^{2+}$ predominates as in aqueous solution [2]. For this chapter, Cu^{2+} will refer to the $Cu(H_2O)_6^{2+}$ ion and Cu will be used to designate to element without consideration of the valence state. In soil environments we are concerned with low concentrations of Cu with approximate averages of 24–55 mg/kg in the Earth's crust and 20–30 mg/kg total Cu in soils. For soils in a normal range of pH, the Cu^{2+} is the species measured as Cu in soils [3]. Organic substances in soils bind Cu. The COO^- groups present in both the solid and liquid phases form stable ligands with Cu. Solid phase ligand formation is considered responsible for Cu deficiencies in organic soils, and the same mechanism is responsible for decreases in Cu toxicity to plants from the addition of peat and other sources of organic matter to high Cu substrates [4]. Since the review by Allaway [5] in 1968, the soil environmental chemistry of Cu has been reviewed extensively [6–14].

Copper, atomic number 29, is the first element of subgroup 1B of the periodic table. The electronic structure of the Cu atom is $1s^2, 2s^2, 2p^6, 3s^2, 3p^6, 3d^{10}, 4s^1$. The single $4s$ electron is outside the filled $3d$ shell and is rather stable. Like all elements of the first transition series (e.g. Cr, Mn, Fe, Co, Ni) and unlike Li, Na, K, and Rb, of Group I, two electrons are removed relatively easily from Cu atoms [3]. While Cu^{2+} is quite stable in water, the second ionisation potential of Cu is sufficiently higher than the first to allow a variety of stable Cu(I) species to exist in the environment. Copper(I) is stable in aqueous solutions containing excess halide ions, acetonitrile, pyridine, or cyanide ions. Parker [3] pointed out from the work of Lindsay [15], that in Cu-rich wet soils of 10^{-6}–10^{-7} M, for example, Cu could

Table 8.1 Common copper minerals in soils and the lithosphere [22, 53]

Name	Formula	Copper content (wt %)	Structure
Chalcite	Cu	100	Cubic close packing, Cu atoms in coordination number 12
Chalcocite, Digenite and other chalcocite-like minerals	$Cu_{1.75-2.0}S$	80	Mostly planar layers of Cu and S in 3-fold coordination
Covelite	CuS	66	2/3 Cu+ and 1/3 Cu^{2+}, Cu^{2+} in planar CuS_3 triangles, Cu^+ in CuS_4 tetrahedra
Bornite	Cu_5FeS_4	63	Cubic antifluorite-type structure, Cu and Fe in tetrahedral coordination, with 18 Cu + Fe atoms randomly distributed among 24 sites
Chalcopyrite	$CuFeS_2$	34	Sphalerite-type structure, Cu in tetrahedral coordination with cubic close-parked S atoms
Cubanite	$CuFe_2S_3$	23	
Enargite	Cu_3AsS_4	48	Wurtzite-like structure, with Cu and As in tetrahedral coordination
Famatinite	Cu_3SbS_4	43	
Tennanite	$(CuFe)_{10}(FeZnCu)_2As_4S_3$	52	
Tetrahedrite	$(CuAg)_{10}(FeZnCu)_2SbAs_4S_{13}$	46	Framework structure with Cu (+1) S_4 tetrahedra. Large holes in framework contain 4 Cu (+1) and 2 Cu (+2) in 3-fold planar coordination, also 4 Sb (+3) or As (+3) in 3-fold pyramidal coordination
Cuprite	Cu_2O	89	O atoms are centre and vertices of cube, Cu in linear coordiantion between two O atoms
Tenorite	CuO	80	Cu coordinated to four O by planar dsp^2 bonding
Malachite	$Cu_2(OH)_2(CO_3)_2$	57	Cu in distorted octahedra, two O and two OH in planar coordination with remaining two axial positions occupied by distant O atoms for half the Cu and OH groups for the other half
Azurite	$Cu_2(OH)_2(CO_3)_2$	55	2/3 of Cu in rectangular coordination with two O and two OH, 1/3 of Cu in pyramidal coordiantion with three O and two OH atoms
Chrysocolla	$CuSiO_3.5H_2O$	25	Infinite SiO_3 chains (like pyroxene), linked by Cu coordinated with 4 O atoms

Table 8.1 (Continued)

Name	Formula	Copper content (wt %)	Structure
Atacamite	$Cu_2(OH)_3Cl$	74	Centres of OH and Cl in cubic close packing, Cu in 6-fold coordination of two types: four OH plus two Cl, five OH plus one Cl
Brochantite	$Cu_4(OH)_6SO_4$	56	Octahedral Cu coordination: four OH groups in a plane and one O atom and one OH group at axial positions
Antlerite	$Cu_3(OH)_4SO_4$	54	
Chalcanthite	$CuSO_4.5H_2O$	25	

be present at 1×10^{-7} M Cu^{2+} and 3×10^{-7} M Cu^{+}. This relationship results from the squared $[Cu^{+}]$ term in the reaction in which Cu^{+} ions readily disproportionate:

$$2Cu^{+}_{(aq)} \rightarrow Cu^{2+}_{(aq)} + Cu_{(s)} \qquad K = 10^{6}M^{-1} \text{ at } 25°C \qquad (1)$$

At solution Cu concentratios of 10^{-2}–10^{-3} M, very little Cu(I) is present. Following a discussion of atomic and physical properties of Cu and the chemistry of Cu ions in different environments using the concept of free energy of transfer of single ions between different solvents, Parker [3], concluded that the aquated Cu(II) ion, $Cu(H_2O)_6^{+4}$ (Cu^{2+} in this chapter) is the most relevant copper species for consideration by soil scientists. However, the flooding of soils may cause Cu^{+} and in some cases Cu^{0} to be more thermodynamically stable than Cu^{2+} [15]. A list of common Cu minerals and their properties are presented in Table 8.1. Chalcopyrite is by far the most abundant Cu mineral. It is found widely dispersed in rocks and concentrated in the largest Cu ore deposits [16]. In addition to the common minerals, Cu in nature occurs in dispersed forms in ordinary rocks, sediments and soils.

8.3 Origin of copper in soils

8.3.1 Soil parent materials

Soil Cu has been discussed by McBride [17]. Aubert and Pinta [7], published a 17 page table including the Cu abundance in parent rocks, the soil types developed and their total Cu content. They report parent materials of crystalline schists (garnet, granulite) containing an intrusion of norites with 1000 mg/kg Cu, and the C horizon of the soils contains up to 100 mg/kg Cu. Typical Cu contents of major rock types are summarised in Table 8.2 [7, 18–20].

From the literature published prior to 1975, Baker and Chesnin [18] reported that Cu in the lithosphere averages 70 mg/kg, the range in soils was given as 2 to

Table 8.2 Typical copper contents of major rock types (mg/kg) [6, 13, 67, 72]

	Range	Average
Basic igneous (basalts)	30–160	90
Acid igneous (granites)	4–30	15
Ultramafic (pyroxenites)	10–40	15
Shale and clay	30–150	50
Black shales	20–200	70
Volcanic rocks	5–20	
Agrillaceous sediments	40–60	
Limestones	5–20	
Sandstones	5–20	
Lithosphere		70
Earth's crust	24–55	
Soil	2–100	20–30

100 mg/kg with a selected average of 20 mg/kg. In 1979, Cox [16] cited 24 to 55 mg/kg as the range of Cu in the earth's crust, and Bowen [8] cited 50 mg/kg. Lindsay [15] reported 70 mg/kg Cu as the average for the lithosphere, but like Bowen, he chose 30 mg/kg Cu as an average value for soils. Parker [3] reported after Hodgson [21] the value of 70 mg/kg Cu as the average amount in the earth's crust and 20 mg/kg for soils. From the values reported, it may be concluded that the average Cu concentration in the earth's crust ranges from 24–55 mg/kg and the average Cu range for soils is 20–30 mg/kg. The abundance of Cu in soil and plants is less than that of Zn unless the soil has been contaminated with an industrial source of Cu.

The abundance of Cu in basaltic rocks is greater than for granitic rocks, and is very low in carbonate rocks [22]. Gabbro and basalt rocks have the highest Cu contents, and granodiorite and granite the lowest. The abundance of Cu in igneous rocks is partly controlled by the process of differentiation during crystallisation. Early-formed crystals separate from silicate melts and settle to the lower part of the magma chamber. In general, Mg silicates, having higher melting points, crystallise first, causing a relative enrichment of other components. Iron and Calcium minerals follow, leaving a residuum rich in low-melting alkali aluminum silicates and quartz. As crystallisation proceeded, the remaining liquid became saturated in sulphide and an immiscible Cu-rich sulphide separated. The Cu present was fixed as bornite and chalcopyrite [16]. Strong covalent bonds are formed between Cu^{+} and sulphide (S^{2-}) atoms. In silicate clays, and mafic (high Mg^{2+} and Fe^{2+}) rocks, Cu^{2+} can replace, by isomorphous substitution, those metals with six-fold coordination (Mg^{2+}, Fe^{2+}, Zn^{2+}, Ni^{2+}, and Mn^{2+} (1.8) and Mg^{2+} (1.3) limits isomorphous replacement of these elements by Cu in soil parent materials [17]. Typical Cu concentrations found in soils derived from different parent materials are presented in Table 8.3 [19].

Table 8.3 Typical total copper concentrations in soils on various parent materials [67]

Peat (Histosols)	15–40
Sandy soils on drift (Arensols, Podzols)	2–10
Sandy soils on granite	10
Silty clay loams on shales (Gleysols, Cambisols, etc.)	40
Clays developed on clay rocks (Gleysols)	10–27
Loams developed on basalt rock (Cambisols and others)	40–150
Humic loams on chalk	7–28
Organic-rich loams on loess (Chernozems) (av 30)	1–100
Soils developed on pumice (Lithosols/Arenosols)	3–25
Tropical soils (Ferralsols)	8–128

8.3.2 Agricultural materials

Soil management of Cu for crop production has been studied carefully since Grossenbacher and Floyd [cited in 23] showed in 1917 that beneficial results could be obtained by application of $CuSO_4.5H_2O$ to the soil and Bordeaux mixture as a foliage spray for control of exanthema, or die-back, a widespread disorder of citrus in Florida, USA. Sources of Cu used for crop production are listed in Table 8.4. A wide range of materials and compounds can be used to supply Cu to soils [24]. The usual Cu fertiliser source is $CuSO_4.5H_2O$ (blue-stone), although other compounds, mixtures and chelates are also used. Hydrated $CuSO_4$ is compatible with most fertiliser materials. Anthropogenic inputs of Cu to land are very diverse. Soil levels of Cu are affected by soil and crop treatments including fungicides, fertilisers not used to rectify Cu deficiencies, livestock manures, sewage sludges and atmospheric deposition. For example, pig slurry containing up to 1990 mg/kg Cu has been reported [12, 25–27].

8.3.3 Atmospheric deposition

Atmospheric inputs of Cu to soils from both rain and dry deposition varies considerably according to the proximity of industrial emissions containing Cu and the type and quantities of wind-blown dust. In the UK, the total annual deposition of Cu from dust was found to vary between 100 g/ha and 480 g/ha. While crop removal was estimated at 50–100 g/ha, the deposition dust was not sufficient to correct Cu deficiency in crops and livestock [19]. The lack of response to the deposition was attributed to the form of the Cu or to soil adsorption. Results for a recent investigation of the long-term contamination of soils with Cu, Zn, Pb, and Cd by zinc smelters in Pennsylvania, USA, indicated that edible lettuce plants were grown in contaminated soils containing 254 mg/kg Cu; 12800 mg/kg Zn; 222 mg/kg Cd and 1106 mg/kg Pb. Except for gardens managed carefully with respect to organic matter and lime additions over decades to maintain a pH above 7.0, soils with more than 200 mg/kg Cu, 400 mg/kg Zn, 25 mg/kg Cd and 500 mg/kg Pb are not suitable for production of food crops [28].

Table 8.4 Sources of copper for use as fertilisers [29]

Source	Formula	%Cu	Water solubility
Cu metal	Cu	100	Insoluble
Cuprite	Cu_2O	89	Insoluble
Tenorite	CuO	75	Insoluble
Covellite	CuS	66	Insoluble
Chalcocite	Cu_2S	80	Insoluble
Chalcopyrite	$CuFeS_2$	35	Insoluble
Malachite	$CuCO_3.Cu(OH)_2$	57	Insoluble
Azurite	$2CuCO_3.Cu(OH)_2$	55	Insoluble
Chalcanthite	$CuSO_4.5H_2O$	25	Soluble
Copper sulphate monohydrate	$CuSO_4.H_2O$	35	Soluble
Basic copper sulphate	$CuSO_4.3Cu(OH)_2$ (General formula)	13–53	Insoluble
Copper nitrate	$Cu(NO_3)_2.3H_2O$		Soluble
Copper acetate	$Cu(C_2H_3O_2)_2.H_2O$	32	Slightly
Copper oxalate	$CuC_2O_4.0.5H_2O$	40	Insoluble
Copper oxychloride	$CuCl_2.2CuO.4H_2O$	52	Insoluble
Copper ammonium phosphate	$Cu(NH_4)PO_4.H_2O$	32	Insoluble
Copper chelate	Na_2Cu EDTA	13	Soluble
Copper chelate	NaCu HEDTA	9	Soluble
Copper polyflavanoids	–	5–7	Soluble
Copper–sulphur frits	–	Varies	Varies
Copper–glass fusions	–	Varies	Varies
Sewage sludges	–	0.04–1.0	Slightly
Animal manures (no Cu supplements)	–	0.002–0.00	Slightly
Animal manures (with Cu supplements)		0.06–0.19	Slightly

The total quantity of Cu that has been emitted to the atmosphere since 3800 BC has been estimated at 3.2×10^6 t (about 1% of the 307×10^6 t produced) [12]. This amount is about three orders of magnitude greater than the present-day atmospheric Cu burden. Because of the relatively short residence time for airborne Cu aerosols, it is doubtful if there can be a substantial build-up of Cu in the atmosphere. However, the atmosphere is the important medium for the transmission of pollutant Cu to the most remote areas of the Earth. Analysis of moss samples and polar ice cores reveals a substantial increase in the airborne Cu at locations far from emission sources. About 80% of the total world production of Cu has been made in the twentieth century, and it has been estimated that 30% of the all-time figure has been produced during the 1970s [12]. Most of this enormous quantity of Cu is apparently wasted on land even though the turnover time of artifacts may be considerable. The hastening effects of acid gases released to the environment and the disposal of sewage sludges and other wastes could have a substantial effect on soils and associated flora and fauna. The total Cu production of 307×10^6 t is about twice the Cu content of the top 2-cm layer of soils of the world and is about an order of magnitude greater than the annual Cu demand for all living land biota [12].

Table 8.5 Composition data reported for Cu, Fe, Zn, and Mn in soils, plants and sewage sludges [9,13, 39, 57]

Element	Soil range (mg/kg)	Plant conc. (mg/kg)			Sewage sludges (mg/kg)			
		Mean	Range	Max.	Range	Median		Max. safe
						a	b	
Cu	10–80	20	7–30	150	84–17000	800	1230	1000
Zn	10–300	50	21–70	300	101–49000	1700	2780	25000
Cd	0.01–0.7	0.1	0.05–0.2	3	1–3410	15	31	25
Mn	20–3000	850	31–100	300	32–9870	260	–	–
Fe	10000–100000	–	21–70	750	–	–	–	–

8.3.4 Sewage sludges

Values reported from Cu, Mn, Fe and Zn for soils, plants and sewage sludges are compared in Table 8.5. Data for 8 h composite samples from a sewage treatment plant in Burlington, Ontario, Canada indicate an average concentration for Cu of 0.31 mg/kg in raw sewage, 0.21 mg/kg in primary effluent and 0.08 mg/kg in the final effluent [29]. If one ignores the volume changes, it becomes apparent that approximately 75% of the Cu in sewage remains in the sewage sludge. The analogous data for Zn and Pb were 77% and 93%, respectively. Data for a Dallas, Texas, USA plant indicated removals of 33% Cu, 65% Zn, 56% Pb, and 39% for Cd. Results for Zurich, Switzerland indicate a Cu removal of about 55%. From these data and results presented in Table 8.5, it may be concluded that sewage sludges are relatively abundant in Cu, Zn and Cd compared with soils and plants. While there is some risk associated with land application of sewage sludges because of Cu, Zn, Cd and other metals, sludges are frequently applied to land, both as a method of disposal and as a source of added soil organic matter, N and P. In addition to the median values reported in Table 8.5, several other median values for Cu have been reported including 800 mg/kg for 42 dry sludges in England and Wales, 560 mg/kg for 93 samples from Sweden; 700 mg/kg for 57 samples from Michigan, USA and 1200 mg/kg for 16 samples for 16 cities in USA [30].

In Pennsylvania and other states in the Northeastern, USA, it is not recommended that sewage sludges containing more than 1000 mg/kg Cu, 2500 mg/kg Zn, 1000 mg/kg Cr, 1000 mg/kg Pb, 200 mg/kg Ni, 25 mg/kg Cd, 10 mg/kg Hg or 10 mg/kg PCBs on a dry weight basis be applied to cropland [31]. Recently in Pennsylvania; USA it has been found that many sludges from waste water treatment plants cannot meet the 1000 mg/kg limit for Cu, but can meet a 1200 mg/kg Cu limit. Sludges more contaminated with metals will effectively exceed the capacity of soils to retain metals with little benefit from added organic matter, N and P. Maximum cumulative metal loadings of soils from sludges are regulated in Pennsylvania, USA. The regulated level for Cu of 60 ppm in the surface plough layer in the field is intended to prevent long-term adverse effects on biological processes essential for soil fertility as well as possible adverse health effects in sheep and cattle on treated pastures [31, 32]. Because of the relatively high concentrations of Cu and other metals compared to non-polluted soils, sewage sludges should be analysed and used as fertilisers as indicated in Table 8.5. Under no conditions is it considered practical to use sewage sludge as a soil conditioner without consideration of the metal and pathogen concentrations in the sludge and the resulting loadings of Cu and other metals in the soil and the crops to be grown.

Adding a Cu-rich sludge (6000 ppm Cu on a dry basis with 2300 ppm extractable in 0.005 M EDTA) to land at 125 t/ha in the first year and 31 t/ha in each of 3 subsequent years, raised the total soil Cu from 30 to 600 mg/kg and the 0.05 M EDTA-extractable copper from 8 to 280 mg/kg , of which 180 mg/kg was still EDTA-extractable after 4 years [30]. High levels of extractable soil Cu were reflected by increased uptake by vegetable crops and timothy (*Phleum pratense*). Numerous workers have published data on the addition of Cu in sewage sludge to soils and the associated uptake by food and other crops [9, 13, 33, 34].

While sewage sludges are capable of substantially increasing the soil levels of Cu, there have been no reports of plant toxicities to sludge Cu when grown in fertile, limed soils. Soil organic matter appears to be the dominant factor controlling Cu retention [34]. Thus, the rate of decompsition of sludge organic matter becomes a prime consideration for sludge treated soils. While toxicity to higher plants is of major concern, soil biological activity may be much more sensitive to sludge Cu inputs. As little as 10 mg/kg DTPA-extractable Cu has been shown to decrease soil enzymatic activity [35], 5 ppm Cu in solution increased chemical denitrification in high pH soils [36], and any added Cu inhibited nitrification [37].

8.4 Chemical behaviour of copper in the soil

Total Cu in soils includes six 'pools' classified according to their physico-chemical behaviour. The pools are soluble ions and inorganic and organic complexes in soil solution; exchangeable Cu; stable organic complexes in humus; Cu adsorbed by hydrous oxides of Mn, Fe and Al; Cu absorbed on the clay–humus colloidal complex; and the crystal lattice-bound Cu in soil minerals. When dilute salt solutions such as 0.01 M $CaCl_2$ are used to extract soils, the soil solution and the exchangeable Cu are considered as one fraction [19]. While it has not been possible to separate Cu retained by organics, Fe and Mn oxides, and silicate minerals [38] the results for soils and sediments indicate that these components rank in the order [33]:

Organics > Fe/Mn oxides >>> clay minerals

As little as 20, 65 and 75% of the Cu adsorbed by organic matter, oxide materials and clays, respectively, was found to be isotopically exchangeable with ^{65}Cu within 24 h [39].

In surface soils, the total Cu concentration in the soil solution is normally only 0.01 to 0.6 μM, owing to its high affinity for sorption by organic and inorganic colloids [19, 26]. Levels over 1.5 to 4.5 mg/kg damage or kill the roots of growing plants [26]. The labile metal measured by isotopic exchange, the amount removed by a chelating agent like EDTA or DTPA, or extracted with an exchange resin is considered as the plant available portion and includes the Cu in solution and part of that adsorbed on the solid phase of soils. However, the chemical activity of this Cu is not constant; and, therefore, the plant uptake of Cu is not well correlated with either the amounts in the various pools or the amounts removed by various extractants. Since most soils contain only 20 to 30 mg/kg total Cu, it is likely that Cu exists in most soils as $Cu(H_2O)_6^{2+}$ ions adsorbed at mineral surfaces, occluded or co-precipitated by silicate and non-silicated clays and as organically complexed ions [15, 17, 51, 41]. While the Cu^{2+} ion is the dominant ion in solution phase of acid soils, soil solutions over a range in pH contain several forms of Cu including Cu^{2+}, $CuSO_4^0$, $Cu(OH)_2^0$, $CuCO_3^0$, Cu^+, $CuCl^0$, $Cu(Cl_2)^-$, and numerous organic complexes of Cu [42].

Most Cu-containing minerals are too soluble to control the very low activities of Cu^{2+} in soil solution [15, 17]. McBride [17] concluded that simple precipitation of Cu as $Cu(OH)_2$, CuO, or $Cu_2(OH)_2CO_3$ are not generally responsible for the 'adsorption' of Cu^{2+} added to soils. Since most soils contain only 20 to 30 mg/kg total Cu, it is likely that Cu exists in most soils as Cu^{2+} ions adsorbed at mineral surfaces, occluded or co-precipitated by silicate and non-silicate clays and as organically complexed ions [15, 17, 40, 41].

8.4.1 Solubility of Cu(II) minerals and soil copper

The activity of Cu^{2+} in soil has been estimated by Lindsay [15] from the equation:

$$\text{Log } Cu^{2+} = 2.8 - 2\text{pH} \qquad (2)$$

For a soil solution pCu of 6, the pH of soil would be 4.4 and at pCu of 14, the soil pH would be 8.4. Most Cu minerals are more soluble than soil–Cu. For crystalline solids (c) and mineral forms, the order of decreasing solubility is

$$CuCO_3\text{ (c)} > Cu_3(OH)_2(CO_3)_2\text{ (azurite)} > Cu(OH)_2\text{ (c)}$$
$$> Cu_2(OH)_2CO_3\text{ (malachite)} > CuO\text{ (tenorite)}$$
$$> CuFe_2O_4O_4\text{ (cupric ferrite)} > \text{soil Cu.}$$

From these solubility relationships, Lindsay [15] concluded that 'soil–Cu may indeed be cupric ferrite'. McBride [17], however indicated that the tetrahedral coordination of Fe^{3+} required for this structure may inhibit its formation under soil conditions. For soils containing appreciable quantities of Fe and Al oxides with high specific surface areas, Cu^{2+} could exist as chemisorbed or occluded ions rather than as separate phases, which renders a lower activity of Cu^{2+} in solution than that predicted for pure Cu^{2+} minerals. Since Cu^{2+} can exist in octahedral coordination in silicate clays and as occluded carbonate, some soil Cu^{2+} can only be removed by dissolution of the clays [41, 43]. Non-diffusible Cu in calcareous soils that is soluble in dilute acid is probably present as an impurity within carbonate minerals and is another form of non-labile Cu immbolised in other soil minerals [17].

8.4.2 Copper adsorption by soils

Reactions of Cu with inorganic and organic components of soils have been reviewed [17, 30, 44] and courses in principles of surface chemistry and instrumentation are available in chemistry and physics departments at universities. Harter has reviewed and made available important publications on adsorption phenomena [34]. However, there is a need for more insight into soil properties and their management for crop production and environmental quality. In most agricultural soils at pH above 5.5, Cu is present as 'specifically' adsorbed ions and complexes. While non-specifically adsorbed ions are either in the diffuse Gouy–Chapman layer or at the outer Helmholtz plane and separated from the solid surface by at least one water molecule, specifically adsorbed ions on a solid surface may be regarded as a growth of an existing solid phase or the forming of new solid phases in aqueous solution. Thus specifically adsorbed Cu^{2+} is not removed by cation exchange when excessive amounts of Cu^{2+} or other exchangeable ions are added to the soil. Characteristics of Cu adsorption have been the subject of several authors [17, 39, 40, 44–46]. The relatively slow rate at which Cu^{2+} becomes specifically adsorbed by soil makes interpretation of data from adsorption studies difficult to relate to soil processes which operate in natural environments. Cupric Cu can undergo hydrolysis, inorganic complex formation, and/or organic complex formation when added to soil.

Stability constants and solubility constants act independently, to control the concentrations of Cu^{+} or Cu^{2+} in solution. Hydrolysis constants for Cu^{2+} at 25°C have been summarised by Leckie and Davis [47]:

$$Cu^{2+} + H_2O = CuOH_2 + H^+ \qquad *B_1$$
$$Cu^{2+} + 2\,H_2O = Cu(OH)_2^0 + 2H^+ \qquad *B_2$$
$$2Cu^{2+} + 2H_2O = Cu_2(OH)_2^{2+} + 2H^+ \qquad *B_{22}$$

The values reported for $-\log *B_1$ range from 7.2 to 8.1, for $-\log *B_2$ from 13.7 to 17.3, and for $-\log *B_{22}$ from 10.3 to 10.95. The hydrolysis of Cu^{2+} is a function of the total metal in solution and the pH. The respective constants assigned by Lindsay [15] are 7.70, 13.78 and 10.68. The hydrolysis products formed in solution regulate the concentrations of Cu through their initial adsorption or precipitation. While Cu^{2+} is the major ion in solution below pH 6.9, $Cu(OH)_2^0$ is the major solution species above pH7 [15]. In addition to important hydrolysis reactions, Leckie and Davis [47], presented solubility data for 11 Cu compounds, and formation equilibria for 23 inorganic Cu complexes, and 36 organic complexes. Mattigod and Sposito compiled association constants for transition metal cations with inorganic ligands [42]. The binding of Cu and other metallic cations by soil organic matter has a profound influence on the physical and chemical properties of soils [48]. Quantification of the plant availability of Cu in soil suspensions requires as a minimum the determination of labile Cu (ions in the solid and solution phases in equilibrium with free Cu forms in the soil solution) and the activity of Cu^{2+} in soil solution for acid soils and $Cu(OH)_2^0$ in neutral or alkaline soils.

Two mechanisms have been proposed to explain the specific adsorption of Cu^{2+} and other ions with six-fold coordination by silicate clays [49]. The first involves the extension of the octahedral layer of 2:1 lattice clays. The projecting portion of the octahedral layer constitutes a hydroxide phase in equilibrium with the metallic ions in solution. While the solubility would not be identical with the corresponding macrocrystalline hydroxide, it could approach it if sufficient hydroxide were available to bridge the gaps between successive layers to form domains of hydroxide with three dimensional crystallinity. The other case involves the reconstitution of the octahedral layer within the confines of the silica tetrahedral layer of clay minerals. Mineral acids cause a differential dissolution of layer silicates in which the octahedral layer passes into solution more rapidly that the silica layers. When new conditions favour the formation of hydroxides, reconstitution of the original octahedral layer occurs. If Cu^{2+} is added in solution at a pH above that of the clay in suspension, these mechanisms are operative [49, 50].

The adsorption of Cu^{2+} by weight of clays is in the order kaolinite > fireclay-type kaolinite > illite > smectite. This relationship indicates that Cu^{2+} adsorption or fixation increases as the silica to sesquioxide ratio of the soil decreases. Since this composition relationship is analogous to an increase in octahedral coordination of metals, the two mechanisms seem valid for explaining the specific adsorption of Cu^{2+} by soil clays. In addition, they serve to explain why specific adsorption of Cu^{2+} by Na-montmorillonite decreases rapidly after Cu^{2+} equivalent to about 30% of the Na^+ has been added [51]. Specific adsorption or occlusion of Cu via octahedral coordination, suggests that the adsorption of Cu^{2+} is not directly dependent on the CEC of the clay, but is strongly related to pH [52] and therefore to base saturation of the exchange complex and to the abundance of octahedrally coordinated ions on exposed surfaces.

While adsorption of Cu^{2+} has been shown to obey the Freundlich adsorption isotherm [6, 53], the mechanisms discussed above suggest that a constant pH, adsorption of Cu^{2+} by clays should be more in agreement with that predicted by the Langmuir adsorption isotherm for chemisorption [37, 45, 46, 54]. The adsorption of Cu^{2+} at a solution activity above about 10^{-6} M, the maximum Cu^{2+} activity tolerated by the most Cu tolerant higher plants, is not of importance in soil management. The minimum concentration of metals for specific adsorption has been defined as that concentration of an ion in solution which causes a reversal of charge (zeta potential) of colloids at infinite dilution. For Cu^{2+} this concentration has been reported as 2.5×10^{-4} for colloidal SiO_2 at pH 6.5 and 5×10^{-5} for kaolinite at pH 5 [44]. These concentrations of Cu^{2+} would be toxic to most plant species.

Since the solution activity of Cu^{2+} increases with the amount of specifically adsorbed Cu, labile Cu will also increase; but the relationship among the three fractions varies with soil pH, organic matter quantity and composition, with soil mineral abundance and composition. It is for this reason that peats and mineral soils with high contents of organic matter are the most prone to Cu deficiency. Even in soils with normal levels of organic matter (1–8%) most of the Cu not in the minerals is organically bound [30]. A fractionation study of the Cu in representative samples of 24 different English soil series, showed that the bulk of the metal was held in the lattice of weatherable minerals (over 50%), about 30% was bound by organic matter and 15% was adsorbed by hydrous oxides of Fe and Mn. The combined amounts of water soluble and exchangeable Cu, which relate to the activity of Cu^{2+} in the soil solution, were very small (1–2%) [45].

The presence of organic matter may be used to explain the observation that clays separated from different soils tended to have similar capacities for metal ion adsorption, which was greater (about 35 meq/kg) than that observed with pure clay minerals. The constant derived from Langmuir adsorption plots has been used to compare the relative bonding strengths of cations with respect to a kaolinitic soil clay at pH 6 [55, 56]. The order derived was:

$Cr^{3+} > Fe^{3+}, Al^{3+} > Ga^{3+} > Cu^{2+} > Pb^{2+} > Y^{3+},$
$La^{3+} > Mn^{2+} > Ni^{2+}, Co^{2+} > Zn^{2+} > Sr^{2+}, Mg^{2+} > NH_4^+, K^+$

and with respect to A-horizon clay colloids from the same soil with its humus coatings the only change in order was Y^{3+}, $La^{3+} > Pb^{2+} > Cu^{2+}$. The intrinsic affinity of the heavy metals on goethite (FeOOH) is in the order Cu > Pb > Zn > Co > Cd, and the pH for 50% retention from solution (10^{-5} M) is Cu (5.2), Pb (5.6), Cd (7.7) [51].

There is sufficient evidence to support the concept that the chemical form of Cu is important in controlling geochemical and biological processes. The proposed mechanisms involving specific adsorption of Cu by silicate, hydrous oxide clays and soil organic matter seem reasonable for soils subjected to alternate wetting and drying. Modelling of adsorption processes by both inorganic and organic soil components requires that speciation of the Cu, the organic fraction and the clay minerals be accurately known. The use of selected, well characterised components, estimated association constants for inorganic Cu ligands [42] and organic Cu ligands [30, 47] are recommended. Using this approach with appropriate concentrations and activities of Cu^{2+} and other components could enable the development of more accurate estimates of the behaviour of Cu in soils. Many more studies using well defined, simple systems will be required before soils can be characterised with respect to structure of solid phases and solution activities of ions and other species of Cu.

While the basic knowledge of Cu retention by clays, soils and sediments continues to accumulate at a steady rate, soil management decisions can be made from measurements of total sorbed Cu, labile Cu and the activity or chemical potential of Cu^{2+} in the soil solution when the soil pH is in the range desired for crops [26, 50, 51, 57–59]. The book by Dragun [60] serves as a model for the applications of soil chemistry to soil management for environmental quality.

8.5 Soil–plant relationships

8.5.1 Copper in crop nutrition

Copper deficiency data has been summarised to the extent possible by Shorrocks and Alloway [19]. Soils deficient in Cu are found in many areas of the world. A total of 23 countries reported Cu deficiencies in wheat, 12 in oats, 12 in pasture, 12 in maize, 12 in barley and 9 in rice. Deficiencies occur most often on organic soils (Histosols) followed by podzols (Spodosols) which are derived from sands and lack desirable capacities to hold either water or nutrients; Solonetz (Natrixerolls) high-Na soils with an associated high pH, highly dispersed clay, and poor internal drainage; and Kastanozems (Ustolls) which are chestnut soils of the sub-arid steppes with an accumulation of organic matter in the surface with high base status and high pH which results in high specific adsorption of Cu, Zn, Mn and Fe. The total areas covered by these five soil groups were calculated for each country. The estimate of the cultivated areas under arable and permanent crops (not including grassland) was done by prorating the areas of soil according to the percentage of the total area covered by arable and permanent crops on the assumption that all soil types are equally represented in the cultivated area. Countries with more than 5×10^6 hectares of cultivated land low in available Cu include Poland with 6.085, Argentina with 5.665, Canada with 12.915, USA with 42.115, Australia with 8.290, and India with 8.254. Other countries with more than 10^6 ha of low available Cu soils include Denmark, Finland, Germany, UK, Brazil, Mexico, China, Indonesia, Pakistan, Turkey, Nigeria and South Africa. Since the above summary includes soils with adverse conditions besides Cu deficiency, a more important summary is included in Table 8.6 [19] which includes data for crop response to Cu. Many of the more economically worthwhile yield increases have been obtained under conditions of subclincal deficiencies where crop yields are reduced without the occurrence of marked symptoms. An increase of 22.4% in spring barley was obtained with 1 litre/ha of foliar-applied $CuOH_2$ (before the end of tillering) on a brown sand. The yield increase was due to a significant increase in the number of grains per head or ear. Foliar applied $CuSO_4.5H_2O$ gave a 13.5% increase in wheat yields on an organic rendzina where no symptoms of deficiency had been observed. Sugar beets on a brown sand gave an 18% increase in root yield where 50 kg/ha of $CuSO_4$ had been applied to the seedbed when no signs of deficiency were apparent. Mean responses of up to 30% were obtained between 1966 and 1968 in 21 trials where Cu at 10 kg/ha was applied to soils with the highest rate of N [19].

The quality of crop products is affected by Cu deficiency. Obvious factors including size, shape and colour of fruits and vegetables as well as composition and associated storage losses, nutritional value and product acceptability are of economic importance and deserve attention. Some examples of adverse effects of Cu deficiency include unattractive appearance and smaller size for citrus fruits,; discoloration and spongy texture in onions; discoloration in carrots; chlorosis and wilting in lettuce; reduced protein, altered proportions of amino acids and Cu content of wheat and barley; high concentrations of amino acids in juice extracted from sugar beets; high incidence of cork spot and superficial cork in pears; and low Cu content of grass resulting in Cu deficiency in ruminant animals especially where the soil has an elevated Mo level [61].

These experiences indicate the great need to manage soil availabilities of all

Table 8.6 Examples of the yield effects caused by copper deficiency in crops as indicated by field experiments with copper treatments.

Crop	Soil type* and Country†	Yield with applied Cu (t/ha)	Yield response due to applied Cu (%)
Wheat	Peat (USA)	2.42	47.2
	Peat (USA)	2.82	95–100
	Sand (SA)	2.75	75–90
	Calc. sand (S. Aus)	0.45	91–100
	Sand (S. Aus)	0.70	67
	Lateritic podz. (S. Aus)	1.84	11–91
	Sand (W. Aus)	0.80	29–91
	Clay (Aus. Qnsld.)	0.80	100
	Org. rendzina (UK)	7.72	13.5
	Rendzina (France)	7.01–7.83	37–124
	Sandy loam (France)	3.50	22.4
	Calc. loam (UK)	3.51	5–33
	Silt loam (USA)	1.23	47–100
	Sandy loam (India)	1.87	14
	Clay loam (Kenya)	1.90	24–45
	Sands, loams and peat (21 sites in western Germany)	3.3–4.7	9.8–29.8
Barley	Peat (USA)	1.83	14.7
	Peat (UK)	5.84	275
	Sand on glacial drift (UK)	2.00–3.21	17.5–22.4
	Loamy sand (UK)	2.26–2.32	63–96
Barley and oats	Sands on fluv/glac. drift (20 sites in Scotland) (UK)	2.70	50–100
Maize	Sand (USA)	8.31	3–15
Rice	Sand (Brazil)	9.5	70
	Sand (Pakistan)	3.95	8.6
Sugar beet	Sand on glacial drift (UK)	8.74 (sugar)	15.5
Cassava	Peat (Malaysia)	12	66.7
Groundnut	Peat (Malaysia)	0.27	96.8
Sorghum	Peat (Malaysia)	2.2	100

* Soil types (mostly be texture) but FAO/UNESCO equivalent classes are: peat = histosol, sand = arensol or podzol, and lateritic soils = ferralsol. Rendzina and podzol (podz) = same. Parent materials: fluv/glac = fluvio-glacial deposits, and calc. sand = calcareous sand (cascic arenosol soils).
† SA = South Africa, S & W Aus. = South and West Australia, Qnsld = Queensland, UK = United Kingdom.
Table taken from Shorrocks and Alloway [67].

essential and toxic elements simultaneously. Crops and livestock too often suffer from hidden hunger of hidden toxicities. A plant response to one element or nutrient can not be expressed when growth and development is limited by a deficiency of toxicity of another element [57, 58].

8.5.2 Absorption and translocation of copper

Rates of absorption of Cu are among the lowest of the essential elements and there are large genetic differences among plant species and cultivars within species [62, 63]. The Cu in the root environment of rhizosphere is almost all organically complexed by root exudates or soil humus, but uptake and translocation is a function of the activity of Cu^{2+} in true solution at the active absorption sites [64]. While root absorption of Cu includes specific adsorption on cell walls of the root free space, the limiting step of transport across the plasmalemma involves the electrochemical gradient relating the activity of Cu^{2+} in solution outside the root to that of the cytoplasm of the cortical cells. Because the plant component of the absorption process is controlled by the plant species and cultivars, it becomes obvious that Cu uptake is a function of the activity of Cu^{2+} at the absorption site outside of the plasmalemma [50, 58, 59, 64].

The activity of Cu^{2+} involved in uptake by plant roots is a function of the soil solution activity of Cu^{2+} as modified by mycorrhizae effects. The uptake of Cu, Zn and P is enhanced by fungi associated with roots, known as vesicular–arbuscular mycorrhizae, whose hyphae penetrate the root at one end with the other end extending several cm into the soil [65]. It is not known whether the beneficial effect(s) of mycorrhizae results from an increased activity of ions or to a greater effective root surface area. Uptake of Cu^{2+} and Zn^{2+} is a metabolically active process at least at normal soil levels and is supported by findings that absorption is reduced by metabolic inhibitors. Since higher activities of either Zn^{2+} or Cu^{2+} in soil or solution cultures is antagonistic to the uptake of the other ion, it is generally recognised that both ions are absorbed in the same way. Other ions which reduce Cu absorption are Ca^{+}, K^{+} and NH_4^{+} (not (NO_3^{-}). However, it is likely that these ions affect uptake of Cu^{2+} via differential complexing, and other surface effects which determine soil solution activities of Cu^{2+} and membrane permeability [66].

All parameters relating to Cu^{2+} absorption by plant roots are affected by pH of the soil solution. A decrease in pH enhances Cu^{2+} absorption because increased solution activity of Cu^{2+} associated with dissolution of minerals, reduces organic complexation and solid phase adsorption of Cu.

The translocation of Cu^{2+} within plants occurs in both the xylem and phloem where the metal is bound by organic nitrogen compounds such as amino acids. Concentrations of 1.5–2.0 μM Cu^{2+} in the xylem and 3–140 μM in the phloem have been reported [66]. Much of the Cu^{2+} associated with plant roots may not be translocated into the shoot even when a deficiency occurs in the aerial parts. In the shoot, N metabolism appears to control the binding and transport of Cu^{2+}. Copper is a relatively immobile element in plants. Immobile elements in plants include Cu, Fe, Mn and other elements which do not move from old leaves to new growing points. Green leaves may accumulate high concentrations of Cu^{2+} and subsequently not release it to younger leaves and other tissues, such as inflorescence, despite their deficiency. With heavy rates of nitrogen where the Cu^{2+} activity and/or its quantity is marginal, the plant will probably produce normal vegetative growth at first but the failure to redistribute Cu from the older leaves, together with the decrease in uptake as root growth slows, produces chlorosis and tip necrosis of the new leaves which are characteristic deficiency symptoms [26, 59, 67].

8.6 Polluted soils

The utilisation of metallic Cu for tools, weapons, utensils, and ornaments has marked important periods in human civilisation. Copper was an important item of trade tracing back to the Old World, to the 'Old Copper Culture' of the Upper Great Lakes in the USA and Canada, and to the Copper Eskimos in the Northwest Territories of Canada and Alaska [68]. The heating of Cu nuggets in an open fire to increase their malleability has been replaced over time by methods which require high inputs of energy derived directly or indirectly from the coal reserves of the world.

The fly ash captured from the burning of coal for electricity generation is also a potential source of Cu pollution. While soil additions of coal ash are more frequently limited by their high concentrations of B and Se, a range in composition of 14–2800 mg/kg Cu has been reported [69]. Land enrichment with Cu^{2+} from air emissions of fly ash from burning of coal is expected to be negligible. However, as utilisation and disposal of fly ash on land becomes a common practice, Cu and other elements will require continued monitoring.

From the combustion of wood products, fossil fuels and waste incineration within urban areas, soils from urban areas are 5–10 times as high in extractable Cu as those of adjacent rural areas. High tension electricity cables can contaminate a strip of land up to 20 m wide, and Cu pollution of soils along highways from dusts has been reported, but the effects are considered minor [12].

From the discussion above, it is obvious that industrial and other uses of Cu combined with the associated burning of coal, oil, wood, and certain wastes contribute to point and non-point source pollution of the environment with Cu. Even where soils have been highly enriched with Cu, movement of Cu will be mainly in the solution phase within the soil and with the soil lost as a result of erosion. The amounts of Cu lost by leaching is very small. Except for areas near smelters, land application of sewage sludges, land application of manures from poultry and swine fed $CuSO_4$, the use of Cu-containing fungicides, algicides, etc, and the use of Cu in plating and chemical industries are the most frequent contributors to high and excessive levels of Cu^{2+} in soils.

Major sources of pollution such as smelters, usually yield the highest concentrations in soils within 1–3 km of the stack with concentrations decreasing exponentially with distance. For the Cu–Ni complex at Sudbury, Ontario, Canada, for example, most of the Cu is deposited within 32 km but soils in areas closer than 7.5 km frequently contain well over 1000 mg/kg. In addition to the effect of air emissions of Cu and other metals on soils and vegetation, the metals from mining areas tend to enter streams via effluents and erosion causing high metal accumulations in fish and other species as well as the transport of contaminants to alluvial soils downstream [68]. Smelter emissions are complex and involve elements other than Cu as well as SO_2. In some locations, smelter emissions of SO_2 and particulates are controlled but no limit has been set on the emissions of specific metals. Copper smelters and brass foundries also release large quantities of As and Zn. Zinc smelters release substantial amounts of Cu, Zn,Cd, Pb, and Ni. Multi-element effects have been assessed but rarely in terms of the relative effects of the constituent elements. In Pennsylvania, USA, emissions from Zn smelters are evident for distances up to 19 km from soil abundance of Zn, Cu, Pb and Cd [70]. At what is considered excessive soil levels of all four elements, many home gardens provide from 25–50% of the vegetables used by families [28]. A recent assessment of possible adverse health effects indicates that with liming and frequent additions of humus to the soils, vegetables can be grown; and they are safe for human consumption [28]. Thus, human health has been protected by phytotoxic effects of Zn or the sum of all metals and soil acidity which prevents excess ingestion of Cd and Pb that are toxic to humans. Also, the acceptable levels of body accumulations of Cd and Pb may have resulted from the protective action of high Zn in air emissions and garden vegetables. Possible adverse effects of Cu on human health were not considered significant. Even with high soil levels of Cu, deer and horses in the area suffered lesions in joints that are characteristic of Cu deficiency. These symptoms were attributed to excessive levels of Zn and Cd in forages [28].

The application of practices which lead to deliberate non-point pollution of agricultural soils is a greater threat to sustained food and fibre production than is point source pollution. Since the application of sewage sludges, municipal composts, pig and poultry slurries, and the use of fungicides, etc. to agricultural land is user-initiated, their extent and impact can be more prevalent and less obvious until plant growth becomes visually retarded and shows symptoms of metal toxicity [19, 26]. Chemical fertilisers rarely exceed 100 mg/kg Cu and even with long-term use, fertilisers do not contribute substantially to soil Cu contamination [71].

The highest concentrations of Cu in agricultural soils have been associated with the use of fungicide sprays in orchards where they reach the soil surface year after year directly or indireclty as leaf litter [71]. It has been reported that 7×10^7 kg of Cu as Bordeaux mixture is sprayed per year on vineyards, orchards, bananas, citrus, and other crops. The accumulated Cu in these soils is most often found to be toxic to the new plantings of the same or a different crop. However, these sprays are quite effective and their widespread use is likely to continue. Soil concentrations ranging from 110–1500 mg/kg have been associated with this practice compared with 20–30 mg/kg for background levels in agricultural soils. In Australia, 30000 ha of apple and pear orchard soils have been converted to other uses [71]. The effects of these soils on crop production, health of grazing livestock, especially sheep and cattle, have not been fully investigated, but the hazards from a Cu-induced Mo deficiency may be anticipated [32].

Sewage sludges even with control of discharges by industries are relatively high in Cu and Zn generated from their frequent use in household plumbing (Table 8.5). Sewage sludge is added to agricultural soils primarily as a source of organic matter and for its fertiliser value in terms of N and P. The significance of the Cu content in sludges can only be evaluated in conjunction with the accompanying concentrations of Ni, Zn, and Cd. Soil levels of metals are regulated on the basis of their abundance in both soils and sludges. *Criteria and Recommendations for Land Applications of Sludges in the Northeast* [31] limits land loadings of metals based on soil texture. The allowable loading ranges for sands to clays are 25–150 kg ha^{-1} for Cu, 2–4.5 for Cd, 50–300 for Zn, 10–60 for Ni, and 100–600 for Pb and Cr [31]. For the majority of sludge and soil combinations found in Pennsylvania, USA, Cu is the element which most often limits the lifetime sludge loadings of the soils. Because the availability of Cu^{2+} is so greatly dependent upon soil properties affecting Cu^{2+} adsorption and pH, total soil loadings for one region may not be applicable in another region. The 1983 USEPA [72] safe metal loading for soils of 280 kg/ha is considered appropriate and operational. The new, 1993, USEPA regulations for land application of sewage sludges [73] allow soil loadings of Cu and other metals in soils to reach concentrations considered appropriate for 'Sludge Bio-Solids', which includes soils and wastes added to soils. Sludges, other wastes and soils are allowed to be polluted with Cu to 1500 mg/kg, Zn to 2800 mg/kg, Ni to 420 mg/kg, Cr to 1500 mg/kg, and Cd to 39 mg/kg. A site in Pennsylvania, USA, now supports no vegetative cover after one very heavy application of sludge meeting the criteria for 'Bio-Solids' some 30 years ago contaminated the soil with 805 mg/kg Cu, 636 mg/kg Zn, 144 mg/kg Ni, 312 mg/kg Cr, and 12.3 mg/kg Cd as total sorbed metals. It is important that land managers avoid their opportunity to add metals from manures and sludges to enrich soils to the current USEPA approved concentrations, otherwise long-term, if not permanent, sterilisation of the soil is possible [73].

8.7 Copper in soil–plant–animal relationships

Copper deficiency in grazing ruminants has been widely reported from many parts of the world and may be attributed either to low soil and forage levels or, more commonly, to an excess or high relative level of Mo and/or an interaction between Cu, Mo and S [20,32, 74, 75]. Copper in forage and pasture crops depends on soil availability of Cu, plant species, stage of growth, time of year and lime and fertiliser applications. Legumes tend to take up larger amounts of Cu than grasses. In some cases, non-crop species or weeds may also increase dietary intake of Cu by grazing livestock. Mixed pasture herbage rarely contains more than 20 mg/kg Cu in the dry matter and usually less than 10 mg/kg [61].

In many areas of the world, pig manure slurries and poultry manures are enriched with Cu resulting from feeding $CuSO_4$ to improve rates of gain and feed efficiency [76–78]. Baker [26] presented an evaluation of copper in soil, water, plant relationships in relation to 'Ecological Problems in High Level Nutrient Feeding' at the 57th Annual Meeting of the Federation of American Societies for Experimental Biology. He estimated that the practice would require an application of approximately 6.7 kg Cu/ha per year, which could be followed for only 5 to 10 years on some sandy soils and for about 60 years on soils with a high cation exchange capacity. It is not possible to accurately predict the role of the organic component of manures and sludges because the stability of copper–organic complexes in soils has not been quantified. Manures with 800–1300 mg/kg Cu are being added to fields repeatedly as required for manure disposal [19, 26, 76–78]. Three annual applications of copper-enriched pig manure at a rate of 12.3 kg Cu/ha, applied shortly after cutting the pasture to avoid risk of grazing animal ingestion of manure contaminated herbage, increased the Cu concentration of pasture grasses, but not to a level that was hazardous to grazing animals [79]. Yield decreases occurred only where plants were smothered by the manure application.

The practice of supplying pigs with growth stimulating levels of copper has been followed in European countries for over 30 years [79–81]. The feed is fortified with 125–255 mg/kg Cu as $CuSO_4$. Sheep on pastures receiving these manures are at some risk because of their particular sensitivity to Cu. This risk is especially great when the manure is applied directly to pastures without soil incorporation.

What has been considered to be high levels of supplemental Cu in swine rations increases both rates of weight gain and feed efficiency [74–76]. Field, greenhouse and laboratory experiments were initiated by Martens, Kornegay and others at Virginia Polytechnic Institute and State University, USA, in 1972 and continued for 15 years through 1992 to determine the effect of Cu-enriched swine manure applications on plant growth and composition [76–78]. They reported equivalent corn grain yields and composition from five annual applications of Cu-enriched swine manure at 22.9 kg Cu/ha/year compared with untreated, control areas. Anderson *et al.* [77] published results for three long-term field experiments established in 1978 and continued through 1988 to evaluate corn (*Zea mays L.*) response to high Cu levels from Cu rich swine manure and $CuSO_4$ applications. The three experiments were conducted on soils with diverse properties. The safe Cu loading rate using the US EPA, 1983 criteria [72] was 280 kg/ha, but a higher amount of Cu was applied to the three soils from either the Cu-rich manure or $CuSO_4$ over the 11 year period. Copper applications of over 300 kg/ha either from pig manure or $CuSO_4$ did not reduce corn grain yields at soil pH 6.7–7.2. However, extractable P concentrations of the soils became very high based on calibration data which indicates that soil test levels > 56 mg/kg are very high for the dilute $HCl–H_2SO_4$ method [83]. With the existing emphasis on water quality and eutrophication of surface and ground water, it is concluded that no adverse effect to plant growth or water quality may be predicted from these data if applications of manures are made which do not result in an equilibrium P concentration (EPC) greater than 200 ppb in the soil solution [84, 85]. In conclusion, the enriched manures may be used as a source of N and/or P for crop production until the soil becomes optimum, but not polluted with N and/or P. Using soil loadings for P to control soil loadings with swine and poultry manures, the cumulative loadings of Cu in soils are not expected to cause forage levels above 20–30 mg/kg Cu, so the forage is safe for sheep and cattle when the soil level of Mo is in the range of 2–5 mg/kg and the soil pH is 6 or above. However, animals on pasture may ingest up to ten times more Cu in the form of soil than in the herbage. Soil ingestion commonly ranges from 1–10% of the dry matter intake of grazing cattle and up to 30% for sheep [20].

Water from wells and other sources for human consumption is almost always below the 1 mg/kg interim Cu limit, which is based on taste considerations [86]. Because of Cu contamination of domestic water supplies by plumbing, several approaches are used by various countries to establish standards for Cu in drinking water. Daily intake of Cu by humans is almost always above the recommended daily requirement of 2 mg/day. The Western diet is between 2 and 5 mg Cu/day, and it is difficult to construct a diet with less than 1 mg Cu [2, 59].

Hereditary Cu toxicosis is known as Wilson's disease. Copper accumulates in the liver as affected infants grow, and evidence of histological damage can be seen in early infancy. However, clinical illness is usually not observed before the age of 5 years [59]. Wilson's disease is rare and can be arrested and prevented by drug therapy. For most individuals homeostatic mechanisms for Cu prevent excessive accumulations.

8.8 The Baker soil test for soil diagnostics

The Baker Soil Test™ became an official method of the American Association for Testing and Materials (ASTM) in 1993. The method designation is D5435-93 'Standard Diagnostic Soil Test Method for Plant Growth and Food Chain Protection'. The method uses macro cations and the chelator DTPA weakly buffered at pH 7.3 to render a small exchange of the available elements. From the amounts of each element extracted, the Baker Soil Test™ computer software program calculates quantity and intensity (Q/I) parameters for each element.The graphical interpretations are based on results of research and data for thousands of user samples from many soils including sludge treated fields and reclaimed lands [28, 57, 58, 64, 65, 86–92].

While the copper concentration in soil solution is in the normal range of 0.01–0.6 μM, when the total soil content of Cu is 25–40 mg/kg, the solution of Cu^{2+} and other ions including Al^{3+}, and Zn^{2+} may be toxic to many higher plants when the soil pH is below about 5.5. This relationship has caused many scientists to conclude that soil pH is the most important factor controlling the plant availability of Cu^{2+}, Cd^{2+}, and Zn^{2+}. However, for polluted soils, the availability of these elements may require an unusually high soil pH to overcome excessive levels in plants and the food chain. As discussed previously, because of the effect of organic matter as well as pH on Cu^{2+} availability, it is not possible to use total or some measure of 'extractable' Cu to predict the solution activity of Cu^{2+} affecting plants. Existing methods for Cu and other trace metals analysis have been reviewed [93]. Since plant root growth is affected more than the aerial plant tissue, plant analysis is not

™Baker Soil Test is a registered Trademark of Land Mangement Decisions, Inc. 1429 Harris Street, State College, PA, to protect the users of the name and computer software for data interpretation.

reliable as a predictor of polluting levels of copper in soils [26, 50]. Inherent differences among plant species and varieties within species cause a greater range in plant leaf Cu than can be induced without creating a Cu^{2+} deficiency or toxicity. Investigations of the author leading to the above conclusions provided the impetus for the development of the Baker Soil Test.

Soil test extracting solutions are in use all over the world. For over 60 years it has been recognised that the available quantity of an element in soil cannot be predicted from the total composition of the soil [94]. Many soil test extractants have been developed to provide a measure of the 'available amount' by removing a portion of the labile pool of respective elements in soils [67]. Where the extractants have been used and calibrated with crop performance in relatively small areas with similar soils, climate and vegetation, the soil test results correlate with crop performance expecially for P, K, and Mg. The DTPA method of Lindsay and Norvell [95] is widely used for determining the relative levels of available metals in soils. However, for Cu and other trace metals, plant uptake is a function of soil pH and other soil composition factors which determine the activity of the metal in the soil solution at the sites of its absorption by plant roots [26, 57, 58, 62, 64]. The pH effect is reflected in the pCu value calculated for the Baker Soil Test.

Both diffusion within the soil solution from zones of high activity to zones of lower activity as well as root absorption are determined by the relative partial molar free energy of an ion in the soil solution which relates indirectly to the quantity or labile quantity of Cu^{2+} in the soil [64, 86]. These relationships for Cu are expressed as follows:

$$(\overline{G} - \overline{G}^0)_{Cu} = RT\ln\overline{A}_{Cu} = RT\ln\overline{\lambda}_{Cu}\overline{C}_{Cu}$$

where $(\overline{G} - \overline{G}^0)_{Cu}$ is the relative partial molar free energy of Cu^{2+} in solution, R is the molar gas constant and T is the temperature in °K, ln is the natural logarithm, and $\overline{A}_{Cu}$ is the activity of Cu^{2+} in the soil solution or the soil suspension if determined by use of a specific ion electrode $\overline{\lambda}_{Cu}$ and $\overline{C}_{Cu}$ are the average activity coefficient and concentration of available Cu^{2+} in the system. Because pCu, which is analogous to pH, is proportional to $(\overline{G} - \overline{G}^0)_{Cu}$, it is obvious that measurements of pH, pCu, pZn, pCd, etc are desirable for predicting plant availability of these ions in soil solutions and soil suspensions. The appropriate expression for the diffusive flux, J_{Cu}, is:

$$J_{Cu} = -AB\pi_{Cu}\frac{d\overline{G}_{Cu}}{dX}$$

where A is the cross-sectional area of the diffusion path, B is the geometry factor that corrects for the porosity and tortuosity of the soil medium, π_{Cu} is the average mobility of the ion, X is the distance in the direction of diffusion, and the other symbols have the meaning assigned above. Whether ion uptake is limited by diffusion or a carrier mechanism, the conclusion that uptake is limited by $\overline{G}_{Cu}$ should hold [64].

Using soil test calibration data developed from the concepts that cation activities and their uptake by plants are related to their mole or equivalent fraction on the soil exchange complex and the ratio-law or energies of exchange concepts of Schofield and Woodruff, respectively, it was calculated that pK, pCa and pMg were in the range determined appropriate for plant nutrition by Marshall [49]. These results served as an impetus for the development of the Baker Soil Test which relates ion availability to both intensive and extensive parameters for several ions [86].

The Baker Soil Test includes one of many combinations of ion activities in solution which could be used to compare soils. The method has been used to monitor macro and trace elements in soils treated with sewage sludge, fly ash, mine spoils and mixtures treated and seeded for vegetative stabilisation of disturbed sites. The approach includes the characterisation of the materials with respect to pH and lime requirements, the preparation of soil–spoil–ash mixtures to be utilised. If the mixtures are satisfactory with respect to pH, soluble salts, B, Se and Mo tested by use of a saturation extract, the Baker Soil Test is applied to test for macronutrient balance (chemical element activity ratios) and ion activities (pH, pK, pCa, pMg, pFe, pMn, pCu, pZn, pAl, pNi and pCd) as well as labile and total sorbed quantities of each metal. The approach is required for management of land use in Pennsylvania, USA. The Baker Soil Test was applied to a US Environment Protection Agency clean-up of an area contaminated by a zinc smelter that has been the subject of much research [28]. The data collected for each sample is evaluated by computer computations and printing of the results [28, 75, 58].

For the clean-up assessment of contaminated gardens near a zinc smelter in Pennsylvania, USA, the statistical regression results for relating plant (*PT*) and soil amounts extracted by Baker Soil Test (*BST*) and the chemical potentials of the respective ions in equilibrium with the DTPA for H, Cu, Zn, Cd and the free DTPA ligand (*LI*) were as follows:

$$PT\mathrm{Cu} = 231.6 - 0.111BST\mathrm{Cu} - 8.97p\mathrm{Cu} + 3.095p\mathrm{H} - 7.68pLI \quad R^2 = 0.38$$
$$PT\mathrm{Zn} = -3798 + 0.88BST\mathrm{Zn} + 182.6p\mathrm{Zn} - 168.1p\mathrm{H} + 201.8pLI \quad R^2 = 0.54$$
$$PT\mathrm{Cd} = -263.3 + 0.99BST\mathrm{Cd} + 9.90p\mathrm{Cd} - 2.1p\mathrm{H} + 12.19pLI \quad R^2 = 0.85$$

While the regressions were all highly significant, the coefficients decrease as the complexing tendency of the ions increase. It is postulated that a substantial amount of Cu is chelated by soluble organics in solution. Since it must be assumed that all Cu in the test solution is associated with DTPA, natural organics in solution will cause an error in the calculated pCu values. While the approach of the Baker Soil Test is considered superior to other existing methods, the use of specific ion electrodes for determinations of the chemical potentials of Cu and other trace metals offers a theoretically desirable approach. While some progress has been made, the change in the standard potential of the specific ion electrode and the lack of sensitivity at normal soil levels of Cu^{2+} limits specific ion electrode applications [50].

8.9 Concluding comments

Soil management of Cu in most areas of the world is somewhat analogous to that for Al. If the pH is not maintained at a desirable range (6 – 7 for many crops), the availability of Cu, Zn, Mn and other elements is more likely to be a problem. The chemistry of Cu in soils is also somewhat like that of Pb in that both are specifically adsorbed or 'fixed' in soils. More accurate methods are needed to measure the plant availability of Cu where the soil abundance is increased several fold from the disposal of wastes. Emphasis needs to be placed in the following areas:

(i) Specific ion electrodes and other methods are needed to precisely and accurately measure the ionic activity of Cu^{2+} in the soil solution, saturation extracts of soils or in dilute $Ca(NO_3)_2$ extracts of soils. The concentration of Cu^{2+} in solution is affected too much by inorganic and organic complexation of Cu^{2+} to permit accurate prediction of Cu availability. The Baker Soil Test is considered the most reliable approach at the present time for predicting Cu availability to plants. Using soil samples from locations where plant growth responses had been obtained, it was found that soils with a pCu_{DTPA} greater than 14.5 and less than 0.2 mg/kg extractable Cu are generally deficient in Cu for optimum plant growth.

(ii) Land disposal of high Cu containing wastes including sewage sludges, some swine and poultry manures and other sources of Cu, should be evaluated for the effects of Cu on the nitrogen cycle. This indirect effect of Cu may perhaps be the more appropriate approach for regulating land disposal of Cu-containing wastes.

(iii) The analytical methods currently used and others being developed to determine mechanisms by which Cu is held by clays, organic matter and mixtures should be continued with greater emphasis on systems of known components in equilibrium with solution activities of Cu below 1 μM. Much more data are needed on the chemistry, mineralogy and biological effects of Cu when the equilibrium solution activity is in the range of 10^{-12}–10^{-6} molar. Once the mechanisms of Cu adsorption are well defined in this activity range, knowlege of dissolution kinetics will become more important.

References

1. Arnon, D.I. and Stout, P.R., *Plant Physiol.* **14** (1939), 371–375.
2. Clementz, D.M., Pinnavaia, T.J., and M.M. Mortland, *J. Phys. Chem.* **77** (1973), 196–200.
3. Parker, A.J., in *Copper in Soils and Plants*. eds. Loneragan, J.F., Robson, AD. and Graham, R.D. (1981), 1–22.
4. Soltanpour, P.N. and Schwab, A.P., *Commun. Soil Sci.* Plant Anal. **8** (1977) 195–207.
5. Allaway, W.H., *Adv in Agron*. **20** (1968) 235–274.
6. Adrinao, D.C., *Trace elements in the terrestrial environment.* Springer-Verlag, New York (1986).
7. Aubert, H. and Pinta, M. *Trace Elements in Soils*. Elsevier, Amsterdam (1977).
8. Bowen, H.J.M., *Environmental chemistry of the elements.* Academic Press, New York. (1979).
9. Davis, B.E., *Applied Soil Trace Elements*. Wiley, New York. (1980).
10. Jeffrey, D.W., *Soil–Plant Relationships. An Ecological Approach.* Croom Helm, London & Sydney and Timber Press, Portland, Oregon (1987).

11. Longeragean, J.F., Robson, A.D. and Graham, R.D. Eds. *Copper in Soils and Plants*. Academic Press, New York (1981).
12. Nriagu, J.O., ed. *Copper in the Environment. Part I. Ecological Cycling*. Wiley, New York (1979).
13. Kabata-Pendias, A.K. and Pendias, H., *Trace Elements in Soils and Plants*. CRC Press, Inc. Boca Taton, Florida (1984).
14. Purves, D., *Trace element contamination of the environment*. Elsevier, New York. (1985).
15. Lindsay, W.L., *Chemical Equilibria in Soils*. John Wiley & Sons, New York (1979).
16. Cox, D.P., in *Copper in the Environment, Part I: Ecological Cycling*. ed. Nriagu, J.O., Wiley, New York (1979), 19–42.
17. McBride, M.B., in *Copper in Soils and Plants*, Loneragan, J.F., Robson, A.D. and Graham, R.D., eds. Academic Press, New York (1981), 25–45.
18. Baker, D.E. and Chesnin, L., *Adv. in Agron*. **27** (1975), 305–374.
19. Shorrocks, V.M. and Alloway B.J., *Copper in Plant, Animal and Human Nutrition*. Copper Developement Assn., Potters Bar (1987).
20. Thornton, I., in *Copper in the Environment. Part I: Ecological Cycling*. ed. Nriagu, J.O., John Wiley, New York (1979).
21. Hodgson, J.F., *Adv. in Agron*. **15** (1963), 119.
22. Krauskopf, D.B., in Geochemistry of micronutrients. *Micronutrients Agriculture*. eds. Mortvedt, J.J., Giordano, P.M. and Lindsay, W.L., eds. *Soil Sci. Soc. Am*., Madison. Wis., (1972), 7–40.
23. Reuther, W. and Labanauska, C.K., in *Diagnostic Criteria for Plants and Soils*, ed. Chapman, H.D., 830 S. Univ. Dr., Riverside, CA, (1965), 157–179.
24. Gartrell, J.W., in *Copper in Soils and Plants*, eds. Lonegran, J.F., Robson, A.D. and Graham, R.D., Academic Press, New York (1981), 313–349.
25. Alloway, B.J., Gregson, J.M., Gregson, S.K., Tanner, R. and Tills, A., in *Management and Control of Heavy Metals in the Environment*, CEP Consultants Ltd, Edinburgh, 545–548.
26. Baker, D.E., *Fed. Proc. Am. Soc. Exp. Biol*. **33** (1974), 1188–1193.
27. Follett, R.H., Murphy, L.S. and Donahue, R.L., *Fertilizers and Soil Amendments*. Prentice–Hall, Englewood Cliffs, NJ, (1981).
28. Baker, D.E. and Bowers, M.E., in *Trace Substances in Environmental Health*, part XXII, ed. Hemphill, D.D., Univ. of Missouri, Columbia, MO, (1988), 281–295.
29. Netzer, A. and Beszedits S., in *Copper in the Environment. Part I: Ecological Cycling*, ed. Nriagau, J.O. (1979), 123–169.
30. Stevenson, F.J. and Fitch, A., in eds. Lonegran, J.F., Robson, A.d. and Graham, R.D., *Copper in soils and plants*. Academic Press, New York (1981), 69–95.
31. Baker, D.E., Bouldin, D.R., Elliott, H.A. and Miller, J.R., eds., *Criteria and Recommendations for Land Application of Sludges in the Northeast*. Pennsylvania State Univ., Agri. Exp. Sta. Bull. 851. (1985).
32. Hornick, S.B., Baker, D.E. and Guss, S.B., in *Molybdenum in the Environment. Volume 2*, eds. Chappell, W.R. and Petersen, K.K.Marcel Dekker, New York (1976), 655–684.
33. Adediran, S.A. and Kramer, J.R., *Appl. Geochem*. **2** (1987), 213–216.
34. Harter, R.D., ed., *Adsorption Phenomena*. Van Nostrand Reinhold, New York (1986).
35. Mathur, S.P. and Sanderson, R.B., *Soil Sci. Soc. Am. J*. **44** (1980), 750–755.
36. Buresh, R.J. and Maragtian, J.T., *J. Environ. Qual*. **5** (1976).
37. Chang, F.H. and Broadbent, F.E., *J. Environ. Qual*. **11** (1982), 1–4.
38. Cavallaro, N. and McBride, M.B., *Soil Sci. Soc. Am. J*. **48** (1978), 1050–1054.
39. McLaren, R.G. and Crawford, D.V., *J. Soil Sci*. **25** (1974), 111–119.
40. McLaren, R.G. and Crawford, D.V., *J. Soil Sci*. **24** (1973), 172–181.
41. Shuman, L.M., *Soil Sci*. **127** (1979), 10–17.
42. Mattigod, S.C. and Spositio, G., *Soil Sci. Soc. Am. J*. **41** (1977), 1092–1097.
43. Kline, J.R. and Rust, R.H., *Soil Sci. Soc. Am. Proc*. **30** (1966), 188–192.
44. James, R.O. and Barrow, N.J., in *Copper in Soils and Plants*, eds. Loneragan, J.F., Robson, A.D. and Graham, R.D., Academic Press, New York (1981), 47–68.
45. McLaren, R.G. and Crawford, D.V., *J. Soil Sci*. **24** (1973), 443–452.
46. Harter, R.D., *Soil Sci. Soc. Am. J*. **43** (1979), 943.
47. Leckie, J.O. and Davis, J.A. III, in *Copper in the Environment. Part I: Ecological Cycling*, ed. Nriagu, J.O., John Wiley, New York (1979), 89–121.
48. Bloom, P.R., in *Chemistry in the Soil Environment*. eds. Dowdy, R.H., Ryan, J.A., Volk, V.V. and Baker, D.E. ASA Spec. Publ. no. 40, Madison, WI (1981), 129–150.

49. Marshall, C.E., *The Physical Chemistry and Mineralogy of Soils. Volume 1: Soil Materials.* John Wiley, New York (1964).
50. Dragun, J. and Baker, D.E., *Soil Sci. Soc. Am. J.* **46** (1982), 921–925.
51. Pickering, W.F., in *Copper in the Environment. Part I: Ecological Cycling.* ed. Nriagu, J.O., John Wiley, New York (1979), 217–253.
52. Kuo, S. and Baker, A.S., *Soil Sci. Soc. Am. J.* **44** (1980), 969–974.
53. Sidle, R.C. and Kardos, L.T., *J. Environ. Qual.* **6** (1977) 313–317.
54. Ellis, B.G. and Knezek, B.D. in *Micronutrients in Agriculture*. eds. Mortvedt, J.J., Giordano, P.M. and Lindsay, W.L., Soil Sci. Soc. Am., Madison, WI (1972), 59–78.
55. Wakatsuki, T., Furukawa, H. and Kawaguchi, K., *Soil Sci. Plant Nutr.* **20** (1974), 353–362.
56. Wakatsuki, T., Furukawa, H. and Kawaguchi, K., *Soil Sci. Plant Nutr.* **21** (1975), 351–360.
57. Baker, D.E., *Soil Sci. Soc. Am. Proc.* **37** (1973), 537–541.
58. Baker, D.E. and Amacher, M.C., *The Development and Interpretaion of A Diagnostic Soil Testing Program*. The Pennsylvania State Univ. Exp. Sta. Bul. 826 (1981).
59. Gupta, U.C., in *Copper in the Environment. Part I: Ecological Cycling*, ed. Nriagu, J.O. John Wiley, New York (1979), 255–288.
60. Dragun, J., *The Soil Chemistry of Hazardous Materials*. Hazardous Materials Control Research Institute, Silver Spring, MD. (1988).
61. Kubota, J. *Agron. J.* **75** (1983), 913–918.
62. Baker, D.E. in *Proc. of Workshop on Plant Adaptation to Mineral Stress in Problem Soils*, ed. Wright, M.J. Conell Univ., Ithaca, NY (1976), 127–149.
63. Gilkes, R.J. in *Copper in Soils and Plants*, eds. Loneragan, J.F., Robson, A.D. and Graham, R.D. Acadmeic Press, NY (1981), 97–117.
64. Baker, D.E. and Low, P.F., *Soil Sci. Soc. Am. Proc.* **34** (1970), 49–56.
65. Lambert, D.H., Baker, D.E. and Cole, H. Jr., *Soil Sci. Soc. Am. J.* **43** (1970), 976–980.
66. Loneragan, J.F., in *Copper in Soils and Plants*, eds. Loneragan, J.F., Robinson, A.D. and Graham, R.D. Academic Press, New York 91987), 165–188.
67. Viets, F.G., *J. Agric. Food Chem.* **10** (1962), 174–177.
68. Hutchinson, T.C. in *Copper in the Environment Part I: Ecological Cycling*, ed. Nriagu, J.O., John Wiley, New York (1979), 451–502.
69. Page, A.L., Elseewi, A. and Straughan, I., *Residue Rev.* **71** (1979), 3–120.
70. Buchaeur, M., *J. Environ. Sci. Technol.* **7** (1973), 131–135.
71. Tiller, K.G. and Merry, R.H. in *Copper in Soils and Plants*, eds. Loneragan, J.F., Robinson, A.D. and Grahan, R.D. Academic Press, New York (1981), 119–137.
72. US Environment Protection Agency, *Process Design Manual for Land Application of Municipal Sludge*, USEPA, Cincinnati, OH. EPA-625/1-83-016. (1983).
73. US Environment Protection Agency, *Standards for the Use and Disposal of Sewage Sludge*, USEPA, Cincinnati, OH. (40 CFR Part 503). Final Rule. (1993).
74. Nriagu, J.O., ed., *Copper in the Environment. Part II: Health Effects*, John Wiley, New York, (1979).
75. Thorton, I. and Webb, J.S., in *Applied Soil Trace Elements*, ed. Davies, B.e., John Wiley, New York (1980), 381–439.
76. Kornegay, E.T., Hedges, J.D., Martens, D.C. and Kramer, C.Y., *Plant Soil* **45** (1976), 151–162.
77. Anderson, M.A., McKenna, J.R., Martens, D.C., Donohue, S.J., Kornegay, E.T. and Lindemann, M.D., *Commun. Soil Sci. Plant Anal.* **22** (1991), 993–1002.
78. International Copper Assn., Ltd., *Crop Response to High Levels of Copper Application. Final Report. ICA Project No. 292 (N)*, (1993).
79. Batey, T.E., Berryman, C. and Line, C., *J. Brit. Grassld. Soc.* **27** (1972) 139–143.
80. Lexmond, Th.M. and de Haan, F.A.M., in *Proc. Internatl. Seminar on Soil Environ. and Fertility Management in Intensive Agric.*, Tokyo, Japan, (1977) 383–393.
81. Braude, R., in *Copper in Animal Wastes and Sewage Sludge*, eds. L'Hermite, P. and Dehandtschutter, J., D. Reidel Publ. Co., Boston, Ma (1981), 3–15.
82. Madsen, A. and Hansen, V., in Copper in Animal *Wastes and Sewage Sludge*, eds. L'Hermite, P. and Dehandtschutter, J., D. Reidel Publ. Co., Boston, Ma (1981), 42–49.
83. Olsen, S.R. and Sommers, L.E., in *Methods of Soil Analysis. Part 2. Chemical and Microbiological Properties. 2nd ed.*, eds. Page, A.L., Miller, R.H and Keeney, D.R. (1982), 418–419.

84. Wolf, A.M., Baker, D.E., Pionke, H.B. and Kunishi, H.M., in *Proceedings of Natural Resources Modeling Symposium*, Pingree Park, CO (1983), 164–169.
85. Baker, D.E., in *Proceedings of International Phosphorus Symposium*. CSIR COnference Center, Pretoria, South Africa (1988), 198–208.
86. Baker, D.E., *Soil Sci. Soc. Am. Proc.* **37** (1973) 537–541.
87. Baker, D.E., in *Soil Testing: Correlating and Interpreting the Analytical Results*, Am. Soc. Agron., Madison, WI (1977), 55–74.
88. Dragun, J. and Baker, D.E., *Soil Sci. Soc. Am. J.* **46** (1982) 921–925.
89. Baker, D.E., *Proc. Int. Conf. Soil Testing & Plant Analysis*. Fresno, Calif. Comm. Soil Sci. and Plant Analysis 21 (1990), 981–1008.
90. Baker, D.E., Pannebaker F.G., Senft, J.P. and Coetzee, J.P., in *Trace Elements in Coal and Coal Combustion Residues*, eds. Keefer, R.F. and Sajwan, K.S. Lewis Publ. (1993) 119–133.
91. Senft, J.P. and Baker, D.E., in *Application of Agricultural Analysis in Environmental Studies*, eds. Hoddinott, K.B. and O'Shay, T.A., ASTM STP 1162, Am. Soc. Test. Mat., Philadelphia (1993), 151–159.
92. Senft, J.P. and Baker, D.E., *Proceedings. Second Int. Conf. on the Abatement of Acidic Drainage. Tome I.* Montreal, Canada (1991) 209–220.
93. Knezek, B.D., and Ellis, B.G., in *Applied Soil Trace Elements*, ed. Davies, B.E.W., John Wiley, New York (180), 259–286.
94. Schreiner, O. and Anderson, M.S., in *Soils and Men, Yearbook of Agriculture*. U.S. Department of Agriculture. (1938), 469–486.
95. Lindsay, W.L. and Norvell, W.A., *Soil Sci. Soc. Am. J.* **42** (1978), 421–428.

9 Lead

B.E. DAVIES

9.1 Introduction

Lead is neither an essential nor a beneficial element for plants or animals. It is well known, however, for being poisonous for mammals and there are fears that human body burdens below those at which clinical symptoms of Pb toxicity appear may cause mental impairment in young children. Soil and dust are important sources of Pb for young children and Pb in blood can be related directly to lead in soil [1]. Many investigations have been carried out on Pb in envrionmental materials, including soil, over the last twenty years and we now have a detailed understanding of its environmental chemistry and ecological and health significance. Lead is present in uncontaminated soils at concentrations < 20 mg/kg but much higher concentrations have been reported in many areas as a consequence of anthropogenic emissions, often over many years. Nriagu [2] should be consulted for an account of lead use and poisoning in antiquity. When Pb is released into the environment it has a long residence time compared with most other pollutants. Lead and its compounds tend to accumulate in soils and sediments where, due to their low solubility and relative freedom from microbial degradation, they will remain bioavailable far into the future.

9.2 Chemistry and geochemistry of lead

9.2.1 *Applied chemistry*

Lead is a member of Group IVB of the Periodic Table of the elements. Two oxidation states (Pb(II) and Pb(IV)) are stable but the environmental chemistry of the element is dominated by the plumbous ion, Pb^{2+}. Elemental Pb is a dense (11.3 pg/cm^3) blue–grey coloured metal which smelts at 327°C and boils at 1744°C. This low melting point has allowed it to be smelted, melted and worked even in primitive societies. The metal is very soft and tends to creep or flow under sustained pressure: it is therefore readily cut and shaped and has long been used on roofs or as pipes. Metallic Pb is relatively opaque to ionising radiation, and makes a valuable shield material in X-ray and radioisotope work. Lead readily alloys with other metals: the Pb/Sb alloy is chiefly used to make battery plates but is also used in shotgun pellets and Pb/Sn alloys are often used as solder. Lead metal, in combination with PbO_2, is used to fabricate the lead–acid accumulator battery. Other inorganic compounds are widely used, e.g., the yellow chromate is used in road markings, and many paints contain Pb oxides or Pb soaps to promote polymerisation. There is an extensive organic chemistry of Pb(IV) compounds, especially tetra-alkyl and tetra-aryl compounds [3].

9.2.2 Lead contents of rocks and soils

The ionic radius of Pb^{2+} is 124 pm and it isomorphously replaces K^+ (133 pm) in silicate lattices or Ca^{2+} (106 pm) in carbonates and apatites. Lead also has a strong affinity for S, i.e., it is chalcophilic. It therefore concentrates in S phases in rocks: the major ore mineral is galena (PbS).

There is general agreement that the abundance of Pb in the average crustal rock is approximately 16 mg/kg. Nriagu [4] calculated the mean Pb content of gabbro as 1.9 mg Pb/kg, andesite as 8.3 mg Pb/kg and granite as 22.7 mg Pb/kg: these data illustrate the tendency for Pb concentrations to rise with increasing silica contents i.e., from ultrabasic to acid igneous rocks.

Although 95% of crustal rocks are of igneous origin, sedimentary rocks account for 75% of surface exposures and are therefore the most widespread soil parent material. The most common sedimentary rocks are shales and mudstones (80%) which have an average Pb content of 23 mg/kg. Black shales are rich in organic matter and sulphide [5], and tend to have higher Pb contents. Sandstones constitue 15% of sedimentary rocks and contain an average of 10 mg Pb/kg while limestones and dolomites (5% of sedimentary rocks) contain about 71 mg Pb/kg.

Estimates of Pb in uncontaminated soils vary. Nriagu [4] has reported the mean as 17 mg Pb/kg and Ure and Berrow [6] have reported 29 mg Pb/kg. A statistical survey of soil data from Wales and England [7] indicated that the normal surface (0–15 cm) soil Pb content lies between 15 and 106 μg/g with a geometric mean of 42 μg/g. Reaves and Berrow [8] examined the distribution of lead in 3944 samples from 896 Scottish soil profiles. The geometric mean content of all mineral samples was 13 mg Pb/kg and all organic samples was 30 mg Pb/kg. Severson *et al.* [9] reported the geometric mean Pb content of surface soils in the Frisian islands of Germany as 7.9 mg/kg and a range of < 4–11 mg/kg. It is probable that in remote or recently settled areas soil Pb contents are less than 20 mg/kg but elsewhere a general, low-level contamination has raised concentrations overall to 30–100 mg/kg. Statistical analyses of data sets have supported the hypothesis that much of the observed Pb in soil in many areas has originated from anthropogenic emissions [10, 11].

9.3 Sources of lead in soils

Soil is a sink for anthropogenic Pb and there are several well-recognised major sources, namely, mining and smelting activities, manures, sewage sludge usage in agriculture and contamination from vehicle exhausts. Lead arsenate ($PbHAsO_4$) has been applied to orchard trees to control insect pests and orchard soils may therefore contain elevated concentrations of Pb [12, 13]. Commercial use of these sprays is now infrequent since they have been replaced by organic pesticides.

9.3.1 Lead derived from vehicle exhausts

As the internal combustion engine developed during the early decades of this century there was a growing demand for petrol of higher octane ratings to avoid uneven combustion in the engine cylinders, i.e. knocking or pinking. In the early 1920s it was discovered that Pb alkyls (tetraethyl and tetramethyl Pb) when added to petrol helped overcome the problem. The first leaded petrol was sold in 1923 and its use rapidly became standard.

Warren and Delavault [14] reported that soil and vegetation samples collected near roads contained unusually high Pb contents and they remarked that the contributions of Pb from petrol fumes merited attention. Subsequently, Cannon and Bowles [15] demonstrated that grass was contaminated by Pb within 152 m downwind of roads in Denver, Colorado, USA and the relationship between Pb content and distance was exponential. This important source of Pb for soil was confirmed by others in the USA [16, 17]. In England, elevated soil Pb concentrations have been reported outside Birmingham, and levels in soil and grass decreased with distance from the road [18]. There are similar reports from other countries (e.g., Switzerland [19], New Zealand [20], Australia [21], Japan [22], Egypt [23], Venezuela [24], Belgium [25], Italy [26], Greece [27], Hong Kong [28]). The literature suggests that there is a zone of about 15 m wide on either side of most roads in which the concentration of Pb exceeds local background levels and contamination of the roadside environment must be a worldwide consequence of the use of leaded petrol.

The most convincing direct evidence for car exhausts as a source of Pb is derived from studies of Pb isotope rations. Gulson *et al.* [29] interpreted their Pb isotope evidence to conclude that vehicle exhausts were the main soil contaminant near Adelaide, Australia. Earlier, Chow [30] collected soil and grass from near two roads in Maryland, USA. He reported the ratios of ^{204}Pb, ^{206}Pb, ^{207}Pb and ^{208}Pb in soil, grass and petrol additives and concluded that the excess of Pb observed in surface soil could be attributed to lead from car exhaust fumes. Lead in grass was derived from surface soil and not deeper layers. Rabinowitz and Wetherill [31] reported a similar investigation for Missouri, USA, contaminating Pb may be derived from leaded petrol or from milling and smelting Pb ores. The $^{206/204}Pb$ isotope ratios were different for the two sources. In four samples of petrol the mean ratio was 18.49 whereas for 131 samples of the local ores the mean ratio was 20.81. The ratios for three soil samples taken from alongside roads were 18.43, 18.58 and 18.82. Contaminating Pb appeared to have come primarily from vehicle exhausts.

Wheeler and Rolfe [32] modelled the deposition of Pb by a double exponential function of the following form:

$$\mathrm{Pb} = A_1 \exp(-kD) + A_2 \exp(k'D)$$

where A1 and A2 are linear functions of average daily traffic volumes and the two exponents are assumed to represent two families of particles of different sizes.

Table 9.1 Lead contents of soil (mg Pb/g soil) at three distances from typical main roads, calculated from [33].

Distance from road (m)	Geometric mean	95% probability range	Number of samples
<10	192	18–2017	20
15	161	50–511	6
>30	53	14–203	17

The larger particles were deposited within 5 m of the road and the smaller particles within 100 m.

It is difficult to compare satisfactorily the results of published studies because of differences in soil sampling depth, distance from the road and methods of analysis. Smith [33] collated literature reports up to 1976. Assuming that soil trace element concentrations are log-normally distributed, the collated data [Table I in Ref. 33) were logtransformed and the geometric means and deviations calculated for soils within 10 m of a road, at approximately 15 m and beyond 30 m. The 95% probability range was then calculated to give a low and a high lead content (Table 9.1). The values demonstrate that Pb contamination does not extend appreciably beyond 30 m from the road.

9.3.2 General atmospheric additions to soil and vegetation

Lead aerosols emitted to the atmosphere from vehicles exhausts or general industrial fumes can be carried over long distances [34–37]. Steinnes [38] has shown how the Pb contents of surface soils in Norway decrease from an average of < 120 mg Pb/kg in the south of the country to < 10 mg Pb/kg in the far north and he ascribed the lower values to remoteness from western European industries.

Elias *et al.* [39] investigated deposition of Pb aerosols in a remote sub-alpine ecosystem in the Yosemite National Park, California, USA. Deposition on to horizontal plates ranged from 92–270 pg Pb/cm^2/day (mean 158) over at two-year study period. These rates were around two orders of magnitude below those found in urban environments. Lindberg and Harris [40] measured atmospheric deposition of Pb and other metals in a deciduous forest, in eastern Tennessee, USA within 20 km of three coal-fired power plants. The deposition rate of Pb to inert flat surfaces ranged from 3 to 15 μg/m^2/day (approximately 0.1–0.6 μg Pb/cm^2/yr). Lindberg and Harris [40] also listed information on total annual atmospheric deposition of Pb measured by other workers at remote, rural and industrial locations in different parts of the world. Total atmospheric deposition of Pb ranged from 3.1 to 31 mg/m^2/yr in remote and rural locations and from 27 to 140 mg/m^2/yr in suburban and industrial areas.

Sposito and Page [41], using published data for air concentrations of Pb and other metals, have calculated deposition of metals from air on to land in various remote regions as 0.4 g/ha/yr at the South Pole, 7.2 g/ha/yr in NW Canada and 6.3 g/ha/yr in northern Michigan, USA. In rural, industrialised and metropolitan Europe calculations of deposited Pb ranged from 87 to 536 g/ha/yr (median 189) and in North America from 71 to 20498 g/ha/yr (median 4257). The calculated high enrichment in Pb was attributed to a 'substantial contribution from vehicle exhaust'. Normalising air concentration of metals to those of Al (derived from a natural source), the authors showed Pb to have high enrichment factors (> 100) in remote, rural and metropolitan regions of Europe and North America and assumed this was due to anthropogenic inputs.

Williams [42] compared the Pb concentrations in soil collected in the Rothamsted Experimental Station, England, in 1972 with samples collected in the late nineteenth century. Increases ranged from 17 to 46%. Direct roadside contamination of the samples was unlikely, and the rise in levels reflected a general environment contamination by atmospheric Pb during the twentieth century. Jones *et al.* [43] have also analysed archived soil samples from the Rothamsted Experimental Station. In one set (Hoosfield) Pb appeared to have remained constant at 38 mg/kg, whereas in the Broadbalk samples the mean Pb content of the pre-1900 samples was 39 mg/kg compared with 46 mg/kg in more recent samples.

Many countries have reduced the Pb content of their petrol or even phased it out completely [44]. As a consequence the Pb content of herbage appears to be decreasing [46–47]. Sippola and Makela-Kurtto [48] have reported that total Pb in Finnish soils has decreased by 4% between 1974 and 1987.

9.3.3 Lead in urban soils

heavy metals in soilsAlthough soils in urban areas have little agricultural significance they should not be dismissed as unimportant. Vegetables grown in urban gardens make a contribution to domestic consumption [49] and children play in parks and gardens and ingest contaminated soil from dirty fingers [1].

Purves [50] and Purves and Mackenzie [51] were the first workers to report elevated levels of trace elements in urban soils which contained about four times as much Pb as rural soils in southeast Scotland. Warren and co-workers [52] reported the Pb contents in soils derived from Liverpool England, and several Canadian cities. City soils collectively, tended to show varying degrees of contamination and the enhancement of values was of the same order of magnitude as those found in some mining and smelting areas.

These early reports have been confirmed in many countries. Fleming and Parle [53] in Dublin found up to 540 mg Pb/kg soil although most values were in the range 70–150 mg/kg. They suggested that Pb in urban soils could be derived from vehicle fumes, abraded tyre material, coal, plastics and rubber factories, insecticides and car batteries. Close to buildings, old paint could also provide a source of Pb.

Czarnowska *et al.* [54] analysed 760 soil samples in Warsaw, Poland and reported a range of 2.0–551 mg Pb/kg soil (20% HCl extracts) and 28% samples contained 50–550 mg Pb/kg. Preer *et al.* [55] analysed 95 soil samples from gardens in Washington DC, USA and reported a range of 44–5300 mg Pb/kg soil with a median of 480 mg/kg. Tiller [56] reported a range of 2–160 mg Pb/kg for 160 samples of rural soils in Australia and up to 212 mg/kg in Adelaide.

In London, Davies *et al.* [57] reported a range of 42–1840 mg Pb/kg (EDTA-extractable). The values could be zoned according to proximity to the city centre and there was an exponential decline of soil Pb content with increasing distance from the centre. A similar curvilinear relationship was found [58] in the Birmingham, England area where soil Pb and distance from the city centre were related by the polynomial regression

$$y = 272.5 - 275x + 0.82x^2 \quad r^2 = 0.528$$

where y is the distance in km and x is the soil Pb content. In New Zealand, Fergusson and Stewart [59] have also shown that the depositional flux of Pb followed an approximately exponential decay curve away from Canterbury. Tiller *et al.* [60] have reported that the rural landscape is contaminated for about 50 km downwind of Adelaide, Australia.

Davies [61] reported the Pb content (EDTA-extractable) of soils from 87 British gardens in both rural and urban environments: 63% of the soils were contaminated when compared with rural arable norms. Although the highest values were derived from Pb mining areas and city gardens, older rural gardens were also mildly contaminated. In two villages in Devon and Cornwall significant correlations were found between the age of the house or garden (i.e., duration of occupacy) and the soil Pb content. Culbard *et al.* [62] have reported data for a nationwide survey of metals in British urban soils and dusts. For 4126 garden soils, the geometric mean Pb content was 266 mg/kg with a range of 13–14 100 mg/kg; for 578 samples form London boroughs the mean content was 654 mg/kg and the range was 60–13 700 mg/kg.

9.3.4 Contamination from mining and smelting

During the years 1908–1913 Griffith [63] investigated frequent complaints of soil infertility from farmers in the Aberystwyth area of Wales and concluded that the primary cause was the presence in soil of appreciable quantities of Pb which had originated from Pb mining in the nineteenth century. Later, Alloway and Davies [64] described a comprehensive investigation of Pb-contaminated soils in Wales. They found that alluvial soils in the Ystwyth river valley contained 90–2900 (mean = 1419) mg Pb/kg compared with 24–56 (mean = 42) mg Pb/kg soil in a neighbouring, control valley. Colbourn and Thornton [65] found high concentrations of Pb in agricultural soils of the Peak District of Derbyshire. Within 100 m of an old smelter they reported mean soil Pb = 30090 mg/kg, and within 100 m of an old ore washing floor a mean of 19400 mg Pb/kg. In the southwest of England, Davies [66] concluded that there was extensive contamination of pasture fields in the Tamar Valley. Davies and Roberts [67] published computer-drawn Pb isoline maps for an area in northeast Wales and calculated that there are 171 km^2 of land containing > 100 mg Pb/kg soil and 47 km^2 have soils with Pb contents ranging from 1000 > 10000 mg Pb/kg. A similar approach was used in a baseline survey in Missouri, USA [68]. The maximum soil Pb concentration was found to be 2200 mg/kg, but 95% of the samples contained < 355 mg Pb/kg. Isopleth plots of the soil Pb contents showed a strong association between concentrations > 355 mg Pb/kg soil and old mine workings, dressing floors and railway ore loading places.

The large scale of the modern Pb industry inevitably leads to accumulations of Pb in soils around the works despite strict environmental controls. In addition, some of the modern mines and smelters are located on the sites of older, dirtier works which have left a legacy of environmental contamination. Avonmouth, England has been a centre for Pb smelting since 1919 and modern production utilises a furnace which was commissioned in 1967. Burkitt *et al.* [69] reported that soils within 0.32 km of the complex contain 600 to 2200 mg/kg, but 95% of the samples contained < 355 mg Pb/kg. Isopleth plots of the soil Pb contents showed a strong association between concentrations > 355 mg Pb/kg soil and old mine workings, dressing floors and railway ore loading places. Levels decline rapidly away from the works but significant quantities of Pb could be detected 11 km away.

In surveys around other smelters, maximum Pb accumulations occur close to the stack. There is a rapid decline away, and the distance–decline curve is usually exponential [70–72]. It is difficult to be categorical about the soil Pb levels likely to be encountered around metal smelters. However, using the data reported above it is concluded that within 3 km of a well-established smelter, soils are likely to contain about 1500 be 2200 mg/kg, but 95% of the samples contained < 355 mg Pb/kg. Isopleth plots of the soil Pb contents showed a strong association between concentrations > 355 mg Pb/kg soil and old mine workings, dressing floors and railway ore loading places, i.e., an enhancement of 15 times the background level. Accumulations of Pb may also be expected in the soils around secondary smelters and other Pb using industries.

9.3.5 *Lead derived from agricultural materials*

Farmyard (barnyard) manure has long been recognised as a desirable soil amendment material. It is a useful short-term and long-term source of the essential plant nutrients N, P and K as well as a source of essential plant and animal micronutrients [73]. Unfortunately, as mixed farming has been replaced by specialised arable farming, supplies of manure at an economic price have decreased. Farmers have therefore turned to the use of alternative materials. In Holland, household wastes are increasingly composted and the Pb content of these has attracted attention [74].

In response to legislation in many countries which forbids the disposal of sewage direct to waterways, city authorities have invested in treatment works which are designed to retain the organic components of sewage. The organic sludge is fermented to destroy pathogens and then is often dewatered to provide an organic material with a variable but useful N and P content [75–77]. This residue is available to farmers as an alternative to farmyard manure. According to Davis [78] 30 million wet tonnes of sewage sludge are produced annually in the UK and 40% is applied to land.

Unfortunately, sewage sludge can have an undesirable chemical property, namely, a high content of heavy metals. Coker and Matthews [79] state that the typical concentration of Pb in human faeces is 11 mg/kg dry solid whereas the median content of sludge from a non-industrial area is 121 mg/kg. The extra Pb is derived from Pb water-pipes and road run-off. Furthermore, it is common practice for industrial effluents to be passed into the foul sewer and the organic residues effectively remove metal ions by complexation. Thus, metal contaminants are retained in sewage sludge. The implications of this were not realised until Le Riche [80] reported metal contamination of soil following long-term application of high rates of sewage sludge. Since that date there has been a proliferation of reports concerning metals in sludges and in crop plants grown on sludge-amended soils.

There are several published reports of metal concentrations in sewage sludges. Berrow and Webber [81] analysed 42 sludges from rural and industrial towns in England and Wales. The Pb contents of the dried materials ranged from 120–3000 mg/kg (mean 820 and median 700, mg Pb/kg respectively). Sommers [76] reported a range of 545–7431 mg Pb/kg for sludges from eight Indiana, USA cities. For 189 samples Pb ranged from 13–19700 mg/kg (mean 1360, median 500 mg/kg, respectively). There is a general agreement that the Pb content of sludges is very variable but the typical sludge contains < 1000 mg Pb/kg. (See Chapter 3.)

The observation that metals accumulate in sludge-amended soils to an extent that Zn phytoxicity symptoms may be seen [80], together with the possibility that Pb might accumulate in crop plants to dangerous levels has led several countries to formulate guidelines for sludge usage. The Council of the European Community in 1986 adopted a Council Directive regulating the use of sewage sludge in agriculture. Repeated applications of sludge should not lead to an accumulation in soil in excess of 50 to 300 mg/kg. Based on a ten year average, no more than 15 kg/ha/yr of Pb may be added from sludge. More details of the behaviour of Pb and other heavy metals in sludge amended soils can be found in Alloway and Jackson [82].

Lead in other agricultural materials has received little attention. Because of the similarity of the ionic sizes of Pb^{2+} and Ca^{2+} the former may proxy for the latter in calcite. In a survey of British agricultural liming materials Chater and Williams [83] reported a mean of 2 mg/kg. Davies *et al.* [84] noted that limestone wastes from Pb/Zn mining have been used as a convenient form of agricultural lime in some areas. Although their use does cause Pb to accumulate in soils, agricultural problems will probably not arise from moderate use.

9.4 Lead in soil profiles

Lead appears to accumulate naturally in surface horizons of soil. Lounamaa [85] reported this phenomenon for soils collected from remote sites in Finland. He observed that

> Lead differs from... other... elements in that soil samples from the different outcrops invariably contain more of it than do the rocks. The strong enrichment of Pb into the soils overlying silicic rocks is quite remarkable.

Wright *et al.* [86] observed that Pb showed the greatest proportional accumulation in surface horizons of soils (podzolics and grey-brown podzols) in eastern Canada. Other authors have also reported Pb accumulations in surface horizons of uncontaminated soils [13, 87–92]. Colbourn and Thornton [65] used the ratio of Pb concentration in topsoil (0–15 cm) to that in subsoil enhancement (RTE). In normal agricultural areas of the UK, RTE ranged from 1.2 to 2.0, whereas in locations affected by mining and smelting operations, values between 4 and 20 were common.

There is little evidence that Pb is readily lost from soil profiles by leaching. Most heavy metals, including Pb, remain in an insoluble or stable form in surface layers after application of sewage sludge [93–95]. Zimdahl and Skogerboe [96] demonstrated that soils have rather large capacities for the immobilisation of Pb, that the organic fraction was largely responsible for the observed fixation of the metal and Pb from vehicle exhausts tends to remain near the soil surface. Korte *et al.* [97] leached 11 soils with a natural leachate spiked with Pb and other trace elements. Lead was immobile in all soils except for one classed as an Ultisol with a loamy sand texture and a very low CEC (2 $cmol_c$/kg). Hooghiemstra-Tielbeek *et al.* [98] were concerned with finding methods to reduce the Pb content of surface soils and thereby reduce the amount of Pb available for direct ingestion by children. For an effective treatment soil pH had to be reduced to 2.5–3 by leaching with M HCl or by applying solid or dissolved $FeCl_3$ to the soil surface.

9.5 The chemical behaviour of lead in soil

In contrast with work on the identification of the sources and assessments of the bulk contents of soil Pb, our knowledge of the chemistry and speciation of Pb in soil is limited. Measurements of total Pb have much value in indentifying and describing areas where contamination of the soil mantle has occurred. But, by analogy with the behaviour of nutrient elements, it can be assumed that only part of the total Pb is available for uptake by plants and Pb compound entering the soil become partitioned among several soil compartments. The main compartments for Pb in soil are the soil solution, the adsorption surfaces of the clay–humus exchange complex, precipitated forms, secondary Fe and Mn oxides and alkaline earth carbonates, the soil humus and silicate lattices.

9.5.1 Fate of Pb and Pb compounds entering the soil

Due to the importance of Pb from anthropogenic sources some workers have attempted to study the fate of Pb in soil when introduced as various inorganic compounds. For example, Santillan-Medrano and Jurinak [99] equilibrated lead acetate solutions with soil. They concluded on theoretical grounds that the solubility of Pb in non-calcareous soils was regulated by $Pb(OH)_2$, $Pb_3(PO_2)_2$, $Pb_5(PO_4)_2OH$ and in calcareous soils $PbCO_3$ was important.

The chemical changes which occur during transport of Pb-rich particles from motor vehicles to soil or plant surfaces have been studied in some detail. When exhaust particles are emitted they consist mainly of PbBr, PbBrCl, Pb(OH)Br, $(PbO)_2PbBr_2$ and $(PbBO)_2PbBrCl$ [100]. After 18 h about 75%, of the Br has gone, 30–40% of the Cl compounds are also lost and Pb carbonates, oxycarbonate and oxides form. In samples of air from Phoenix, Arizona, USA [101] the most abundant species (33%) was α-2PbBrCl NH_4Cl. Other compounds indentified included PbBrCl and $(PbO)_2PbBrCl$. No particles containing only Br and Pb were identified.

Olson and Skogerboe [102] studied the fate of Pb compounds from exhaust fumes after they had entered the soil. More than 75% of the Pb in samples from two areas was associated with soil density > 3.32 g/ml, and $PbSO_4$ was the primary compound present. However, Harrison *et al.* [103] concluded that $PbSO_4$ is not of any significance in roadside soils and Pb was associated with Fe–Mn oxides and organic phases and, to a lesser extent, with carbonates.

Occasionally, Pb may enter soil in the metallic form. Jorgensen and Willems [104] showed that Pb shotgun pellets in Danish soils were transformed into hydrocerussite ($Pb_3(CO_3)_2(OH)_2$, cerrusite ($PbCO_3$) and anglesite ($PbSO_4$). Complete transformation of the pellets would happen in 100–300 years but could be as little as 15–20 years.

9.5.2 *Lead in soil fractions*

Another approach has been to infer the chemical forms of soil Pb through the use of various fractionation procedures. Zimdahl and Skogerboe [96] studied the fixation of Pb added to soils as $PbNO_3$. It took 24–48 hours for ionic Pb to become distributed among the soil compartments, and in the range 20–40°C temperature did not affect the equilibration. From a statistical study of their data they concluded that pH and CEC were the main soil properties involved in the immobilisation of Pb and that soil organic matter was more important in this process than precipitation as the carbonate or sorption by hydrous oxides. Khan and Frankland [105] added $PbCl_2$ or PbO to contaminated soils and noted that it took only about an hour for much of the water-soluble Pb to change to a less soluble form (EDTA-extractable).

Garcia-Miragaya [24] concluded that in soils from near roads < 0.7% Pb was in readily exchangeable forms and most of the Pb was associated with the organic and residual fractions or specifically bonded to inorganic sites: the amount occluded in iron oxides was low. Berrow and Mitchell [106] noted a tendency for Pb concentrations to increase with a decrease in particle size for profile samples in Scotland. Lead released by weathering appeared to be largely adsorbed by silt and clay.

Hildebrand and Blum [107] reported that soil humus, especially high-molecular-weight humic acids contributed most to the immobilisation of Pb added to the soil and that Pb was bound through coordinated binding by free electron pairs. In contrast, Harter [108] concluded from adsorption studies modelled by the Langmuir isotherm that organic matter content had no significant relationship to Pb adsorption by surface horizons.

9.5.3 *Pb in the soil solution*

Soil solution Pb is the immediate source for plant roots and a dynamic equilibrium can be assumed between the soil solution and the other compartments. The very low concentration of Pb in the soil solution presents analytical difficulties which have limited studies of this form of element.

Kabata-Pendias [109] found the concentration of heavy metals in solutions obtained by centrifuging soil to be of the order of 10^{-7} M. The soil solution can also be extracted directly by preparing a 'saturation' extract of soil. Bradford *et al.* [110] collected 68 soil sample in California, USA and found the concentration of Pb in the saturation extract to be about 10^{-9} M.

Davies [7] has proposed that the normal Pb content of uncontaminated soils is about 40 mg/kg (0.2×10^{-6} M) and if the concentration in the soil solution is assumed to be of the order of 10^{-8} M then soil solution Pb in uncontaminated soils represents less than 0.005% of the total.

Gregson and Alloway [111] extracted soil solution from contaminated soils by a centrifugation technique and fractionated the solution by chromatography with Sephadex gels. Table 9.2 provides selected data from the authors' results. The total Pb concentrations are approximately 100 to 1000 times the normal concentration of 40 mg/kg [7] and the concentrations in the soil solution are 10^{-6} M (approximately), or 1000–100000 times the normal soil solution concentrations quoted above. The proportion of the total metal content represented by the soil solution ranges from 0.005% to 0.13% appears to be somewhat higher than in uncontaminated soils and poses the interesting question of whether Pb in polluted soils is more plant-available.

Tills and Alloway [112] used ion-exchange chromatography to determine the speciation of Pb in the soil solution. In acid soils Pb was found to be mainly in a cationic form with some organic complexation. In calcareous soils neutral complexes were dominant, together with some cationic lead species. Similarly,

Table 9.2 Lead in soils and soil solutions (from data of Gregson and Alloway [111])

Total lead (mg/kg)	Soil solution lead (mg/l)	Solution lead as % total
49900	112	0.05
2820	18	0.13
45800	11	0.005
1890	4	0.04
3830	4	0.02

Gregson and Alloway [111] reported results of fractionating soil solutions using gel permeation chromatography. In heavily polluted soils part of the Pb was present as a high-molecular-weight organo–Pb complex, and the proportion represented by this form was greater in soils of higher pH.

9.6 Soil–plant relationships

9.6.1 Uptake of Pb from soil and translocation to shoots

In broad terms there is a positive relationship between the concentration of Pb in the soil and that in the plant [113–115]. There is also general agreement that only a small proportion of the Pb in soil is available for uptake by plants. The uptake of Pb and its transportation to the shoots of perennial ryegrass was studied by Jones *et al.* [116] in conventional and flowing culture solutions. Lead was absorbed rapidly even when the shoot was removed or if the root was dead. The amount of Pb translocated to the shoot was small, being only 3.5–22% of total uptake after several days.

Lane and Martin [117] used histochemical techniques to investigate the distribution of Pb in radish seedlings. The metal accumulated at the endodermis which acts as a partial barrier to Pb translocation to the shoot, but the barrier is only partial. Koeppe [118] has reviewed the uptake and translocation of Pb in plants. He concluded that translocation is highly dependent on physiological status. Under conditions of optimal growth Pb precipitates on root cell-walls in an insoluble, amorphous form which, in maize, has been identified as a Pb phosphate.

Dollard [119] used the radioisotope ^{210}Pb to investigate the redistribution of lead applied to foliage. He estimated that for radish about 35% of the internal Pb burden of the storage tissue could be accounted for by transport from the leaf. For carrot, this pathway contributed about 3%. Dollard considered that even in the case of radish the extra pathway for Pb uptake did not go far towards explaining the often large discrepancies between concentration factors derived during field studies and tracer experiments.

The uptake of Pb by plant roots and translocation to the shoots varies seasonally. Mitchell and Reith [120] first drew attention to this phenomenon when they reported that the Pb content of grass shoots increased during autumn and winter. During the period of active growth the concentration of Pb in grass was 0.3–1.5 mg/kg dry matter; by late autumn it had risen to 10 mg/kg and to 30–40 mg/kg by late winter.

Baker [121] and Simon and Ibrahim [122] have discussed whether the relationship between plant uptake and substrate concentration is linear, curvilinear or has some other shape. Some plants may be described as excluders in that metal concentrations in shoots are maintained at a low and constant value over a range of soil concentrations until some threshold value is exceeded, when uptake becomes unrestricted. For accumulator species the relationship between uptake and substrate concentration is curvilinear: uptake becomes progressively less at high concentrations. Simon and Ibrahim [122] extended the concept to the elements themselves and concluded that Pb (they used ^{210}Pb) was an accumulator metal. They described the uptake by the following equation, in which s is the substrate Pb concentration

$$\text{plant Pb} = 0.74\,(1 - \exp(-1.45s) + 0.16s$$

Although Pb is not especially toxic to plants high substrate concentrations do result in stunted growth or death. Some plants, especially grasses can evolve tolerance to high soil Pb and can be used to revegetate contaminated land [123–125]. In other plants hyperaccumulation has been observed [126, 127].

9.6.2 The problem of airborne Pb and foliar deposition

Tjell and co-workers [128] used ^{210}Pb as a tracer in rural Denmark and showed very low uptake of Pb into grass from the soil and they deduced that 90–99% of Pb in the leaf material was due to foliar uptake. Crump and Barlow [129] concluded that 45–80% of the total Pb in grass growing near a highway was due to airborne deposition. Dalenberg and van Dreel [130] similarly concluded that $> 90\%$ of the Pb in grass and carrot leaves could be attributed in airborne Pb but only 5.7% of the Pb in carrot roots.

Chamberlain [131] has reviewed a number of papers where the authors have compared Pb levels in both plants and soil. Using the reported data he calculated a concentration factor (*CF*):

$$\text{CF} = \text{Pb in plant } (\mu\text{g/g dry weight})/\text{Pb in soil } (\mu\text{g/g dry weight})$$

Chamberlain then examined data published by other authors. Results from field surveys of radish and potatoes yielded *CF* values of 0.05–0.2, and work in greenhouses using filtered air for several crops produced similar *CF* values. Data from Karamanos *et al.* [113] where the uptake of both stable Pb and ^{210}Pb by lucerne and bromegrass was investigated yielded *CF* values of 0.090–0.19. In contrast, however, when ryegrass was grown in soil labelled with ^{210}Pb the *CF* value was only $1\text{–}3 \times 10^{-3}$ and a similar value was calculated for data from Solgaard *et al.* [132] for cereals.

This foliar effect has confused many studies of Pb uptake in the field and there is still doubt on the relative contributions of foliar and root accumulation to the observed Pb content of plant tissue.

9.6.3 Soil factors controlling Pb uptake

Several soil parameters have been shown to influence the uptake of Pb. The modifying effect of soil pH is not large and plants growing on calcareous soils may still contain appreciable Pb [133]. Lagerwerff *et al.* [70] grew lucerne and maize on soils to which lead had been added as $PbCl_2$ and lime was used to raise soil pH from 5.2 to 7.2. Plant Pb contents were reduced by 9–21%. Dijkshoorn *et al.* [134] found little change in the uptake of Pb by grass with changing pH but interpretation was made difficult by the very low concentrations of Pb. Davies [135] grew radish in various field locations and used multivariate statistics to interpret the uptake of metals and the interaction of soil parameters. Only weak associations were observed between Pb contents of the hypocotyl and soil pH. Jopony and Young [136] also found uptake of Pb by radish or red fescue to be poorly correlated with soil pH.

Major nutrient elements and other heavy metals have also been observed to affect the plant uptake of Pb. But it is difficult to draw firm conclusions from the literature and there is little convincing evidence that other agricultural practices, such as the use of manures or fertilisers, are particularly effective.

9.7 Lead and soil microorganisms

Some heavy metals can influence microbial activities in soil and thereby reduce the soil productivity. Liang and Tabatabai [137] found that Pb could inhibit N mineralisation rates in an acid soil in Sweden, Pb had no significant influence. Chang and Broadbent [139] reported that Pb did inhibit N mineralisation and nitrification; the order of decreasing inhibition was Cr > Cd > Cu > Zn > Mn > Pb. Cornfield [14] studied the depression of CO_2 release when an acid sandy soil was incubated with additions of different heavy metals. Over a two-week incubation period there was no inhibitory effect by 10 mg Pb/kg soil, and only 14% by 100 mg Pb/kg; over an eight-week period the two levels of Pb gave 6% and 25% depressions in CO_2 evolution. In comparison, the highest rate of Ag over eight weeks caused a 72% reduction in CO_2 release, and the highest rate of Hg caused a 55% repression. Hattori [141] also found Pb to be the least active inhibitor of CO_2 evolution among the heavy metals.

Al-Khafaji and Tabatabai [142] examined the effects of trace elements on arylsulphatase activity in soils, and the average percentage inhibition by Pb was 11% compared with 95% for Ag and 94% for Hg. Khan and Frankland [105] have examined the effects of heavy metals on the decomposition of cellulose in soils. After 30 days 43.6% of added cellulose was decomposed in the control soil and when Pb (in the form of $PbCl_2$) was incorporated in the soil at 100, 500 and 1000 μg/g levels the proportion of cellulose decomposed was 40.0, 37.1 and 33.8% respectively. McNeilly *et al.* [143] studied a Pb–Zn-contaminated pasture in an old metal mining area of N. Wales. They concluded that plant productivity is less sensitive to metal contamination than the decomposition processes which follow its death. The sensitivities to metals of the components of the decomposition could be summarised as productivity < litter accumulation < soil organic matter breakdown < soil humus decomposition.

9.8 Concluding comments

Lead is a widespread soil contaminant in all areas except those remote from populous regions or where human settlement is very recent. It enters the soil from

many sources, and this ubiquity mirrors the widespread use of the metal. Its residence time is so long that it can be regarded as permanent in soils. It is still accumulating in many areas, and it is available for plant uptake. Fortunately, its solubility and mobility, and hence biovailability are low. Nonetheless, in many environments concentrations are sufficiently high as to pose a potential risk to health, especially near major lead-using industries and in major cities.

References

1. Wixson, N.G. and Davies, B.E., *Environ. Sci. Technol.*, **28** (1994), 26A–31A.
2. Nriagu, J.O., *Lead and Lead Poisoning in Antiquity.* John Wiley & Sons, New York (1983).
3. Greninger, D., Kollonitsch, V. and Kline, C.H., *Lead Chemicals.* International Lead Zinc Research Organisation, New York (1974).
4. Nriagu, J.O., in *The Biogeochemistry of Lead.* ed. Nriagu, J.O. Elsevier Biomedical Press, Amsterdam (1978), 18–88.
5. Meyers, P.A., Prattt, L.M. and Nagy, B., *Chem. Geol.* **99** (1992), 7–11.
6. Ure, A.M. and Berrow, M.L., in *Environmental Chemistry, Volume 2.* ed. Bowen, H.J.M., Royal Society of Chemistry, London (1982), 94–204.
7. Davies, B.E., *Geoderma* **29** (1983), 67–75.
8. Reaves, G.A. and Berrow, M.L. *Geoderma* **32** (1984), 1–8.
9. Severson, R.C., Gough, L.P. and Boom, G.U.D., *Water, Air and Soil Pollut.*, **61** (1992), 169–184.
10. Davies, B.E. and Wixson, B.G., *Water, Air and Soil Pollut.*, **33** (1987), 339–348.
11. Dudka, S., *Sci. Total Environ.*, **121** (1992), 39–52.
12. Frank, R., Ishida, K. and Suda, P., *Can. J. Soil Sci.*, **56** (1976), 181–196.
13. Merry, R.J., Tiller, K.G. and Alston, A.M., *Austr. J. Soil Sci.*, **21** (1983), 549–561.
14. Warren, H.V. and Delavault, R.E., *Trans. Roy. Soc. Can.*, **54** (1960), 1–20.
15. Cannon, H.L. and Bowles, J.M., *Science*, **137** (1962), 765–766.
16. Singer, M.J. and Hanson, L., *Soil Sci. Soc. Am. Proc.*, **33** (1969), 152–153.
17. Lagerwerff, J.V. and Specht, A.W., *Environ. Sci. Technol.*, **4** (1970), 583–586.
18. Davies, B.E. and Holmes, P.L., *J. Agric. Sci. (Camb.)*, **79** (1972), 479–484.
19. Quinche, J.P., Zuber, R. and Bovay, E., *Sonderduck aus Phyopathologische Zeitschrift*, **66** (1969), 259–274.
20. Ward, N.I.. and Brooks, R.R., *Environ. Pollut.*, **6** (1974), 149–158.
21. David, D.J. and Williams, C.H., *Austr. J. Exp. Agric. and Anim. Husb.*, **15** (1975). 414–418.
22. Minami, K. and Araki, K., *Soil Sci. Plant Nutr.*, **21** (1975), 185–188.
23. Belal, M. and Saleh, H., *Atmos. Environ.*, **12** (1978), 1561–1562.
24. Garcia-Miragaya, J., *Soil Sci.*, **138** (1984), 147–152.
25. Albasel, N. and Cottenie, A., *Water, Air and Soil Pollut.*, **24** (1985), 103–109.
26. Favretto, L., Marletta, G.P. and Favretto, L., *J. Sci. Food Agric.*, **37** (1986), 481–486.
27. Fytianos, K., Vasilikiotos, G. and Saminidov, V., *Chemosphere* **14** (1985), 271–277.
28. Ho, Y.B., *Sci. Total Environ.*, **93** (1990), 411–418.
29. Gulson, B.L., Tiller, K.G., Mizon, K.J. and Merry, R.M., *Environ. Sci. Technol.*, **15** (1981), 691–696.
30. Chow, T.J., *Nature (London)*, **225** (1969), 295–296.
31. Rabinowitz, M.B. and Wetherill, G.W., *Environ. Sci. Technol.*, **8** (1972), 705–709.
32. Wheeler, G.L. and Rofe, G.L., *Environ. Pollut.*, **18** (1979), 265—274.
33. Smith, W.H., *J. Air Pollut. Control Assoc.*, **26** (1976), 753–766.
34. Nriagu, J.O. and Pacyna, J.F., *Sci. Total Environ.*, **333** (1988), 134–139.
35. Nriago, J.O., *Nature (London)*, **338** (1989), 47–49.
36. Radlein, N. and Heumann, K.G., *Int. J. Environ. Anal. Chem.*, **48** (1992), 127–150.
37. Kral, R., Mejstrik, V. and Velicka, J., *Sci. Total Environ.*, **111** (1992), 125–133.
Steinnes, E., in *Pollutants in Porous Media*, eds. Yaron, B., Dagan, G. and Goldshmid, J.
38. Springer-Verlag, Berlin (1984), 115–122.
39. Elias, R.W. and Davidson, C., *Atmos. Environ.*, **14** (1980), 1427–1432.

40. Lindberg, S.E. and Harris, R.C., *Water, Air and Soil Pollut.*, **16** (1981), 13–31.
41. Sposito, G. and Page, A.L., in *Metal Ions in Biological Systems*, ed. Siegel, H. Marcel Dekker, New York (1984), 287–232.
42. Williams, C., J. *Agric. Sci. (Camb.)*, **82** (1974), 189–192.
43. Jones, K.C., Symina, C.J. and Johnson, A.E., *Sci. Total Environ.*, **61** (1987), 131–144.
44. Nriagu, J.O., *Sci. Total Environ.*, **92** (1990), 13–28.
45. Jones, K.C., *Environ. Pollut.*, **69** (1991), 311–325.
46. Jones, K.C. and Johnston, A.E., *Environ. Sci. Technol.*, **25** (1991), 1174–1178.
47. Jones, K.C., Symon, C., Taylor, P.J.L., Walsh, J. and Johnston, A.E., *Atmos. Environ.* **25A** (1991), 361–369.
48. Sippola, J. and Makela-Kuritito, R., *Int. J. Environ. Anal. Chem.*, **51** (993), 201–203.
49. Van Lune, P., *Neth. J. Agric.*, **35** (1987), 207–210.
50. Purves, D., *Trans. 9th Intern. Cong. Soil Sci.*, **2** (1968), 351–355.
51. Purves, D. and Mackenzie, E.J., *J. Soil Sci.*, (1969), 288–290.
52. Warren, H.V., Delavault, R.E. and Fletcher, K.W., *Can. Min. Metal. Bull.*, July (1977), 1–12.
53. Fleming, G.A. and Parle, P.J., *Irish J. Agric. Res.*, **16** (1977), 35–48.
54. Czarnowska, K., Gworek, B., Janowska, E. and Kozanecka, T., *Polish Ecological Studies*, **9** (1983), 81–95.
55. Preer, J.R., Akintoe, J.O. and Martin, J.L., *Biological Trace Elements*, **6** (1984), 79–91.
56. Tiller, K.G., *Austr. J. Soil Res.*, **30** (1992), 937–957.
57. Davies, B.E., Conway, D. and Holt, S., *J. Agric. Sci. (Camb.).*, **93** (1979), 749–752.
58. Davies, B.E. and Houghton, J.J., *Urban Ecol.*, **8** (1984), 285–294.
59. Fergusson, J.E. and Stewart, C., *Sci. Total Environ.*, **121** (1992), 247–269.
60. Tiller, K.G., Smith, L.H., Merry, R.H. and Clayton, P.M., *Austr. J. Soil Res.*, **25** (1987), 155–166.
61. Davies, B.E., *Sci. Total Environ.*, **9** (1978), 243–262.
62. Culbard, E.B., Thorton, I., Watt, J., Wheatley, M., Moorcroft, S. and Thompson, M., *J. Environ. Qual.*, **17** (1988), 226–234.
63. Griffith, J.J., *J. Agric. Sci. (Camb.)*, **4** (1919), 367–394.
64. Alloway, B.J. and Davies, B.E., *Geoderma*, **5** (1971), 197–208.
65. Colbourn, P. and Thornton, I., *J. Soil Sci.*, **29** (1978), 513–526.
66. Davies, B.E., *Oilos*, **22** (1971), 366–372.
67. Davies, B.E. and Roberts, L.J., *Sci. Total Environ.*, **4** (1975), 249–261.
68. Davies, B.E. and Wixson, B.G., *J. Soil Sci.*, **36** (1985), 551–570.
69. Burkitt, A., Lester, P. and Nickless, G., *Nature (London)*, **238** (1972), 327–328.
70. Lagerwerff, J.V., Brower, D.L. and Biersdorf, G.T., in *Trace Substances in Environmental Health*, ed. Hemphill, D.D., University of Missouri, Columbia, Mo. VI (1973), 71–78.
71. Ragaini, R.C., Falston, H.R. and Roberts, N., *Environ. Sci. Technol.*, **11** (1977), 773–781.
72. Kobayashi, J., in *Trace Substances in Environmental Health*, ed. Hemphill, D.D. University of Missouri, Columbia, Mo., **V** (1972), 117–120.
73. Atkinson, J.J., Giles, G.R. and Desjardins, J.G., *Can. J. Agric. Sci.*, **34** (1954), 76–80.
74. Lustenhouwer, J. and Hin, J., *Sci Total Environ.*, **128** (1993), 269–278.
75. Sommers, L.E., Nelson, D.W. and Yost, K.J., *J. Environ. Qual.*, **5** (1976), 303–306.
76. Sommers, L.E., *J. Environ. Qual.*, **6** (1977), 225–232.
77. de Haan, S., *Phosphorus in Agriculture*, **78** (1980), 33–41.
78. Davies, R.D., *Water Sci. Technol.*, **19** (1987), 1–8.
79. Coker, E.G. and Matthews, P.J., *Water Sci. Technol.*, **15** (1983), 209–225.
80. Le Riche, H.H., *J. Agric. Sci.*, **71** (1968), 205–208.
81. Berrow, J.L. and Webber, J., *J. Sci. Food Agric.*, **23** (1972), 93–100.
82. Alloway, B.J. and Jackson, A.R., *Sci. Total Environ.*, **100** (1991), 151–176.
83. Chater, M. and Williams, R.J.B., *J. Agric. Sci. (Camb.)*, **82** (1974), 193–205.
84. Davies, B.E., Paveley, C.F. and Wixson, B.G., *Soil Use and Man.*, **9** (1993), 47–52.
85. Lounamaa, J., *Ann. Bot. Soc. Vanama*, **29** (1956), 1–196.
86. Wright, J.L., Levick, R. and Atkinson, H.J., *Soil Sci. Soc. Am. Proc.*, **19** (1955), 340–344.
87. Swaine, D.J. and Mitchell, R.L., *J. Soil Sci.*, **11** (1960), 347–368.
88. Archer, F.C., *J. Soil Sci.*, **14** (1963), 144–148.
89. Presant, E.W. and Turner, W.M., *Can. J. Soil Sci.*, **45** (1965), 305–310.
90. Bradley, R.I., Rudeforth, C.C. and Wilkins, C., *J. Soil Sci.*, **29** (1978), 258–270.

91. Berrow, M.L. and Mitchell, R.L., *Trans. Roy. Soc. Edin.*, **71** (180), 103–121.
92. Friedland, A.J., Johnson, A.H. and Siccama, T.G., *Water, Air and Soil Pollut.*, **21** (1984), 161–170.
93. McGrath, S.P., *J. Agric. Sci. (Camb.)*, **103** (1984), 25–35.
94. Ganze, C.W., Wahlstrom, J.S. and Turner, D.C., *Water Sci. Technol.*, **19** (1987), 19–26.
95. Davis, R.D., Carlton-Smith,C.H., Stark, J.H. and Campbell, J.A., *Environ. Pollut.*, **49** (1988), 99–115.
96. Zimdahl, R.L. and Skogerboe, R.K., *Environ. Sci. Technol.*, **11** (1977), 1202–1207.
97. Korte, H.E., Skopp, J., Fuller, W.H., Niebla, E.E. and Alesh, B.A., *Soil Sci.*, **122** (1976), 350–359.
98. Hooghiemstra-Tielbeek, M., Keizer, M.G. and de Haan, F.A.M., *Neth. J. Agric. Sci.*, **31** (1983), 189–199.
99. Santillan-Medrano, J. and Jurinak, J.J., *Soil Sci. Soc. Am. Proc.*, **39** (1975), 851–856.
100. Ter Haar, G.L. and Bayard, M.A., *Nature (London)*, **232** (1971), 553–554.
101. Post, J.E. and Buseck, P.R., *Environ. Sci. Technol.*, **19** (1985), 682–685.
102. Olson, K.W. and Skogerboe, R.K., *Environ. Sci. Technol.*, **9** (1975), 227–230.
103. Harrison, R.M., Laxen, D.P.H. and Wislon, S.J., *Environ. Sci. Technol.*, **15** (1981), 1378–1383.
104. Jorgensen, S.S. and Williams, M., *Ambio*, **16** (1987), 11–15.
105. Khan, D.E. and Frankland, B., *Environ. Pollut. (A)*, **33** (1984), 63–74.
106. Berrow, J.L. and Mitchell, R.L., *Trans. Roy. Soc. Edin.*, **82** (991), 195–209.
107. Hildebrand, E.E. and Blum, W.E., *Zeitschrift für Pfanzen und Bodenkunde*, **3** (1975), 279–294.
108. Harter, R.D., *Soil Sci. Soc. Am. J.*, **43** (1879), 679–683.
109. Kabata-Pendias, A., *Roczniki Gleboznawcze*, **43** (1972), 3–14.
110. Bradford, G.R., Page, A.L., Lund, L.J. and Olmstead, W., *J. Environ. Qual.*, **4** (1975), 123–127.
111. Gregson, S.K. and Alloway, B.J., *J. Soil Sci.*, **35** (1984), 55–61.
112. Tills, A.R. and Alloway, B.J., *Environ. Technol. Lett.*, **4** (1983), 529–534.
113. Karamanos, R.E., Bettany, J.R. and Stewart, J.W.B., *Can. J. Soil Sci.*, **56** (1976), 485–496.
114. Hemkes, O.J., Kemp, A. and Van Broekhoven, L.W., *Neth. J. Agric. Sci.*, **31** (1983), 227–232.
115. Korcak, R.F. and Fanning, D.S., *Soil Sci.*, **140** (1985), 23–24.
116. Jones, L.H.P., Clement, C.R. and Hopper, M.J., *Plant and Soil*, **38** (1973), 403–414.
117. Lane, S.D. and Martin, E.S., *New Phytol.*, **79** (1977), 281–286.
118. Koeppe, D.E., *Sci. Total Environ.*, **7** (1977), 197–206.
119. Dollard, G.J., *Environ. Pollut. (B)*, **40** (1986), 109–119.
120. Mitchell, R.L. and Reith, J.W.S., *J. Sci. Food Agric.*, **17** (1966), 437–440.
121. Baker, A.J.M., *New Phytol.*, **106** (1987), 93–111.
122. Simon, S.L. and Ibrahim, S.A., *J. Environ. Radioactivity*, **5** (1987), 123–142.
123. Oxbrow, A. and Moffat, *J., Plant and Soil*, **52** (1979), 127–130.
124. Atkins, D.P., Trueman, I.C., Clarke, C.B. and Bradshaw, A.D., *Environ. Pollut. (A)*, **27** (1982), 233–241.
125. Wong, M.H., *Environ. Res.*, **29** (1982), 42–47.
126. Reeves, R.D. and Brooks, R.R., *Environ. Pollut. (B)*, **31** (1983), 277–285.
127. Baker, A.J.M. and Brooks, R.R., *Biorecovery*, **1** (1989), 81–126.
128. Tjell, J.C., Hovmand, E.M.F. and Mosbaek, H., *Nature (London)*, **280** (1979), 25–26.
129. Crump, D.E. and Barlow, P.J., *Sci. Total Environ.* **15** (1980), 269–274.
130. Dalenberg, J.W. and Van Driel, W., *Neth. J. Agric. Sci.*, **38** (1990), 369–379.
131. Chamberlain, A.C., *Atmos. Environ.*, **17** (1983), 693–706.
132. Solgaard, P., Aarkrog, A., Flyger, H., Fenger, J. and Graabaek, A.M., *Nature (London)*, **272** (178), 346–347.
133. Thornton, I. and Webb, J.S., in *Trace Substances in Environmental Health*, ed. Hemphill, D.D. University of Missouri, Columbia, Mo, **9** (1976), 14–25.
134. Dijkshoorn, W., Lampe, J.E.M. and Van Broekhoven, L.W., *Neth. J. Agric. Sci.*, **31** (1983), 181–188.
135. Davies, B.E., *Water, Air Soil Pollut.*, **63** (1992), 331–342.
136. Jopony, M. and Young, S., *Plant and Soil*, **151** (193), 273–278.

137. Liang, C.N. and Tabatabai, M.A., *Environ. Pollut.*, **12** (1977), 141–147.
138. Tyler, G., *Nature (London)*, **255** (1975), 701–702.
139. Chang, F.H. and Broadbent, F.E., *J. Environ. Qual.*, **11** (1982), 115–119.
140. Cornfield, A.H., *Geoderma*, **19** (1977), 199–203.
141. Hattori, H., *Soil Sci. Plant Nutr.*, **38** (1992), 93–100.
142. Al-Khafaji, A.A. and Tabatabai, M.A., *Soil Sci.*, **127** (1979), 129–133.
143. McNeilly, T., Williams, S.T. and Christian, P.J., *Sci. Total Environ.*, **38** (1984), 183–198.

10 Manganese and cobalt

K.A. SMITH AND J.E. PATERSON

10.1 Introduction

Manganese and cobalt are elements which are important to our industrial civilisation, and also to the maintenance of life within it. Manganese is mined in large quantities, and finds important applications in products such as engineering steels and electrical storage batteries. Cobalt, although much less abundant (and correspondingly more expensive) is also used for the manufacture of special steels, and has been used for centuries in the manufacture of blue pigments and glass. Both elements have essential roles in living organisms: Mn in microorganisms and higher plants, Co in some microorganisms, and both in animals. Environmental pollution problems are relatively insignificant compared with those associated with some other heavy metals; the most common biological toxicity effects are found in plants exposed to excess natural Mn in the soil, particularly in flooded conditions.

Scientific interest in Mn in soils is primarily related to its role in plant and animal systems. The fact that Mn was an essential element was first recognised as long ago as 1863, apparently, when Raulin demonstrated that it was needed for the growth of the fungus *Rhizopus* (*Ascophora*) *nigricans* [cited in 1]. The role of the metal as a micronutrient in higher plants was first established by McHargue [2] in his work with oats, soya bean and tomato, and later confirmed by Samuel and Piper [3]. Shortly after, Mn was shown to be essential in mammalian systems also [4], and later it became known that Mn(II) activates enzymes and is involved in the synthesis of glycoproteins [5] and is present in metalloenzymes such as arginase [6] and pyruvate carboxylase [7]. Manganese is involved in fatty acid synthesis [8] and in bone development in poultry [9]. There are a number of important functions for which Mn is essential in plants. For example, it is present in the NAD malic enzyme system found in the leaves of C_4 plants [10]. It also appears to be a specific constituent of the photosynthetic oxygen-evolving system in chloroplasts [11].

At present Mn is the trace element most commonly deficient in cereal crops in the UK [12], and soya bean crops in the USA have been found to be notably sensitive to Mn deficiency [13]. The problem may assume considerable economic importance in the farming areas where it is prevalent; these are located mainly in cool temperate regions, on soils of high pH, high organic matter and carbonate contents, and low content of readily reducible Mn [14].

The primary interest in Co as a constituent of soils lies in its essential roles in ruminant animals and microorganisms, and because of deficiencies rather than

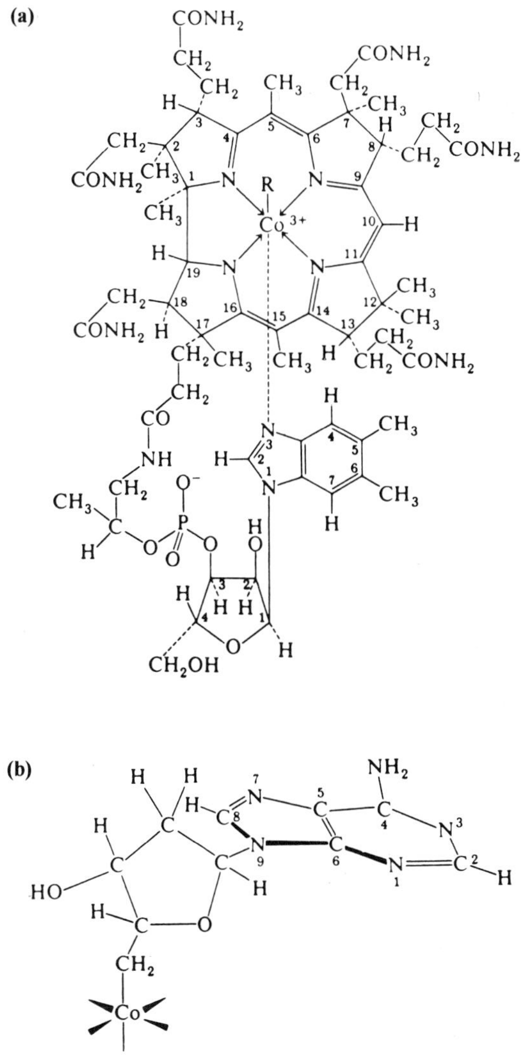

(a)

(b)

Figure 10.1 Structure of vitamin B_{12} coenzyme (Purcell and Kotz [21]). R in structure (a) is 5′-deoxyadenosine, shown in structure (b).

excesses. For centuries, farmers in different parts of the world had found some pastures to be unsuitable for grazing sheep and cattle. Even on apparently rich pastures, these animals lost appetite, became weak and emaciated, suffered severe anaemia and eventually died. In the 1930s the symptoms were attributed to low Co concentrations in the herbage, and both top dressing of the pastures with Co salts and oral administration of Co were used to alleviate the condition [15–17].

In 1948 an 'anti-pernicious–anaemia factor' containing 4% Co was isolated from liver [18,19]. This substance became known as vitamin B_{12}, and was shown to cure pine in lambs [20]. This work proved conclusively the importance of Co deficiency in ruminants, and showed that the resulting disorders were, in fact, due to a deficiency of vitamin B_{12} (or, more precisely, of the closely related molecule vitamin B_{12} coenzyme) [21]. Vitamin B_{12} and its coenzyme (Figure 10.1) were subsequently found to be complex molecules containing Co(III) at the centre of a tetrapyrrole macrocyclic structure, which are synthesised by microorganisms in the rumen.

Interest in Co in plant biology stems particularly from its essential role in biological N fixation. A requirement for the element was demonstrated both for *Rhizobium* bacteria forming symbiotic associations with legume roots [22, 23] and for free-living N-fixing bacteria, e.g *Azotobacter* spp. [24]. Cobalt appears to be essential also for blue–green algae [25]. Although to date it has not been conclusively demonstrated that there is a comparable essential requirement by higher plants, evidence has been obtained of growth benefits resulting from an enhanced supply of the element, e.g. in increased cereal yields [26].

10.2 Geochemical occurrence

All the rocks of the Earth's crust contain Mn in concentrations which are generally much higher than those of other micronutrients, apart from Fe. The highest concentrations (up to several thousand mg/kg) are found in basic igneous rocks such as basalt and gabbro. This is because Mn is mainly present as a substitute for Fe^{2+} in octahedral sites in the structures of ferromagnesian silicate minerals [27]. Contents vary widely in acid igneous rocks (granite, rhyolite, etc) and metamorphic rocks such as schists, but are generally in the range 200–1000 mg/kg. In sedimentary rocks, the concentration range in limestones is about 400–600 mg/kg, but much lower contents are commonly found in sandstones (20–500 mg/kg) [27–29].

Manganese can exist in all oxidation states from Mn(II) to Mn(VII), of which the (II), (III) and (IV) states occur in minerals in combination with O, carbonate and silica. Examples are the ore minerals pyrolusite (MnO_2), rhodochrosite ($MnCO_3$) and rhodonite ($MnSiO_3$) [30], and many other oxides and oxyhydroxides (Table 10.1), in which substitution of Mn(II) and Mn(III) for Mn(IV) occurs extensively [27, 31]. The Mn ions in these oxides may be oxidised or reduced without changing position, and when the valency of a sufficient number has been changed the structure becomes mechanically unstable, and rearranges to that of a new phase [32, 33].

A number of the oxides of Mn are known by more than one name. Synthetic forms, although they may be identical to those occurring in nature, have usually been given different names (Table 10.1). Birnessite refers to a range of different forms of MnO_2, some of the synthetic forms of which have been called δ-MnO_2, or manganous manganite. It occurs widely in mineral deposits, and is

Table 10.1 Manganese oxides and hydroxides occurring in soils. After Gikes and McKenzie [27]

Mineral	Formula	Occurrence
Pyrolusite	MnO_2	Rare
Ramsdellite	MnO_2	Rare
Nsutite	$(Mn^{2+}, Mn^{3+}, Mn^{4+})(O, OH)_2$	Rare
Hollandite	$Ba_2Mn_8O_{16}$	Medium
Cryptomelane	$K_2Mn_8O_{16}$	Medium
Coronadite	$Pb_2Mn_8O_{16}$	Medium
Romanechite	$(Ba, K, Mn, Ca)_2Mn_5O_{10}$	Medium
Todorokite	$(Na, Ca, K, Ba, Mn^{2+})_2Mn_5O_{12}.3H_2O$	Medium
Birnessite	$(Na_{0.7}, Ca_{0.3})\ Mn_7O_{14}.2.8H_2O$	Common
Vernadite	$MnO_2(?)$	Common
Ranceite	$(Ca, Mn)Mn_4O_9.3H_2O$	Medium
Buserite	$Na_4Mn_{14}O_{27}.9H_2O$	Rare
Lithiophorite	$(Al, Li)MnO_2(OH)_2$	Medium
Manganite	$MN^{3+}O\ OH$	Rare
Hausmannite	$Mn^{2+}Mn_2^{3+}O_4$	Rare

one of the two most common forms of mineralised Mn in soils, the other being vernadite [27].

Cobalt ore minerals include cobaltite (CoAsS–FeAsS) and skutterudite ($CoAs_3$–$NiAs_3$). Apart from these deposits, it is most abundant in relatively unstable ferromagnesian minerals such as olivine, pyroxenes, amphibole and biotite, which are concentrated in basic and ultrabasic igneous rocks [28, 29] (Figure 10.2). These minerals contain as their principal cations Mg^{2+} (ionic radius 7.8 nm) and Fe^{2+} (ionic radius 8.3 nm), and ions such as Co^{2+}, with an ionic radius of 8.2 nm, are incorporated into the crystal lattice by isomorphous substitution. In contrast, silica-rich acidic igneous rocks such as granite, which do not contain ferromagnesian minerals, are low in Co [28].

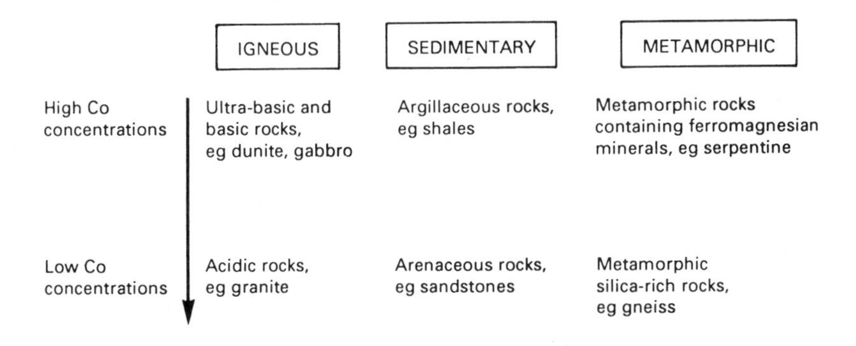

Figure 10.2 General relationships between rock types and total Co content (after J.E. Paterson, Ph.D. thesis, University of Glasgow, 1988).

Ultrabasic rocks such as dunite and peridotite, and the product of their metamorphism, serpentinite, contain 100–200 mg/kg of Co. Basic rocks such as basalts and andesite contain 30–45 mg/kg, whereas acid or neutral types such as granite and rhyolite contain only 5–10 mg/kg [29].

The Co content of sedimentary rocks reflects the composition of the material from which they were originally derived. Thus shales formed from the easily weathered ultrabasic and basic rocks are relatively rich in Co: 10–50 mg/kg [28], while sandstones originating from silica-rich acidic rocks have much lower concentrations.

10.3 Origin of manganese and cobalt in soils

Soils derive all their Mn content from the parent materials, and the concentrations found in mineral soils reflect the composition of these parent materials (Table 10.2). In soils Mn commonly occurs as the oxide minerals birnessite and vernadite [27]. Many other oxide forms exist, however (Table 10.1), and a further seven of these have been reported in soils from several localities [27]. Non-stoichiometric oxides of mixed valency are the normal outcome of oxidataion of Mn(II) or reduction of Mn(IV) [36]. Dubois [37] collated references to 150 materials which had a composition between $Mn_{1.2}$ and $Mn_{2.0}$. The various oxides of Mn show a strong tendency to form mixed crystals with other transition metals, including Co.

Birnessite is the first oxidation product of weathered carbonate rocks [36], and occurs in concretions as well as in more finely divided material. Lithiophorite is a double-layer manganite containing substantial amounts of Li and Al (Table 10.1). In well-aerated soils, in neutral, or slightly alkaline conditions, the Mn(IV) oxidation state is the most stable, and the composition of the manganese oxides ranges from Mn_3O_4 to MnO_2, but in reducing conditions the Mn(II) state is favoured [38].

The manganese oxides vary in their solubility and ease of reduction, and this may have a relationship with their relative capacities to supply Mn to plant roots. The results of individual experimental studies demonstrate this variability, but conflict with each other in their conclusions as to which oxide is the most effective source. For example, Jones and Leeper [39] found that oats utilised Mn from pyrolusite and manganite but not from hausmannite, whereas Page [40] reported that the availability of Mn from these three compounds was in the order pyrolusite < manganite < hausmannite.

Apart from the natural mineralogical sources, the only other significant source of Mn in soils is the result of application of the element to deficient crops. Application to the soil is normally in the form of $MnSO_4$, MnO, or as an addition to macronutrient fertilisers. This topic has been reviewed by Walter [41]. Quantities applied may range from < 10 to > 100 kg Mn/ha. Those made directly to plant leaves (a proportion of which inevitably enters the soil) are at the lower end of this range, and represent only a very small addition to the total quantity present.

The only significant sources of Co in soils are (i) the parent materials from which the soils are derived, and (ii) the deliberate applications of Co salts or Co-

Table 10.2 Manganese and Co contents of some Scottish and Welsh soils formed on different parent materials. Adapted from Aubert and Pinta [29] (original data of Swaine and Mitchell [34] and Archer [35])

Location	Parent rock and parent rock content (ppm)	Soil type	Soil classification 7th approximation (USDA)	Total Mn content in soil (mg/kg)	Total Co content in soil (mg/kg)
Scotland	Serpentine till	Brown podzolic soil, freely drained	Haplorthod	1500	40–200
	Olivine gabbro till	Brown podzolic soil with gleyed B and C horizons imperfectly drained	Haplorthod	7000	40
	Andesitic moraine	Brown forest soil freely drained	Ochrept	1000	10–20
	Granitic till	Peaty gleyed podzol with iron pan poorly drained	Ferrod or sideraquod	50	< 1–3
	Granitic gneiss till	Podzol freely drained	Spodosol	1000	10–20
	Quartz mica schists till	Podzol freely drained	Spodosol	3000	25–40
	Silurian slate till	Non-calcareous gley soil poorly drained	Haplaquept	300	10–30
	Sandstone till	Peaty podzol with iron pan freely drained	Placaquod	80	< 3–4
Wales	Rhyolite			1000–1200	3
	Mixed drifts	Soils on these parent rocks		3000	30
	Dolerite			2500–3000	20–30
	Pumice tuff			1000–11,500	20–25
	Mixed drifts			2500–3000	8–15

treated phosphate fertilisers to topsoils to overcome deficiencies which are causing problems with ruminant nutrition or the cultivation of legumes.

Total Co contents of soils vary widely: from 0.05 to 300 mg/kg, with an average content in the range 10–15 mg/kg [29]. The contents vary mainly in relation to the parent materials from which they were derived, even though there are also differences with depth in the soil profile and between different soil types derived from a common parent material due to pedological processes. Thus, for example, in Scottish soils formed on serpentine, andesite and granite the respective Co concentrations were 40–200, 10–20 and < 1–3 mg/kg (Table 10.2); and under very contrasting tropical conditions, in the Central African Republic, comparable trends have also been found; the concentrations in grey ferruginous soils were 20–100 mg/kg in soils formed on amphibolite, 39–60 mg/kg in soils formed on migmatite, and < 4 mg/kg in those formed on granite [29].

Within a given soil profile, Co is generally concentrated in those horizons rich in organic materials and clays. In podzols Co accumulates in the illuvial B horizon, whereas the eluvial Ea horizon is generally depleted. In contrast, in chernozems and vertisols the distribution of Co is relatively uniform throughout the profile [29]. The oxide, hydroxide and carbonate of Co are all very insoluble, thus in alkaline conditions the element is immobile. In acid conditions, on the other hand, dissolution and leaching are more likely to occur. This results in generally greater concentrations of total Co in alkaline than in acid soils.

10.4 Manganese and cobalt chemistry affecting plant availability

Uptake of Mn and Co by plants is a function of the concentration of these elements in ionic form in the soil solution and the concentration present on the exchange sites of the cation exchange complex, i.e. the 'available' or labile pool. For Mn and Co, as for other nutrients, much effort has been applied to solving the problem of estimating the size of this labile pool, in order to predict the likely uptake of these elements by plants from a given soil. Plants take up Mn as Mn(II), and the factors that affect the capacity of a soil to provide an adequate supply of Mn for plant nutrition are those that have the greatest influence on the reduction of Mn from its higher oxidation states to the more labile divalent form. Cobalt is also taken up as the divalent cation, the only oxidation stated commonly found in soil minerals. However, the factors which are important in determining the availability of Mn also apply to Co, since the latter is generally found in association with the Mn oxide minerals.

One concept of the different forms in which Mn occurs in soils is shown in Figure 10.3 and Figure 10.4 illustrates the distribution of Mn between various

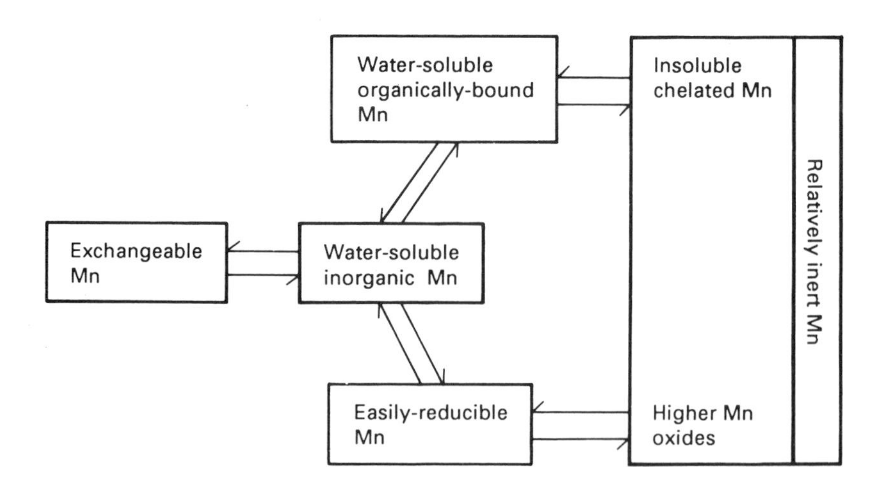

Figure 10.3 Inter-relationships between different forms of Mn in soils (from Ghanem *et al.* [42])

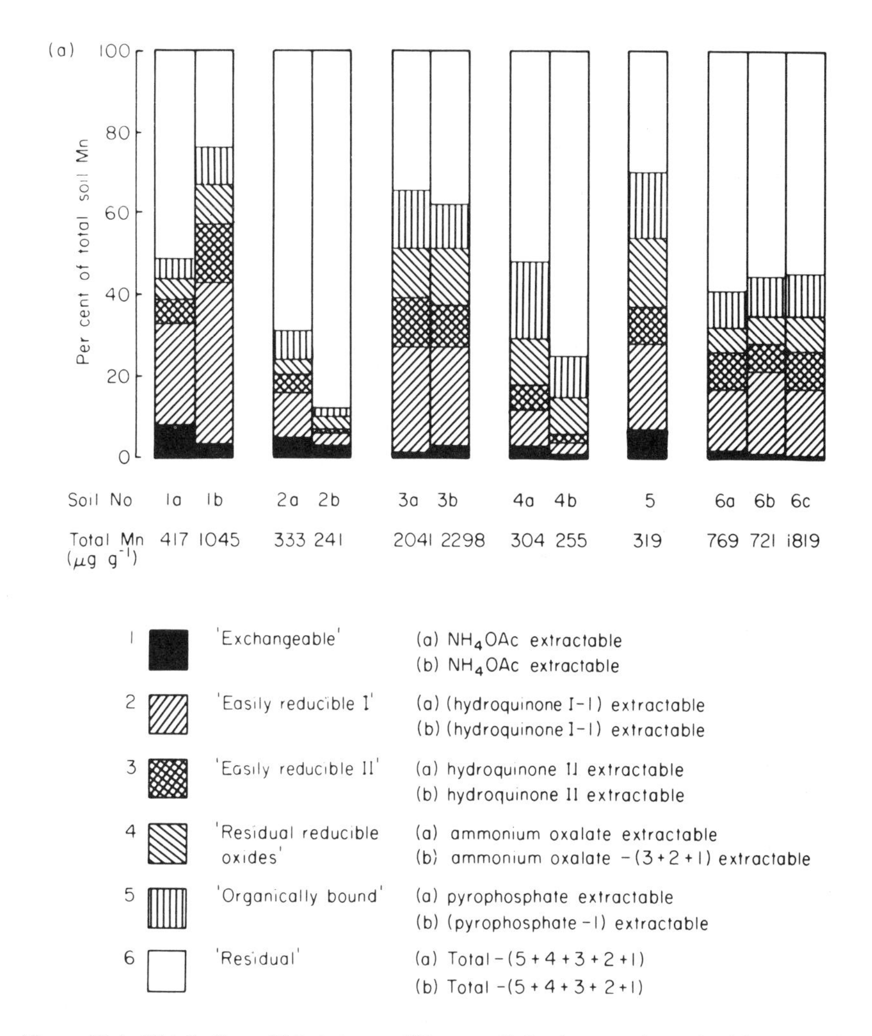

Figure 10.4 Distribution of Mn between different soil fractions, as determined by successive extractions (from Jarvis [38])

fractions in an acidic soil [38]. The availability of Mn is largely governed by the supply of H^+ ions and electrons, which reduce the higher valency states of Mn^{2+}. To take one example, the release of Mn^{2+} from pyrolusite occurs by the reaction [43]

$$MnO_2 + 4H^+ + 2e^- = Mn^{2+} + 2H_2O$$

In general, the relationship between pMn^{2+} and pH in well-aerated soils is likely to reflect the varying contribution to the reduction from complex formation as well as those from H^+ ions and electrons. However, the empirical straight-line

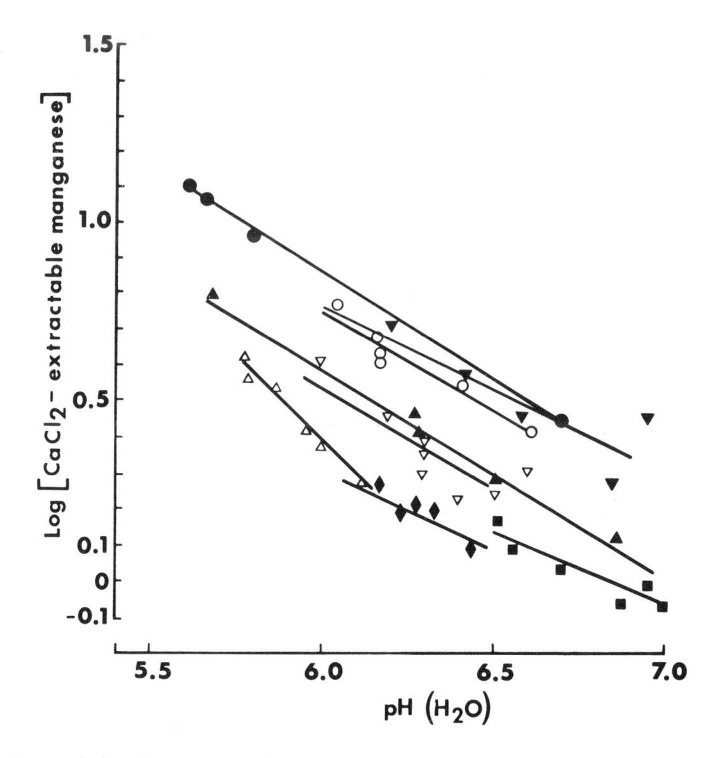

Figure 10.5 Relationship between soil pH and $CaCl_2$-extractable Mn in soils from eight farms in S.E. Scotland from ([44])

plots of $CaCl_2$-extractable Mn against pH that can be obtained from field studies (Figure 10.5) are likely to prove useful for predicting the likely effects of pH on Mn availability, at least within a limited range of soils.

In flooded soils, the reduction of higher valency states of Mn is both chemical and biochemical [45]. In the thermodynamic sequence of soil reduction reactions (Table 10.3), the reduction of MnO_2 follows that of NO_3^- and precedes that of Fe (III). According to Ponnamperuma [45], acid soil high in Mn and organic matter may have the concentration of water-soluble Mn^{2+} increase to as much as 90 mg/l, at temperatures of 25–35°C, within 1–2 weeks after flooding (Figure 10.6). Alkaline soils and soils low in Mn rarely contain more than 10 mg/l of water-soluble Mn at any stage of submergence [45]. When high concentrations of Mn^{2+} are released, toxic effects on plants are often observed. Plant contents may reach > 300 mg/kg dry matter in such situations.

The decrease in water-soluble Mn, after the initial rise following flooding, is the result of precipitation as $MnCO_3$ [45]. The activity of Mn^{2+} after the peak is given by

$$pH + ½log Mn^{2+} + ½log P_{CO_2} = 4.4$$

Table 10.3 Themodynamic sequence of soil reduction. From Ponnamperuma [45]

System	E_0^7*	pE_0^7†
$O_2 + 4H^+ + 4e^- = 2H_2O$	0.814	13.80
$2NO_3 + 12H^+ + 10e^- = N_2 + 6H_2O$	0.741	12.66
$MnO_2 + 4H^+ + 2e^- = Mn^{2+} + 2H_2O$	0.401	6.80
$CH_3COCOOH + 2H^+ + 2e^- = CH_3CHOHCOOH$	−0.158	−2.67
$Fe(OH)_3 + 3H^+ + e^- = Fe^{2+} + 3H_2O$	−0.185	−3.13
$SO_4^{2-} + 10H^+ + 8e^- = H_2S + 4H_2O$	−0.214	−3.63
$CO_2 + 8H^+ + 8e^- = CH_4 + 2H_2O$	−0.244	−4.14
$N_2 + 8H^+ + 6e^- = 2NH_4^+$	−0.278	−4.69
$NADP^+ + H^+ + 2e^- = NADPH$	−0.317	−5.29
$NAD^+ + H^+ + 2e- = NADH$	−0.329	−5.58
$2H^+ + 2e^- = H_2$	−0.413	−7.00
Ferredoxin (Fe^{3+}) + e^- = Ferrodoxin (Fe^{2+})	−0.431	−7.31

*E_0 corrected to pH 7.0
†pE_0 corrected to pH 7.0

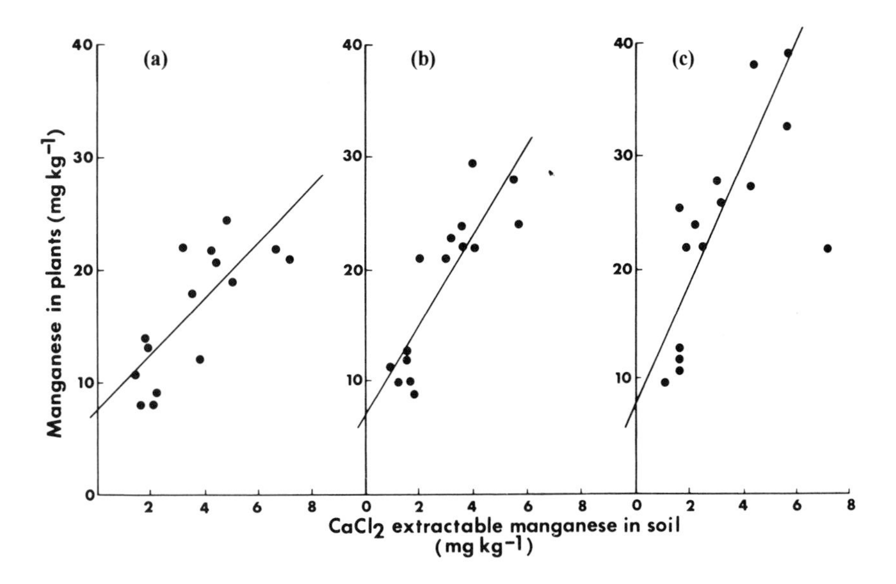

Figure 10.6 Kinetics of water-soluble Mn in five submerged soils (from [45])

Although reactions occurring in the bulk soil are important, the changes in manganese solubility taking place within the rhizosphere may have a greater effect in determining the availability of Mn to the plant roots [46]. Rhizosphere pH can differ by up to two units from the rest of the soil [47]. For example, when ammonium N fertiliser is used the rhizosphere becomes more acid than the bulk soil due to

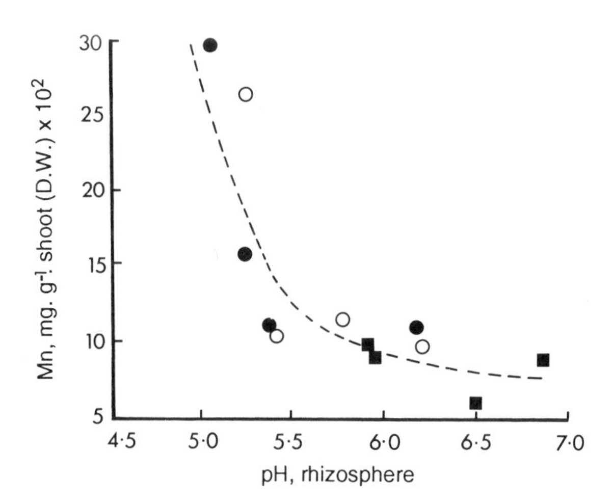

Figure 10.7 Effect of rhizosphere pH (as obtained by using the following sources of nitrogen: choline dihydrogen phosphate ●, ammonium dihydrogen phosphate ○, calcium nitrate ■) on shoot Mn levels. (From Sarker and Wyn Jones [51]).

(i) the increased excretion of H^+ ions by the roots as a result of greater NH_4^+ uptake [48, 49], and/or (ii) the microbial transformations of ammonium to nitrate [48]. Godo and Reisenauer [50] found considerably greater amounts of Mn extractable with 0.01 M $CaCl_2$ in soil from the rhizosphere of wheat roots than in the bulk soil, and also found that the rate of increase with decreasing pH went up considerably below about pH 5.5. This sharp increase at pH 5.5 was also observed by Sarker and Wyn Jones [51] for French beans and is illustrated in Figure 10.7. However, Linehan *et al.* [52, 53] measuring the fluctuation over the growing season of Mn (and also Cu and Zn) in the soil solution of a soil cropped with barley found a large peak in Mn concentrations during the summer months, accompanied by a slight increase in pH (Figure 10.8). Linehan *et al.* suggested that low-molecular weight organic compounds produced by roots during this time of greatest growth may act as complexing agents increasing Mn availability. This effect would appear to override the effect of the increased pH.

Bromfield [54, 55] produced the first experimental evidence that MnO_2 can be solubilised by root exudates. Plants such as white lupins develop rhizospheres which not only lower pH but also contain more reductants and chelating agents than the bulk soil. Leachates taken from white lupin rhizospheres have been found to dissolve ten times more MnO_2 than leachates from bulk soil [56, 57].

Roots exudates is a broad term covering a range of organic carbon compound produced or released from plant roots and include low molecular weight organic solutes, mucilage and sloughed-off cells and tissues [46]. The low molecular weight forms are the principal forms involved in complexing Mn. They are composed largely of organic acids, and one of these, malic acid, has been shown to be

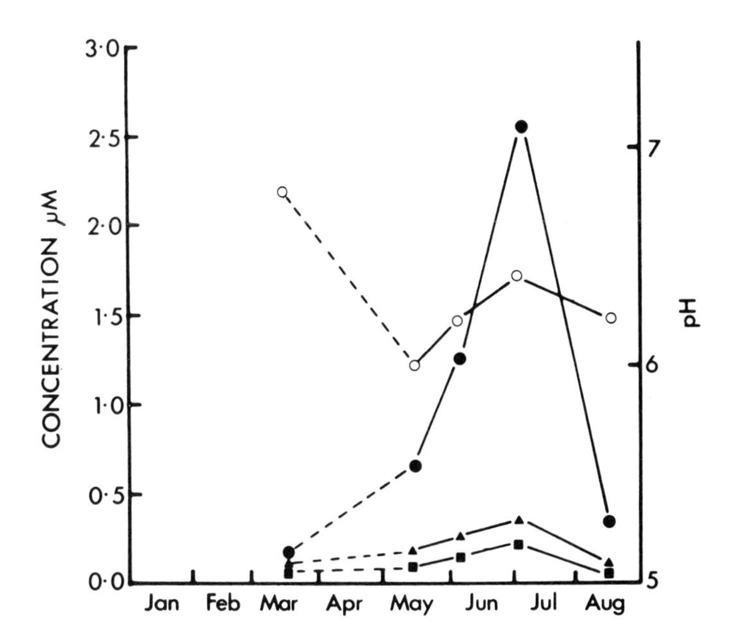

Figure 10.8 Changes in Mn, Cu and Zn concentrations in soil solutions and in pH, with time. In March, solutions isolated from soil prior to seed sowing, subsequently solution isolated from rhizosphere soils. ○, pH; ●, Mn; ■, Cu; ▲, Zn (from [52])

particularly effective in solubilising MnO_2 [47, 58]. Phenolic acids, such as sinapic acid and ferulic acid, although present in much smaller concentrations, are resistant to microbial degradation and have been found to be very effective in MnO_2 reduction [59].

Since the rhizosphere contains such an abundant supply of organic carbon, the microbial population tends to be higher (up to 50 times) than in the bulk soil [46]. Such increased microbial activity will affect Mn availability but whether this is enhanced or decreased will depend on, for example, the relative activity of Mn-oxidising or reducing bacteria and the changes in E_h occurring due to microbial respiration [46]. Mn-oxidising bacteria are not very sensitive to acidity but are particularly active in the pH range 6 to 7.5; Bromfield [60] found the Mn concentrations of oat plants were lower for those plants treated with the Mn-oxidising bacteria *Arthrobacter* spp than those left uninoculated, for treatments where the pH was above 6.0. However, in a study in which sodium azide was used to inhibit microbial activity in an acid soil, it was found that the organisms responsible for MnO_2 oxidation were well adapted to acid conditions and the rate of oxidation fell as the pH was increased by liming from 5.0 to 6.5 [61].

Mn toxicity symptoms have been found in a wide range of crops, including soya bean, cotton, tobacco and upland rice grown in soils with high available Mn. Reported toxic concentrations for these and other species range from 80 to 5000 mg/kg [62]. The toxicity is commonly associated with acidic soils and warm

Table 10.4 Relationship between soil drainage status, extractable Co and plant uptake. From Mitchell *et al.* [63]

Drainage status	Acetic acid-extractable Co (mg/kg)	Co content of mixed herbage (mg/kg)
Freely drained	1.0	0.12
Poorly drained	2.7	0.86

climates. Water-soluble soil Mn appears to be a better guide to the likely occurrence of toxicity than the amounts of exchangeable or reducible Mn, but actual values appear to be applicable only to local circumstance [62].

Uptake of Co by plants is a function of the concentration of Co in the soil solution and on the exchange sites of the cation exchange complex i.e. the 'available' or labile pool. Soil drainage status has a major influence on the amount of Co available for plant uptake. In poorly drained soils the amount of extractable Co is generally greater than in adjoining areas which are well drained [63–66], and plant uptake is significantly increased (Table 10.4). Mitchell *et al.* [63] attributed this effect to differences in the rates of weathering of ferromagnesian minerals and differences in types of clay mineral and organic complexes formed under the different drainage conditions.

Next to drainage, soil pH is the most important soil factor determining the availability of Co [67]. Evidence has been obtained that the uptake of Co increases as the soil pH decreases [68–71]. Mitchell [72] found that increasing the pH from 5.4 to 6.4 almost halved herbage Co concentrations while other workers found that increasing the pH above 6.0 had little further effect [71, 73]. Various suggestions have been put forward to explain this phenomenon, including changes in the forms of Co present at different pH values [69], stronger and irreversible exchange reactions on both clay and sesquoxide surfaces [74] and the precipitation of various cobalt salts [75].

Cobalt can be adsorbed onto Mn minerals such as the dioxides, and then become more strongly bound with time as it is oxidised and replaces Mn in the crystal structure [27, 76]. The point of zero charge (PZC) of some synthetic manganese dioxides has been shown to vary from pH 1.5 for birnessite to 7.3 for pyrolusite [27, 31, 77], increasing with the degree of crystallinity. For synthetic oxides of the types occuring in soils, the range of PZC is from pH 1.5 to 4.6 [31]. Thus birnessite would be expected to be negatively charged in all soils, and most others also to be negatively charged in all but the most acid soils. These materials provide effective sites for adsorption of other cations, therefore, particularly for Fe, Co, Ni, Cu, Pb and Zn [78–80]. The adsorption is strongly pH-dependent (Figure 10.9).

Taylor and McKenzie [31, 81, 82] have studied the association of other trace elements with the manganese dioxides in a number of Australian soils. The oxide minerals were extracted with acidified H_2O_2, and the ratio of each trace element

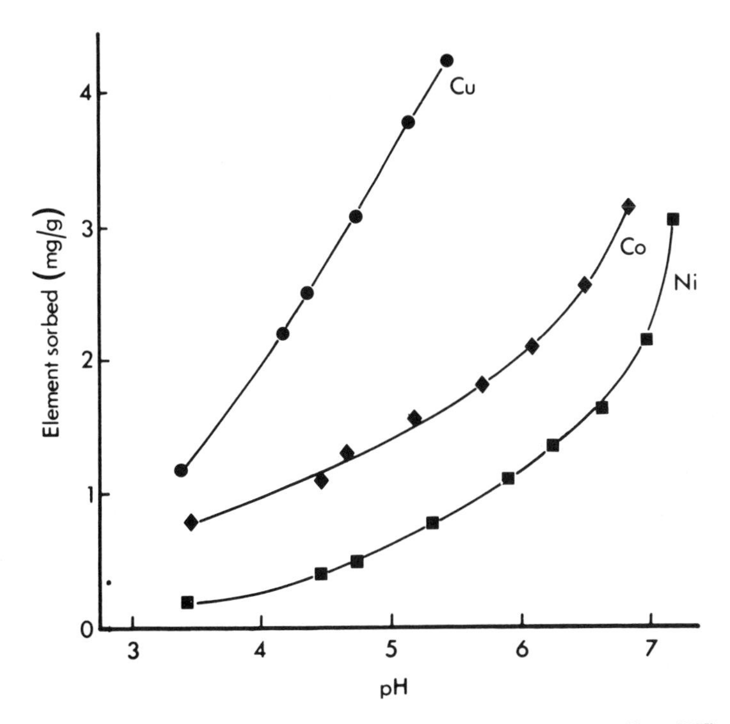

Figure 10.9 Effect of pH on the sorption of Co, Cu and Ni by MnO_2 (from [78])

extracted to the total present in the soil was plotted against the corresponding ratio for Mn. These plots are shown in Figure 10.10, which demonstrates a remarkable association of Co with the Mn. It was concluded that most of the Co (79%, on average) in the soils was contained in or associated with Mn in mineral form. Taylor [83] showed that the bulk of the Co in soils from Europe, Bermuda and the Middle East was also associated with Mn.

Other studies [84, 85] have indicated the importance, from the point of view of Co availability for plant uptake of this association with soil Mn minerals. Adams *et al.* [84] found that the uptake by clover of Co applied to some Australian soils in pot and field experiments was inversely related to the total Mn content of the soil. On moderately acid soils with a low Mn content (*ca.* 100 mg/kg), the basal dressing of 1.1 kg/ha of $CoSO_4.7H_2O$ appeared to be adequate; on soils with Mn contents in the range 100–1000 mg/kg Co deficiency could be alleviated by dressing with Co, although higher applications were necessary; but on soils with > 1000 mg/kg of Mn the use of Co as an amendment was not a practical solution. The experiments showed that the fixation of Co by the manganese dioxides was relatively rapid, and once fixed the Co was unavailable for plant uptake.

According to Jarvis [86], the younger soils of the temperate regions show more variation in the forms in which Mn occurs than the soils of Australia, because of

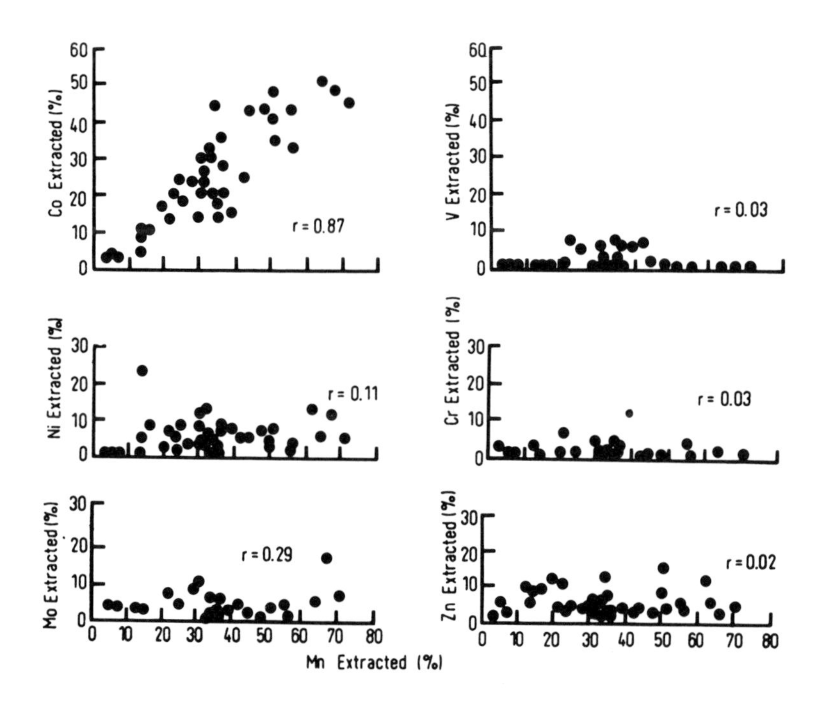

Figure 10.10 Association of Co (in comparison with five other trace elements) with Mn oxide minerals in soil. Soils extracted with acidified hydrogen peroxide, and ratio of element extracted to total content plotted against corresponding data for Mn (from [31])

fluctuations in redox potential and possibly of pH. This author investigated therelationship between Co and the most reactive form of oxidised Mn, i.e. the easily reducible Mn as extracted by hydroquinone, in some acidic grassland soils from England. He found that there was a general tendency for total Co to increase with Mn, particularly with the easily reducible fraction. The relationship was described by the equation

$$y = 6.23 + 0.01x \ (r = 0.759)$$

where y is total Co and x is easily reducible Mn in mg/kg.

By using ^{58}Co tracer methods McLaren *et al.* [87] were able to investigate the sorption of Co by individual components from solutions containing Co concentrations within the range found in natural soil solutions. Soil-derived oxides sorbed by far the greatest amounts of Co, although substantial amounts were also sorbed by organic materials. Clay minerals and non-pedogenic Fe and Mn oxides sorbed relatively little Co. The Co sorbed by soil oxides was not readily desorbed, and rapidly became non-isotopically exchangeable with solution Co, whereas the Co sorbed by humic acid was relatively easily desorbed, and a large proportion remained isotopically exchangeable (Figure 10.11).

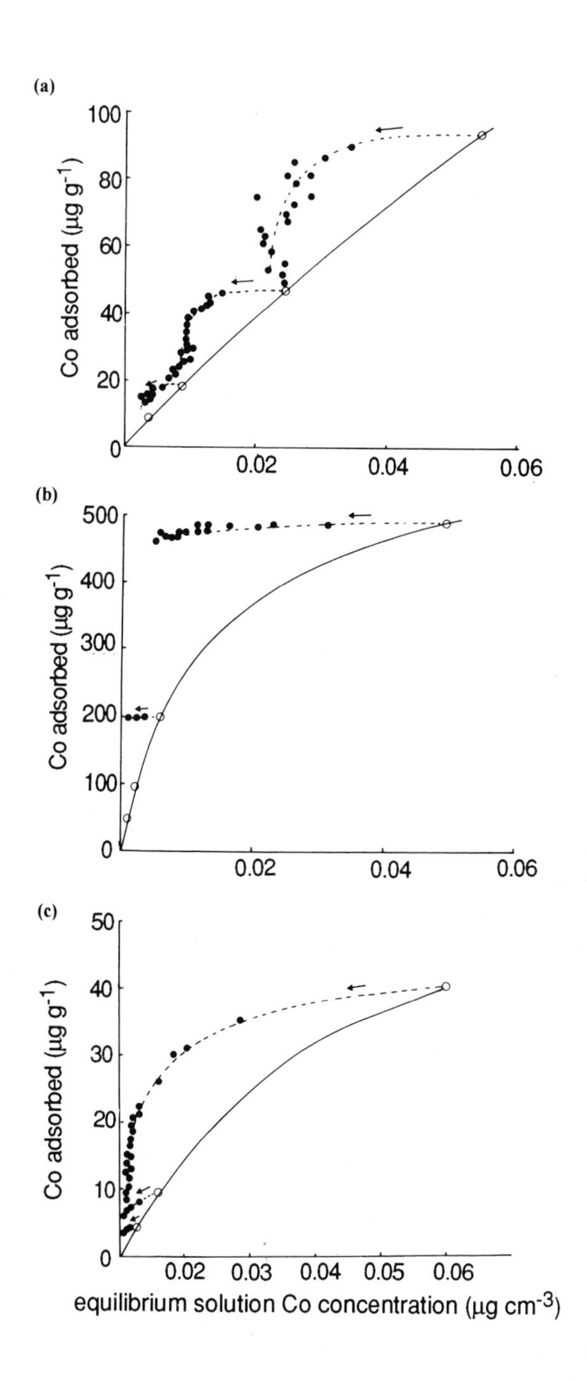

Figure 10.11 Desorption of Co from soil components; ○, sorption; ●, desorption (from [87])

10.5 Estimates of plant availability

10.5.1 Manganese

Although soil analysis is an unreliable predictor of Mn availability, it is still widely used for this purpose. Reisenauer [88] lists four main groups of extractants for soil Mn in current use: water, and dilute neutral salt solutions; 1 M ammonium acetate at pH 4 or 7 with and without reducing agents; acids; and solutions of chelating agents. Fusion methods are also employed. However, Reisenauer concluded that soil analysis was not correlated with plant uptake, either in the greenhouse or the field. One reason for this may be that changes in the rhizosphere (as discussed in Section 10.4) that are not reflected in bulk soil measurements may lead to an increase or decrease in Mn solubility and hence, acquisition by the roots.

10.5.2 Cobalt

As for manganese, it has been concluded that most of the widely used reagents fail to correlate with plant uptake under natural or near-natural conditions. The reagents used include those (e.g. dilute acetic acid) which alter significantly soil pH, one of the two most important factors governing Co availability. Thus soils of relatively high pH, giving characteristically low plant concentrations, appear to be less deficient than they really are. The converse also applies: neutral 1 M ammonium acetate will buffer an acid soil at a pH above that which it exhibited prior to extraction, and consequently may reduce the amount of Co passing into solution.

A few years ago, it was experimentally very difficult to measure the concentrations of Co at the lower end of the range found in soil extracts, with the techniques then available. Had an extractant such as dilute $CaCl_2$ solution been used, concentrations would have often been below the limit of detection, and there was a natural tendency to use a reagent that extracted enough to be measurable. However, the outcome of the use of extractants that lower the pH so drastically, when studying availability in a range of soils which are themselves of widely differing pH, has been to normalise the conditions to those of the most acid soils. Thus the natural variations in availability caused by pH-dependent factors such as difference in solubility of Co oxides, oxyhydroxides and/or carbonates have been obscured.

10.5.3 Assessment using isotopic exchange

Estimations of available Mn by isotopic exchange have shown generally higher values than those obtained by straightforward extraction by 0.05 M $CaCl_2$ (Table 10.5 and [89]). One possible explanation is that some of the soil Mn exists in forms that are exchangeable with solution Mn^{2+} but not with Ca^{2+} ions. The complexing agent DTPA gives higher E-values than those obtained with $CaCl_2$, and Goldberg and Smith [90] have shown that the former reagent apparently dissolves some of the Mn in mineral fractions as well as removing it from exchange sites.

Table 10.5 Comparison of values for exchangeable Mn as determined by isotopic exchange and by simple extraction. Adapted from Goldberg and Smith [90]

Isotopically exchangeable Mn (mg/kg)		Extractable Mn (mg/kg)	
$CaCl_2$	DPTA	$CaCl_2$	DTPA
51.7		37.1	
71.0		50.4	
165	264	37.2	263
101	157	75.2	155
13.5		13.1	
1.5		1.3	
16.4	35	6.4	35
31.1		15.7	
42.6	106	40.9	106
135	327	34.1	325

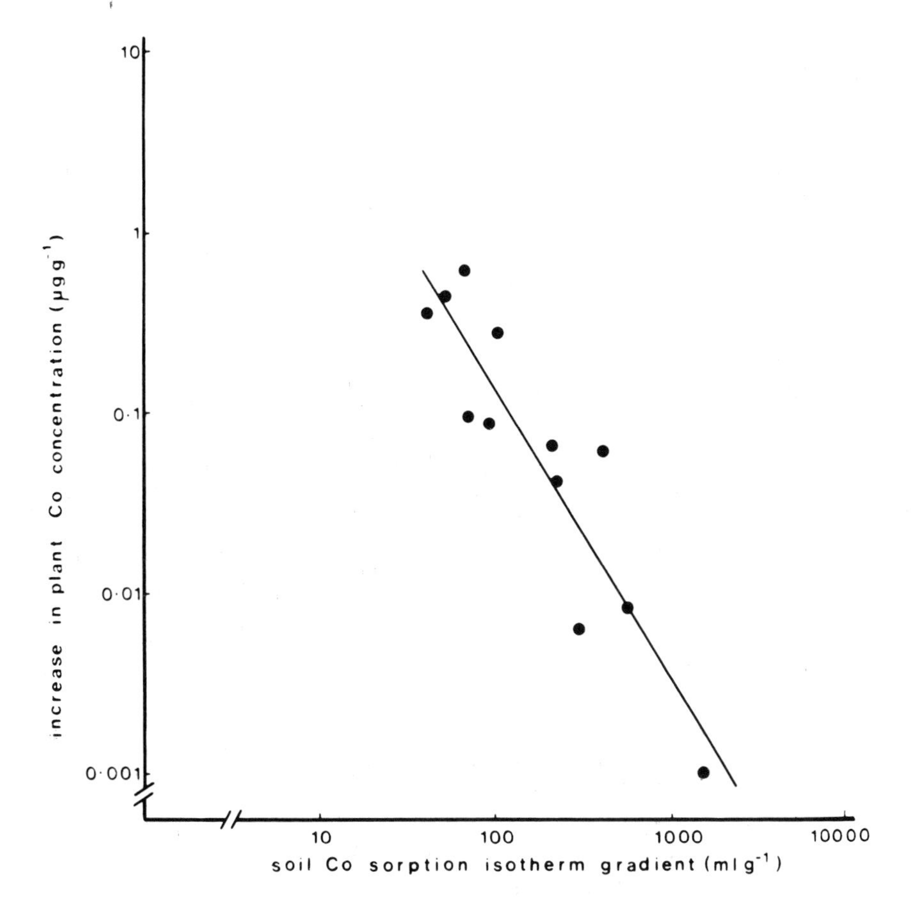

Figure 10.12 Relationship between increase in plant Co concentrations and soil Co sorption isotherm gradients (from [71])

Tiller *et al.* [91], using the long-lived isotope ^{60}Co, found that the istopically exchangeable Co in 25 Australian soils ranged from 0.16 to 5.4 mg/kg, and that the values were highly correlated with the sum of ammonium acetate- and hydroquinone-extractable Co. They found that the bonding of Co was much stronger at low percentage saturation of adsorption sites, and suggested that the initial gradient of the Co sorption isotherm might provide a measure of the intensity factor of Co availability to plants. Subsequent work with 20 Scottish soils by McLaren *et al.* [71] has provided support for this concept; Figure 10.12 shows the increase in plant concentrations with decreasing Co sorption isotherm gradient observed by these workers.

10.6 Concluding comments

The essential requirement for Mn by higher plants and for Co by ruminants and N-fixing microorganisms has stimulated a great amount of research into the forms, quantities and chemical behaviour of these two elements in soils. There is now a good understanding of their soil inorganic chemistry, and factors affecting their rate of release from soil minerals and concentrations in the soil solution, such as the role of organic substances exuded from roots or microorganisms, are becoming better understood. However, more work is needed in this area, particularly to investigate the effectiveness of a more diverse range of plant species (and their associated microflora) in mobilising the two elements.

References

1. Hewitt, E.J. and Smith, T.A., *Plant Mineral Nutrition*. English Universities Press, London (1974).
2. McHargue, J.S., *Agric. Res.* **24** (1923), 781
3. Samuel, F., and Piper, C.S., *J. Agric. S. Aust.* **31** (1928) 696 and 789 (cited in [1]).
4. Kemmerer, A.R., Elvehjem, C.A. and Hart, E.B., *J. Biol. Chem.* **92** (1931), 623.
5. Bowen, H.J.M., *Environmental Chemistry of the Elements*. Academic Press, London (1979).
6. Hirsch-Holb, H., Kolb, H.J. and Greenberg, J.M., *J. Biol. Chem.* **246** (1971), 395.
7. Scrutton, M.C., Utter, M.F. and Mildvan, M.S., *J. Biol. Chem.* **241** (1966), 3480.
8. Tietz, A., *Biochim. Biophys. Acta.* **25** (1957), 303.
9. Leach, R.M., *Poultry Sci.* **47** (1968), 828.
10. Hatch, M.D., and Kagawa, T., *Arch. Biochim. Biophys.* **160** (1974), 346.
11. Cheniae, G.M. and Martin, I.F., *Biochim. Biophys. Acta.* **197** (1970), 219.
12. Berndt, G.F., *J. Sci. Fd. Agric.* **45** (1988), 119.
13. Masagni, H.J. and Cox, F.R., *Soil Sci. Soc. Am. J.* **49** (1985), 382.
14. Reuter, D.J., Alston, A.M., and McFarlane, J.D., in *Manganese in Soils and Plants*, eds. Graham, R.D., Hannam, R.J., and Uren, N.C., Kluwer, Dordrecht (1988), Chapter 14.
15. Underwood, E.J. and Filmer, J.F., *Aust. Vet. J.* **11** (1935), 84.
16. Askew, H.O. and Dixon, J.K., *N.Z. J. Sci. Technol.* **18** (1937), 688.
17. Corner, H. and Smith, A.M., *Biochem. J.* **32** (1938), 1800
18. Rickes, E.L., Brink, N.G., Koniosky, F.R., Wood, T.R. and Folkers, K., *Science* **108** (1948), 134.
19. Smith, E.L., *Nature (Lond)*, **162** (1948), 144.
20. Smith, K.E., Koch, B.A. and Turk K.L., *J. Nutr.* **44** (1951), 455.
21. Purcell, K.F. and Kotz, J., *Inorganic Chemistry*. W.B. Saunders, Philadelphia (1977).
22. Ahmed, S. and Evans, H.J., *Biochim. Biophys. Res. Comm.* **1** (1959), 271.

23. Reisenauer, H.M., *Nature (Lond).* **198** (1960), 375
24. Iswaran, V. and Sundara Rao, W.V.B., *Nature, (Lond).* **203** (1960), 549.
25. Young, R.S., *Cobalt in Biology and Biochemistry.* Academic Press, London (1979).
26. Busse, M., *Planta* **53** (1959), 25.
27. Gilkes, R.J. and McKenzie, R.M., in *Manganese in Soils and Plants*, eds. Graham, R.D., Hannam, R.J. and Uren, N.C. Kluwer, Dordrecht (1988), Chapter 2.
28. Mitchell, R.L., in *Chemistry of the Soil*, ed. Bear, F.E., Van Nostrand Reinhold, New York (1964), Chapter 8.
29. Aubert, H. and Pinta, M., *Trace Elements in Soils*, Elsevier, Amsterdam (1977).
30. Krauskopf, K.B., in *Micronutrients in Agriculture*, ed. Mordvedt, J.J., Giordano, P.M. and Lindsay, W.L., Soil Sci. Soc. Am., Madison, Wis. (1972), Chapter 2.
31. McKenzie, R.M., *Z. Pfanzen. Bodenk.* **131** (1972), 221.
32. Feitnecht, W., Oswald, H.P. and Feitnecht-Steimann, V., *Helv. Chim. Acta.* **48** (1960), 1947.
33. Wadsley, A.D. and Walkley, A., *Rev. Pure Appl. Chem.* **1** (1951), 203.
34. Swaine, D.J. and Mitchell, R.L., *J. Soil Sci.* **11** (1960), 347.
35. Archer, F.C., *J. Soil Sci.* **14** (1963), 144.
36. Ponnamperuma, F.N., Loy, T.A. and Tianco, E.M., *Soil Sci.* **108** (1969), 48.
37. Dubois, P., *Ann. Chim.* **5** (1936), 411.
38. Jarvis, S.C., *J. Soil Sci.* **35** (1984), 421.
39. Jones, L.H.P. and Leeper, G.W., *Pl. Soil* **3** (1951), 141.
40. Page, E.R., *Pl. Soil* **17** (1962), 99.
41. Walter, K.H., in *Manganese in Soils and Plants*, eds. Graham, R.D., Hannam, R.J. and Uren, N.C. Kluwer, Dordrecht (1988), Chapter 15.
42. Ghanem, I., El-Gabaly, M.M., Hassan, M.N. and Trados, V., *Pl. Soil* **34** (1971), 653.
43. Lindsay, W.L., *Chemical Equilibria in Soils.* Wiley, New York (1979).
44. Goldberg, S.P., Smith, K.A. and Holmes, J.C., *J. Sci. Fd. Agric.* **34** (1983), 657.
45. Ponnamperuma, F.N., in *Flooding and Plant Growth*, ed. Kozlowski, T.T, Academic Press, Orlando, Fla (1984), Chapter 2.
46. Marschner, H., in *Manganese in Soils and Plants*, eds. Graham, R.D., Hannam, R.J. and Uren, N.C. Kluwer, Dordrecht (1988), Chapter 13.
47. Marschner, H., Römfield, V. and Kissel, M., *J. Plant Nutr.* **9** (1986), 695.
48. Smiley, R.W., *Soil Sci. Soc. Am. Proc.* **38** (1074), 795.
49. Marschner, H., Römfield, V., Horst, W.J. and Martin, P., *Z. Pflanzen. Bodenk.* **149** (1986), 441.
50. Godo, G.H. and Reisenauer, H.M., *Soil Sci. Soc. Am. J.* **44** (1980), 993.
51. Sarkar, A.N. and Wyn Jones, R.G., *Pl. Soil* **66** (1982), 361.
52. Linehan, D.J., Sinclair, A.H. and Mitchell, M.C., *Pl. Soil* **86** (1985), 147.
53. Linehan, D.J., Sinclair, A.H. and Mitchell, M.C., *J. Soil Sci.* **40** (1989), 103.
54. Bromfield, S.M., *Pl. Soil* **9** (1958), 325.
55. Bromfield, S.M., *Pl. Soil* **10** (1958), 147.
56. Gardner, W.K., Parbery, D.G. and Barber, D.A., *Pl. Soil* **68** (1982), 19.
57. Gardner, W.K., Parbery, D.G. and Barber, D.A., *Pl. Soil* **68** (1982), 33.
58. Juaregui, M.A. and Reisenauer, H.M., *Soil Sci. Soc. Am. J.* **46** (1982), 314.
59. Lehmann, R.G., Cheng, H.H. and Harsh, J.B., *Soil Sci. Soc. Am. J.* **51** (1987), 352.
60. Bromfield, S.M., *Pl. Soil* **49** (1978), 23.
61. Sparrow, L.A. and Uren, N.C., *Soil Biol. Biochem*, **19** (1987), 143.
62. Page, E.R., *Pl. Soil* **16** (1962), 247.
63. Mitchell, R.L., Reith, J.W.S. and Johnson, I.M., *J. Sci. Fd. Agric.* **8** (1957), s.51.
64. Walsh, T., Fleming, G.A. and Kavanagh, T.J., *J. Eire Dept. Agric.* **52** (1956), 56.
65. Adams, S.N. and Honeysett, J.L., *Aust. J. Agric. Res.* **15** (1964), 357.
66. Berrow, M.L. and Mitchell, R.L., *Trans. Roy. Soc. Edinburgh, Earth Sci.* **71** (1980), 103.
67. West, T.S., *Phil. Trans. Roy. Soc. Lond. B.* **294** (1981), 19.
68. Finch, T. and Rogers, P.A.M., *Ir. J. Agric. Res.* **17** (1979), 107.
69. Sanders, J.R., *J. Soil Sci.*, **34** (1983), 315.
70. Lawson, D.M., Unpublished Ph.D. Thesis, University of Edinburgh (1983).
71. McLaren R.G., Lawson, D.M. and Swift, R.S., *J. Sci. Fd. Agric.* **39** (1987), 101.
72. Klessa, D.A., Dixon, J. and Voss, R.C., *Res. Dev. Agric.*, **6** (1989), 25.
73. Mitchell, R.L., *Geol. Soc. Am. Bull.* **83** (1972), 1069.
74. Wild, A., *Russell's Soil Conditions and Plant Growth* (11th ed.), Longman Scientific, Harlow (1988).

75. Young, R.A., *Soil Sci. Soc. Am. Proc.* **13** (1949), 122.
76. Murray, J.W. and Dillard, J.G., *Geochim. Cosmochim. Acta.* **43** (1979), 781.
77. Morgan, J.J. and Stumm, W.J., *Colloid Sci.* **19** (1964), 347.
78. McKenzie, R.M., *Aust. J. Soil Res.* **5** (1967), 235.
79. Murray, D.J., Healy, T.W. and Furstenau, D.W., *Adv. Chem.* **79** (1968), 577.
80. McKenzie, R.M., *Aust. J. Soil Res.* **16** (1980), 61.
81. Taylor, R.M. and McKenzie, R.M., *Aust. J. Soil Res.* **4** (1966), 29.
82. McKenzie, R.M. and Taylor R.M., *Trans. 9th Int. Congr. Soil Sci.* **2** (1968), 577.
83. Taylor, R.M., *J. Soil Sci.* **19** (1968), 77.
84. Adams, S.N., Honeysett, J.L., Tiller, K.J. and Norrish, K., *Aust. J. Agric. Res.* **7** (1969), 29.
85. McLaren R.G., Lawson, D.M. and Swift, R.S., *J. Sci. Fd. Agric.* **37** (1986), 223.
86. Jarvis, S.C., *J. Soil Sci.* **35** (1984), 431.
87. McLaren R.G., Lawson, D.M. and Swift R.S., *J. Sci. Fd. Agric.* **37** (1986), 413.
88. Reisenauer, H.M., in *Manganese in Soils and Plants*, eds. Graham, R.D., Hannam, R.J. and
89. Uren, N.C. Kluwer, Dordrecht (1988), Chapter 6.
90. Bromfield, S.M., *Pl. Soil* **190** (1958), 147.
91. Goldberg, S.P. and Smith, K.A., *Soil Sci. Soc. Am. J.* **48** (1984), 559.
92. Tiller, K.G., Honeysett, J.L. and Hallsworth, E.G., *Aust. J. Soil Res.* **7** (1969), 43.

11 Mercury

E. STEINNES

11.1 Introduction

There is archaeological evidence that Hg has been known and used by humans for at least 3500 years [1]. The ancient Egyptians apparently knew how to make amalgamations with Sn and Cu as early as the sixth century BC, and uses of Hg metal and cinnabar in medicine existed in China and India at the same time. The Greeks were also familiar with techniques for the extraction of Hg from ores and with medicinal applications of the element. The Romans inherited most of this knowledge and greatly extended the commerical applications of the metal. Most of the Hg consumed by the Romans went into production of the red pigment vermillion, but Hg was also used to treat a variety of diseases.

After the fall of the Roman Empire, Hg consumption was chiefly restricted to medicinal and pharmaceutical uses until the invention of scientific instruments such as the barometer in 1643 by Torricelli and the Hg thermometer in 1720 by Fahrenheit involved the introduction of Hg into scientific research. Later on the metal has found an increasing number of applications in industry such as in the Hg-cathode electrolytic cell used in the chlor-alkali process.

11.1.1 Current uses of mercury

The principal uses of Hg have undergone great changes in recent decades. By the turn of this century, main uses of the metal were in the recovery of Au and Ag and the manufacture of fulminate and vermilion [1]. These uses have essentially disappeared, but a series of new applications has appeared over the years. The trend is illustrated in Table 11.1, showing the development in annual consumption of Hg among the major areas of use in the USA during the period 1950–1984 [2].

While applications in agriculture, pharmaceuticals and general laboratory practice have declined over the last 20 years, the uses in electrical apparatus, chlor-alkali plants and dental fillings are still substantial and have shown no clear tendency to decline, and in the USA gross demands for Hg have been projected to stay at the same level for several decades to come [3]. Increasing concern about the environmental hazards of Hg may, however, invalidate this estimate.

11.1.2 Release of mercury to the environment

At the present time the most significant anthropogenic activities giving rise to emission of Hg to land, water and air appear to be the following:

Table 11.1 Main areas of Hg consumption in the USA during the period 1950–1984 (76 lb flasks) [2].

Areas of use	1950	1955	1960	1965	1970	1975	1980	1984
Agriculture	4504	7399	2974	3116	1811	600	—	—
Chloralkali industry	1309	3108	6211	8753	15011	15222	9470	7342
Dental preparations	1458	1409	1783	1619	2286	2340	1779	1432
Electrical apparatus	12049	9268	9630	14764	15952	16971		
Batteries							27829	29700
Other							4098	3210
General laboratory use	646	976	1302	2827	1806	335	363	217
Measuring and contol apparatus	5385	5628	6525	4628	4832	4598	3049	2812
Pharmaceuticals	5996	1578	1729	3261	690	445	—	—
Paints								
Antifouling	3133	724	1360	255	198	—	—	—
Mildew-proofing	—	—	2861	7534	10149	6928	8621	4651
Redistilled*	7600	9583	9678	12257	—	—	—	—
Other	7135	14983	7472	17440	8768	3399	3774	5238
Total	49214	54656	51525	76454	61503	50838	58983	54602

*Contributing to the above groups, mainly dental, electrical, and measuring/control apparatus.

(i) Mining and smelting of ores, in particular Cu and Zn smelting.
(ii) Burning of fossil fuels, mainly coal.
(iii) Industrial production processes, in particular the Hg cell chlor-alkali process for production of Cl and caustic soda.
(iv) Consumption-related discharges, including waste incineration which is catching up rapidly in some countries.

The magnitude of the inputs into the atmosphere, hydrosphere and pedosphere may be illustrated by Andren and Nriagu's estimates of world-wide Hg production and release to the environment [4]. According to their figures, the annual anthropogenic release of Hg on a global basis was about 3×10^6 kg around the year 1900, and had increased to about three times that amount during the 1970s. Around 45% was emitted to air, 7% to water, and 48% to land. The continental degassing of Hg was estimated at 1.8×10^7 kg annually, which means that human activity is now interfering substantially with the natural Hg cycle. 1973 figures for the USA [4] indicated that over 30% of the atmospheric burden was anthropogenic. In Europe the atmospheric discharge in 1975 exceeded the natural degassing [2].

11.1.3 Harmful effects of mercury

No essential biological function of Hg is known. On the contrary, Hg is among the most toxic elements to man and many higher animals. All chemical compounds of Hg are toxic to humans [5], although Hg^0 may have to be oxidised to ionic forms in order to show toxic effects. Mercury salts show a high acute toxicity, with a variety of symptoms and damages. Some organomercurials, in particular low-molecular-weight alkyl mercury compounds, are considered even more hazardous to humans because of their high chronic toxicity with respect to various, largely irreversible, defects of the nervous system. Methyl mercury is particularly significant in this respect because it is produced by microorganisms from Hg^{2+} in different natural environments. Methyl mercury appears to show strong teratogenic effects, and carcinogenic and mutagenic activity has also been implied.

Occupational Hg poisoning, e.g. in miners, has been documented during the last few centuries. Environmental Hg poisoning, however, is of a much more recent date. The first known case occurred in Japan during the late 1950s when the inhabitants of the small town of Minamata were poisoned by the consumption of fish containing high levels of methyl mercury. Several cases of wildlife poisoning from seeds treated with methyl mercury were documented in Sweden during the period 1948–1965 [6]. The investigations initiated by these events and carried out during the subsequent decade showed that elevated methyl mercury levels in fish were widespread globally. Thus methyl mercury is the dominant toxic Hg species in the environment, consumption of fish and crustacea being the main hazard to humans and higher animals.

On the other hand, Hg does not seem to represent a major problem with respect to phytotoxicty [7]. The levels at which toxicity symptoms are apparent are far above those encountered under normal conditions. In general, the availability of soil Hg to plants is low, and the roots serve as a barrier to Hg uptake.

Hg^{2+} is one of the metal ions which are the most toxic to soil biota. In a number of experiments mercury added as Hg^{2+} appeared to be highly toxic to soil respiration [8], sometimes at levels of the same order as the indigenous amounts of the soil. In order cases significantly higher Hg concentrations were required to bring about measurable effects. In any case the critical level is likely to depend quite strongly on the properties of the soil.

11.2 Geochemical occurrence

Although more than 20 principal Hg minerals are known in nature, commercial production of Hg is made almost entirely from cinnabar, HgS [1]. Mercury also occurs in the Earth's crust as complex sulphides with Zn, Fe and other metals, but only to a small extent as the native metal. The major Hg deposits may have been formed from hydrothermal solutions which transported the Hg as sulphide or chloride complexes. The crustal average of Hg is of the order of 50 ng/g [9] or perhaps even lower. A further discussion of Hg abundances in different rock types is given in section 11.3

Cinnabar is resistant to the normal processes of oxidation and weathering and is extremely insoluble in water, and therefore enters the geochemical cycle mainly in the form of mechanically degraded particulate material. A far more important source for the release of Hg from crustal rocks is degassing of elemental Hg. Some gaseous Hg is also contributed by volcanic emissions, from vegetation and from the ocean. The most significant chemical species of Hg participating in the geochemical cycle of this element may be classified as follows [6]:

Volatile compounds:	Hg^0; $(CH_3)_2Hg$
Reactive species:	Hg^{2+}; HgX_2, HgX_3^- and HgX_4^{2-} with X = OH^-, Cl^- and Br^-; HgO on aerosol particles; Hg^{2+} complexes with organic acids.
Non-reactive species:	Methyl mercury (CH_3Hg^+, CH_3HgCl, CH_3HgOH) and other organomercuric compounds; $Hg(CN)_2$; HgS; Hg^{2+} bound to S atoms in humic matter.

Typical background concentrations in air [6] appear to be about 3 nm/m^3 over land and slightly lower over the sea, mostly in the form of Hg^0. In aquatic systems representative concentration ranges may be considered to be 0.5–3 ng/l in the open ocean and 1–3 ng/l in rivers and lakes. Most of this is probably in the form of inorganic Hg(II) species.

11.3 Origin of mercury in soils

The original Hg sources common to all soils are the minerals constituting the rocks forming the soil parent material. In the case of surface soils, atmospheric deposition is also a very significant Hg source, which has become even more important with the increasing contribution from anthropogenic activity to atmospheric Hg. In the case of agricultural soils the use of fertilisers (commercial fertiliser, manure, sewage sludge), lime and Hg-containing fungicides may sometimes increase substantially the Hg load.

11.3.1 Soil parent materials

Data for Hg in rocks have been produced for more than 50 years, and more data are available than for most other trace elements present in rocks at similar concentration levels. Possibly some of the data reported in the literature are erroneously high, in particular in some of the older work. Nevertheless there is reasonable consistency in at least one respect, notably that the Hg content in rocks not containing appreciable organic matter is very low, frequently below 30 ng/g.

In Table 11.2 data from some investigations assumed to be representative with respect to the levels of Hg in major igneous and sedimentary rock types [10–13] are presented. For igneous rocks the Hg contents appear to be very low in general, and 10 ng/g may be estimated to be a representative level. Concerning sedimentary rocks, sandstones and limestones fall within the same general range as the igneous rocks, whereas shales cover a very wide range of Hg contents. The organic matter content is obviously an important factor in this respect [11, 13], but other factors may also be significant [11]. Rocks derived from volcanic debris or deposited in volcanic areas have more Hg. The stable dissolved Hg species, as determined by the E_h and pH of the sedimentary environment, may be important. Rocks deposited in a reducing environment, as indicated by the presence not only of organic material

Table 11.2 Some selected abundance data for Hg in common rock types (ng/g)

	Connor and Shacklette [10]	NcNeal and Rose [11]	Henriques [12]	Cameron and Jonasson [13]
Igneous rocks				8.4*
Basalt				
Mean	—	—	3.9	(—)
Range			0.2–17.7	
Granite				
Mean	—	—	3.5	(—)
Range			1.4–28.1	
Sedimentary rocks				
Sandstone				
Mean	12	7	25	
Range	< 10–150		0.8–6.0	
Limestone				
Mean	49	9	6.0	
Range	<10–290		0.8–31.2	
Shale				
Mean	45	23	5.9	513; 129; 42‡
Range	<10–190		0.9–33.5	
Mean	340**		234†	
Range	< 40–1500		31.9–340	

*Average for igneous rocks (basalts, andesites, rhyolites)
**Bituminous shales
†Black shales
‡Averages for three groups of Precambrian shales

but also of S, have more Hg. Rocks formed in oxidising environments tend to have more Hg when they contain newly precipitated Fe and Mn oxides. Diagenetic and thermal effects may also be important.

It seems probable from the data presented above that the crustal average of Hg is closer to 20 ng/g than to the values of 50 ng/g [9] and 80 ng/g [14] estimated in previous reviews.

11.3.2 Atmospheric deposition

Some of the Hg present in soil and water may slowly be transformed to volatile species that are emitted to the atmosphere [6], probably mainly as Hg^0 and $(CH_3)_2Hg$, which can both be formed by biochemical processes. It is generally assumed that most of the gaseous Hg in the atmosphere is Hg^0. Measurements over the European continent and the Atlantic have indicated that an appreciable amount of $(CH_3)_2Hg$ is emitted initially, but is relatively rapidly degraded to Hg^0 in the air. The residence time of Hg^0 in the atmosphere is long, possibly as much as one year [6].

It is suggested [15] that Hg^0 is eventually subjected to atmospheric oxidation processes which yield water-soluble forms subsequently scavenged by wet or dry deposition. In industrialised areas air pollutants such as ozone may significantly influence the formation rate of oxidised Hg species [16]. The extent of Hg removal from the atmosphere by dry deposition is not well known [6]. Typical concentrations of Hg in precipitation at remote sites appear to be of the order of 2–10 ng/l, while levels in more polluted areas may typically be of the order of 5 times higher, excluding areas in the vicinity of Hg-emitting industries. With an annual precipitation of 1000 mm and a Hg content of 20 ng/l the wet deposition is 20 μg/m^2/yr, which is slightly lower than the 30–60 μg/m estimated by Andersson [17] for background areas and 30–200 μg/m suggested on the basis of data from Danish peat bogs [6].

It has generally been assumed that most of the Hg emitted to the atmosphere even from anthropogenic processes is in the elemental vapour form [18]. For high-temperature combustion processes at least, this seems questionable. Brosset [19] found a fraction of the order of 50% of the Hg emission from a coal combustion plant was present as forms of Hg[II]. If this is representative, the deposition of Hg from coal-fired power plants or waste incinerators could turn out to be more significant than generally assumed in the past.

Recent investigations in Scandinavia [20, 21] and USA [22] indicate that long-range atmospheric transport from anthropogenic sources is a much more significant source of Hg to surface soils than previously anticipated, at least for humus-rich natural soils. In Sweden more than 50% of Hg in the mor layer is supposed to be of pollution origin [20], partly from a small number of strong domestic point sources, partly from source regions elsewhere in Europe. Implications of this surface accumulation of Hg in soil are discussed in Section 11.6.

11.3.3 Agricultural materials

Considerable amounts of Hg may be added to agricultural land along with ferilisers, lime, and manures, as evident from the literature figures summarised by Andersson [17]. Most commercial fertilisers have Hg contents below 50 ng/g, but considerably higher values occur in P fertilisers. The Hg may be derived partly from the phosphate rocks and partly from the H_2SO_4 used for their dissolution. Samples of lime tend to show values below 20 ng/g, while manure typically exhibits Hg levels of the order of 100 ng/g.

Since the beginning of the twentieth century, Hg compounds have been utilised in agriculture as fungicides and seed disinfectants. During most of the time organic mercurials have been used, in particular during the period 1945–1970. The supply of Hg to the soil by seed dressing can be as high as 1 mg/m^2. The supply usually does not occur every year, hence a reasonable average seems to be 100–200 mg/m^2/yr for cereal-producing soils [17]. This is the same order as the total amount of Hg already present in the upper 20 cm of the soil, and considerably more than what may normally be supplied by atmospheric deposition.

11.3.4 Sewage sludge

Sludge from wastewater treatment plants is frequently applied as fertiliser on arable land. Often restrictions are imposed on the use of sludge because of the high content of some heavy metals found in this material, including Hg. It appears from the examples of literature values selected by Andersson [17] that 5–10 μg/g is a typical Hg level in sludge, and occasionally values of 100 μg/g or more are reported. More recent literature values [23, 24] support this trend. Assuming that 50 t/ha of sludge are applied to the soil, the added amount of Hg is typically of the order of 50 mg/m^2, which is a considerable increment with respect to a normal background level.

11.4 Chemical behaviour of mercury in soil

11.4.1 Occurrence and stability of inorganic mercury species in soil

Depending on the redox conditions, Hg may occur in three different valency states, namely as Hg^0, Hg_2^{2+} and Hg^{2+}, of which Hg^0 and Hg^{2+} are the states normally encountered in soil. In addition to the redox potential, pH and Cl^- concentration are key parameters in determining the speciation of Hg in the soil solution and the chemical transformations occurring. In addition to chemical reactions, transformations may also be mediated by microbial activity. Knowledge of the speciation and transformation reactions of Hg is important in order to explain the retention and mobility of this element in the soil, equilibria between the solid phases and the soil solution, and availability for plant uptake.

Due to its strong ability to form complexes, Hg^{2+} rarely occurs in the free ionic form under natural conditions. In acid solutions Hg^{2+} is stable at a redox potential above 0.4 V, and normally occurs as the $HgCl_2^0$ complex. Above pH 7 the complex $Hg\ (OH)_2^0$ is the corresponding stable form. Hg^{2+} also forms strong complexes with humic matter, but the character of these complexes is not very well known.

Another important property of Hg is the ability to bind strongly to the sulphide ion. Under strongly reducing conditions, Hg^0 is stable in the presence of H_2S or HS^-, but at increasing redox potential, HgS will precipitate, or in strongly alkaline soils the soluble HgS_2^{2-} ion will be formed. Further increase will bring about oxidation of sulphide to sulphate, but at this point the potential is still not high enough to prevent reduction to Hg^0. Further rise in redox potential to the level normally found in surface soil will finally transform Hg into the +2 state. A more detailed treatment of the physical chemistry of Hg in aqueous solutions is given by Andersson [17, 25].

Only a very minute fraction of Hg^{2+} occurs in the soil solution, the major fraction being either bound in soil minerals or adsorbed on solid surfaces, inorganic and organic. Since $HgCl_2^0$ is only weakly retained by mineral matter it can be assumed that Hg^{2+} is mainly attracted to organic matter in acid soils, whereas in neutral and slightly alkaline soil, mineral compounds are also active. A recent review of the literature on the behaviour of mercury in soil is given by Schuster [26], with emphasis on complexation and adsorption processes.

11.4.2 Volatilisation of mercury from soil

A number of investigators have observed loss of Hg added to soils in the form of inorganic salts [27–30]. Organic matter tended to enhance the loss. The volatilisation seems to be mediated by microorganisms [30], but significant losses were also reported from sterilised soils [27]. These studies and others were normally carried out with Hg concentrations far above the naturally occurring ones, and it is hard to evaluate the significance of the demonstrated reactions and processes for the Hg turnover under natural conditions [17].

Some of the experimental evidence seems conflicting with regard to the influence of pH and organic matter on the volatilisation of Hg from soil [29]. For soils low in clay and humus, more was vaporised from a neutral than from an acidic soil, whereas for two soils with 4–5% humus content and 15–17% clay the opposite was observed. In acid soil, a higher humus content inhibited the vaporisation of Hg at the 1 μg/g level in the surface layer, which at 50 μg/g the loss was higher at higher humus content. A probable explanation [17] seems to be that at low Hg content the humus exhibits sufficient complexing capacity to allow only a very small fraction to be present in the soil solution and thus be available for reduction and subsequent loss by volatilisation. At higher Hg content a significant fraction may be present in the solution and eventually be lost after reduction.

Under natural conditions the release of Hg^0 and possibly also other volatile Hg compounds from soil is probably very significant in the cycling of Hg. It is a fact that anomalously high concentrations of Hg^0 in air are frequently observed in the vicinity of Hg-bearing ores. It has also been suggested that depletion of Hg in the A horizon of natural soils relative to deeper horizons might be explained by evaporation loss [31]. Heating experiments on some of these soils, which were all low in Hg, indicated incipient volatilisation loss even at about 50°C. On the other hand, sorption of Hg^0 vapour in natural surface soils was shown in radiotracer experiments [32]. The majority of the sorbed Hg was found to volatilise at 100–200°C, and extraction experiments with various agents indicated that the predominant part of the tracer was present in the soil as an organomercury species.

11.4.3 Leaching of mercury from soil

The strong ability of soil to fix Hg^{2+} species means that the removal of Hg from soils by leaching is in most cases insignificant [28, 29, 33, 34]. Hogg *et al.* [33] conducted an experiment where radioactively labelled Hg compounds were applied to soil columns which were then leached with sewage effluent. It appeared that neither $HgCl_2$, methyl mercuric chloride nor phenyl mercuric acetate was found in measurable amounts below 20 cm depth in the soils. Lodenius *et al.* [34] studied the leaching of Hg^{2+} in peat lysimeters and found that additions of chloride, fertiliser or sterilant did not affect the leaching of Hg. The only treatment that had an effect was drying of the column, which resulted in cracks presumably allowing Hg bound to humus colloids to penetrate the column. Similar transport might occur laterally in surface soil layers in periods with surface runoff. The annual transport of Hg on humus colloids from forest soils to lakes in Sweden is estimated to be less than 1% of the amount stored in the humus layer of the soil [35].

11.4.4 Retention of mercury in soil material

The adsorption process is apparently dominant in providing the retention of Hg species in soil. Adsorption of Hg depends on a number of factors, including the chemical form of Hg introduced, the grain size distribution of the soil, the nature and amount of inorganic and organic soil colloid, the soil pH, and the redox potential. In addition Hg^{2+} may be fixed in the form of low-solubility precipitates, in particular the sulphide and the selenide.

The retention of Hg^{2+} may occur as a result of ion exchange, but stronger bonds are probably involved to a large extent, such as hydroxoligands in the case of sesquioxides and various ligands in the case of humic substances. Andersson [25] found the following sequence for the retention of inorganic Hg under neutral conditions: $Al(OH)_3$ < kaolinite < montmorillonite < illitic clay soil < lateritic soil < organic soils < $Fe_2O_3.x\ nH_2O$. Below pH 5.5 where $HgCl_2^0$ is the dominant species in solution, organic matter will be mainly responsible for the sorption of Hg. The retention of Hg in the organic soil is not significantly reduced until pH < 4, where it slowly decreases. In neutral soils (pH > 5.5) iron oxides and clay minerals participate more efficiently in the adsorption of Hg^{2+}. The maximum adsorption occurs around pH 7, where $HgOHCl^0$ is the dominant species [17]. Also, organomercury compounds such as methylmercuric chloride and phenylmercuric acetate are strongly adsorbed in soils in the pH range around neutrality [33]. Clay minerals appear to be efficient in this respect [36], but only within a narrow pH range and when the compounds are present in low enough concentrations that they appear mainly in dissociated form.

11.4.5 Methylation of mercury in soil

Since the observation that microorganisms in natural lake sediments could methylate Hg [37], a number of studies confirming the production of mono- and dimethyl mercury in the aquatic and the terrestrial environment have been carried out. The most relevant literature with respect to the methylation of Hg in soils is reviewed by Adriano [7]. It appears that methyl mercury may be formed in soils under a variety of conditions, including purely abiological methylation [38] possibly associated with the fulvic acid fraction of the soil. Westling [39], studying different forms of Hg in run-off from peatlands in southern and central Sweden, found a higher proportion or methyl mercury in the case of drained (3.5–14.2%) than in undrained mires (2.0–5.7%).

11.4.6 Concentration and distribution of mercury in soils

In most soils the Hg content varies with depth in particular in virgin soils, which means that the sampling depth is important. In cultivated soils the ploughed layer

(0–20 cm) is homogenised due to the soil management. Comparison of data from cultivated and non-cultivated soils is therefore difficult unless the homogenised sampling depth is the same for both. In the case of comparisons between different virgin soils, similar sampling depths or the same horizons should be used [17].

In Table 11.3 some selected literature data showing typical Hg levels in presumably uncontaminated soils are listed. Most of the data are from North America, the UK and Scandinavia. Very few published data on Hg in soils seem to exist for southern and eastern Europe, Asia, Africa, Latin America and Australia. Some of the investigations listed in Table 11.3 may have included agricultural soils treated with organic mercurials. By and large, however, the data are thought to be representative for areas with no appreciable influence from anthropogenic sources giving rise to elevated Hg levels in the surface soil.

Organic soils commonly have higher average Hg contents than mineral soils [47, 48]. Låg and Steinnes [53] found a highly significant correlation between Hg and organic matter content in the top layer of forest soils. Andersson [25] in a study of Hg in profiles of virgin soils demonstrated a very close correlation between Hg and organic matter content in acidic soils, whereas in neutral soils ($pH > 6$) where the dominating species are HgOHCl and $Hg(OH)_2$ rather than $HgCl_2$, the co-variation between Hg and Fe was stronger than that of Hg and organic matter. It seems to be a possibility that the distinctly higher Hg content observed in soils in the eastern compared to the western USA [40] could in part be explained by differences in soil organic matter.

In some studies, however, such as those of Dudas and Pawluk investigating chernozemic and luvisolic soils in Alberta [39, 31], Hg appeared to be depleted in the surface layer relative to lower horizons. In soils with low organic matter contents and $pH > 6$ this trend would be very likely to be expected. However, McKeague and Kloosterman [46], in an extensive survey of virgin soils from all the provinces of Canada, observed that in more than half of the samples, particularly podzolic and gleysolic soils the highest Hg concentrations were present in the surface layer.

In cultivated soils the Hg level is normally quite constant within the plough layer. Below this layer there is a gradual decrease downwards to the background levels in the parent material [17, 59]. In some cases however the Hg content in the surface horizon is similar to that of the subsoil [60].

Ombrotrophic peat bogs offer a unique opportunity to study the transfer of Hg between the atmosphere and organic soils virtually independent of the underlying mineral material [50, 54]. In a study of Hg in peat cores from 20 Norwegian bogs [54] a consistently higher level was found in the surface layer than at 50 cm depth. It is not evident to what extent this is due to either recent increases in atmospheric deposition of Hg or to natural redistribution of Hg within the peat profile. It appears, however, that the largest enrichment in the surface peat was found in the southern part of the country where the supply of large-range transported pollutants is much more prominent than in areas farther north, which indicates surface accumulation of anthropogenic Hg. Similar findings were reported by Jensen and Jensen [61] in a study of dated peat cores from Scandinavian bogs.

Table 11.3 Selected literature data for the Hg contents of surface soils

Country	Soil type	Number of samples	Hg (ng/g) Range	Hg (ng/g) Average*	Reference
USA	Surface soils (entire country)	912	< 10–4600	112	40
	Surface soils (Western states)	492	< 10–4600	83	
	Surface soils (Eastern states)	420	10–3400	147	
	Surface horizon (Missouri)	1140	<10–800	39	9
	B horizon (Missouri)	300	10–1500	72	
	Surface horizon (Colorado)	168	< 10–420	35	
	B horizon (Eastern states)	420	10–3400	96	
	B horizon (Western states)	492	<10–4600	55	
	Urban and and agricultural (0–5 cm)	264	—	110	41
	Cropland and non-cropland (0–7.5 cm)	379	< 50–1060	(70–80)	42
	Cultivated (0–15 cm)	200	—	30	43
	Cultivated (0–7.5 cm)	96	50–360	120	44
Canada	Cultivated and non-cultivated	27	5–36	22	45
	Virgin soils, A horizon	65	< 5–660	64	46
	Cultivated and virgin, A horizon	170	13–741	102	47
	Cultivated (0–15 cm)	290	10–1140	110	48
	53 profiles	173	5–100	59	49
	Peat (0–40 cm)	11	10–110	60	50
Austria	Cultivated and non-cultivated (0–20 cm)	40	5–340	95	51
Sweden	Cultivated and non-cultivated (0–20 cm)	273	4–920	60	52
Norway	Forest soil, A_0 horizon	700	20–550	188	53
	Ombrogenous peat (0–5 cm)	20	35–255	119	54
	Ombrogenous peat (50 cm)	13	15–40	31	
UK	Cultivated and non-cultivated (0–15 cm)	51	10–1780	32	55
	Topsoils and profile soils	354	< 10–1710	130	56
	Cultivated (0–15 cm)	53	8–190	40	57
	Cultivated (0–15 cm)	305	20–400	90†	58
East and Central Africa	Surface horizons	14	11–41	23	52
India	Surface soils	12	3–689	20	

*Some investigators report range and average after the emission of anomalous values
†Median

In the survey presented by Ure and Berrow [56], data for the Hg content of 3049 soils averaging 98 ng/g were reported. No similar calculation is present for the data listed in Table 11.3, because the format of reporting is not uniform. Some investigators include all data obtained on a given material, while others exclude anomalous values before calculating the average. What seems to be evident, however, is that the Hg content of surface soils normally exceeds to a substantial degree that of the mineral material from which it was derived. Much of this excess Hg is probably present in some form that permits its participation in exchange processes within the soil and between atmosphere and soil.

11.5 Mercury in the soil–plant system

The uptake of Hg in terrestrial plants has been the subject of numerous investigations, and reviews of this work are available [7, 62]. Most of the work has been performed on agricultural crops under controlled experimental conditions, and almost exclusively with Hg loads far above those encountered under normal conditions. In general, the availability of soil Hg to plants is low, and there is a general tendency to Hg accumulation in the roots, indicating that the roots serve as a barrier to Hg uptake [63, 64]. Experiments on the uptake of Hg^{2+} in higher plants from a solution culture [65] showed a certain transport of Hg to the shoots at levels exceeding 0.1 mg/kg Hg in the external medium. The fraction of Hg retained in the roots was about 20 times that observed in the shoots. Lindberg *et al.* [64] studying the plant uptake of Hg from agricultural soils near a Hg mine found that the relative Hg content in roots was closely related to the NH_4OAc-extractable Hg in the soils. The Hg concentration in above-ground parts of plants, on the other hand, appeared to be largely dependent on foliar uptake of Hg^0 volatilised from the soil. Several authors report that Hg compounds applied to some above-ground parts of plants can be readily translocated into others [7]. The Hg content in crop plants grown on soils low in Hg are reported to be in the same range as for the soils [45]. In the case of cereals the content in grain was about 3–10 times lower than that of the straw. Other work indicated still lower Hg levels (~1–2 ng/g) in grains of barley and wheat [66]. Even at these very low levels it may be assumed that foliar uptake of Hg^0 from the air plays a significant role.

11.6 Investigations of mercury-polluted soils

Present anthropogenic activities giving rise to Hg emission to the atmosphere may significantly contaminate surface soils in the neighbourhood of the emission sources. These activities include the smelting of non-ferrous metals, burning of fossil fuels, industrial activity such as the chlor-alkali process, and waste incineration. Soil contents of the order of 1–10 mg/kg Hg were found at distances < 0.5 km from a chloralkali plant [67]. In the vicinity of mercury mines, values of the order of 100 mg/kg Hg have been observed in agricultural soils [64, 68].

The high content of methyl mercury in fish in several regions in Europe and North America has focused great attention on the widespread atmospheric deposition of Hg even very far from the source regions. Perhaps the most extensive study on the pathways of Hg from air via soil to fresh waters has been carried out in Sweden [69, 70]. At present deposition rates about 80% of the supplied Hg is accumulated in the soil humus layer [71]. The transport of Hg into surface waters occurs mainly with humic matter, and is not appreciably affected by factors such as acidification [72]. In order to decrease Hg content in fish to acceptable levels, the atmospheric wet deposition of Hg must be reduced to about 20% of the present level [73].

The most significant problem of Hg pollution of agricultural soils still seems to be that of organic mercurials added in the form of seed dressing (now discontinued in many countries) and of foliar sprays against plant diseases. The behaviour of organomercurials in soils is discussed by Andersson [17]. It appears that these compounds may be fixed both to organic and inorganic soil colloids, but the mineral component, in particular clay minerals, dominates the retention. Different organomercurials show different retention characteristics and decomposition rates in soils. In cases where organic Hg compounds are added at rates commonly used in agricultural practice they are likely to contribute a fraction of the annual supply greater than that from precipitation or normal fertilisation [17].

11.7 Global budgets of mercury

The release of volatile Hg compounds from the land surface and the atmospheric deposition of Hg acting in the opposite direction are important processes not only for the Hg chemsitry in soils but also for the global circulation of Hg. Several attempts have been made in the literature to present a model for the global budget of Hg. In Table 11.4 some estimates of fluxes that are important for the Hg budget of soils from two different publications [4, 6] are presented. The two budgets are based on the assumption that release and supply of Hg balance each other, and they agree on the fact that the total annual flux seems to be of the order of 10^{10} g Hg on a global basis, but considerable uncertainty exists, mainly because of the difficulties involved in making reliable measurements. Andren and Nriagu [4] estimate the global Hg store in soils to be 2.1×10^{13} g. If we accept their value for natural Hg emission to the atmosphere, the mean residence time for Hg in soils is of the order of 1000 years, provided that the removal of Hg by leaching is small compared to the volatilisation.

Table 11.4 Some estimates of global fluxes relevant to the cycling of Hg. Unit: 10^9g Hg/yr

Process	Andren and Nriagu [4]	Lindqvist *et al.* [6]
Present anthropogenic emissions	10	2–10
Present background emissions	21	< 15
Total present emissions	31	2–17
Wet deposition	—	2–10
Dry deposition	—	< 7
Total present deposition	31	2–17
Pre-industrial deposition (and emission)	—	2–10

References

1. Nriagu, J.O., in *The Biogeochemistry of Mercury in the Environment*, ed. Nriagu, J.O. Elsevier, Amsterdam (1979), Chapter 2.
2. *Minerals Yearbook* (1950–84), U.S. Bureau of Mines, Washington, D.C.
3. Watson, W.D., in *The Biogeochemistry of Mercury in the Environment*, ed. Nriagu, J.O. Elsevier, Amsterdam (1979), Chapter 3.

4. Andren, A.W. and Nriagu, J.O., in *The Biogeochemistry of Mercury in the Environment*, ed. Nriagu, J.O. Elsevier, Amsterdam (1979), Chapter 1.
5. Greenwood, M.R. and Von Burg, R., in *Metalle der Umwelt*, ed. E. Merian, Verlag Chemie, Weinheim (1984), Chapter II, 18.
6. Lindqvist, O., Jernelöv, A.,, Johansson, K. and Rodhe, H., *Mercury in the Swedish Environment*. National Swedish Environment Protection Board, Report SNV PM 1816 (1984).
7. Adriano, D.C., in *Trace Elements in the Terrestrial Environment*, Springer-Verlag, New York (1986), Chapter 9.
8. Rundgren, S., Rühling, Å., Schlüter, K. and Tyler, G., *Mercury in Soil—Distribution, Speciation and Biological Effects*. Nordic Council of Ministers, Report Nord 1992:3 (1992).
9. Bowen, H.J.M., in *Environmental Chemistry of the Elements*, Academic Press, London (1979), Chapter 3.
10. Connor, J.J. and Shacklette, H.T., *Background Geochemistry of Some Rocks, Soils, Plants and Vegetables in the Conterminous United States*. U.S. Geol. Survey Prof. Paper 574-F (1975).
11. McNeal, J.M. and Rose, A.W., *Geochem. Cosmochim. Acta* **38** (1974), 1759.
12. Henriques, A., unpublished work, cited in R. Fern and J.E. Larsson, *Kvicksilver*–användning, kontroll och miljöeffekter, National Swedish Environment Protection Board Report SNV PM 421 (1973).
13. Cameron, E.M. and Jonasson, I.R., *Geochim. Cosmochim. Acta* **36** (1972), 985.
14. Vinogradov, A.P., *Geochemie seltener und nur in Spuren vorhandener chemisher Elemente im Boden*. Akademie-Verlag, Berlin (1954), 203.
15. Brosset, C., *Water, Air, Soil Pollut.* **16** (1981), 253.
16. Iverfeldt, Å. and Lindqvist, O., *Atmos. Environ.* **20** (1986), 1567.
17. Andersson, A., in *The Biogeochemistry of Mercury in the Environment*, ed. Nriagu, J.O. Elsevier, Amsterdam (1979), Chapter 4.
18. Matheson, D.H., in *The Biogeochemistry of Mercury in the Environment*, ed. Nriagu, J.O. Elsevier, Amsterdam (1979), Chapter 5.
19. Brosset, C., *Water, Air, Soil Pollut.* **34** (1987), 145.
20. Håkanson, L., Nilsson, Å. and Andersson, T., *Water, Air, Soil Pollut.*, **50** (1990), 311.
21. Steinnes, E. and Andersson, E.M., *Water, Air Pollut.*, **56** (1991), 391.
22. Nater, E.A. and Grigal, D.F., *Nature*, **358** (1992), 139.
23. Lester, J.N., Sterritt, R.M. and Kirk, P.W.W., *Sci. Total Environ.* **30** (1983), 45.
24. Chaney, R.L., in *Proc. Pan American Health Organization Workshop on the International Transportation, Utilization or Disposal of Sewage Sludge* (1984).
25. Andersson, A., *Gundförbättring* **23** (1970) No. 5, 31. (In Swedish, English Summary).
26. Schuster, E., *Water, Air, Soil Pollut.*, **56** (1991), 667.
27. Frear, D.E.H. and Dills, L.E., *J. Econ. Entomol.* **60** (1967), 970.
28. Gilmour, J.T. and Miller, M.S., *J. Environ. Qual.* **2** (1973), 145.
29. Wimmer, J., *Bodenkultur* **25** (1974) 369 (In German, English summary).
30. Rogers, R.D. and McFarlane, J.C., *J. Environ. Qual.* **8** (1979), 255.
31. Dudas, M.J. and Pawluk, S., *Can. J. Soil Sci.* **56** (1976), 413.
32. Landa, E.R., *Geochim. Cosmochim Acta* **42** (1978), 1407.
33. Hogg, T.J., Stewart, J.W.B. and Bettany, J.R., *J. Environ. Qual.* 7 (1978), 440.
34. Lodenius, M., Seppänen, A. and Autio, S., *Chemosphere* **16** (1987), 1215.
35. Johansson, K., Lindqvist, O. and Timm, B., *Kvicksilvers förekomst och omsättning i miljöen*. National Swedish Environment Protection Board, Report 3470 (1988).
36. Inoue, K. and Aomine, S., *Soil Sci. Plant Nutr.* **15** (1969), 86.
37. Jensen, S. and Jernelöv, A., *Nature, Lond.* **223** (1969), 753.
38. Rogers, R.D., *J. Environ. Qual.* **6** (1977), 463.
39. Westling, O., *Water, Air, Soil Pollut.*, **56** (1991), 419.
40. Shacklette, H.T., Boerngen, J.G. and Turner, R.L., *Mercury in the Environment—Surficial Materials in the Conterminous United States*. U.S. Geological Survey Circular 644 (1971).
41. Klein, D.H., *Environ. Sci. Technol.* **6** (1972), 560.
42. Wiersma, G.B. and Tai, H., *Pesticid. Monit. J.* **7** (1974), 214.
43. Sell, J.L., Deitz, F.D. and Buchanan, M.L., *Arch. Environ. Contam. Toxicol.* **3** (1975), 278.
44. Gowen, H.A., Wiersma, G.B. and Tai, H., *Pesticid. Monit. J.* **10** (1976), 111.
45. Gracey, H.I. and Stewart, H.B.W., *Can. J. Soil Sci.* **54** (1974), 105.
46. McKeague, J.A. and Kloosterman, B., *Can. J. Soil Sci.* **54** (1974), 503.

47. John, M.K., van Laerhoven, C.J., Osborne, V.E. and Cotic, I., *Water, Air, Soil Pollut.* **5** (1975), 213.
48. Frank, R., Ishida, K. and Suda, P., *Can. J. Soil Sci.* **56** (1971), 181.
49. McKeague, J.A. and Wolynetz, M.S., *Geoderma* **24** (1980), 299.
50. Glooschenko, W.A. and Capoblanko, J.A., *Environ. Sci. Technol.* **16** (1982), 187.
51. Wimmer, J. and Haunold, E., *Bodenkultur* **24** (1973), 25.
52. Andersson, A., *Grundförbättring* **20** (1967), 95. (English summary).
53. Läg, J. and Steinnes, E., *Acta Agric. Scand.* **28** (1978), 393.
54. Hvatum, O.Ø. and Steinnes, E., unpublished work.
55. Davies, B.E., *Geoderma* **16** (1976), 183.
56. Ure, A.M. and Berrow, M.L., *In Environmental Chemistry, Vol. 2*, ed. Bowen, H.J.M. Royal Society of Chemistry, London (1982), Chapter 3, 155.
57. Archer, F.C., *Ministry of Agriculture, Fisheries & Food Reference Book 326* (1980), 184.
58. Archer, F.C. and Hodgson, I.P., *J. Soil Sci.* **38** (1987), 421.
59. Whitby, L.M., Gaynor, J. and MacLean, A.J., *Can. J. Soil Sci.* **58** (1978), 325.
60. Mills, J.G. and Zwarich, M.A., *Can. J. Soil Sci.* **55** (1975), 295.
61. Jensen, A. and Jensen, A., *Water, Air, Soil Pollut.* **56** (1991), 769.
62. Kaiser, G. and Tölg, G., *in The Handbook of Environmental Chemistry,* Vol. 3, Part A, ed. Hutzinger, O. Springer-Verlag, Berlin (1980), 1.
63. Gracey, H.I., and Stewart, J.W.B., in *Proc. Intl. Conf. on Land Waste Management*, Ottawa (1974), 97.
64. Lindberg, S.E., Jackson, D.R., Huckabee, J.W., Janzen, S.A., Levin, M.J. and Lund, J.R., *J. Environ. Qual.* **8** (1979), 572.
65. Beauford, W., Barber, J. and Barringer, A.R., *Physiol. Plant.* **39** (1977), 261.
66. Läg, J. and Steinnes, E., *Scientific Reports of the Agricultural University of Norway* **57** (1978), No. 10.
67. Bull, K.R., Roberts, R.D., Inskip, M.J. and Goodman, G.T., *Environ. Pollut.* **12** (1977), 135.
68. Morishita, T., Kishino, K. and Idaka, S., *Soil Sci. Plant Nutr.* **28** (1982), 523.
69. Lindqvist, O., Johansson, K., Aastrup, M., Andersson, A., Bringmark, L., Hovsenius, G., Häkanson, L., Iverfeldt, Å., Meili, M. and Timm, B., *Water, Air, Soil Pollut.*, **55** (1991), 1.
70. Meili, M., *Water, Air, Soil Pollut.*, **56** (1991), 333.
71. Aastrup, M., Johnson, J., Bringmark, E., Bringmark, L. and Iverfeldt, Å., *Water, Air, Soil Pollut.*, **56** (1991), 155.
72. Johansson, K. and Iverfeldt, Å., V*erh. Internat. Verein. Limnol.*, **24** (1991), 2200.
73. Johansson, K., Aastrup, M. Andersson, A., Bringmark, L. and Iverfeldt, Å., *Water, Air, Soil Pollut.* **56** (1991), 267.

12 Selenium

R.H. NEAL

12.1 Introduction

Recent experiences with Se in a number of locations worldwide, including the western USA, China and Australia, have resulted in renewed interest in this element in both toxic and deficient capacities. Selenium, a member of Group VI of the Periodic Table (atomic number 34), has considerable chemical similarities to S, resembling the latter in many forms and compounds. Such similarities result in a number of biological interrelations, and these have been evidenced by toxic and deficient responses in both humans and animals. Globally, Se does not occur in any large deposits which might be considered as economically viable, but it is obtained commercially usually as a by-product of the metal electrolytic refining processes involving Cu, Zn and Ni ores. Uses of Se include the manufacture of electronic components, glass, plastics and ceramics, and use by chemical industries including the manufacture of pigments and lubricants [1]. In addition, selenium sulphide is used in shampoos for the treatment of certain scalp conditions, and Se in combination with vitamin E can be found on the shelves of many health food stores in the form of a dietary supplement.

In animal nutrition, the amount of Se present in the diet may result in either deficient or toxic responses. The occurrence of Se in native vegetation in amounts toxic to animals has been of historical interest [2], and although most highly seleniferous indicator plants are unpalatable to livestock, ingestion of plant material containing such levels has been reported to induce chronic conditions of 'alkali disease' and 'blind staggers' [3]. Although rare, acute Se poisoning resulting in death has been observed in both sheep and cattle [4]. Insufficient quantities of Se in animal feeds, however, can lead to a disorder known as 'white muscle disease'. This affliction may be alleviated through the addition of Se in conjunction with vitamin E to feeds [3]. The range between sufficient and toxic levels of Se in animal nutrition is very narrow, with a minimal dietary level of 0.10 mg/kg considered sufficient to prevent white muscle disease [2], dependent on the availability of vitamin E. Selenium toxicity in livestock is generally considered to occur at concentrations > 3 or 4 mg/kg material eaten. As a result of this limited range of essentiality, Se uptake by plants, and its inclusion in both human and animal diets is of considerable interest in many scientific areas.

Although Se is an important element in animal nutrition, there is little evidence to suggest its essentiality for plant growth. Concentrations of Se in plants are usually a good indicator as to whether an area may be classed as Se-deficient or as having high levels of available Se in the soil. Generally, plants growing on seleniferous soils exhibit selenosis, in most cases indicated by chlorosis of the leaves and a pink coloration of the roots. Selenium accumulators, such as *Astragalus* sp., take up many orders of magnitude more Se than most agricultural crops without any signs of poisoning [3], and thus may be a source of poisoning to animals. In addition, some plants, such as *Munroa squarrosa*, are non-accumulators and can grow on high-Se soils without the accumulation of more than a few mg Se/kg [5]. Conversely, accumulator plants grown on Se-deficient soils may provide higher Se concentrations in the edible portions compared to plants with a lower uptake capacity, and as such could be useful in reducing Se deficiency in animals. A more detailed discussion of the role of Se in soil–plant relationships is given in a later section.

Whilst it is not the purpose of this chapter to include work involving the nutritional aspects of Se, readers are encouraged to consult recent reviews of Se of a more general scope published by Adriano [1], Ihnat [6], and Gissel-Nielsen [7]. In addition, a number of excellent reviews covering many aspects of the role of Se in the environment are available, and these give much greater detail to certain aspects discussed in this chapter than can be included in the space available [3, 8–10].

12.2 Geochemical occurrence

The abundance of Se in the earth's crust is reported to be about 0.05–0.09 mg/kg. Selenium is a chalcophile element [11], and as such is associated with sulphide ore deposits, having about 1/6000 the abundance of S [12]. Although Se is inconsistently dispersed through geological deposits, it may be detected in most earth materials, and is frequently enriched in black shales in concentrations ranging up to 675 mg Se/kg [1]. Elevated amounts of Se in shales and in the carbonaceous debris of sandstones may be attributed to deposition under reducing conditions, the latter induced by the presence of organic matter at the time of formation. When reduction occurs in the presence of a sulphide ion, precipitation results [13], and a similar mechanism could be responsible for the formation of selenides. Elevated concentrations of Se are associated with some types of phosphatic rocks, with concentrations between 1 and 300 mg/kg reported [3]. It appears that there is little correlation between the Se and P_2O_5 content of these deposits, although rocks of darker coloration (low grade) exhibit higher Se concentrations. The occurrence of significant deposits of independent Se minerals in nature has not been reported, and this may be explained by the fact that considerably increased concentrations of the element would be required for their formation. Usually Se is associated with sulphides, where it is incorporated into the sulphide crystal lattice by a process of isomorphous substitution of S [9]. The similarity in the crystallochemistry of these two elements explains why Se substitutes for S and does not occur as Se mineral in pyrite deposits.

Primary sources of Se in nature are volcanic emissions, and metallic suphides associated with igneous activity. An example of volcanic selenite, selenate and natural selenium may be found in tuff deposits in Wyoming, USA. A thorough review of the occurrence of Se in geologic materials has been tabulated and discussed by Berrow and Ure [14], with significant amounts of Se minerals found to occur in silicic rocks at locations in Russia, China, Canada, the US and New Zealand, among others. In addition, considerable concentrations of Se are associated with various fossil fuel types, including coal and oil, although the latter to a lesser extent. Such associations are significant when considering the impact of airborne combustion products on soils and waters. Secondary sources of Se are biological pools in which accumulation has occurred.

12.3 Origin of selenium in soils

12.3.1 Soil parent materials

The Se content of soils generally reflects the weathering of parent materials, although in certain circumstances atmospheric, and more recently, anthropogenic inputs may influence their composition. In the natural environment, elevated concentrations of Se in soils are associated primarily with volcanic materials,

Table 12.1 Concentrations of Se commonly observed in soils and parent materials

Material	Mean (mg/kg)	Range	Reference
Igneous rocks	0.35	0.09–1.08	14
Volcanic rocks, USA			119
(CO, CA, NM, ID, AK)	< 1.0		
(HI)	< 2.0		
Volcanic tuffs (WY)	9.15	12.5–187	3
Sandstones		< 0.01–0.05	16, 103
Carbonates	0.08		14, 16
Marine carbonates	0.17		
Carbonaceous materials			121
Shales (Western US)		< 1–675	118
Shales (WY)	19.86	2.3–52.0	3
Shales (general)	0.05		16, 122
Mudstones		few–1500	120
Limestone	0.03		122
Phosphate rocks		1–300	120, 121
Coal			
USA	3.36	0.46–10.65	123
Australia	0.79	0.21–2.5	16
Oil		0.01–1.4	14
Soils			
US		< 0.1–5000	16
Selected (CA)	1.5	0.6–1.6	61, 124
UK	0.5	0.2–2.0	17
Selected (Wales and Ireland)		30–3000	18

sulphide ore bodies, black shales and carbonaceous sandstones. The intensity of weathering and leaching processes on these parent materials will define the ultimate Se composition of associated soils. Selenium is apparently readily oxidised during weathering, becoming more mobile with increasing oxidation state [1], and thus in arid alkaline environments, Se is most likely to be present in soils as available selenate. In humid regions, selenite appears to be the predominant species, and under these conditions, Se is less soluble for reasons that will be discussed shortly. In the USA, seleniferous soils are generally concentrated in a broad band covering much of the central continent, and it is because of their widespread occurrence that these soils have been examined in considerable detail [14]. Extensive investigations of soils and subsoils in Wyoming and South Dakota have resulted from the proliferation of seleniferous vegetation (i.e. plants with > 2 mg/kg Se) in this region, which pose a hazard to livestock. Recent identification of seleniferous soils in California has led to a resurgence in investigations of these soil types. Commonly observed concentrations of Se in various parent materials and a range of soils are illustrated in Table 12.1. A number of investigations have determined soil Se concentrations in a variety of soils, illustrating the diversity in the distribution of this element. The amount of Se present in most soils is detectable but highly variable, ranging from trace amounts (< 0.1 mg/kg) up to areas of acute toxicity (up to 8000 mg/kg) [14, 16]. Archer and Hodgson [17] investigated the total and extractable trace element concentrations of soils in England and Wales. Total (i.e. perchloric/nitric acid digestible) Se concentrations were determined in 229 soil samples collected from the upper 15 cm of a variety of agricultural soils. Values determined ranged between 0.02 and 2.0 mg Se/kg, with a median valve of 0.5 mg Se/kg. In contrast to the seleniferous soils commonly found in semi-arid regions such as those of the western USA, the occurrence of toxic concentrations of Se under humid conditions has also been reported [18]. Leaching of Se-enriched carboniferous rock formations in various regions of Ireland has resulted in elevated amounts of Se in the valley soils high in organic matter.

12.3.2 Agricultural materials

The use of Se as an agrochemical encompasses a variety of areas including its use in pest control, and as a dietary supplement for livestock grazed in Se-deficient areas. As a pesticide, Se dissolved in potassium ammonium sulphide ($[K(NH_4)S]_5Se$) has been used to control the occurrence of mites and insects on citrus, grapes and ornamentals [19]. An apparent resistance to this chemical appeared in some types of mites, and the use of this material has since declined. Selenium has also been used in foliar sprays and in the soil application of selenates, with the intention of increasing Se concentrations in plant tissues rendering them toxic to certain insects. The use of this type of pesticide on edible crops was discontinued prior to 1961, due to concerns about human health effects. The main agricultural use for Se is without doubt, as a dietary supplement in animal feeds, in order to prevent problems associated with Se deficiency.

12.3.3 Atmospheric deposition

The major anthropogenic source of atmospheric Se is derived from the combustion of coal in electric power plants, and in industrial, commercial and residential burners. Nriagu and Pacyna [20] reported that of an estimated worldwide maximum emission of Se of 5780×10^3 kg during 1983, 2755×10^3 kg were derived from coal combustion. Other noteworthy sources of atmospheric Se include the combustion of oil (827×10^3 kg/yr) and release of Se during industrial pyrometallurgical processes such as Cu–Ni refining (1280×10^3 kg/yr). A mean global emission rate of Se from natural sources has recently been estimated to be between 6000 to 13 000 t/yr [21], with between 60 and 80% of the total Se emission arising from marine biogenic origin. The inventory of Nriagu and Pacyna [20] clearly suggests that soils also are receiving large quantities of Se from a wide variety of industrial wastes, the two principal sources being the deposition of ash residues from coal combustion and the general wastage of commercial products on land.

12.3.4 Sewage sludges

There is little information concerning concentrations of Se in sewage sluges or sewage sludge-amended soils. In the USA, a study of sewage sludges from 16 metropolitan areas reported Se contents ranging between 1.7 and 8.7 mg/kg [22]. A limited evaluation of sludges from the greater Los Angeles area indicates a range of values between 2 and 6 mg Se/kg [23], whilst a median value of 1.1 mg Se/kg was calculated for sludges generated in a survey of 40 cities. Data are available which indicate that Se is generally removed from municipal wastewater at the primary stage, and as such is concentrated into sewage sludge. Primary and secondary effluent contain as little as 0.2% of the original Se, most of which is probably drawn from the manufacturing industries discussed earlier.

Compared with other metals, the Se content of most sewage sludges is quite low. Logan *et al.* [24] investigated the impact of Se in soils amended with sewage sludge on barley (*Hordeum vulgare* L. var. Briggs), Swiss chard (*Beta vulgaris* L. var. Fordhook) and radish (*Raphanus sativa* L. var. Cherry Belle). The authors reported that only 13–25% of the sludge-applied Se could be accounted for in soil in the 0–15 cm depth of incorporation, with no measurable evidence of leaching to a depth of 150 cm. This is much lower than for the other elements, such as Cd, Zn, Ni and Pb, determined in these experiments by Chang *et al.* [25], from which $> 90\%$ at 0–15 cm depth were recovered. The authors concluded that losses of Se were due either to leaching or volatilisation. Uptake of Se by all crops were also low, with little evidence for correlation between applied Se and that accumulated by the plants.

12.4 Chemical behaviour in soils

12.4.1 Speciation: Inorganic ions

The oxidation states of Se are II, III, IV and VI, all of which may be found in soils under a variety of conditions. The speciation and distribution of Se is dependent

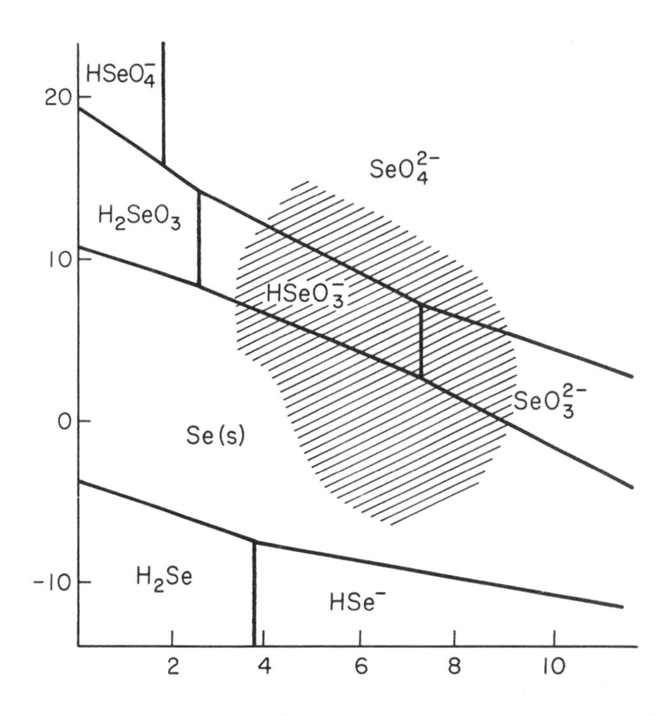

Figure 12.1 pE–pH diagram for the system Se–H_2O under the condition Se_T = 78.96 μg/l (1 mmol/m^3). From Neal *et al.* [38].

on a number of these soil conditions, not the least of which are pH and the oxidation–reduction (redox) potential. Other physicochemical factors which may influence the speciation of Se in soils include the chemical and mineralogical composition, microbial intervention, and the nature of the adsorbing surfaces. Before considering the complex interactions of these various components, however, one must consider the behaviour of Se under simple chemical conditions.

Figure 12.1 illustrates the redox (E_h) speciation of an aqueous system containing 1 mmol Se/m^3 (78.96 μg/kg) as a function of varying pH and E_h conditions. This predominance diagram has been constructed from thermodynamic data compiled by Sposito *et al.* [26], and includes only inorganic complexes (pE = E_h/59.16, where E_h is the conventional electrode potential in mV). The shaded portion of the diagram indicates pE–pH environments found typically in soils, and from this it is evident that under most soil conditions only elemental Se, selenite and selentate should be stable. However, reduction to selenides may occur under suitable anoxic conditions. Selenium (IV) which may exist either as selenite (SeO_3^{2-}) or biselenite ($HSeO_3^-$) depending on pH, and selenate (Se(VI)) are the aqueous oxygenated forms of Se and are primarily responsible for the reactivity of this element in soils. In comparison with S, Se is a more redox-stable element. Of the oxyanions selenite is the favoured redox-stable form over selenate, whereas sulphate is the favoured

Table 12.2 Equilibrium potentials of oxidation-reduction reactions pertinent to soil systems.

Reaction	Eh at pH 7 (V)	Reference
O_2/H_2O	0.82	125
NO^-/NO^-	0.54	125
SeO_4^{2-}/SeO_3^{2-}	0.44	125
MnO_2/Mn^{2+}	0.40	26
SeO_3^{2-}/Se^0	0.27	26
$FeOOH/Fe^{2+}$	0.17	125
SO_4^{2-}/HS^-	−0.16	125
CO_2/CH_4	−0.24	27
$N_2NH_4^+$	−0.28	27

form of sulphur species. It is interesting to consider the relative positions of the Se redox couples, SeO_4^{2-}/SeO_3^{2-} and SeO_3^{2-}/Se^0, compared to others that are commonly found in aqueous systems and soils (Table 12.2). Thermodynamic calculations would predict that SeO_4 should be reduced after NO_3 but before MnO_2 as the pE value decreases in a soil. This sequence was confirmed in a study of kinetic reduction processes in a 1:8 (w/w) suspension of soil in its own water extract, incubated without O_2 supply for up to 250 h [28]. Masscheleyn *et al.* [29] demonstrated a similar sequence in suspensions of Se-containing sediments under controlled pE and pH conditions. Selenate was detected only when NO_3^- was present, and these authors concluded that pE and pH were key factors in the biogeochemistry of Se in the sediments. Other authors have also noted the importance of NO_3 on possibly retarding SeO_4 reduction [28, 30–32], and this effect may be chemical, i.e., the poising of the pE value at a high level because of the abundance of NO_3 [33], or biochemical, via direct inhibition of respiration in SeO_4 reducers [30]. A recently isolated *Pseudomonas* species has been shown to utilise a separate SeO_4 reductase enzyme to catalyse the reduction of SeO_4 to SeO_3 [34], but that the synthesis of this enzyme is inhibited by the presence of NO_3^-. If a sufficiently low Eh level is reached SO_4^{2-} reduction, CH_4 fermentation and NH_4^+ formation will occur.

Concentrations of Se oxyanions in soil, surface and ground waters are also controlled by the solubility of Se minerals in soils. A recent study by Elrashidi *et al.* [35] considered a theoretical development of equilibrium reactions and constants for a wide range of Se minerals and inorganic solution species. These authors indicated that, in general, metal selenate and selenite minerals are too soluble to be stable in most soils, and that the formation of selenides, such as $Cu_2Se(c)$, PbSe(c) and SnSe(c), under reduced conditions would prevent the precipitation of elemental Se. Similar calculations were performed to evaluate Se mineral stability in anoxic soil and sediment pore water systems in which reducing conditions lead to microbially mediated transformations of Se oxyanions [36]. Elemental Se and the formation of FeSe and $FeSe_2$ were determined to control Se solubility, even though Fe selenides were found to be thermodynamically unstable with respect to their sulfur counterparts. In a study of reservoir sediment suspensions under controlled redox (500, 200, 0 and –200 mV) and pH (5, natural, and 7.5) conditions [37], decreasing concentrations of Ni, Cu and Zn upon reduction were seen as evidence that the precipitation of metal selenides under anaerobic conditions could provide a method for the removal of Se from solution. Iron selenites are not expected to exist in acid soils, contrary to earlier suggestions by Lakin [12] and Rosenfeld and Beath [3], although the formation of ferric oxide-selenites should not be excluded as the complexes controlling selenite solubility in alkaline soils. A similar complex involving gypsum ($CaSO_4$) and selenite may also attenuate soluble selenite concentrations [38], but such coprecipitated complexes are difficult to identify or characterise at present.

Calculations of stability constants for selenate, selenite and selenide complexes have shown that only such complexes as $MnSeO_4^0$ and $NiSeO_4^0$, $NaHSeO_3^0$, and $KHSe_0$ and NH_4HSe^0, appear to make any contribution to soluble Se in normal cultivated soils, depending on redox conditions. Under inorganic conditions where a high redox status prevails, SeO_4^{2-} appears to be the major contributing species above pH 2, while in systems of moderate redox both SeO_3^{2-} and $HSeO_3^-$ are significant contributors to soluble Se concentrations. It is most likely, however, that under natural conditions, the presence of biological reducers will have a significant impact on oxyanion stability.

12.4.2 Speciation: Organic ions

There is less information available regarding the behaviour of organic Se compounds in soils. A significant amount of data pertaining to the chemistry of organic Se has been accumulated, with recent emphasis in the area of seleno-carbohydrates, seleno–aminoacids, selenopeptides and other Se containing molecules of biochemical interest [39]. However, compounds found in soil systems are often unidentifiable since they form either as products and by-products of microbial processes, or through direct reaction with organic materials such as humic and fulvic acids. Various soluble organic Se species have been identified [40], including nonvolatile organic selenides, seleno–amino acids, a dimethylselenium ion and the volatile methylated forms dimethylselenide (DMSe) and dimethyldiselenide (DMDSe). Dimethylselenide, a member of the largest class of identified organic Se compounds, is one of the simplest and perhaps most environmentally significant organic Se compounds in soils. The production of this compound is generally associated with microbial methylation of inorganic Se species, resulting in the generation of a volatile organic species which may be lost to the atmosphere from both soil and water surfaces. Many investigations of Se speciation in soils generally identify soluble organic selenium as a single entity owing to problems encountered in the analytical determination of organic Se species. More recently, however, efforts have been made to fractionate organic Se, as reflected by the work of Abrams *et al.* [41]. Organic Se compounds extracted from a selected topographic sequence of Californian soils with a solution of alkaline pyrophosphate (0.1 M NaOH/0.1 M $Na_4P_2O_7$) were fractionated into humates, and hydrophilic and hydrophobic fulvates. Selenomethionine was also identified in this study, and may be a potential source of Se for uptake by plants. Similar results were found in a study of Se fractions present in soil profiles from three Swedish podzols [42]. Se was shown to be considerably enriched in soil organic matter, most markedly in the hydrophobic fulvate fraction, and it was noted that the distribution of Se among the organic fractions differed significantly from that of sulphur.

In summary, the chemical species in which Se predominates in soils will determine its reactivity, and ultimately, its behaviour. The chemistry and mobility of the inorganic oxyanions, selenate and selenite, differ significantly. Although under inorganic conditions these oxyanions are reported to be stable with respect to changing redox conditions, the presence of organic matter and bacteria suggest that reduction will be more likely to occur. The prevalence of unidentified organic species serves to complicate the picture further, and it would seem to be of considerable importance that a greater understanding of these species and their interactive behaviour should be the focus of future research.

12.4.3 Adsorption

The adsorption of ions on to solid surfaces offers an important mechanism by which a compound may be attenuated by soil, thus affecting its progress through the profile. The difference in behaviour between selenite and selenate may be evaluated by comparing their adsorptive properties. Most researchers agree that selenite is adsorbed by soil and specimen minerals to a greater extent than selenate, and that this difference is most likely due to the adsorption mechanisms involved.

It is widely established that selenite is adsorbed on solid surfaces by a mechanism known as ligand exchange. Such an exchange usually involves a hydroxyl group residing on the surface of clay particles or metal hydrous oxides [43, 44]. This mechanism of specific adsorption is thought to be similar to those of ions such as phosphate and arsenate, and as such is highly dependent on pH, since the presence of hydrogen ions will affect the ability of the surface to attract the negatively charged ion. As a result, the amount of selenite, or indeed any anion, adsorbed by a solid surface will decrease as a function of increasing pH. This behaviour is in direct contrast to that of cations, which are adsorbed to a greater extent with increasing pH.

In contrast, selenate behaves in a manner attributed to non-specifically adsorbed ions such as sulphate and nitrate. This mechanism is believed to involve the formation of outer-sphere surface complexes which incorporate a water molecule between the surface and the adsorbing species [45]. The existence of these mechanisms cannot be proved using macroscale investigations such as the construction of adsorption isotherms, although the latter may be very useful in comparing the behaviour of various ions and surfaces under experimental conditions. However, a recent investigation utilising entended X-ray absorption fine structure (EXFAS) measurements to probe the complexes formed by Se oxyanions at the goethite–water interface have shown that, indeed, selenate does form a weakly bonded, outer-sphere complex, and that selenite is more strongly bonded as an inner-sphere complex [46]. In realistic terms this means that selenite will be retained in soils in favour of selenate and other ions such as sulphate and chloride, but that it may be replaced by phosphate. In a practical sense, this would seem to indicate that the application of phosphatic fertilisers to seleniferous soils may result in increased Se availability.

Adsorption isotherms and pH envelope experiments have been used extensively to examine the behaviour of aqueous ions at solid surfaces. Investigations of selenite adsorption by specimen minerals have given insight into the mechanistic aspects of Se solubility in soils. Hingston *et al.* [43] and Hingston [47] investigated selenite adsorption on goethite and gibbsite and reported an increase in both the pH of the suspension and the negative surface charge on the oxide which they attributed to be a result of specific adsorption. Hamdy and Gissel-Nielsen [48] showed selenite adsorption by Fe oxides to be extensive, rapid and decreasing with pH between pH 3 and 8. These authors also showed selenite sorption by clay minerals to be dependent on pH rather than layer type, although evidence for greater sorption capacity by 1:1 minerals, such as kaolinite, than 2:1 minerals vermiculite and montmorillonite, was found. Bar-Yosef and Meek [49] attributed this difference to the presence of a more effective negative electric field near the surface of montmorillonite than kaolinite, restricting the accessibility to adsorption sites. Scott [50] observed selenite adsorption at goethite and birnessite surfaces to occur within a few minutes, and reach equilibrium within a few hours. Redox products (Mn(II) and Se(VI) were observed when selenite was added to aqueous suspensions of birnessite, although the appearance of measurable selenate occurred slowly as a function of the amount of selenite adsorbed.

Significant adsorption of selenate on specimen minerals has been reported [49, 51–54]. Extensive ion removal has been found at low pH (e.g. 80% of 10 mmol/m^3 selenate pH 4.5) by Fe oxides [51, 52] and kaolinite [49]. Although some evidence exists postulating the formation of inner-sphere surface complexes [53], selenate adsorption is generally considered to be a non-specific process, and, unlike selenite, is sensitive to changes in ionic strength [46].

It is unlikely that weathered heterogeneous adsorbents will behave identically to specimen minerals, since adsorption characteristics of soils depend on the extent of pedochemical weathering [45]. In addition, the relative proportions of adsorbing surfaces, such as those of oxides and clay minerals, vary in soils together with the absolute mass of adsorbent present. John *et al.* [55], in a study of 66 New Zealand soils, concluded that in general selenite adsorption increased as a function of soil weathering. Although there is little evidence in the literature for selenate adsorption by soils under experimental conditions [56, 57], what there is, is conflicting. Singh *et al.* [58] and Goldhamer *et al.* [59] both reported at least 15% of 5–150 mmol SeO_4^{2-}/m^3 adsorbed by soils of varying composition, whereas Neal and Sposito [56] and Goldberg and Glaubig [57] were unable to detect any significant adsorption, even at low pH (< 5.0). Alemi *et al.* [6] observed selenate adsorption on soil in batch experiments, but this phenomenon was not evident in soil column studies. Similarly Fio *et al.* [61] were also unable to detect any adsorption of selenate in soil column studies, but in both instances adsorption of selenite was observed [60, 61].

In addition to the physical and chemical nature of the adsorbent, studies have shown that solution composition may be of considerable importance with respect to differentiating adsorption mechanisms. Changes in ionic strength, for example, may indirectly affect the charge distribution on a solid surface and in the surrounding aqueous solution, thereby altering the attractive or repulsive interaction between adsorbing anions and the surface [62]. Experimental studies by a number of researchers, however, indicate that variation of ionic strength has little effect on the extent of selenite sorption [38, 43, 46], whereas selenate adsorption varies significantly. Changes in ionic strength could be considered strictly as increases (or decreases) in the concentration of competing anions. In the case of chloride, this competition is expected to be minimal, but certain anions, notably phosphate and sulphate, may be considered as competitors with both selenite and selenate in aqueous colloidal systems. A distillation of data from a variety of sources suggests the following affinity sequence for selected anion adsorption on soils and specimen minerals [43, 44, 54, 56, 63–65].

phosphate > arsenate ⩾ **selenite** ⩾ silicate >> sulphate ⩾ **selenate**
> nitrate > chloride

An example of the relative behaviour of four of these anions is illustrated in Figure 12.2. Selenate (0.74 mg SeO_4^{2-}/l), selenite (0.24 mg SeO_3^{2-}/l), sulphate (96 g SO_4^{2-}/l) and chloride (1.7 g.l) were equilibrated with an alluvial soil (fine

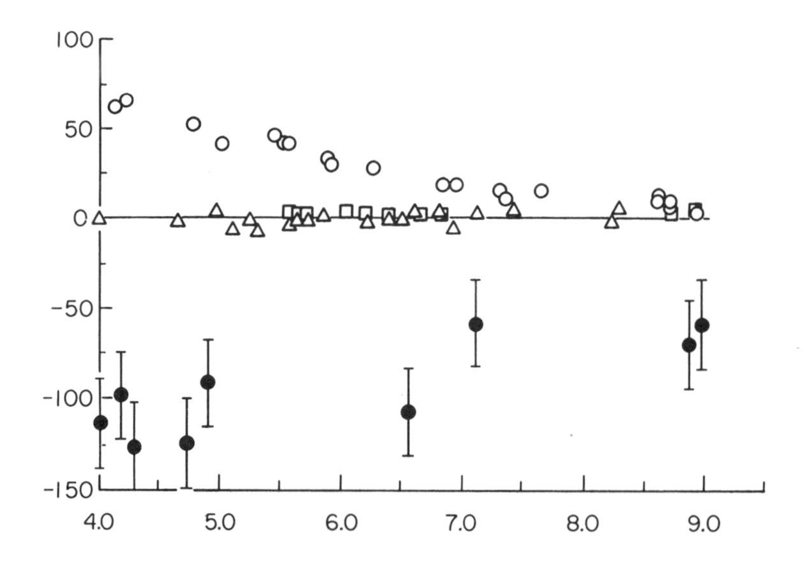

Figure 12.2 Adsorption of four anions on Panhill soil. ○, selenite; △, selenate; □, sulfate; ●, chloride. (From Neal and Sposito [56].)

loamy mixed thermic, Typic Haplargids) from the Panhill soil series, California, USA, and the amount adsorbed expressed as a percentage of the initial anion concentration [56]. The adsorption of selenite exhibits a phenomenon known as a pH envelope, the typical response of an anion which is adsorbed by inner-sphere surface complexation whereby increasing pH results in decreased attenuation. Selenate and sulphate, almost indistinguishable in their response, showed no adsorption under the conditions of this experiment, while chloride exhibits the phenomenon of anion exclusion. The latter may be described mechanistically as the repulsion of an ion by a charged surface of like charge [45]. If phosphate were to be included in this figure, then it would probably exhibit a response similar to that of selenite.

12.4.4 Biological transformations

In the soil environment, most transformations of Se appear to be microbially mediated, and these transformations may fall in either one or more of three defined categories: oxidation and reduction; immobilisation and mineralisation; and methylation. Increased availability and toxicity that arises from the presence of the more oxidised forms of Se have led researchers to focus their attention on the microbial reduction of these species, to its elemental form or through the incorporation into organic compounds. Many fungi, bacteria and actinomycetes that occur in soils are capable of reducing inorganic salts of Se either to elemental forms that appear as red intracellular deposits [66], or to both volatile and non-volatile organic compounds, within a relatively short time [67]. Under anaerobic conditions, the evolution of hydrogen selenide (H_2Se) [68], from soils amended with selenate, selenite, and elemental Se has been reported [68], and the formation of metal selenides and/or elemental Se in O_2-containing zones postulated.

Immobilisation of Se, i.e. processes which effectively reduce the availability to plants, animals or microorganisms, involves the formation of many organic compounds, including amino acids, proteins, selenium compounds and selenides [66]. The chemical similarity of Se to S results in the production of many compounds analogous to those bearing sulfur, and the metabolism of organic Se compounds is considered to follow similar pathways as those of S analogues also [69]. Recent investigations of methods to purify agricultural wastewater containing Se have resulted in the isolation of two organisms which together reduce selenate to elemental Se. One organism, *Pseudomonas stutzeri*, anaerobically respires selenate to selenite, while the second, a strict anaerobe, reduces selenite to Se [70, 71]. However, the inhibition of selenate reduction by these organisms in the presence of nitrate suggests that they use nitrate respiratory systems to reduce selenate. In contrast, *Thauera selenatis* has been shown to reduce selenate and nitrate concomitantly [71] using acetate as a preferred electron donor. Similarly, metabolism of organic Se compounds is considered to follow similar pathways as those of S analogues [69].

One of the more prominent effects of biological mediation on Se in soils is its conversion to volatile organic forms. Volatile organic species evolving from soils include dimethylselenide (DMSe)CH_3SeCH_3, dimethyldiselenide (DMDSe) $CH_3SeSeCH_3$ and dimethylselenone $CH_3SeO_2CH_3$ [72], and these are considered to be produced predominantly by soil fungi [67, 73, 74, 75] and bacteria [66, 76]. The addition of certain C sources to facilitate elevated volatilisation rates by these microbial populations has been under increased scrutiny in recent years, since conversion of potentially toxic selenate to a volatile species may be an important method for controlling the mobility and toxicity of Se in soils [77], pectin [76] and certain proteins, including egg albumen [67], have been found to increase Se volatilisation in both soil and water samples, respectively. In soils, Karlson and Frankenberger [75] have identified the fungi *Acremonium falciforme*, *Penicillium citrinum*, and *Ulocladium tubercalatum* as active converters of both selenite and selenate, although the latter to a far lesser extent, and have shown accelerated volatilisation rates can be achieved through the addition of a C source such as pectin. Similar findings are reported by Thompson-Eagle and Frankenberger [67] for Se volatilisation by *Alternaria alternata* in seleniferous waters.

12.4.5 Selenium in ground waters

The extent to which Se is present in groundwaters depends on a number of factors, including the depth to groundwater, and the type of material through which Se-bearing water must percolate. In areas of shallow groundwater, such as much of the San Joaquin Valley, California, USA, Se enrichment is considered to have occurred as a result of groundwater evaporation [78]. In high water tables of this type, Se is found predominantly as selenate, at concentrations between < 1 and 3800 μg/l, owing to alkaline oxic conditions. As this species, Se is most mobile, and may, therefore, be easily transported in the groundwater. It is unlikely that deeper groundwaters underlying acid soils will contain much Se. In view of the highly oxic conditions required for Se mobility, any Se present in these waters is most likely to be in the form of a reduced immobile species.

12.4.6 Contrasting behaviour in marine waters and sediments

The amount of Se present in ocean waters would reach hazardous concentrations, were it not for adsorption and precipitation mechanisms involving hydrous oxides, organic matter and Fe sulphides [3]. As a result, marine sediments are usually enriched in Se, of both mineral and biological origin, the latter resulting from decayed organisms which incorporated Se during their life cycle. The distribution of Se in marine waters indicates increasing concentrations with increasing depth [79]. Selenate is the predominant species in surface waters ($\sim$ 20 μg/l), with an increasing contribution from selenite as a function of depth, and increased Se concentrations are seen at points of oxygen depletion. The behaviour of Se in sediments appears conflicting, however, since both oxidation of selenite to selenate, and the reverse reaction, have been reported [80]. Both oxidation and reduction processes in freshwater sediments taken from Canadian lakes were attributed to either biotic processes or the presence of organic materials. The latter, together with sesquioxides, were found to play an important role in the retention of selenite by these sediments.

12.5 Soil–plant relationships

The uptake and accumulation of Se by plants may be influenced by a number of environmental factors, the most significant of which appear to be the concentration and form in which Se occurs in soils. Other factors which contribute to the eventual Se contents of plants, such as pH, soil mineralogical composition, plant species, etc., may combine under certain conditions to enhance or attenuate Se translocation from soil to plants. Excellent reviews of these topics have been addressed in recent years, and provide greater detail than it is possible to present within the confines of this chapter [1, 3, 81].

The previous sections of this chapter discussed extensively the importance of speciation in the behaviour of Se in soils. It is apparent, therefore, that the predominant species of Se present in a soil will have a marked effect on its uptake and accumulation by that plant. Selenate, the most mobile, soluble and least well sorbed of the inorganic species, will be the dominant form of Se in aerated alkaline soils, such as those of semi-arid regions, and as such will be readily available for plant uptake. However, irrigation of such soils could leach the Se below the root zone and effectively render Se unavailable to plants. It is unlikely that this process would be effective in regions where excessive evaporation of high subsurface waters results in enhanced salt concentrations, including Se, at the soil surface. In neutral and acid soils, selenite is the predominant aqueous species (Figure 12.1), and is for the most part unavailable due to adsorption on clays and hydrous oxides (Figure 12.2). Experiments investigating Se uptake by plants as a function of pH have substantiated these findings. Alfalfa (*Medicago sativa*) grown in soils of both acid and neutral pH have been shown to contain higher levels of Se in plants grown on the soils of higher pH [82]. The addition of lime amendments to soils resulted in increased Se concentrations in plants relative to those grown on the unlimed soil [83]. Similarly, the uptake of Se by wheat (*Triticum aestivum*) and rape (*Brassica napus*) grown in a sandy soil at pH 5 and pH 7 was greatest at the higher pH [84].

Whilst pH may affect the uptake of Se through surface interactions in soils, the extent to which such phenomena will affect plant uptake will depend on soil texture and mineralogy. Soil texture has been found to affect the uptake of Se by a variety of plants including alfalfa, wheat and rape [84–86]. Plants grown in sandy soils have been found to contain significantly more Se compared to those grown on silt loam-, clay-, and organically amended soils. In each instance negative correlations between soil clay content and Se concentrations in plant tissues were established, and in some instances the uptake of selenate was as much as 10 times that of selenite. Such behaviour is predictable if the adsorptive effects of selenite versus selenate on clay surfaces are considered. The interaction between Se and organic material, on the other hand, is less well understood, although its effects on Se availability have been measured. Studies have shown a decrease in Se availability as a result of elevated organic matter content in soils [58, 87], whilst Davies and Watkinson [88] reported the opposite effect when they considered the uptake of added selenite from peat soil versus mineral soils. The complexity surrounding the issue of organoselenium compounds in soils prevents a clear explanation of these phenomena, since there is little information concerning the availability of these forms.

The earlier discussion of adsorption phenomena revealed the adsorption of Se by soil to be affected by the presence of other cations and anions. Anions such as phosphate will readily displace selenate ions from surfaces and compete with selenite for inner surface complexation sites on soil surfaces. Sulphate, on the other hand, does not affect selenite adsorption, but will compete with selenate. As a result, the presence of anions such as these in soils will influence the extent to which Se is adsorbed by a soil, and, depending on their relative concentrations, its availability to plants. Interactions between Se and phosphate in soils have been reported to affect Se accumulation by plants [88, 90]. Increasing applications of P to soils resulted in increasing concentrations of Se in clover, while increasing amounts of soil-applied selenite resulted in increased levels of plant P. Similarly, the depressive effect of P on selenate accumulation by alfalfa (*Medicago sativa* L. cv Germain WL 512) grown in sand culture irrigated with modified Hoagland's solution [91] was found to be highly significant, whereas the presence of arsenate stimulated Se uptake under identical conditions. Whilst competition for surface sites on soils may be one mechanism which could explain these results, the effect of greater root growth resulting from increased P fertilisation may also be an important consideration, thereby providing the plant with a larger root–soil contact area and a larger volume of soil from which Se could be extracted [89]. Data generated from studies involving plant essential elements such as P require careful consideration, since additional effects, such as effective dilution, alleviation of plant deficiencies, and Se contamination from the use of rock phosphate, may make significant contributions to the availability of Se.

The interaction between Se and sulphate in soils is well known. Several studies have shown that sulphate has a much greater effect on plant uptake of selenate than selenite [92–95]. In general, increasing concentrations of sulphate in soil solution results in decreased accumulation of Se by plants [95, 96], with much less Se accumulated in the edible tissues of alfalfa (*Medicago sativa* L.), barley (*Hordeum vulgare* L.), beet (*Beta vulgaris* L.) and tomatoes (*Lycoperiscon esculentum* Mill) than in the edible portions. Mikkelsen *et al.* [97] investigated the effect of both pH and Se oxidation state on Se uptake by alfalfa grown in nutrient solutions containing between 0 and 3.0 mg Se/l as either selenate or selenite. Both tissue concentrations of Se and shoot growth were affected regardless of Se oxidation state, whilst root growth was reduced only in the presence of selenate. Additions of selenate applied to acid soils from the southeastern USA were shown to have a greater toxic effect on the growth of a forage species (*Sorghum vulgare*) than comparable amounts of added selenite, resulting in reductions in plant biomass and a significant increase in Se tissue concentrations [85]. A study by Mikkelsen *et al.* [98] investigated the effect of flooding conditions, such as those employed in rice cultivation, on Se uptake. Selenium accumulated in all rice (*Oryza sativa* L. cv. M101) plant parts, although plants grown in a flooded soil with added organic matter had tissue Se concentrations 20% less than that of rice grown under both unamended-flooded and non-flooded conditions. It is probable that the inclusion of organic material resulted in the enhanced reduction of selenate to less plant available forms, such as selenite, organoselenium compounds and selenides, reducing both Se phytotoxicity and yield loss.

Selenium is absorbed by plants in both organic and inorganic form. The uptake of selenite and selenate into roots does not appear to follow the same pathway. Uptake of selenate has been shown to be an energy-required process, whilst selenite assimilation is energy-independent [90, 100]. Roots immersed in a solution containing selenite do not accumulate Se to concentrations greater than the external bathing solution, whereas selenate may be actively accumulated to concentrations far exceeding the concentration in which the roots reside. Once present in the root, a differential preference for the translocation of selenite and selenate to the plant shoot is exhibited. Selenate moves unchanged through the plant [101] and may reach concentrations many times in excess of the external solution [102]. Selenite, on the other hand, is rapidly converted to selenate and organic Se compounds. Conversion of selenate into organic compounds is believed to occur in the leaves [103]. The primary compounds formed include Se analogues of amino acids which may be incorporated into plant proteins in some plants and result in toxicity [104]. Selenium accumulator plants have developed a mechanism which prevents inclusion of Se-containing compounds in proteins, and thus are able to avoid Se-induced phytotoxicity.

The capacity of different plants to accumulate Se varies widely. Primary Se accumulators, which include several species of *Astragalus* (Leguminosae), *Conopsis* (Compositae), *Stanleya* (Cruciferae) and *Xylorhiza* (Compositae), possess the ability to extract and/or accumulate large amounts of Se (10^3 to 10^4 mg Se/kg dry wt) from soils that contain only several mg/kg of Se [105], and store it primarily in organic form [1]. Secondary indicator plants, including species of *Aster*, *Atriplex Mentizelia* and *Sideranthus* accumulate much lower concentrations of Se, usually only a few hundred mg/kg, which is generally present as selenate and organic selenium compounds. In general, most crop plants, grains and native grasses do not accumulate Se to levels above 50 μg/kg. A number of investigations concerning Se accumulation by crop species indicate that although there are differences in the potential of various plant species to concentrate Se, all plants and plant parts can accumulate Se when grown in soil containing moderate levels of water-soluble Se [81]. The species wild brown mustard *(Brassica juncea* Czern L.), Old Man saltbush (*Atriplex nummularia* Lindl L.), creeping saltbush (*Atriplex semibaccata* R. Br. L.), *Astragalus incanus*, and a tall fescue (*Festuca arundinacea* Schreb L.) have been indentified as capable of accumulating Se when grown in either selenate or selenite treated soil, although each species accumulated large amounts of Se from the selenate treated soil [105]. The effect of sulphate on Se uptake by primary accumulators such as *Astragalus bisculcatus* ((Hook.) A. Gray) have shown preferential accumulation of Se [106]. This finding is contrary to the competitive behaviour of S and Se reported for most non-accumulator plants, such as alfalfa, (*Medicago sativa* L.), and suggests that primary Se-accumulators have a unique ability to accumulate selenate inspite of competition from sulphate ions [106].

The volatilisation of Se from plants has been shown to be a significant pathway for the removal of selenium from soil [107, 108, 109]. Species such as *Astragalus bisulcatus*, broccoli (*Brassica oleracea* Waltham 29 Midseason variety), rice (*Oryza sativa*), and cabbage (*Brassica oleracea capitata* L.) have been reported to have the highest Se volatilisation rates, and studies of a wider range of crop species have shown that the rate of Se volatilisation is highly correlated with the concentration of Se in plant tissues. However, it has been suggested that sulphate salinity in soils such as those found on the west side of the San Joaquin Valley, California, USA, may diminish substantially the rate of Se volatilisation by plants not specifically adapted for such an environment [109].

12.6 Polluted soils: case histories

12.6.1 California

One of the most recent experiences with Se pollution occurred in the San Joaquin Valley of California, USA. The area affected is one of the world's most productive agricultural areas, and as such has attracted a great deal of attention regarding the environmental impact, and the instigation of 'clean-up' programmes. A number of recent publications are available in the literature which give detailed discussions of wide-ranging aspects of this problem.

The location of the San Joaquin Valley in California is illustrated in Figure 12.3. Approximately 400 km in length, this wide, almost flat, valley produces a variety of vegetable produce, and has an average of only 10 frost nights in the entire growing season per year. The importance of this region in supplying food for the 'dinner tables of America' cannot be over-emphasised; its commercial significance explains the intensity of purpose with which this problem has been addressed. The Se which exists in the soils of the western San Joaquin Valley (generally those to the west of the San Joaquin River) is believed to be of natural origin, and to have originated through the following processes. During the Jurassic and Cretaceous periods, the deposition of sediments comprising sandstones, shales and conglomerates, are thought to have included seleno-sulphides of Fe. The subsequent uplifting of these sediments to form what are now known as the Coast Ranges exposed these sediments to oxidative conditions, and Se was released as selenite and selenate salts, while sulphur appeared as sulphates [110]. Erosive processes on both the east and west sides of the valley have given rise to the wide, flat fertile valley of today. However, the characteristics of the two sides are quite different because of the nature of the parent material. Alluvium on the east side is derived from the granitic rocks of the Sierra Nevada mountains, and contains few native salts or trace elements. In contrast, however, the finer textured west-side alluvial soils are derived from the sedimentary Coast Range, and contain significant quantities of soluble mineral salts and trace elements including Se, As, B, Cd, etc. [11]. Irrigation, used to supplement precipitation and increase crop yield, has resulted in the increased salinisation of these soils, and encouraged the use of excessive amounts of water to flush the salts below the root zone to maintain crop yields. The presence of impermeable strata combined with continuous leaching has resulted in the elevation of the groundwater table and the salinisation of the root zone. In response to this problem a network of subsurface collector drains and

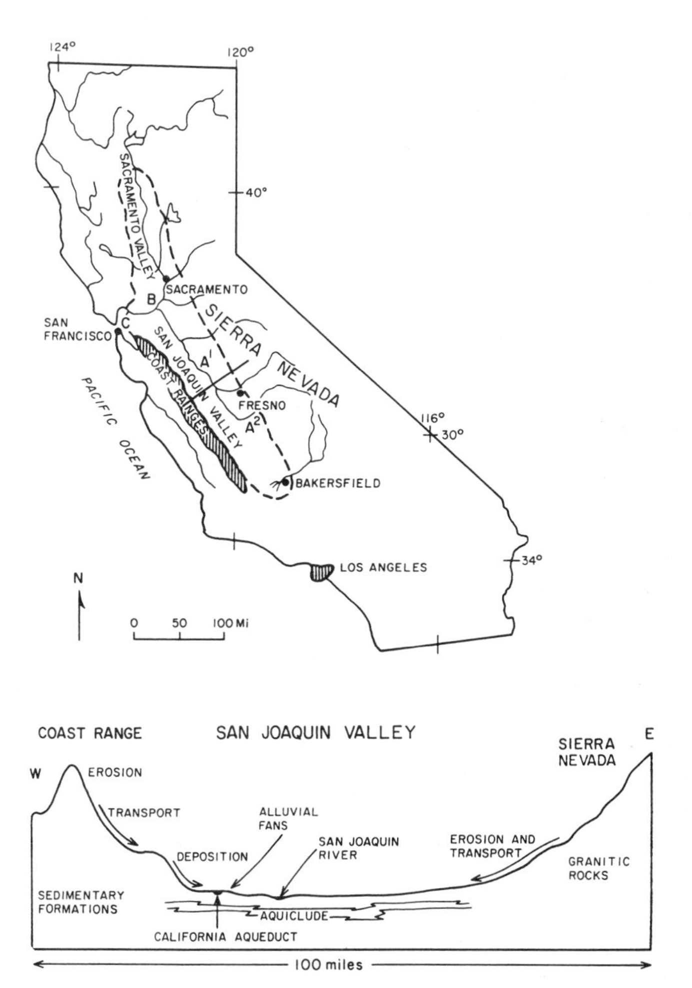

Figure 12.3 Location of the San Joaquin Valley, California, USA.

open canals were installed and the drainage water from the western San Joaquin Valley directed north with the intention of discharging it into the Sacramento delta. However, in 1973 environmental concerns halted this construction of the primary exit drain (the San Luis Drain) and this resulted in the diversion of drainage runoff to a holding reservoir (the Kesterson Reservoir). In 1978, Kesterson Reservoir began receiving subsurface drainage waters, but not until the early 1980s were adverse effects noted in the form of wildfowl embyro death and deformity. The detection of elevated concentrations at the Kesterson Reservoir, which have since been attributed to the subsurface drainage waters collected from the alluvial soils of the western San Joaquin Valley, have given rise to considerable concern not only with the immediate problem of cleaning up Kesterson Reservoir, but of controlling Se-containing drainage waters. Dissolved Se concentrations in drainage waters vary widely, as do those in shallow groundwaters. The latter have been reported to range from undetectable to several thousand μg/l [112]. The highest concentration measured in the San Joaquin River was 2 μg/l, a level well below the public health standard of 10 μg/l.

Despite the termination of inputs of drain waters into the Kesterson Reservoir, a substantial residual inventory of Se, other trace elements, and salts was left in the previously ponded soils. While great variability exists, the average total Se concentration in the surface 0.15 m of soil is about 2 mg/kg, although below this depth concentrations are generally less than 2 mg/kg [113]. This partitioning is the result of dissolved selenate in the pond waters entering the anoxic pond bottoms and undergoing microbial reduction to the relatively unavailable forms of selenite, elemental Se, and various forms of organic Se [114]. Thus, during ponding, a large proportion of Se in the infiltration waters is selectively immobilised within the surface soils. However, upon re-exposure to aerated conditions with the cessation of pond operations, the previously insoluble and adsorbed Se inventories of the surface soils are susceptible to re-oxidation, and thus may become available for translocation and entry into the food chain [113]. One of the site management options considered for the Kesterson Reservoir entailed the excavation of surface soils to a depth of 0.15 m or more in many of the former ponds, which would then be contained indefinitely in a nearby lined landfill. However, the distribution of Se below this depth combined with the yearly rise of the shallow water table was found to be counterproductive to attaining the clean-up goal of 4 mg Se/kg soil. Thus, problems associated with present day clean-up and future management practices have resulted in diverse investigation of Se behaviour in the environment. Among the remedial approaches that are being considered are the use of Se-accumulator plants such as saltgrass and *Astragalus*, to remove Se from the soil [105, 106, 108, 109, 115]. Specifically, plants might be grown such that maximal Se is accumulated in their aerial parts which would then be harvested, removed from the site, and disposed of elsewhere. Complications to this approach arise from the ability to select plants that not only accumulate Se but can also tolerate soil salinity and high concentrations of boron, such as *Astragalus bisulcatus* and *A. racemosus* [115]. Volatilisation of Se as alkylselenides by soil microbes has also been recommended as a suitable remedial approach, especially when the process may be enhanced by amendment with organic C sources [116]. In addition, the stimulation of Se biomethylation from evaporation pond water through the addition of specific amendments such as proteins [67] is being considered. Future disposal of drainage water perhaps remains the most challenging problem, although reuse for the irrigation of salt-tolerant crops is under investigation, and could reduce the amount of drainage water requiring disposal [110]. Where such an approach would not be suitable, the effective use of selenate respiring bacterium in a biological reactor system for the removal of selenium oxides in drainage water has shown promise [71].

12.6.2 China

Although it is a matter of considerable concern to the population of central California, little evidence for selenosis in humans arising from the occurrence of Se pollution described above has been reported in this area. This may well be attributed to the variety in diet of the local population and the effective dilution of foodstuffs that may contain elevated levels of Se in the general marketplace. In contrast, an endemic disease was discovered in the People's Republic of China in 1961 that was directly attributed to the diet of people living in Enshi County, Huebi Province [117]. During the years of the highest prevalence morbidity was almost 50% in the most heavily affected areas, and its cause was determined to be Se intoxication. Symptoms of loss of hair and nails were the most common signs of poisoning. Dietary intake ranged between 3.20 and 6.69 mg Se/day, approximately 1000 times those levels associated with the Se-deficient Keshan disease which is also found in China, located in a wide area running from the northeastern province of Heilongjiang to the southwestern Yunnan province. The ultimate environmental source of Se was determined to be coal, from which Se had been mobilised through processes of weathering, leaching and biological action, and released into the soil. The traditional use of lime fertiliser had elevated soil pH and further increased the availability of Se for uptake by crops. In addition, drought conditions led to the failure of the rice crop, resulting in an increased consumption of local vegetables as a dietary supplement. This example of Se poisoning illustrates how a number of soil processes can be exacerbated by environmental and anthropogenic conditions, and give rise to toxic concentrations of Se in the human diet.

12.7 Concluding comments

In this chapter, an attempt has been made to summarise the wide range of information that is available concerning Se in soils and its bioavailability. The dual nature of Se as both an essential and potentially toxic element has resulted in considerable interest in aspects of both human and animal nutrition, together with its behaviour in soils and plants. Although there are many environmental properties which govern the reactivity of Se in soils, the principal factors may be summarised as follows:

(i) The reactivity and mobility of Se in soils is dependent on the chemical speciation. Selenate, which prevails under alkaline oxic conditions, is more mobile, soluble and less well adsorbed than selenite. Organic Se species are known to exist, and have been associated with organic fractions in soils, but identification of these compounds is still in the preliminary stages.

(ii) In acidic soils, the presence of clay, Fe oxides and organic matter will affect Se availability to a greater extent than in soils where selenate is the predominant species. This is due to differences in the adsorptive behaviour of the two inorganic oxyanions. Transformations between Se species are pH- and redox-dependent and are often facilitated by microbes. In addition, microbial activity results in the formation of volatile species which can be lost to the atmosphere.

(iii) Plant availability of Se is many times greater for selenate than for selenite. Organic Se is also accumulated by plants, although there is little information available as to its effects. The mechanisms involving uptake of the inorganic species are different, and may result in different levels of phytotoxicity. Changing pH and the presence of other anions, such as phosphate and sulphate will affect apparent Se accumulation by plants. Phytotoxicity occurs as a result of the incorporation of Se analogues of essential S compounds in plant tissues, and the ability of different plant species to incorporate and tolerate Se varies widely.

In conclusion, however, it is apparent that there are still many aspects of Se chemistry which require further understanding if a complete picture of its role in the environment is to be obtained. Fundamental concepts, including kinetic interactions under varying soil conditions and the nature and behaviour of indeterminate organic materials, are not well understood. Further research in these areas will be necessary if a comprehensive response to potential problems involving Se toxicity or deficiency is required.

Acknowledgment

We would like to thank Linda Bobbit for drawing the figures presented in this chapter.

References

1. Adriano, D.C., in *Trace Elements in the Terrestrial Environment*. Springer-Verlag. New York (1986), Chapter 12.
2. Allaway, W.H., in *Trace Substances in Environmental Health*. ed. Hemphill, D.D., Univ. of Missouri, Columbia, Mo. (1968), 181.
3. Rosenfeld, I. and Beath, O.A., *Selenium: Geobotany, Biochemistry, Toxicity and Nutrition*. Academic Press. New York (1964).
4. Beath, O.A., Draize, J.H and Eppson, H.F., *Wyoming Agr. Expt. Sta. Bull. No. 189* (1932), 1.

5. Miller, J.T. and Byers, H.G., *J. Agric. Sci.*, **42** (1937), 182.
6. Ihnat, M., *Occurrence and Distribution of Selenium*. CRC Press (1989).
7. Gissel-Nielsen, G., Gupta, U.C., Lamand, M. and Westermarck, T., *Advances in Agronomy*, **37** (1984), 397.
8. Moxon, A.L. and Olson, O.E., in *Selenium*. eds. Zingaro, R.A. and Cooper, W.C., Reinhold, New York (1974), Chapter 12.
9. Cooper, W.C., Bennett, K.G. and Croxton, F.C., in *Selenium*. eds. Zingaro, R.A. and Cooper, W.C., Reinhold, New York (1974), Chapter 1.
10. Combs, G.F. and Combs, S.B., *The Role of Selenium in Nutrition*. Academic Press, New York (1986).
11. Goldschmidt, V.M., *Geochemistry*. Oxford University Press, NJ, (1954).
12. Lakin, H.W., *Geol. Soc. Am. Bull.*, **83** (1972), 181.
13. Krauskopf, K.B., *Introduction to Geochemistry 2nd Ed.* MacGraw-Hill, Tokyo, (1979).
14. Berrow, M.L. and Ure, A.M., in *Occurrence and Distribution of Selenium*. ed. Ihnat, M. CRC Press (1989). Chapter 9.
15. Byers, H.G., Miller, J.T., Williams, K.T. and Lakin, H.W., in *USDA Tech. Bull. 601*. USDA-ARS, Washington, DC, (1938), 74.
16. Swaine, D.J., in *Trace Substances in Environmental Health-XII*. ed. Hemphill, D.D., Univ. Missouri-Columbia, MO. (1978), 129.
17. Archer, F.C. and Hodgson, I.H., *Soil Sci.* **38** (1987), 421.
18. Walsh, T. and Fleming, G.A., *Trans. Intern. Soc. Soil Sci. Comm. II and IV.* **2** (1952), 178.
19. National Research Council. *Selenium*. National Academy of Sciences, Washington DC. (1976).
20. Nriago, J.O. and Pacyna, J.M., *Nature*, **333** (1988), 134.
21. Mosher, B.W. and Duce, R.A., *J. Geophys. Res.* **92** (1987), 13289.
22. Furr, A.K., Lawrence, A.W., Tong, S.S.C., Grandolfo, M.C., Hofstader, R.A., Bache, C.A., Gutenmann, W.H. and Lisk, D.J., *Env. Sci. Technol.* **10** (176), 683.
23. Los Angeles/Orange County Metropolitan Area Project. *Technical Report*. P.O. Box 4998, Whittier Calif. 90607, (1977).
24. Logan, T.J., Chang, A.C., Page, A.L. and Gange, T.J., *J. Environ. Qual.* **16** (1987), 349.
25. Chang, A.C., Warneke, J.E., Page, A.L. and Lund, L.J., *J. Environ. Qual.* **13** (1984), 87.
26. Sposito, G., Neal, R.H., Holtzclaw, K.M. and Traina, S.J., in *1985–86 Technical Progress Report, Univ. Calif. Salinity Drainage Task Force*, UC Davis, CA (1986) 81.
27. Stumm, W. and Morgan, J.J., *Aquatic Chemistry 2nd Ed.* Wiley-Interscience, NY, (1981), Chapter 7.
28. Sposito, G., Yang, A., Neal, R.H., and Mackzum, A., *Soil Sci. Soc. Am. J.*, **55** (1991), 1597.
29. Masscheleyn, P.H., Delaune, R.D. and Patrick Jr., W.H., *Environ. Sci. Technol.* **24** (1989), 91.
30. Oremland, R.S., Hollibaugh, J.T., Maest, A.S., Presser, T.S., Miller, L.G. and Culbertson, C.W., *Appl. Environ. Microbiol.*, **55** (1989), 2333.
31. Oremland, R.S., Steinberg, N.A., Maest, A.S., Miller, L.G., Hollibaugh, J.T., *Environ. Sci. Technol.*, **24** (1990), 1157.
32. Steinberg, N.A. and Oremland, R.S., *Appl. Environ. Microbiol.*, **56** (1990), 3550.
33. Weres, O., Bowman, H.R., Goldstein, A., Smith, E.C., Tsao, L. and Harnden, W., *Water Air Soil Pollut.*, **49** (1990), 251.
34. Macy, J.M., in *UC Salinity/Drainage Task Force 1989–1990 Tech. Progress Rep.*, (1990), 95.
35. Elrashidi, M.A., Adriano, D.C., Workman, S.M. and Lindsay, W.L., *Soil Science*, **144** (1987), 141.
36. Masscheleyn, P.H., Delaune, R.D. and Patrick Jr., W.H., *Environ. Sci. Health*, **26** (1991), 555.
37. Masscheleyn, P.H., Delaune, R.D. and Patrick Jr., W.H., *J. Environ. Qual*, **20** (1991), 522.
38. Neal, R.H., Sposito, G., Holtzclaw, K.M. and Traina, S.J., *Soil Sci. Soc. J.*, **51** (1987), 1161.
39. Irgolic, K.J. and Kudchadker, M.V., in *Selenium*. eds. Zingaro, R.A. and Cooper, W.C., Reinhold, New York (1974), Chapter 8.
40. Cooke, T.D. and Bruland, K.W., *Environ. Sci. Technol.* **21** (1987), 1214.
41. Abrams, M.M., Burau, R.G. and Zasoski, R.J., *Soil Sc. Soc. Am. J.*, **54** (1990), 979.
42. Gustafsson, J.P. and Johnsson, L., *J. Soil. Sci.*, **43** (1992), 461.
43. Hingston, F.J., Posner, A.M. and Quirk, J.P., *Adv. Chem. Ser.*, **79** (1968), 82.
44. Parfitt, R.L., *Adv. Agron.*, **30** (1978), 1.

45. Sposito, G., *The Surface Chemistry of Soils*. Oxford University Press, (1984), Chapter 4.
46. Hayes, K.F., Roe, A.L., Brown Jr., G.E., Hodgson, K.O., Leckie, J.O. and Parks, G.A., *Science*, **238** (1987), 783.
47. Hingston, F.J., *Specific Adsorption of Anions on Goethite and Gibbsite*. Ph.D. diss. Univ. of W. Australia, Nedlands, (1970).
48. Hamdy, A.A. and Gissel-Nielsen, G., *Z. Pfanzenernaehr. Bodenkd.*, **140** (1977), 63.
49. Barr-Yosef, B. and Meek, D., *Soil Sci.*, **144** (1987), 11.
50. Scott, M.J., *Kinetics of Adsorption and Redox Processes on Iron and Manganese Oxides: Reactions of As(III) and Se(IV) at Geothite and Birnessite Surfaces*. EQL Rept 33, Calif. Inst. Technol. Pasadena, Calif. (1991).
51. Benjamin, M.M., *Environ Sci. Technol.*, **17** (1983) 686.
52. Davis, J.A. and Leckie, J.O., *J. Colloid Interface Sci.*, **14** (1980), 32.
53. Harrison, J.B., and Berkheiser, V.E., *Clays and Clay Minerals*, **30** (1982), 97.
54. Balistrieri, L.S. and Chao, T.T., *Soil Sci. Soc. Am. J.*, **51** (1987), 1145.
55. John, M.K., Saunders, W.M.H. and Watkinson, J.H., *N.Z. J. Agric. Res.*, **19** (1976), 143.
56. Neal, R.H. and Sposito, G., *Soil Sci. Soc. Am. J.*, **53** (1989), 70.
57. Goldberg, S. and Glaubig, R.A., *Soil Sci. Soc. J.*, **52** (1988), 954.
58. Singh, M., Singh, N. and Relan, P.S., *Soil Sci.*, **132** (1981), 134.
59. Goldhamer, D.A., Nielsen, D.R., Grismer, M. and Biggar, J.W., in *1985–86 Technical Progress Report*, Univ. Salinity Drainage Task Force, UC Davis, Calif. (1986), 55.
60. Alemi, M.H., Goldhamer, D.A. and Nielsen, D.R., *J. Environ. Qual.*, **20** (1991), 89.
61. Fio, J.L., Fujii, R. and Deverel, S.J., *Soil Sci. Soc. Am. J.*, **55** (1991), 1313.
62. Hingston, F.J., in *Adsorption of Inorganics at Solid/Liquid Interfaces*. eds. Anderson, M.A., and Rubin, A.J., Ann Arbor Science, Ann Arbor Mich. (1981), 51.
63. Neal, R.H., Sposito, G., Holtzclaw, K.M. and Traina, S.J., *Soil Sci. Soc. Am. J.*, **51** (1987), 1165.
64. Marsh, K.B., Tillman, R.W. and Syers, J.K., *Soil Sci. Soc. Am. J.*, **51** (1987), 318.
65. Ryden, J.C., Syers, J.K. and Tillman, R.W., *Soil Sci.*, **38** (1987), 211.
66. Doran, J.W., in *Advance in Microbial Ecology: Vol. 6.* ed. Marshall, K.C., Plenum Publishing Corp. (1982), 1.
67. Thompson-Eagle, E.T., and Frankenberger Jr., W.T., *J. Environ. Qual.*, **19** (1990), 125.
68. Doran, J.W., and Alexander, M., *Sci. Soc. Am. J.*, **40** (1977), 687.
69. Shrift, A., in *Organic Selenium Compounds: Their Chemistry and Biology*. eds. McElroy, W.D. and Gunther, W.H.H., John Wiley and Sons, New York, (1973), 763.
70. Macy, J.M., in *1990–91 Technical Progress Report, Univ. Calif. Salinity Drainage Task Force*, UC Davis, Calif. (1991), 120.
71. Macy, J.M., Lawson, S., DeMoll, H. and Rech, S., in *1991–92 Technical Progress Report, Univ. Calif. Salinity Drainage Task Force*, UC Davis, Calif. (1992), 66.
72. Reamer, D.C., and Zoller, W.H., *Science*, **208** (198), 500.
73. Fleming, R.W. and Alexander, M., *Appl. Microbiol.*, **24** (1972), 424.
74. Barkes, L. and Fleming, R.W., *Bull. Environ. Contam. Toxicol.*, **12** (1974), 308.
75. Karlson, U. and Frankenberger Jr., W.T., *Soil Sci. Soc. Am. J.*, **53** (1989), 749.
76. Thompson-Eagle, E.T., Frankenberger Jr, W.T. and Karlson, U., *Appl. Environ. Microbiol.* **55** (1989), 1406.
77. Craig, P.J., in *Organometallic Compounds in the Environment*. ed. Craig, P.J., Longman Group Ltd. Harlow, Essex (1986), 1.
78. Deverel, S.J. and Fujii, R., *Water Resources Research*, **24** (1988), 516.
79. Measures, C.I. and Burton, J.D., *Earth and Planetary Science Letters*, **46** (1980), 385.
80. Lipinski, N.G., Huang, P.M., Liaw, W.K. and Hammer, U.T., *Can. Tech. Rep. Fish Aquat. Sci.*, (1986), 166.
81. Mikkelsen, R.L., Page, A.L. and Bingham, F.T., in *Selenium in Agriculture and the Environment*. ed. Jacobs, L.W., Soil Science Society of America Special Publication 24, ASA Madison Wis. (1989).
82. Cary, E.E. and Allaway, W.H., *Soil Sci. Soc. Am. Proc.*, **33** (1969), 571.
83. Gupta, U.C., McRae, K.B. and Winter, K.A., *Can. J. Soil Sci.*, **62** (1982), 145.
84. Johnsson, L., *Plant and Soil*, **133** (1991), 57.
85. Carlson, C.L., Adriano, D.C. and Dixon, P.M., *J. Environ. Qual.*, **20** (1991), 363.
86. Cary, E.E., Wieczorek, G.A. and Allaway, W.H., *Soil Sci. Soc. Am. Proc.*, **31** (1967), 21.

87. Ylaranta, T., *Ann Agric. Fenn.*, **22** (1983), 29.
88. Davies, E.B. and Watkinson, J.H., *N.Z. J. Agric. Res.*, **9** (1966), 641.
89. Carter, D.L., Robbins, C.W. and Brown, M.J., *Soil Sci. Soc. Am. Proc.*, **36** (1972), 624.
90. Singh, M. and Malhotra, P.K., *Plant Soil*, **44** (1976), 261.
91. Khattak, R.A., Page, A.L., Parker, D.R. and Bakhtar. D., J. Environ. Qual., **20** (1991), 165.
92. Pratley, J.E. and MacFarlane, J.D., *Aust. J. Exp. Agric. Anim. Husb.*, **14** (1974), 533.
93. Gissel-Nielsen, G., *J. Sci. Food Agric.*, **24** (1973), 649.
94. Mikkelsen, R.L., Hagnia, G.H. and Page, A.L., *Plant and Soil*, **107** (1988), 63.
95. Mikkelsen, R.L., Hagnia, G.H., Page, A.L. and Bingham, F.T., *J. Environ. Qual.*, **17** (1988), 85.
96. Wan, H.F., Mikkelsen, R.L. and Page, A.K., *J. Environ. Qual.*, **17** (1988), 269.
97. Mikkelsen, R.L., Hagnia, G.H. and Page, A.L., *J. Plant Nutrition*, **10** (1987), 937.
98. Mikkelsen, R.L., Mikkelsen, D.S. and Abshahi, A., *Soil Sci. Soc. Am. J.*, **53** (1989), 122.
99. Shrift, A., and Ulrich, J.M., *Plant Physiol.*, **44** (1969), 893.
100. Ulrich, J.M. and Shrift, A., *Plant Physiol.*, **43** (1968), 14.
101. Peterson, P.J., Benson, L.M. and Zieve, R., in *Effects of Heavy Metal Pollution on Plants: Volume 1*. ed. Lepp, N.W., Applied Sci. Publ. London (1981), Chapter 8.
102. Asher, F.C., Butler, G.W. and Peterson, P.J., *J. Exp. Bot.*, **28** (1977), 279.
103. Brown, T.A. and Shrift, A., *Biol. Rev.*, **57** (1982), 59.
104. Anderson, J.W. and Scarf, A.R., in *Metals and Micronutrients: Uptake and Utilization by Plants*. eds. Robb, D.A. and Pierpoint, W.S., Academic Press, NY, (1983), 241.
105. Banuelos, G.S. and Meek, D.W., *J. Environ. Qual.*, **19** (1990), 772.
106. Bell, P.F., Parker, D.R. and Page, A.L., *Soil Sci. Soc. Am. J.*, **56** (1992), 1818.
107. Zieve, R. and Peterson, P.J., *Sci. Total Environ.*, **32** (1984), 197.
108. Duckart, E.C., Waldron, L.J. and Doner, H.E., *Soil Sci.*, **53** (1991), 94.
109. Terry, N., Carlson, C., Raab, T.K. and Zayed, A.M., *J. Environ. Qual.*, **21** (1992), 341.
110. U.C. Salinity/Drainage Task Force. *Principal Accomplishments 1985–1990*. Reagents of the University of California, Division of Agricultural and Natural Sciences, (1991).
111. Letey, J., Robers, C., Penberth, M. and Vasek, C., *An Agricultural Dilemma*. Regents of the University of California, Division of Agricultural and Natural Sciences, (1986).
112. Deverel, S.J., Gilliom, R.J., Fujii, R., Izbicki, J.A. and Fields, J.C., *Water Resources Investigations Report* 84-4319, U.S. Geological Survey, Denver, Col, (1984).
113. Tokunaga, T.K. and Benson, S.M., *J. Environ. Qual.*, **21** (1992), 246.
114. Weres, O., Jaouni, A-R., and Tsao, L., *Appl. Geochem.*, **4** (1989), 543.
115. Parker, D.R., Page, A.L. and Thomason, D.N., *J. Environ. Qual.*, **20** (1991), 157.
116. Frankenberger Jr., W.T. and Karlson, U., *Soil Sci. Soc. Am. J.*, **53** (1989), 1435.
117. Yang, S., Wang, R., Zhou, R. and Sun, S., *Am. J. Nutr.*, **37** (1983), 872.
118. Davidson, D.F. and Lakin, H.W., *U.S. Geol. Survey Res.*, Profess. Paper No. 424-C (1961), 329.
119. Davidson, D.F. and Powers, H.A., *U.S. Geol. Survey Bull.*, No. 1084-C (1959), 69.
120. Davidson, D.F. and Gulbrandsen, R.A., *Geol. Soc. Am. Bull.*, No. 68 (1957), 1714.
121. Bowen, H.J.M., *Environmental Chemistry of the Elements*. Academic Press, London (1979).
122. Kolijonen, T., *Oikos*, **25** (1972), 353.
123. Pillay, K.K.S., Thomas Jr, C.C. and Kaminski, J.W., *Nucl. Applns. and Tech.*, **7** (1969), 478.
124. Fujii, R. and Deverel, S.J., in *Selenium in Agriculture and the Environment*. ed. Jacobs, L.W., Soil Science Society of America Special Publication 24, Asa Madison, Wis (1989).
125. Bohn, H.L., McNeal, B.L. and O'Connor, G.A., *Soil Chemistry*. 2nd ed. Wiley Interscience, New York (1985).

13 Zinc

L. KIEKENS

13.1 Introduction

Zinc is an essential trace element for humans, animals and higher plants. Although the beneficial effect of Zn on the growth of *Aspergillus niger* was discovered by Raulin in 1869–1870, the discovery of the essentiality of Zn for higher plants in 1926 is generally attributed to Sommer and Lipman.

Zinc is also essential for humans and animals, i.e. a deficient intake consistently results in an impairment of a function from optimal to suboptimal, and supplementation with physiological levels of this element, but not of others, prevents or cures this impairment [1]. The recommended safe and adequate dietary intake for adults is around 15 mg/day [2]. Zinc acts as a catalytic or structural component in numerous enzymes involved in energy metabolism and in transcription and translation. Zinc deficiency symptoms in humans and animals are failure to eat, severe growth depression, skin lesions and sexual immaturity. For humans, depression of immunocompetence and change of taste acuity also occur.

Higher plants predominantly absorb Zn as a divalent cation (Zn^{2+}), which acts either as a metal component of enzymes or as a functional, structural, or regulatory cofactor of a large number of enzymes. According to Marschner [3], at least four enzymes contain bound Zn: carbonic anhydrase, alcohol dehydrogenase, Cu–Zn superoxide dismutase and RNA polymerase. Furthermore, Zn is required for the activity of various enzymes, such as dehydrogenases, aldolases, isomerases, transphosphorylases, RNA and DNA polymerases. Because of these functions, Zn is involved in carbohydrate and protein metabolism. Zinc is also required for the synthesis of tryptophan, a precursor for the synthesis of indoleacetic acid (IAA). It is clear that the most pronounced Zn deficiency symptoms, namely stunted growth and 'little leaf' rosette of trees are related to the latter physiological function of Zn.

Crops particularly sensitive to Zn deficiency are the cereals maize and sorghum, flax, hops, cotton, legumes, grapes, citrus and fruit trees (peach, apple). In general, the most permanent symptoms of Zn deficiency are inter-veinal chlorosis (mainly of monocotyledons), stunted growth, malformation of stems and leaves, better known as 'little leaf' rosette of trees; and violet-red points on leaves. Typical names for Zn deficiency are white bud (corn and sorghum), little leaf (fruit trees), mottle leaf (citrus) and sickle leaf (cacao).

The Zn content of plants varies considerably, as a function of different soil and climate factors and also of plant genotypes. As an overall approximation, the following ranges of Zn concentrations and their classification for the mature leaf tissue can be presented as follow: (i) deficient, if less than 10–20 mg/kg dry matter, (ii) sufficient or normal if between 25 and 150 mg/kg, (iii) excessive or toxic if more than 400 mg/kg [4, 5]. A lot of information about Zn levels in various plant species has been compiled by various authors [6–8].

13.2 Geochemical occurrence of zinc

The average total Zn content of the lithosphere is estimated to be approximately 80 mg/kg [9]. The most abundant sources of Zn are the ZnS minerals sphalerite and wurtzite and to a lesser extent minerals such as $ZnCO_3$ (smithsonites), Zn_2SiO_4 (willemite), ZnO (zincite), $ZnSO_4$ (zinkosite), $ZnFe_2O_4$ (franklinite) and $Zn_3(PO_4)_2.4H_2O$ (hopeite) [9].

In magmatic rocks Zn appears to be uniformly distributed. Mean Zn contents vary from 40 mg/kg in acid rocks (granites) to 100 mg/kg in basaltic rocks [10]. In sedimentary rocks, the highest Zn contents are found in shales and clayey sediments (80–120 mg/kg), while sandstones, limestones and dolomites generally have lower contents, ranging from 10 to 30 mg/kg [5].

13.3 Origin of zinc in soils

13.3.1 Soil parent materials

Total Zn content of soils is largely dependent on the composition of the parent rock materials [5, 11, 14]. The common range for total zinc concentrations in soils is 10–300 mg/kg with an average of 50 mg/kg [9]. Kabata-Pendias *et al.* [15] studied the background levels of trace metals in soils of the temperate humid zone of Europe. Zn status showed a relatively uniform distribution among soil units. The lowest Zn concentrations were always found in podzols (28 mg/kg) and luvisols (35 mg/kg), while higher levels were found in fluvisols (60 mg/kg) and histosols (58 mg/kg).

The average Zn concentration for all soils of the area studied was 50 mg/kg, with a range for the soil units from 10 to 105 mg/kg. Similar patterns of variation of Zn levels were also observed for soils in Poland. Angelone and Bini [16] reviewed the literature on trace element distribution in soils of western Europe. The Zn concentrations in most soils of western Europe are consistent with the levels considered as background values, with exception for mined areas. Table 13.1 reports the Zn concentrations in soils of western Europe compared to the means of USA, Canada and world soils.

13.3.2 Atmospheric fall-out

The burning of coal and other fossil fuels and the smelting of non-ferrous metals are the major Zn sources contributing to air pollution. An estimation of the global release of Zn as a contaminant into the environment may be based on the world

Table 13.1 Ranges and mean Zn contents of soils (mg/kg) [16]

Country	Range	Mean
Austria	36–8900*	65
Belguim	14–130	57
Denmark	7–76	7
France	5–38	16
Germany	13–492	83
Greece	80–10547	1038
Italy	–	89
Netherlands	9–1020*	72.5
Norway	40–100	60
Portugal	–	58.4
Spain	10–109	59
Sweden	100–318	182
England and Wales	–	78.2
Scotland	0.7–987	58
USA	–	54
Canada	–	74
World soils	–	50

*Indicates data not included in the mean.

mineral and energy consumption and demand. According to Kabata-Pendias [5] the forecast for Zn demand in the year 2000 is estimated at 11×10^6 t. Compared to the consumption in 1975 this means an increase of about 100%. The steady global increase of Zn concentrations in the atmosphere can be illustrated by the following Zn concentrations in air (ng/m^3) from different locations [5]: 0.002–0.05 (South Pole), 10 (Norway), 15 (Shetland Islands) 14–6800 (Japan) and 550–1600 (Federal Republic of Germany). It should be emphasised that some natural sources such as volcanic eruptions and aeolian dusts may also contribute to Zn pollution.

13.3.3 Agricultural use of sewage sludge

Sewage sludge is the by-product of wastewater treatment and contains N, P and organic matter as major beneficial constitutents for plants and soils. However, sewage sludges often contain appreciable amounts of Zn which can affect crop plants. Zinc in sewage tends to be associated with suspended solids and is partitioned into the sludge during treatment. Conventional sewage treatment removes 40–74% Zn from the influent [17].

Sludges exhibit a wide range of Zn concentrations which are generally higher than the background levels found in soils [18]. As an example the following ranges of Zn concentrations in sludge have been reported: 700–49000 mg/kg [19], 600–20000 mg/kg [17], 101–27800 mg/kg [20], 91–28766 mg/kg [21]. The corresponding mean Zn concentrations are respectively 4100, 1500, 2790 and 1579 mg/kg dry matter.

It is clear that the uncontrolled utilisation of sewage sluge on agricultural land will lead to accumulation of Zn and other heavy metals in the soil and consequently constitute a permanent risk for plants and crops. In this context many efforts have been made to minimise these risks by regulating the sewage sludge organisation on agricultural land. Guidelines for sludge use have been adopted in many countries, and they are generally based on the following assumptions [18]:

(i) That Zn concentrations in sludge may not exceed defined limits. The Commission of the European Community (CEC) directive gives as recommended and mandatory concentrations in sludge respectively 2500 and 4000 mg/kg Zn. For most countries the maximum permissible total Zn concentrations in sludges are in the range 1000–5000 mg/kg Zn, with an average of 2500–3000 mg/kg Zn.
(ii) That Zn concentrations in soils may not exceed defined limits. In most countries a maximum level of 300 mg/kg Zn is adopted.
(iii) That heavy metal loadings to agricultural land may not exceed defined limits. The CEC directive proposes values of 175–550 kg/ha as maximum Zn loading, whereby it is assumed that the normal soil background Zn concentration is 80 mg/kg and that a 1 mg/kg increase in soil Zn concentration for the upper 20 cm horizon results from the application of 2.5 kg/ha Zn.
(iv) That Zn is less likely to cause problems if it is added to the soil in several small increments during an extended period of time rather than in one or a few large increments. In this context the CEC suggest an annual Zn loading of 30 kg/ha based on a 10-year average.

13.3.4 Agrochemicals

In addition to aerial sources of Zn and all sewage-derived materials, fertilisers and pesticides may also increase the Zn concentrations of soils. All fertilisers, mineral as well as organic, and soil amendments contain Zn, most often as impurities. Zinc concentrations in inorganic phosphate fertilisers range from 50–1450 mg/kg, in limestones from 10–450 mg/kg while in manure values from 15–250 mg/kg Zn have been reported [5, 22–24]. Some pesticides contain Zn concentrations up to 25%, and hence may increase the Zn concentrations in soils.

Input–output balances of Zn in soils can be calculated, generally based on the earlier-mentioned Zn pollutant sources and exported quantities of Zn by crops. For Belgium, Verloo *et al.* [25] calculated an input/output ratio of 7.7, indicating that a gradual enrichment of agricultural soils with Zn takes place. The relative importance of the different input sources decreased in the following order: manure (70%) > atmospheric fall-out (25%) > mineral fertilisers (4.5%) > compost (0.3%) > sewage sludge (0.2%). The very low input from sewage sludge is due to the fact that the agricultural use of sludge in Belgium is very small compared to the situation in most other countries. Therefore these figures should not be generalised; for other countries quite different input/output ratios are obtained, mainly determined by the use of sewage sludge and industrial activities.

13.4 Chemical behaviour of zinc in soils

13.4.1 Zn fractions in soils

The total amount of Zn in soils is distributed over some more or less distinct fractions. Viets [15] distinguished the following five pools:

(i) Water-soluble pool: the fraction present in the soil solution.
(ii) Exchangeable pool: ions bound to soil particles by electrical charges.
(iii) Adsorbed, chelated or complexed pool: metals bound to organic ligands.
(iv) Pool of clayey secondary minerals and insoluble metallic oxides.
(v) Pool of primary minerals

Only those fractions which are soluble or may be solubilised are bio-available. It is therefore important to distinguish between the total quantities and the amounts which can be transferred into more soluble forms. Figure 13.1 shows the different storage possibilities of Zn^{2+} in the soil.

The total amount of Zn in the soil is distributed among the following forms:

(i) Free ions (Zn^{2+}) and organo-zinc complexes in the soil solution.
(ii) Adsorbed and exchangeable Zn in the colloidal fraction of the soil, composed of clay particles, humic compounds, Fe and Al hydroxides.
(iii) Secondary minerals and insoluble complexes in the solid phase of the soil.

The distribution of Zn among these forms is governed by the equilibrium constants of the corresponding reactions, in which Zn is involved, namely:

(i) precipitation and dissolution,
(ii) complexation and decomplexation,
(iii) adsorption and desorption.

The kind of interaction which will be the most critical in a given system depends on several parameters, such as:

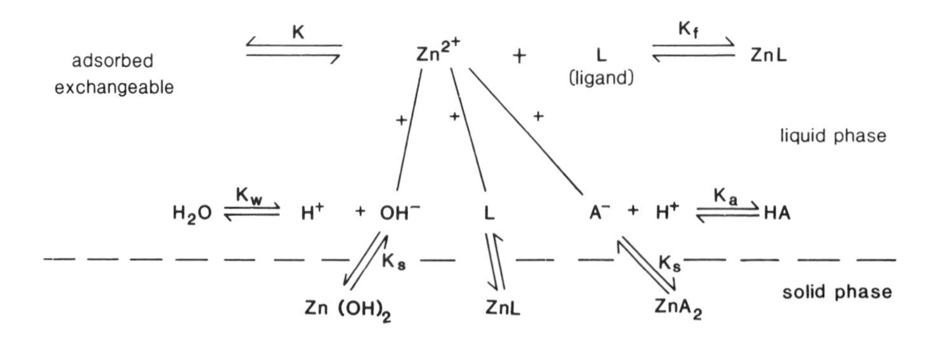

Figure 13.1 Chemical equilibria between Zn and soil components.

(i) The concentration of Zn^{2+} and other ions in the soil solution,
(ii) The kind and amount of adsorption sites associated with the solid phase of the soil,
(iii) The concentration of all ligands capable of forming organo-zinc complexes,
(iv) pH and redox potential of the soil.

Changing one or more of these parameters will result in a shift of the global equilibrium, and a transfer of Zn from one form to another will take place until a new equilibrium is attained. Such equilibrium displacements may occur as a result of plant uptake, losses by leaching, input of Zn by different ways, changing moisture content of the soil, pH changes, mineralisation of organic matter and changing redox status of the soil.

13.4.2 Solubility of Zn in soils

Compared to the average total Zn content of soils (50 mg/kg), the concentration of Zn in the soil solution is very low. Hodgson *et al.* [27] have indicated values ranging from 3×10^{-8} to 3×10^{-6} M. If all of the Zn present in an average soil were in solution at 10% moisture, it would comprise $10^{-2.12}$ M Zn [14]. Kabata-Pendias [5] reports values from the literature ranging from 4 to 270 μg/l, depending on the soil and the techniques used for obtaining the solution. In very acid soils (pH < 4), Zn concentration in solution was reported to average 7137 μg/l.

A free Zn^{2+} ion in the soil solution will precipitate when the solubility product of its compound with any reaction partner R^{m-} is attained. Precipitates with hydroxides, carbonates, phosphates, sulphides, molybdates and with several other anions, including humates, fulvates and other organic ligands may be formed. In this way, a lot of precipitation reactions can take place, and the maximum activity of Zn^{2+} can be calculated.

Theoretical equations with regard to the chemical activity of Zn ions in soils have been proposed [9, 28], based on the solubility products of the different Zn compounds. The solubilities of various Zn minerals that may occur in soils are given by the reactions of Table 13.2 (after Lindsay [9, 28]).

In most of these reactions pH seems to be an important parameter. Reaction 8 in Table 13.2 gives the solubility of soil Zn based on experimental measurement for a number of soils [9, 28–30]. The equilibrium constant for this reaction,

$$\text{soil-Zn} + 2H^+ \rightleftharpoons Zn^{2+} \qquad \log K° = 5.8$$

may be expressed as follows:

$$\log Zn^{2+} = 5.8 - 2\ \text{pH}$$

or

$$\text{pZn} = 2\ \text{pH} - 5.8$$

This equation provides a very useful reference solubility for Zn^{2+} in soils. It shows that the activity of Zn^{2+} in the soil solution is directly proportional to the square of

Table 13.2 Solubilities of Zn minerals. After Lindsay [9, 28]

	Equilibrium reaction	log K^0
	Oxides and hydroxides	
1.	$Zn(OH_2$ (amorph) $+ 2H^+ \rightleftharpoons Zn^{2+} + 2H_2O$	12.48
2.	α–Zn $(OH)_2(C) + 2H^+ \rightleftharpoons Zn^{2+} + 2H_2O$	12.19
3.	β–Zn $(OH)_2(C) + 2H^+ \rightleftharpoons Zn^{2+} + 2H_2O$	11.78
4.	γ–Zn $(OH)_2(C) + 2H^+ \rightleftharpoons Zn^{2+} + 2H_2O$	11.74
5.	ε–Zn $(OH)_2(C) + 2H^+ \rightleftharpoons Zn^{2+} + 2H_2O$	11.53
6.	ZnO (zincite) $+ 2H^+ \rightleftharpoons Zn^{2+} + H_2O$	11.16
	Carbonates	
7.	$ZnCO_3$ (smithsonite) $+ 2H^+ \rightleftharpoons Zn^{2+} + CO_2(g) + H_2O$	7.91
	Soil-Zn and Zn-Fe oxide	
8.	Soil-Zn $+ 2H^+ \rightleftharpoons Zn^{2+}$	5.80
9.	$ZnFe_2O_4$ (franklinite) $+ 8H^+ \rightleftharpoons Zn^{2+} + 2Fe^{3+} + 4H_2O$	9.85
	Silicates	
10.	Zn_2SiO_4 (willemite) $+ 4H^+ \rightleftharpoons 2Zn^{2+} + H_4SiO_4^0$	13.15
	Chlorides	
11.	$ZnCl_2(C) \rightleftharpoons Zn^{2+} + 2Cl^-$	7.07
	Sulphates	
12.	$ZnSO_4$ (zinkosite) $\rightleftharpoons Zn^{2+} + SO_4^{2-}$	3.41
13.	$ZnO.2ZnSO_4(C) + 2H^+ \rightleftharpoons 3Zn^{2+} + 2SO_4^{2-} + H_2O$	19.12
14.	$Zn(OH)_2.ZnSO_4(C) + 2H^+ \rightleftharpoons 2Zn^{2+} + SO_4^{2-} + 2H_2O$	7.50
	Phosphates	
15.	$Zn_3(PO_4)_2.4H_2O$ (hopeite) $+4H^+ \rightleftharpoons 3Zn^{2+} + 2H_2PO_4^- + 4H_2O$	3.80

the proton activity. This means that the solubility of Zn will increase at decreasing pH values of the soil. The solubility of several Zn minerals decreases in the following order:

$$Zn(OH)_2 \text{ (amorph)} > \alpha\text{-}Zn(OH)_2 > \beta\text{-}Zn(OH)_2 > \gamma\text{-}Zn(OH)_2 > \varepsilon\text{-}Zn(OH)_2 > ZnCO_3 \text{ (smithsonite)} > ZnO \text{ (zincite)} > Zn_2SiO_4 \text{ (willemite)} > \text{soil-Zn} > ZnFe_2O_4 \text{ (franklinite)}.$$

These minerals represent a range in Zn^{2+} solubility of 10^8. All of the $Zn(OH)_2$ minerals, ZnO (zincite) and $ZnCO_3$ (smithsonite) are about 10^5 times more soluble than soil Zn. These salts make good Zn fertilisers when applied in the vicinty of the roots. The various soluble Zn species in equilibrium with soil Zn are given in Table 13.3, after Lindsay [9, 28].

When calculating the activities of the different Zn species as a function of pH, it seems that below pH 7.7 Zn^{2+} is the predominant species, while $ZnOH^+$ is more prevalent above this pH. At pH values above 9.11 the neutral species $Zn(OH)_2$ is predominant, while $Zn(OH)_3^-$ and $Zn(OH)_4^{2-}$ are never major soluble species in the pH range of soils. Furthermore, one may calculate that at pH 5 the activity of Zn^{2+} in solution is about 10^{-4} M (6.5 mg/l), whereas at pH 8 it decreases to about 10^{-10} M (0.007 μg/l).

Table 13.3 Soluble Zn Species in equilibrium with soil Zn. After Lindsay [9, 28]

Equilibrium reaction	log K^0
Hydrolysis species	
1. $Zn^{2+} + H_2O \rightleftharpoons ZnOH^+ + H^+$	− 7.69
2. $Zn^{2+} + 2H_2O \rightleftharpoons Zn(OH)_2^0 + 2H^+$	− 16.80
3. $Zn^{2+} + 3H_2O \rightleftharpoons Zn(OH)_3^- + 3H^+$	− 27.68
4. $Zn^{2+} + 4H_2O \rightleftharpoons Zn(OH)_4^{2-} + 4H^+$	− 38.29
Chloride complexes	
5. $Zn^{2+} + Cl^- \rightleftharpoons ZnCl^+$	0.43
6. $Zn^{2+} + 2Cl^- \rightleftharpoons ZnCl_2^0$	0.00
7. $Zn^{2+} + 3Cl^- \rightleftharpoons ZnCl_3^-$	0.50
8. $Zn^{2+} + 4Cl^- \rightleftharpoons ZnCl_4^{2-}$	0.20
Other complexes	
9. $Zn^{2+} + H_2PO_4^- \rightleftharpoons ZnH_2PO_4^+$	1.60
10. $Zn^{2+} + H_2PO_4^- \rightleftharpoons ZnHPO_4^0 + H^+$	−3.90
11. $Zn^{2+} + NO_3^- \rightleftharpoons ZnNO_3^+$	0.40
12. $Zn^{2+} + 2NO_3^- \rightleftharpoons Zn(NO_3)_2^0$	−0.30
13. $Zn^{2+} + SO_4^{2-} \rightleftharpoons ZnSO_4^0$	2.33

Zinc also forms complexes with chloride, phosphate, nitrate and sulphate. From the equilibrium reactions mentioned in Table 13.3, it seems that the complexes $ZnSO_4^0$ and $ZnHPO_4^0$ are the most important and may contribute significantly to total Zn in solution. The activity of $ZnSO_4^0$ species equals that of Zn^{2+} when SO_4^{2-} is $10^{-2.33}$ M. Thus the $ZnSO_4^0$ complex may increase the solubility and mobility of Zn^{2+} in soils. This explains why acidifying fertilisers, like $(NH_4)_2SO_4$, may increase the availibility of Zn.

The $ZnHPO_4^0$ species may contribute to Zn in solution, particularly in neutral and alkaline soils, depending upon phosphate activity.

Summarising, the Zn species contributing significantly to total inorganic zinc in solution can be represented as

$$[Zn_{inorg}] = [Zn^{2+}] + [ZnSO_4^0] + [ZnOH^+] + [Zn(OH)_2^0] + [ZnHPO_4^0]$$

Substituting activities and taking the equilibrium constants of Table 13.3 into consideration, the above equation may be rearranged to give [9, 28]:

$$(Zn^{2+}) = \frac{[Zn_{inorg}]}{\dfrac{1}{\gamma Zn^{2+}} + 10^{2.33}(SO_4^{2-}) + \dfrac{10^{-7.69}}{\gamma ZnOH^+ (H^+)} + \dfrac{10^{-16.80}}{(H^+)^2} + \dfrac{10^{-3.90}(H_2PO_4^-)}{(H^+)}}$$

This equation indicates that Zn^{2+} activity can be calculated when the following parameters are known: total inorganic Zn, pH, ionic strength, SO_4^{2-} and $H_2PO_4^-$ activity.

However, Zn may also be present in the soil solution as organic species, so that the total soluble Zn is split up into inorganic and organic Zn. According to Hodgson *et al.* [27], inorganic Zn can be estimated from total soluble Zn by extracting soils in the presence and absence of charcoal which absorbs the organic Zn complexes.

Other authors have also studied the solubility of Zn in soils. Herms [31] determined experimentally the solubility of Zn as a function of pH and found the following relationship:

$$\log(Zn^{2+}) = -0.5\ pH - 1.02 \quad \text{or} \quad pZn = 0.5\ pH + 1.02$$

This equation is quite different from the one mentioned earlier and reported by Lindsay [9, 28]. This indicates that the solubility of Zn may not only be explained by dissolution–precipitation reactions, but that other mechanisms, such as adsorption–desorption and complexation, may also affect the solubility of Zn in soils.

13.4.3 Adsorption and desorption of Zn in soils

The term adsorption is commonly used for the processes of sorption of chemical elements from solutions by soil particles [5]. The term sorption refers to all phenomena at the solid–solution boundary. The most important soil components contributing to adsorption of Zn are clay minerals, hydrated metal oxides and organic matter; they constitue the so-called colloidal phase of the soil. In general, at normal soil pH values, the surface of the colloidal phase is negatively charged. The negatively charged adsorption sites are compensated by equivalent amounts of positive charges, such as protons and other cations, e.g. Zn^{2+}. Therefore the adsorption of Zn^{2+} from the soil solution by the solid soil particles is generally accompanied by the simultaneous desorption of equivalent amounts of other cations from the solid phase to the soil solution. This process is called ion exchange or equivalent adsorption.

The adsorption of Zn by soils and soil constituents has been extensively studied. From the studies of Zn adsorption [9, 32, 33] it has been shown that clays and organic matter may adsorb Zn quite strongly and that apparently two different Zn adsorption mechanisms occur: one in acid conditions related to cation exchange sites and another in alkaline conditions that are considered to be chemisorption and highly affected by organic ligands.

Tiller and Hodgson [34] found that Zn is adsorbed reversibly by ion-exchange and irreversibly by lattice penetration in clay minerals. The latter phenomenon was also reported by Elgabaly [35]. De Mumbrum and Jackson [36] and Bingham *et al.* [37] stated that montmorillonite, especially at neutral or alkaline pH, fixed Zn in amounts in excess of the CEC. This may be explained by the adsorption of Zn in hydrolysed form and precipitation of $Zn(OH)_2$. According to Misra and Tiwari [38], formation of Zn carbonates is also possible. McBride and Blasiak [39] reported that nucleation of Zn hydroxide on clay surfaces may produce strongly pH-dependent retention of Zn in soils.

Reddy and Perkins [40] studied the adsorption of Zn by bentonite, illite and kaolinite at different pH values, by alternate wetting and drying and after incubation at moisture saturation. They found that Zn is fixed as a result of precipitation, physical entrapment in clay lattice wedge zones or strong adsorption at the exchange sites. Adsorption experiments with clay minerals [41] and with organic matter [42] showed that Zn is more strongly adsorbed at alkaline pH values. Shuman [43] studied the effect of soil properties on Zn adsorption and found that soils high in clay or organic matter had higher adsorption capacities for Zn than sandy soils low in organic matter. A study on the effect of pH on Zn adsorption revealed that at low pH values, Zn adsorption was more reduced for sandy soils than for those high in colloidal-size materials. Abd-Elfattah and Wada [44] found the highest selective adsorption of Zn by Fe oxides, halloysite and allophanes and the lowest by montmorillonite. Thus it seems that clay minerals, hydrous oxides, organic matter and pH are likely to be the most important factors affecting Zn adsorption by soils.

After the observation of the ion-exchange phenomenon in soils by Thomas Way in 1850 [45], a lot of investigators have tried to establish a relationship between the amounts of adsorbed ions and their concentration in the equilibrium solution. In a first phase, this resulted in empirical equations, as experimental observations showed that at low concentrations in solution, the amount of adsorbed ions was proportional with these concentrations. The Langmuir and Freundlich equations belong to this group. Many authors have shown that adsorption of Zn by soils can successfully be described by the Langmuir adsorption equation [43, 46–52].

Table 13.4 shows the calculated Langmuir parameters for the description of Zn adsorption by three soils of different texture [52].

The high values of r and r^2 indicate that Zn adsorption data for the three soils fit the Langmuir equation. The value of the bonding energy coefficient K was highest in the heavy clay soil and lowest in light sandy loam soil. This clearly illustrates the effect of the physicochemical soil parameters on Zn adsorption.

Table 13.4 Calculated Langmuir parameters for Zn adsorption by soils. From Kiekens [52]

Soil texture	Soil characteristics	Adsorption maximum		Bonding energy coefficient K (ppm^{-1})	Correlation coefficient (r)	Coefficient of determination (r^2)
		(mg/kg)	% of CEC			
Light sandy loam	pH = 6.25 % C = 1.3 % clay = 7.8	2000	68	0.013	0.981	0.963
Light loam	pH = 6.50 % C = 1.42 % clay = 13.6	2500	66.5	0.043	0.990	0.980
Heavy clay	pH = 7.0 % C = 1.74 % clay = 39.6	5000	73.9	0.250	0.994	0.988

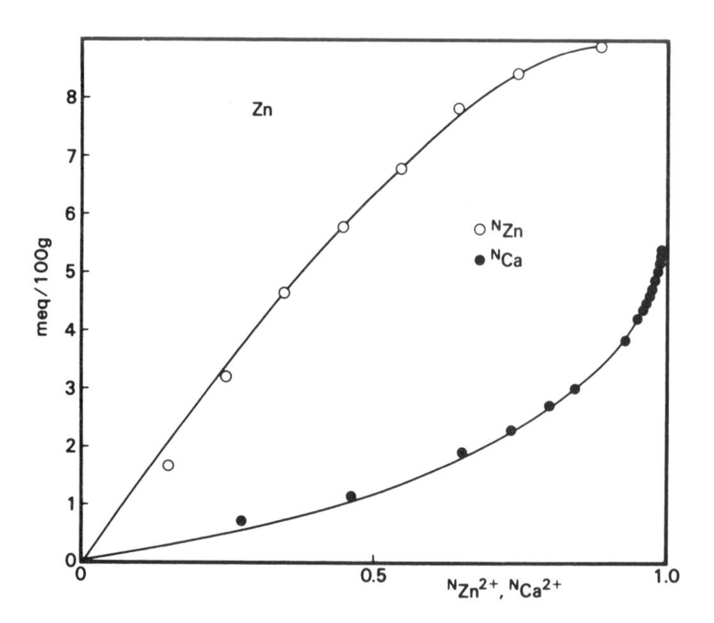

Figure 13.2 Adsorption-desorption hysteresis of Zn. Adsorption (○) and desorption (●) isotherms are plotted in function of respectively the equivalent fractions $N_{Zn^{2+}}$ and $N_{Ca^{2+}}$ in the equilibrium solution. From Kiebens [40].

A more or less important fraction of Zn may be involved in specific adsorption reactions. According to Maes [53] and Van Der Weyden [54], pH-dependent adsorption sites and broken bonds may effectively contribute to selective or preferential adsorption. Kiekens [51] studied the reversibility of the exchange reaction

$$\text{Ca soil} + Zn^{2+} \underset{K_2}{\overset{K_1}{\rightleftharpoons}} \text{Zn Soil} + Ca^{2+}$$

This reaction is reversible if $K_1 = 1/K_2$, while hysteresis occurs if $K_1 \neq 1/K_2$, so that the same equilibrium is not attained in both directions. It was found that an important fraction of Zn was irreversibly fixed by the soil, as illustrated in Figure 13.2. In this figure the cumulative amounts of Zn adsorbed (●) by the soil upon Zn^{2+}/Ca^{2+} exchange (adsorption of Zn^{2+}, desorption of Ca^{2+}), and desorbed (○), upon Ca^{2+}/Zn^{2+} exchange (desorption of Zn^{2+}, adsorption of Ca^{2+}) are plotted in function of respectively the equivalent amount of Zn^{2+} and Ca^{2+} in the equilibrium solution ($N_{Zn^{2+}}$, $N_{Ca^{2+}}$). The equivalent amount of Zn^{2+} is defined as the ratio between Zn^{2+} and the sum of $Zn^{2+} + Ca^{2+}$ ions in the equilibrium solution.

The fraction of Zn corresponding with the amounts between the adsorption and desorption isotherm may be considered as irreversibly fixed by the soil. This hysteresis effect seems to be relatively more important at low occupancy of the adsorption complex with Zn. Hysteresis also indicates that other than regular ion exchange reactions are involved in the immobilisation of Zn in soils.

The following parameters may play a role in the selective adsorption of Zn and in the occurrence of an adsorption–desorption hysteresis effect:

(i) Number of pH dependent adsorption sites
(ii) Interactions with amorphous hydroxides
(iii) Affinity to form organomineral complexes and their stability
(iv) Formation of hydroxy complexes
(v) Steric factors
(vi) Properties of ion:ionic radius, polarisabilty, thickness of the hydration sheet, equivalent conductance, hydration enthalpy and entropy.

The observed hysteresis effect may have important practical consequences and applications. Addition of selective cation exchangers such as macroporous polystyrol resins with weak acidic complexing functional groups, reduced Zn uptake by plants cultivated in a contaminated sandy soil [55]. Similar effects were obtained by addition of a heavy clay soil [52]. Soil acidity also affects Zn adsorption: at normal soil pH values, the adsorbed Zn fraction is much higher than the water-soluble one. Desorption occurs at low pH values, and for Zn the critical pH value is about 5 [56].

13.4.4 Influence of soil organic matter on the behaviour of zinc in soils

Soil organic matter is an important soil constituent, originating from decaying plant and animal products that have been converted to a more or less stable product known as humus. The end-products of the degradation of biota residues are humic substances, organic acids of low and high molecular weight, carbohydrates, proteins, peptides, amino acids, lipids, waxes, polycyclic aromatic hydrocarbons and lignin fragments. The most stable organic compounds in soils are humic substances. On the basis of solubility, they can be divided into humic acids, soluble only in alkaline medium, and fulvic acids, soluble in both alkaline and acid medium. Their chemical configuration is similar, but humic acids generally have higher molecular weights than fulvic acids. Both of these humic substances contain a relatively large number of functional groups (OH, COOH, SH, C=O), having a great affinity for metal ions, such as Zn^{2+}.

Fulvic acids mainly form chelates with Zn ions over a wide pH range, thus increasing the solubility and mobility of Zn. A large part of Zn present in the soil solution and in surface waters appears to be linked to a yellowish compound with fulvic acid properties, Verloo [57] showed that 16% of soluble Zn was present in this way in a soil leachate. Hodgson *et al.* [27] and Geering *et al.* [58], however, found that 60–75% of soluble Zn was present as soluble organic complexes.

The interactions between humic substances and Zn have been studied by many authors [29, 59–66]. All studies reveal that fulvic acids show a selectivity towards metal ion, which is reflected by the stability constants of the metal fulvic acid complexes or chelates.

The stability constant K is defined as the equilibrium constant of a reaction forming a complex or chelate. For the reaction between a moles of a cation M^{x+} and b moles of a ligand L^{y-},

$$aM^{x+} + bL^{y-} \rightleftharpoons M_aL_b^{ax-by}$$

the stability constant is given by

$$K = \frac{(M_aL_b)^{ax-by}}{(M^{x+})^a \cdot (L^{y-})^b}$$

in which parentheses denote activities. Several authors have determined experimentally the stability constant of Zn fulvates. According to Schnitzer and Skinner [67, 68] and Stevenson and Ardakani [62] the stability constant (log K) of Zn fulvates is 1.7 at pH 3.5 and 2.3 at pH 5. Schnitzer and Khan [63] reported log K value of 2.3 at pH 3 and 3.6 at pH 5. Courpron [69] found a log K value of 2.83 for Zn fulvic acid compounds. Note that the complex stability increases with increasing pH value. Verloo [57–64] showed that in fact two stability maxima occur, one at about pH 6 and another at pH 9. This might be attributed to the dissociation of functional –carboxyl and –hydroxyl groups in the fulvic acid molecule. Furthermore, Zn fulvates show only slight colloidal properties. This means that flocculation of fulvic acids occurs only at high electrolyte concentrations. In this respect fulvic acids act in soils as mobilising agents for Zn and other metals, such as Cu, Fe and Pb.

Humic acids show a more complicated interaction and solubility pattern. They are insoluble in acid conditions, and dissolve gradually as pH increases. In alkaline media they are completely soluble, but behave as a colloidal system, which means that they can be flocculated by cations. In neutral and alkaline soils Ca^{2+} and Mg^{2+} are dominating flocculating ions, and in acid soils also Fe^{3+} and Al^{3+} may become very important. In extremely leached soils, humic acids may be peptised and may move in the profile up or downwards depending on the prevailing water regime.

Interactions between Zn and humic acids are strongly affected by pH, because the solubility of Zn and humic acids is very pH dependent and shows opposite characteristics. In acid conditions most of the Zn is soluble, while humic acids are insoluble.

In a detailed study Verloo [57] discussed the behaviour of Zn in purified humic acid–Zn systems. The distribution pattern of Zn humates as a function of pH is represented in Figure 13.3, from which it is quite obvious that the behaviour of Zn is strongly affected by the presence of humic acid. At low pH values most of the Zn is present in cationic form, and humate complexes are formed at increasing pH values. Most of the formed Zn humates are soluble, and at alkaline pH values only a very small fraction is present as Zn hydroxides. The soluble Zn humates may flocculate in the presence of electrolytes.

Finally, it should be emphasised that simple organic compounds, such as amino

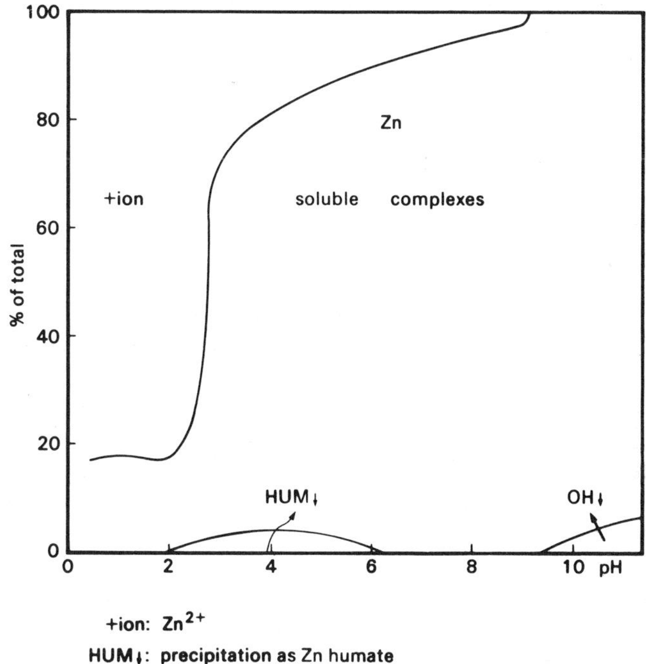

Figure 13.3 Distribution pattern of Zn humates in function of pH, after Verloo [37].

acids, hydroxy acids and also phosphoric acids are also effective complexing or chelating agents for Zn, thus increasing its mobility and solubility in soils [5].

Summarising, one may conclude that soil organic matter is an important factor affecting the behaviour of Zn in soils. The fulvic acid fraction and low-molecular-weight organic acids mainly form soluble complexes and chelates with Zn, thus increasing its mobility. Due to their colloidal nature, Zn humates may be considered as an organic storage pool for Zn.

13.5 Soil–plant relationships

13.5.1 Factors affecting the availability of zinc in soils

Zn^{2+} seems to be the predominating form in which Zn is absorbed by plant roots, but hydrated Zn and several other complexes and Zn organic chelates may also be absorbed [5, 70–72]. Factors affecting the availability and solubility of Zn in soils have been extensively reviewed by several authors [14, 61, 73–75]. Among the factors affecting Zn availability in soils, mainly soil parameters, such as total content, pH, organic matter, adsorption site, microbial activity, moisture regime play an important role. Other factors, such as climatic conditions and interactions between Zn and other macro- and micronutrients at the soil level and on the plant level also largely affect Zn availability. In general, the effect of different soil and other factors on Zn availability may be summarised as follows:

(i) In some highly leached acid soils total Zn levels may be very low, resulting in low available contents.
(ii) Availability of Zn decreases at increasing pH values of the soil, due to the lower solubility of Zn minerals and increasing adsorption of Zn by negatively charged colloidal soil particles.
(iii) In soils low in organic matter content, Zn availability is directly affected by the content of organic complexing or chelating ligands, originating from decaying organic matter or root exudates.
(iv) Low temperatures and light intensities generally decrease Zn availability, mainly because of restricted root development.
(v) High P levels in the soil may decrease Zn availability and uptake by plants. The Zn–P antagonism is one of the best-known nutrient interactions in soil chemistry and plant nutrition. This antagonism mainly seems to be based on chemical reactions in the rhizosphere [76]. According to Smilde *et al.* [77], the Zn–P antagonism can be explained on a plant physiological basis.
(vi) Interactions with other nutrients may decrease Zn availability. Mainly Zn–Fe antagonism, but also Zn–Cu, Zn–N and Zn–Ca interactions are widely known.

13.5.2 Evaluation of zinc availability in soils

Zinc availability in soils is generally evaluated by the extraction of a fraction of the total Zn by means of chemical reagents. Some extracting solutions are used for the selective determination of Zn, while others aim at the simultaneous extraction of a number of nutrient elements.

The choice of extracting solutions has often been based on empirical rather than theoretical considerations. Their value is often judged on the basis of more or less good correlations between the extracted Zn amounts and Zn uptake by plants. The analytical results are moreover influenced by operating modalities such as soil/solution ratio, extraction time, soil properties, chemical form or matrix of Zn present in the soil [78, 79].

As the biologically active fraction of Zn in soils mainly consists of its soluble, exchangeable and complexed forms, it is possible to base Zn extractions on the following phenomena: solubilisation of compounds present in the solid phase of the soil, ion exchange and complexation. Thus an extracting solution offering the greatest likelihood of giving a good correlation with Zn uptake by plants should meet the following criteria.

(i) It should be sufficiently acid to solubilise precipitated Zn compounds that contribute to plant uptake.

(ii) It should contain a displacing cation in order to exchange adsorbed Zn.
(iii) It should have the property of extracting soluble organo-zinc compounds.

Based upon these considerations, a number of extracting solutions have been proposed by several authors.

In this context the FAO European Cooperative Network on Trace Elements has proposed a reference method, using 0.5 N NH_4 acetate + 0.02 M EDTA pH 4.65 as the extracting solution [80].

Other widely used extractants are:

0.005 M DTPA + 0.01 M $CaCl_2$ + 0.1 M TEA pH 7.30 [30]
1 M NH_4HCO_3 + 0.005 M DTPA pH 7.60 [81]
0.1 M HCl [82, 83]
2 N $MgCl_2$ [84]
$(NH_4)_2CO_3$ + EDTA [85]
0.1 M $NaNO_3$ [86]
0.05 M $CaCl_2$ [87]
0.5 M HNO_3 [88]
0.5 N Na acetate + DTPA pH 4.8 [89].

Furthermore, a number of sequential extraction procedures for heavy metals in soils, sludges and sediments have been proposed [90–93]. The main purpose of fractionation is to better understand the distribution of Zn over different forms. Various reagents have been proposed to extract fractions representing soluble, exchangeable, organic, adsorbed and precipitated forms. The great diversity of extraction and fractionation procedures and consequently the difficulty of comparing results obtained by different methods have focused attention on developing a more universal method for assessing the availability of Zn and other heavy metals in soils. The proposed method consists of the determination of the mobility of Zn [94] by step-wise acidification of a soil–water suspension. The mobile fraction of Zn may be defined as the sum of the actual amount present in the soil solution, corresponding to the so-called nutrient intensity, and an amount present in the solid phase which can be transferred to the liquid phase by changing the pH of the soil.

Application of the proposed mobility test to a large number of polluted and non-polluted soils led to the following observations:

(i) Typical mobilisation patterns are obtained, which are largely affected by the physicochemical properties of the soil. The influence of soil texture on the mobility of Zn in a light, medium and heavy soil containing the same total amount of Zn is shown in Figure 13.4.
(ii) The mobility test makes it possible to differentiate between soils with normal and increased levels of Zn [96].
(iii) Very significant relationships have been observed between Zn uptake by plants and the amounts solubilised at different pH values.
(iv) The chemical species in which Zn is present, is also relected in the mobilisation pattern.

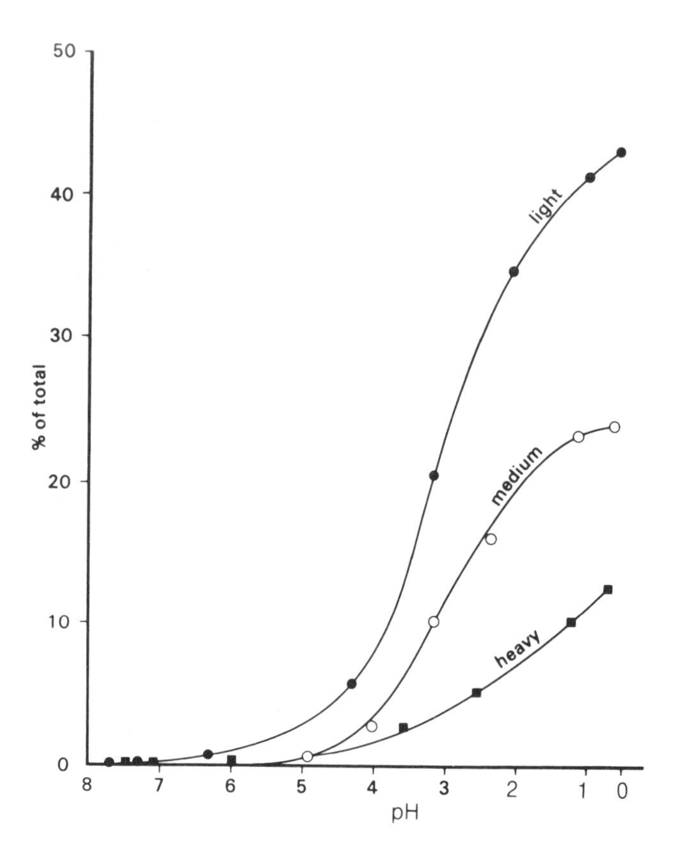

Figure 13.4 Percentage of total Zn solubilised in function of pH in three soils of different texture.

Stepwise or progressive acidification of a soil–water suspension releases amounts of Zn which are indicative for the actual available quantities as well as those that may become available over a longer lapse of time.

13.5.3 Plant factors affecting the uptake of zinc from soil

Zinc absorption and accumulation by plants vary widely between plant species and cultivars. Generally, increases in soil zinc concentrations cause increases in plant tissues. Falahi-Ardakani *et al.* [97] grew six vegetable species (broccoli, cabbage, lettuce, eggplant, pepper and tomato) for 8 weeks in a medium enriched with composted sewage sludge. Zn was accumulated at rates of 4 to 10 mg per week by the plants. Henry and Harrison [98] studied the uptake of metals by turfgrass, tomatoes, lettuce and carrots grown in a control soil, a soil amended with commercial NPK fertiliser, compost and compost/soil mixture at 1:1. The loading rates of Zn in control soil, compost mixture and compost were respectively

232, 239 and 245 kg/ha. Uptake rates of Zn by the plants in this study were in the order lettuce > grass > carrots > tomatoes. Uptake slopes of Zn by plants grown in compost were higher than in soils for lettuce, grass and carrots.

13.6 Zinc polluted soils

In recent years Zn concentrations in some soils have gradually increased, particularly in industrialised countries as a consequence of human activities. Total Zn levels of several hundred and thousands of mg/kg Zn in soils have been reported.

Zinc belongs to the group of trace metals potentially most hazardous to the biosphere. Most of the concern about excessive Zn concentrations in soils relates to its possible uptake by crops and consequent adverse effects on the crops themselves and on livestock and human diets. Together with Cu, Ni and Cr, Zn is principally phytotoxic, so the concern about this metal is mainly directed at effects on crop yield and soil fertility.

The main pollutant sources for Zn in soils are metalliferous mining activities, agricultural use of sewage sludge and composted materials, and the use of agrochemicals such as fertilisers and pesticides.

13.6.1 Contamination from metalliferous mining and smelting

Mining and smelting of Zn containing ores is a major source for Zn pollution. Zn mining during the eighteenth and nineteenth century in the village of Shipham in Somerset, UK resulted in a major increase in the Zn concentration in the soils. For the 320 soil samples analysed from this region the range of Zn concentration was 250–37200 mg/kg, with a median of 7600 mg/kg [99].

In Poland [100] Zn concentrations ranging from 1665 to 4245 mg/kg were found in soils polluted by metal-processing industries. Letunova and Krivitsky [101] report values from 400 to 4245 mg/kg in Russian soils polluted by nonferrous metal mining activities. Zn contamination of flooded soils in Japan was found to be primarily due to irrigation with waste waters discharged from the Ikuno mine in Japan [102] and from various mines in the Yoneshiro River Basin in Japan [103]. The concentrations of Zn in the surface water of the polluted area ranged from 3.40 to 6.20 mg/l and are much higher than those of the allowable concentration of natural irrigation water, i.e. 0.5 mg/l [103]. Asami [102] found Zn levels of 1310 to 1780 mg/kg in polluted flooded soils near the Maruyame river in Japan.

13.6.2 Soil pollution from heavy applications of sewage sludge

Juste and Mench [104] reviewed the effects of long-term application of sewage sludge on metal distribution in the soil profile and on metal uptake by crops, using results from field trials in Europe and the US. Total zinc input to the soil seemed to be the major factor affecting zinc concentration in plant tissue.

In experiments in the US, maximum Zn inputs ranged from 290 to 4937 kg/ha, while for the European long-term trials Zn inputs varied from 746 and 4882 kg/ha.

Obviously, such differences in Zn inputs induced marked differences in Zn behaviour in soils and plants. In the Bordeaux experiment (France) total Zn content of the soil increased from 8.1mg/kg for the control plots to 1074 mg/kg for the highest cumulative sludge application rate. It was also observed that the downward movement of Zn increased with its loading rate.

Alloway *et al.* [105] found soil total Zn contents of up to 1748 mg/kg at sites which had received heavy applications of sewage sludge. Elhassanin *et al.* [106] investigated the extent of Zn pollution of sandy soils of Abu Rawash and El-Gabel El-Asfar areas, Egypt, introducted by long-term use of sewage water for irrigation. Results revealed that prolonging the periods of irrigation was associated with significant increases in the total and available forms of Zn.

It has been recognised for more than 10 years that the microbial biomass of some soils which have received heavy applications of sewage sludge over many years is significantly reduced in comparison with non-sludged soils [107]. Several workers have investigated the causes of this reduced biomass and it is considered that elevated concentrations of several heavy metals, including Zn, are mainly responsible. One important aspect is that *Rhizobia* species, the symbiotic nitrogen fixing bacteria in the roots of legumes, are sensitive to metals. Although Chaudri *et al.* [108] found that the order of metal toxicity to *Rhizobium leguminosum* bv *trifolii*, (in white clover) is Cu > Cd > Ni > Zn, it is considered that Zn constitutes the greatest management problem. The concentrations of Zn in sludges are still relatively high, although the concentrations of almost all metals have decreased in recent years (see Chapter 3). Chaudri *et al.* [109] found that Zn concentrations below the maximum EU guideline limits (< 300 mg/kg) caused significant toxic effects in *R. leguminosum* bv *trifolii*. Although some European countries have lower maximum limits (Germany has 200 mg/kg), the level of 300 mg/kg is used in the UK. As a result of the findings from experiments in both the UK and Germany, recommendations were made in November 1993 to the UK Ministry of Agriculture, Fisheries and Food that the limit should be reduced from 300 to 250 mg/kg (see Chapter 2, section 2.4.9).

Investigations on other *Rhizobia* species have shown that *R. mellilot*i (in roots of alfalfa) from soils contaminated by atmospheric deposition from a Zn smelter is not so sensitive to Zn as *R. leguminosum* bv *trifolii* [110].

13.7 Concluding remarks

Zinc may be considered, together with Cd as a very mobile and bioavailable metal, which may accumulate in crops and human diets. Therefore the knowledge of long- and short-term efffects of atmospheric deposition and sewage sludge applications on soils is urgently needed. Despite results provided by long-term field experiments, more information is needed to evaluate long-term changes in factors controlling Zn bioavailability.

References

1. Mertz, W., *Ann. N.Y. Acad. Sci.* **199** (1972), 191.
2. Mertz, W., *Science* **213** (1981), 1332.
3. Marschner, H., *Mineral Nutrition in Higher Plants*. Academic Press, London (1986).
4. Loué, A., *Les Oligo-éléments en Agriculture*. Agri-Nathan, Paris (1986).
5. Kabata-Pendias, A. and Pendias, H., in *Trace Elements in Soils and Plants*. 2nd Ed. Lewis Publ. Inc. Boca Raton, Florida (1992).
6. Chapman, H.D., *Diagnostic Criteria for Plants and Soils*, University of California, Riverside (1972).
7. Bergmann, W. and Neubert, P., in *Pflanzendiagnose und Pflanzenanalyse*. Fischer Verlag, Jena (1976).
8. Bergman, W., in *Ernährungsstörungen bei Kulturpflanzen, Enststehung und Diagnose*. Fischer Verlag, Jena (1983)
9. Lindsay, W.L., *Adv. Agron.* **24** (1972), 147.
10. Lindsay, W.L., *Soil Sci. Soc. of America*, Madison (1991), Chapter 2.
11. Graham, E.R., *Soil Sci.* **75** (1953), 333.
12. Swaine, D.J. and Mitchell, R.L., *J. Soil Sci.* **11** (1960), 347.
13. Wells, N., *J. Soil Sci.* **11** (1960), 409.
14. Sillanpaa, M., *Trace Elements in Soils and Agriculture*. Soils Bulletin. FAO Rome (1972).
15. Kabata-Pendias, A., Dudka, S., Chipecka, A. and Gawinsowska, T., in *Biogeochemistry of Trace Metals*, ed. Adriano D.C. Lewis Publishers, Boca Raton (1992), Chapter 3.
16. Angelone, M. and Bini, C., in *Biogeochemistry of Trace Metals*, ed. Adriano D.C. Lewis Publishers, Boca Raton (1992), Chapter 2.
17. Davis, R.D., Control of Contamination Problems in the Treatment and Disposal of Sewage Sludge. WRC Technical Report TR 156, Stevenage (1980).
18. Webber, M.D., Kloke, A. and Chr. Tjell, J., in *Processing and Use of Sewage Sludge*, eds. L'Hermite, P. and Ott, H. Reidel, Dordrecht (1984), 371.
19. Berrow, M.L. and Webber, J., *J. Sci. Fd. Agric.* **23** (1972), 93.
20. Dowdy, R.J., Larson, R.E. and Epstein, E., in *Proc. Soil Conservation Society of America*, Ankeny, Iowa (1976), 118.
21. O'Riordan, E.G., *Ir. J. Agric. Res.* **25** (1986), 239.
22. Andersson, A., *Swed. J. Agric. Res.* **7** (1977), 1.
23. Kloke, A., *Gesunde Pflanz.* **32** (1980), 261.
24. Adriano, D.C., *Trace Elements in the Terrestrial Environment*. Springer-Verlag, New York (1986).
25. Verloo, M. and Tack, F., in *Grondontleding en Bemestingsadviezen, Genootschap Planten-produktie en Ekosfeer*, KVIV, Antwerpen (1988), 6.
26. Viets, F.G., *J. Agr. Fd. Chem.* **10** (1962), 174.
27. Hodgson, J.F., Lindsay, W.L. and Trierweiler, J.F., *Soil Sci. Soc. Amer. Proc.* **30** (1966), 723.
28. Lindsay, W.L., *Chemical Equilibria in Soils*. Wiley Interscience, New York (1979).
29. Norvell, W.A., in Micronutrients in Agriculture, eds. Mordvedt, J.J., Giordano, P. and Lindsay, W.L. Soil Sci. Soc. of America, Madison (1991), Chapter 6.
30. Lindsay, W.L. and Norvell, W.A., *Agron. Abstr.* **61** (1969), 84.
31. Herms, U., Untersuchungen zur Schwermetalöslichkeit in kontaminierten Böden und kompostierten Siedlungsabfällen in Abhängigkeit von Bodenreaktion, Redoxbedingungen und Stoffbestand. Doctoral Thesis, Kiel (1982).
32. Farrah, H. and Pickering, W.F., *Aust. J. Chem.* **30** (1977), 1417.
33. Wada, k. and Abd-Elfattah, A., *J. Soil Sci. Plant Nutr.* **24** (1978), 417.
34. Tiller, K.G. and Hodgson, J.F., *Clays and Minerals* **9** (1962), 393.
35. Elgabaly, M.M., *Soil Sci.* **85** (1950), 319.
36. Demumbrum, L.E. and Jackson, M.L., *Soil Sci. Soc. Amer. Proc.* **20** (1956), 334.
37. Bingham, F.F., Page, A.L. and Sims, J.R., *Soil Sci. Soc. Amer. Proc.* **28** (1965), 351.
38. Misra, S.G. and Tiwari, R.C., *Soil Sci.* **101** (1966), 465.
39. McBride, M.B. and Blasiak, J.J., *Soil Sci. Soc. Am. J.* **43** (1979), 866.
40. Reddy, N.R. and Perkins, H.F., *Soil Sci. Soc. Amer. Proc.* **38** (1974), 229.
41. Jurinak, J.J. and Bauer, N., *Soil Sci. Soc. Amer. Proc.* **20** (1956), 446.
42. Randhawa, N.S. and Broadbent, F.E., *Soil Sci.* **99** (1965), 362.

43. Shuman, L.M., *Soil Sci. Soc. Amer. Proc.* **39** (1975), 454.
44. Abd-Elfattah, A. and Wada, K., *J. Soil Sci.* **32** (1981), 271.
45. Way, J.T., *J. Roy. Agr. Soc. Eng.* **11** (1850), 313.
46. Udo, E.J., Bohn, H.L. and Tucker, T.C., *Soil Sci. Soc. Amer. Proc.* **34** (1970), 405.
47. Griffin, R.A. and Shimp, N.F., *Environ. Sci. Technol.* **10** (1976), 1256.
48. Sinha, M.K., Dhillon, S.K., Pundeer, G.S., Randhawa, N.S. and Dhillon, K.S., *Geoderma* **13** (1975), 349.
49. Bolt, G.H. and Bruggenwert, M.G.M., in *Soil Chemistry A. Basic Elements.* Elsevier, Amsterdam (1976).
50. Kiekens, L., *Med. Fac. Landbouww. Rijksuniv. Gent* **40** (1975), 1481.
51. Kiekens, L., *Adsorptieverschijnselen van zware metalen in gronden.* Doctoral Thesis, Gent (1980).
52. Kiekens, L., *Academiae Analecta* **48** (1986), 45.
53. Maes, A., Ion exchange of some transition metal ions in montmorillonites and synthetic faujasites. Doctoral Thesis, Louvain (1973).
54. Van Der Weijden, C.H., Report 221, Reactor Centrum Nederland, Petten (1975).
55. Van Assche, C. and Jansen, G., *Landwirtsch. Forsch.* **34** (1978), 215.
56. Cottenie, A. and Kiekens, L. in *Potassium in Soil*, proc. 9th IPI Colloq. Landshut (1972), 91.
57. Verloo, M.G., Komplexvorming van Sporenelementen met Organische Bodemkomponenten. Doctoral Thesis. Gent (1974).
58. Geering, H.r. and Hodgson, J.f., *Soil Sci. Soc. Amer. Proc.* **33** (1969), 54.
59. Mortensen, J.L., *Soil Sci. Soc. Amer. Proc.* **27** (1963), 179.
60. Wallace, A., *Soil Sci. Soc. Amer. Proc.* **27** (1963), 176.
61. Hodgson, J.F., *Adv. Agron.* **15** (1963), 119.
62. Stevenson, F.J. and Ardakani, M.S., in *Micronutrients in Agriculture*, eds. Mordvedt, J.J., Giordano, P. and Lindsay, W.L., Soil Sci. Soc. of America, Madison (1991), Chapter 5.
63. Schnitzer, M. and Khan, S.U., *Soil Organic Matter.* Elsevier, Amsterdam (1978).
64. Verloo, M., in *Essential and Non-essential Trace Elements in the System Soil–Water–Plant*, ed. Cottenie, A., State University, Gent (1979) 7.
65. Kitagishi, K. and Yamane, I., *Heavy Metal Pollution in Soils of Japan.* Japan Science Society Press (1981).
66. Sillanpää, M., *Micronutrients and the Nutrient Status of Soils: a Global Study.* FAO (1982).
67. Schnitzer, M. and Skinner S.I.M., *Soil Sci.* **102** (1966), 361.
68. Schnitzer, M. and Skinner S.I.M., *Soil Sci.* **103** (1967), 247.
69. Courpron, C., *Ann. Agron.* **18** (1967), 623.
70. Loneragan, J.F., *Trace Elements in Soil–Plant–Animal Systems.* Academic Press, London (1975).
71. Tiffin, L.O., in *Micronutrients in Agriculture*, eds. Mordvedt. J.J., Giordano, P. and Lindsay, W.L. Soil Sci. Soc. of America, Madison (1991), Chapter 9.
72. Weinhberg, E.D., in *Microorganisms and Minerals.* Marcel Dekker, New York (1977).
73. Mitchell, R.L., in *Chemistry of the Soil*, ed. Bear, F.E. Reinhold, New York (1964).
74. Lucas, R.L. and Knezek, B.D., in *Micronutrients in Agriculture*, eds. Mordvedt, J.J., Giordano, P. and Lindsay, W.L. Soil Sci. Soc. America, Madison (1991), Chapter 12.
75. Kiekens, L., in *Proc. 1st Int. ISAMA Symp. Gent* (1985), 4.
76. Olsen, S.R., in *Micronutrients in Agriculture*, eds. Mordvedt, J.J., Giordano, P. and Lindsay, W.L. Soil Sci. Soc. America, Madison (1991), Chapter 11.
77. Smilde, K.W., Koukoulakis, P. and van Luit, B., *Pl. Soil* **41** (1974), 445.
78. Kiekens, L. and Camerlynck, R., *Landwirtsch. Forsch.* **39** (1982), 255.
79. Verloo, M., Cottenie, A. and Van Landschoot, G., *Landwirtsch. Forsch.* **39** (1982), 394.
80. Larkanen, E. E. and Erviö, R., *Acta Agr. Fenn.* **123** (1971), 223.
81. Soltanpour, P.N. and Schwab, A.P., *Commun. Soil Sci. Plant Anal.* **8** (1977), 195.
82. Viets, F.G., Boawn, L.C. and Crawford, C.L., *Soil Sci.* **78** (1954), 305.
83. Nelson, J.L., Boawn, L.C. and Viets, F.G. Jr., *Soil Sci.* **88** (1959), 275.
84. Stewart, J.A. and Berger, K.C., *Soil Sci.* **100** (1965), 244.
85. Trierweiler, J.f. and Lindsay, W.L., *Soil Sci. Soc. Amer. Proc.* **33** (1969), 49.
86. Häni, H. and Gupta, S., in *Environmental Effects of Organic and Inorganic Contaminants in Sewage Sludge*, eds. Davis, R.D., Hucker, G. and L'Hermite, P. Reidel, Dordrecht (1983), 121.

87. Sauerbeck, D.R. and Styperek, P., in *Chemical Methods for Assessing Bio-Available Metals in Sludges and Soils*, eds. Leschber, R., Davis, R.D. and L'Hermite, P. Elsevier, London (1985), 49.
88. Cottenie, A., Verloo, M., Kiekens, L. and Velghe, G., in *Biological and Analytical Aspects of Soil Pollution*, ed. Cottenie, A. State University of Gent (1982).
89. Wolf, B., *Comm. Soil Sci. Plant Anal.* **13** (1982), 1005.
90. Badri, M.A. and Aston, S.R., in *Proc. Int. Conf. Heavy Metals in the Environment, Amsterdam*. CEP Consultants, Edinburgh (1981), 705.
91. Förstner, U., Calmano, W., Conradt, K., Jaksch, H., Schimkus, C. and Schoer, J., in *Proc. Int. Conf. Heavy Metals in the Environment. Amsterdam* (1981), 691.
92. Kiekens, L. and Cottenie, A. in in *Proc. Int. Conf. Heavy Metals in the Environment, Heidelberg*. CEP Consultants, Edinburgh (1983), 657.
93. Tessier, A., Campbell, P.G.C. and Bisson, M., *Anal. Chem.* **51** (1979), 844.
94. Cottenie, A. and Kiekens, L., *Korrespondenz Abwasser* **28** (1981), 206.
95. Kiekens, L., in *Utilization of Sewage Sludge on Land: Rates of Application and Long-term Effects of Metals*, eds. Berglund, S., Davies, R.D. and L'Hermite, P. Redel, Dordrecht (1984), 126.
96. Kiekens, L. and Cottenie, A., in *Processing and Use of Sewage Sludge*, eds. L'Hermite, P. and Ott, H. Reidel, Dordrecht (1984), 140.
97. Falahi-Ardakani, A., Bouwkamp, J.C., Gouin, F.R. and Chaney, R.L., *J. Environ. Hort.*, **5** (1987), 112.
98. Henry, C.L. and Harrison, R.B., in *Biogeochemistry of Trace Metals*, ed. Adriano D.C. Lewis Publishers, Boca Raton (1992), Chapter 7.
99. Sims, D.L. and Morgan, H., *Sci. Total Environ.*, **75** (1988), 1.
100. Faber, A. and Niezgoda, J., *Rocz. Glebozn.*, **33** (1982), 93.
101. Letunova, S.V. and Krivitskiy, V.A., *Agrokhimiya*, **6** (1979), 104.
102. Asami, T., in *Heavy Metal Pollution in Soils of Japan*, eds. Kitagishim K. and Yamane, I., Japan Scientific Societies Press (1981), 149.
103. Homma, S., in *Heavy Metal Pollution in Soils of Japan*, eds. Kitagishim K. and Yamane, I., Japan Scientific Societies Press (1981), 137.
104. Juste, C. and Mench, C., in *Biogeochemistry of Trace Metals*, ed. Adriano D.C. Lewis Publishers, Boca Raton (1992), Chapter 7.
105. Alloway, B.J., Thornton, I., Smart, G.A., Sherlock, I. and Quinn, M.J., *Sci. Total Environ.*, **75** (1988), 41.
106. Elhassanin, A.S., Labib, T.M. and Dobal, A.T., *Water Air and Soil Pollution*, **66** (1993), 239.
107. Brooks, P.C. and McGrath, S.P., *J. Soil Sci.* **35** (1984), 341–346.
108. Chaudri, A.M., McGrath, S.P. and Giller, K.E., *Soil Biol. Biochem.* **24** (1992), 625–632.
109. Chaudri, A.M., McGrath, S.P. and Giller, K.E., Reitz, E. and Sauerbeck, D.R., *Soil Biol. Biochem.* **25** (1993), 301–309.
110. Angle, J.S. and Chaney, R.L., *Water, Air and Soil Pollut.* **57–58** (1991), 597–604.

14 Other less abundant elements of potential environmental significance

R. EDWARDS , N.W. LEPP and K.C. JONES

14.1 Antimony

Antimony is non-essential to plants but can be readily taken up by roots when in soluble forms in soil. It therefore has the potential to act as a significant plant contaminant in industrial areas. Antimony exists at valence states between +3 and +5 and occurs in the lithosphere as antimonosulphides, metal antimonides, and antiminoxides [1]. Antimony becomes enriched in the early stages of magmatic differentiation [2] with the primary mineral stibnite (Sb_2S_3) occurring in polymetallic hydrothermal deposits with other sulphantimonides e.g. pyargyrite (Ag_3SbS_3), tetrahydrite ($CuSbS_3$) and bournonite ($PbCu(Sb, As)S_3$). It also occurs with sphalerite, pyrite, galena and is also found in minor amounts in Hg deposits [2]. The major commercial sources of Sb are stibnite and antimonial Pb ores with minor contributions from antimonoxides. A Pb–Sb alloy, antimonial Pb, is used for storage batteries [3]; other uses include flame retardants, pigments and explosives. Organo-antimony pharmaceuticals are used in trypanocidal and antisyphilitic drugs, but without any likely environmental consequence [4].

Antimony is usually concentrated in soil relative to parent materials. Average concentrations for parent materials are reported in the order of 0.1–0.2 mg/kg for basic rocks and 0.2 mg/kg for intermediate and acidic rocks [5]. Overall igneous rocks average 0.2 mg/kg. Shales average 1–2 mg/kg, with limestones and sandstones containing around 0.2 mg/kg [2]. Stibnite-containing rocks and those with other Sb minerals such as valentite and kermesite can be expected to have higher concentrations. On average soils contain Sb at *ca.* 1 mg/kg [5], but wide ranges have been reported (see Table 14.1).

Antimony may reach agricultural soils by wet and dry deposition, following emissions from incineration and fossil fuel combustion, and by the addition of soil amendments such as chemical fertilisers, sewage sludge and fly-ash. Atmospheric concentrations of Sb are found to be up to 10^3–10^5 times higher adjacent to mining and smelting works than in rural areas, and are elevated in areas of high coal combustion [1]. Antimony levels in UK coal range from 1–10 mg/kg (mean 3.3 mg/kg) [10] and reach 20 mg/kg in Australasian coal [11]. Antimony in fly-ash averages 3.5 mg/kg, and concentrations are inversely related to particle size [12]. Mean air concentrations in the UK over a period from 1957 to 1974 were found to decline as a consequence of clean air legislation [13]. Atmospheric deposition of Sb in the vicinity of Pb–Zn smelters has resulted in soil levels reaching 50–100

Table 14.1 Content of Sb in soils

Location	Sb content (mg/kg)	Reference
Overall	0.05–260	6
USA	2.3–9.5	3
UK	1.1–8.6	7
Scotland	0.29–1.3	8
Canada	1–3	3
Nigeria	1–5	1
Bulgaria	0.8–2.2	1
Netherlands	0.6–2.1	1

mg/kg [14, 15] and as high as 200 mg/kg adjacent to a Cu smelter [16]. Significantly higher levels have been reported in the vicinity of an Sb smelter. Immediately adjacent to the smelter, soil Sb levels reached a maximum of 1489 mg/kg (dry weight) and were still elevated at sampling sites over 1 km away (263 mg/kg) [17]. Dust particulates blown from Au mine spoil materials can also be a significant source for Sb contamination in mining areas since concentrations can reach 50 000 mg/kg in these dumps where As is present [18]. Analysis of 500 soil samples in Norway indicated an association of Sb with humus layers where the Sb was derived from atmospheric inputs from pollution and marine sources, often over a long range [19]. The radioactive isotope ^{125}Sb has been detected in Indian soils as a consequence of nuclear weapons testing [6].

Antimony concentrations in sewage sludge range from 2.6–44.4 mg/kg in the USA [20] and 15–19 mg/kg in the UK [21]. Beckett *et al.* [22] recommended that in addition to monitoring commonly occurring sludge contaminants (e.g. Cu, Zn, Ni) more unusual elements including Sb should be monitored until likely soil accumulations have proved harmless. Data on Sb levels in chemical fertilisers are scarce but superphosphates can in some cases reach 100 mg/kg of Sb and therefore act as a significant source [6].

The weathering reactions of Sb have been poorly investigated. Antiminosulphides are converted to their corresponding oxides, and Sb may be adsorbed by clays and hydrous oxides [2]. As Sb_2O_3, antimony is largely immobile. A profile of the surface 20 cm of a soil contaminated aerially with antimony(III) oxide showed that Sb accumulated at the surface, whilst concentrations decreased with increasing depth. At greater depths increased proportions of Sb were in more available forms which may reflect the downward movement of Sb after conversion to more mobile forms [23, 24]. Antimony is classed as having moderate mobility in soil [25] and probably occurs as antimonate in soluble form [16]. Complexes with soil humates in soil are also possible [16]. It is assumed that a major proportion of Sb remains in a labile form with only low adsorption by soil components over long duration [6]. The isotope ^{125}Sb applied to soil has been shown to move readily down the profile [6]. Experiments on the adsorption processes between pentavalent ^{119}Sb and haematite [26] have shown that strong adsorption occurs below pH 7 with a rapid reduction in adsorption at higher pH. Antimony ions, adsorbed in the range pH 2–5, were not desorbed, with subsequent adjustment to alkaline conditions. The $Sb(OH)_6^-$ form of the ion was reported to predominate and adsorption on haematite could be interpreted in terms of electrostatic attraction.

Methylantimony compounds have been detected in natural waters, but data or sediments and soils are lacking. Biomethylation of Sb has not been categorically proven, despite the similarity of Sb chemistry to that of As and Se. However, strong circumstantial evidence exists for the production of methylated Sb compounds from mould cultures [4]. Application of phenylstibonic acid ($C_6H_6SbO(OH)_2$) or a soluble antimonate salt ($KSbO_3$) to *Penicillium notatum* resulted in the evolution of volatile Sb products. However, reduction to stibnine could also account for a positive result. If biomethylation does occur it is most likely to act as a detoxification mechanism where antimonate ions are converted to less toxic organic forms. The possibility of a direct foliar uptake of methylated Sb released from soil therefore also exists. Other evidence for an interaction between microorganisms and Sb is scarce although the microbial biomass has been shown to contain up to 2.61% of the total Sb content of some ore deposits and spoils [27].

Phytotoxic levels of Sb have been given as 5–10 mg/kg in plant tissue [25]. Antimony toxicity to plants has been described as moderate [28], although there are no reported environmental cases. Antimony toxicity in cabbage was found to be associated with a purple colouring of the leaf veins and midribs of outer leaves at concentrations as low as 2–4 mg/kg [29]. Plants were found to be more tolerant of excess Sb than of As. Normal levels quoted for terrestrial flora range from 2–30 μg/kg [30], 50 μg/kg [28], and 0.0002–5 mg/kg [6]. Coughtrey *et al.* [6] summarised the available data to give figures of 0.1 mg/kg for natural vegetation, 50 μg/kg for pasture grasses, herbage, and herbaceous vegetable, grain and cereal products, and 5 μg/kg for leguminous and root vegetables and garden fruits. Other figures for edible plants have been given as 0.2–4.3 μg/kg with corn grain and tubers at less than 2 μg/kg [25].

Radiological assessments of the translocation of Sb in plants show that Sb is preferentially located in older leaf tissue, lower stems, and particularly roots [6]. Coughtrey *et al.* [6] have calculated soil/plant ratios of Sb at 0.1, 5×10^{-2}, and 5×10^{-3} respectively for these tissues, assuming a mean soil concentration of 1 mg/kg. Translocation from root to shoots is reported to be unlikely to exceed 15% of the total plant uptake [6]. Barley and flax roots have been found to have a reduced uptake of Sb when grown in a peaty soil where leaf concentrations were found to remain the same [25].

Although there is little or no information on the metabolism and toxicity of Sb in plants in contaminated areas, elevated concentrations in tissues have been reported. Concentrations between 0.35 and 2.5 mg/kg in trees and shrubs have been found in mineralised areas of Alaska [31], and levels as high as 110 mg/kg occurred in grasses (unwashed) adjacent to a Pb smelter [15]. In the vicinity of Sb smelters levels of 336 mg/kg [23] and 900 mg/kg [32] have been recorded for grasses. Grasses in the vicinity of a Au mining and refining site have been found to contain up to 15.4 mg/kg exceeding background concentrations by up to a factor of three [17]. Species-specific effects in relation to Sb uptake are seen when plants are grown on soils amended with fly-ash. Onion plants for example show elevated uptake of Sb, but clover may have uptakes which are the same as, or less than an uncontaminated soil [20, 30].

Overall the behaviour of Sb in soils and its uptake and distribution in plants has been poorly investigated. Geochemical studies are needed in order to more fully understand adsorption–desorption processes, mobility in soil, and transfer pathways—especially for Sb derived from atmospheric inputs, and contaminated soil amendments. Similarly, investigations into the uptake and metabolism of Sb in terrestrial vegetation, together with effects on the soil microflora, are needed in order to recognise potential toxicity problems. The verification of Sb biomethylation would also reveal more about a possible Sb soil–microorganism–plant interaction via the release of volatile Sb compounds.

References: Antimony

1. Monitoring and Assessment Research Centre, *Exposure Commitment Assessments of Environmental Pollutants*, Vol. 3. Chelsea College, University of London (1983).
2. Fairbridge, R.W., *The Encyclopedia of Geochemistry and Environmental Sciences*. Encyclopedia of Earth Sciences Series Volume IV A. Van Nostrand Reinhold, New York (1972).
3. Onishi, H., in *Handbook of Geochemistry*, ed. Wedepohl, K.H. II-4/51. Springer-Verlag, Berlin (1978).
4. Craig, P.J., *Organometallic Compounds in the Environment—Principles and Reactions*. Longman, Harlow (1986).
5. Ure, A.M. and Berrow, M.L., in *Environmental Chemistry*, Vol. 2 (H.J.M. Bowen, Senior Reporter). Royal Society of Chemistry, London (1980).
6. Coughtrey, P.J., Jackson, D. and Thorne, M.C., *Radionuclide Distribution and Transport in Terrestrial and Aquatic Ecosystems*, Vol. 3 A.A. Balkema, Rotterdam (1983).
7. Cawse, P.A., *A Survey of Atmospheric Trace Elements in the UK (1972–1973)*. UK Atomic Energy Report AERE-R 7669. HMSO, London (1976).
8. Ure, A.M., Beacon, J.R., Berrow, M.L. and Watt, J.J., *Geoderma* **22** (1979), 1.
9. Greenberg, R.R., Zoller, W.H., Jacko, R.B., Nuenendorf, D.W. and Yost, K.J., *Environ. Sci. Technol.* **12** (1978), 1329.
10. Goetz, L., Sabvbioni, E., Springer, A. and Pietra, R., in *Proc. Internat. Conf. Management and Control of Heavy Metals in the Environment*, London (1979).
11. Swaine, D.J., in *Trace Substances in Environmental Health* XI ed. Hemphill, D.D. University of Missouri (1979).
12. Coles, D.G., Ragaini, R.C., Ondov, U.M., Fisher, G.L., Silberman, D. and Prentice, B.A., *Environ. Sci. Technol.* **13** (1979), 455.
13. Salmon, L., Atkins, D.H.F., Fisher, E.M.R., Healy, C. and Law, D.V., *Sci. Total Environ.* **9** (1980), 161.
14. Lynch, A.J., McQuaker, N.R. and Brown, D.F., *J. Air Pollut. Control Assoc.* **30** (1980), 257.
15. Ragaini, R.C., Ralston, H.R. and Roberts, N., *Environ. Sci. Technol.* **11** (1977), 773.
16. Crecelius, E.A., Johnson, C.J. and Hofer, G.C., *Water Air Soil Pollut.* **3** (1974), 337.
17. Ainsworth, N., Cooke, J.A. and Johnson, M.S., *Environ. Pollut.* **65** (1990), 65.
18. Peterson, P.J. and Girling, C.A., in *Effects of Heavy Metal Pollution on Plants* Vol. 1., *Effect of Trace Metals on Plant Function*, ed. Lepp, N.W. Applied Science Publishers, London (1981).
19. Steines, E., in *Geomedical Aspects in Present and Future Research*, ed. Låg, J. Norwegian Academy of Science, Oslo (1980).

20. Furr, A.K., Lawrence, A.W., Tong, S.S.C., Grandfoli, M.C., Hofstader, P.A., Bachje, C.A., Guntenmann, W.H. and Lisk, D.J., *Environ. Sci. Technol.* **10** (1976), 683.
21. Wiseman, B.F.H. and Bedri, G.M., *J. Radioanal. Chem.* **24** (1975), 313.
22. Beckett, P.H.T., Davis, R.D., Brindley, P. and Chem, C., *J. Water Pollut. Control* **78** (1979), 419.
23. Ainsworth, N., Cooke, J.A. and Johnson, M.S., *Water, Air and Soil Pollut.* **57–58** (1991), 193.
24. Ainsworth, N., Cooke, J.A. and Johnson, M.S., *Environ. Pollut.* **65** (1990), 79.
25. Kabata-Pendias, A. and Pendias, H., *Trace Elements in Soils and Plants.* CRC Press, Boca Raton (1984).
26. Ambe, F., Ambe, S., Odada, T. and Sekizawa, H., in *Geochemical Processes at Mineral Surfaces*, eds. Davis, J.A and Hayes, K.F. American Chemical Society, Washington (1986).
27. Hara, T., Sonada, Y. and Iwai, I., *Soil Sci. Plant Nut.* **23** (1977), 253.
28. Brooks, R.R., *Geobotany and Biogeochemistry in Mineral Exploration.* Harper and Row, New York (1972).
29. Letunova, S.V., Ermakov, V.V. and Alekseeva, S.A., *Agrockhimiya* **4** (1984), 77.
30. Bowen, H.J.M., *Environmental Chemistry of the Elements.* Academic Press, London (1979).
31. Shacklette, H.T., Erdman, J.A., Harms, T.F. and Papp, C.S.E. In *Toxicity of Heavy Metals in the Environment* Part I, ed. Oehme, F.W. Marcek Dekker, New York (1978).
32. Programmatie van het Wetenschapsbeleid, *Report of the National Research and Development Programme on Environment – Air*, Brussels, Belgium (1982), 373.
33. Furr, A.K., Kelly, W.C., Bache, C.A., Gutenmann, W.H. and Lisk, D.J., *J. Agric. Fd. Chem.* **24** (1976), 885.
34. Furr, A.K., Kelly, W.C., Bache, C.A., Gutenmann, W.H. and Lisk, D.J., *Arch. Env. Health*, **30** (1975), 244.

14.2 Gold

Gold has fascinated the human race for millennia, and prospecting for this noble metal has always captured the imagination. Because of its unique properties and considerable economic and political value there has been continued interest in the environmental behaviour of Au, primarily through a desire to exploit and understand its cycling and movement for prospecting purposes. Gold is rare in the Earth's crust, and tends to be concentrated in rich seams and deposits—in other words, it has a rather restricted mobility. During the weathering process the sulphides, sulphosalts, and tellurides with which Au is frequently associated are oxidised, and the Au commonly remains as, or is reduced to , the metallic state. Nonetheless, the presence of Au in plants, stream waters and the oceans indicates that some mechanisms for solution of Au do exist. Efforts have been made by geochemists to understand these processes in recent years, largely because of the implications for Au exploration, prospecting and recovery from low-grade ores or deposits. Attention has been focused in this regard on the behaviour of Au in soils and plants. This increased research effort has been made possible in recent years by improvements in analytical techniques, which now enable determinations routinely at the parts per billion (μg/kg) range commonly found in soils and plants. Two techniques have been used—instrumental neutron activation analysis (INAA) [35–39] and graphite furnace atomic absorption spectrophotometry (GFAAS) [40]. The actual instrumental procedures are now, in general, sufficiently well researched and tested to be used with confidence. INAA, for example typically involves irradiation of samples and standards in a flux of ca. 1×10^{12} neutrons/cm^2/s for 8 h, followed by a period of decay for about 3 days to enable optimum resolution and detection of the gamma emitting ^{198}Au (half-life 2.7 days) at 412 keV. INAA can frequently be used to determine several potential pathfinder elements which often occur along with Au in mineralised areas, such as As, Sb and W [41, 42]. However, considerable care has to be exercised in the collection, preparation and pretreatment of samples for Au analysis. One major problem is the difficulty in obtaining representative soil or sediment samples from the field because Au tends to occur as discrete particles or flakes rather than being well dispersed through the soil [43]. Clifton *et al.* [41] studied the problem of sample size and meaningful Au analysis. They set as a goal 'that it be 95% probable that the true Au content of the deposit be no more than approximately 50% larger than the Au content obtained by chemical analysis of the sample'. They stated further that 'the number of Au particles in the sample is the only factor controlling the precision of the chemical analysis if, in addition to assuming that the Au particles are distributed randomly, it is also assumed that (1). Au particle mass is uniform, (2). Au particles make up less than 0.1% of all the particles, (3). the sample contains at least 1000 particles of all kinds, and (4). analytical errors are absent.' They concluded that 'a sample has the minimum adequate size. . .if it is large enough that it can be expected to contain 20 particles of Au.' This idealised model of random distribution of Au particles of uniform mass and no analytical error then requires a minimum of 20 Au particles for a ± 50% error. The difficulty encountered in soils containing coarse Au is illustrated by the results of the determination of Au in 18 10-g subsamples of a single soil sample [45] (Table 14.2).

Partly because of these analytical problems with *soil* analysis, several workers have advocated the analysis of soil *humus* or *mull* for prospecting studies [39, 45–47]. Lakin *et al.* [45] state that mull 'is often an excellent sampling medium for reconnaissance geochemical exploration for Au.' Gold present in this layer is primarily derived from biogeochemical cycling through vegetation (see below). Dunn [48] has graphically described plants '. . .efficient geochemical sampling tools. A highly corrosive micro-environment occurs around the myriad of roots and rootlets which feed a plant. Elements dissolved and absorbed by the plants include not only those essential to growth, but traces of other elements which can safely be tolerated in cell structures. In glaciated terrain extensive areas are covered with overburden which could be blanketing mineralised bedrock. In such areas

Table 14.2 Determination of Au in soil subsamples (see text for details)

Number of subsamples	Au (mg/kg)
14	<0.04
1	0.06
1	0.12
1	0.25
1	1.0

biogeochemical exploration may be the only feasible method of detecting metals such as Au, that accumulate in low concentrations.' The humus layer then acts as a reductant and concentration medium which retains exceedingly fine Au, or Au held in complexed form with organic matter at the surface that has been passed through and accumulated from a large volume of soil by the vegetation. The advantage of humus analysis over vegetation for prospecting is the way it presents an integrated picture of biologically accumulated Au, avoiding the vagaries of seasonal and species-specific Au uptake by plants. The enrichment of Au in soil surface layers was reported as long ago as 1937 by Goldschmidt [49] in the soil humus layer of an ancient, undisturbed beech and oak forest, and in a later discussion of the geochemistry of Au [50] he concluded that enrichment in the soil surface layers was evidence for Au circulation in living plants. He proposed that with the decomposition of stems, leaves and dead roots the metals, together with the most stable organic compounds, would remain in the humus layer. More recent comprehensive evidence for surface enrichment comes from studies by Lakin *et al.* [45]. Their work on soil profiles and vegetation in 12 Au districts in Colorado, Utah and Nevada, USA, tended to show the enrichment of Au both in the upper, organic-rich horizons and near the bottom of profiles. They concluded that biogeochemical recycling was the primary factor responsible for concentration in the upper layers, and hypothesised that the high concentration of Au near the bottom of the profiles may be due to the release of Au from disintegrating fragments of the vein material or to the downward migration of Au particles (of high density) within the soil profiles during downslope creep. Conventionally 20–50 g of dried, macerated soil humus is formed into a briquette (pellet) for INAA analysis [38].

The chemistry of Au in soils is essentially a chemistry of complex compounds; no simple Au cations exist in aqueous solution [45]. In addition, its chemistry is complicated by the strong interactions with organic matter—about which there are considerable uncertainties. It should also be emphasised that the behaviour of Au in soil can usefully be conceived as a number of distinct steps or processes. Firstly, possible mechanisms for its dissolution from minerals; secondly, the need to invoke mechanisms for its transport and mobility through the soil solution; thirdly, various complex interactions with inorganic and organic soil constituents; and fourthly, the role of all of these factors in determining its availability for uptake by biological systems.

Existing data reveal a rather conflicting picture about possible mechanisms for the dissolution of Au. Gold is known to be relatively immobile, not readily entering the aqueous phase [37, 45]. It has been reported by some workers that Au may be solubilised from minerals and soils by microbial activity [51, 52], whilst the results of others are rather ambiguous [45]. Similarly, there is dispute over the role of organic acids in Au dissolution [53, 54]. One of the most interesting confirmed observations is the solubilisation of Au by cyanide released into soil solution by cyanogenic plants [45, 55, 56]. Some of the explanation for these apparent discrepancies probably arises from the complexity of Au chemistry, and the need to perform dissolution experiments under precise experimental conditions. It can be shown, for example, that the dissolution of Au in soils may vary considerably, depending on the presence of other chemical constituents, and the prevailing pH and redox conditions. Gold may be in contact with oxygenated water for decades without dissolving but if 250 mg/l of cyanide ion (CN^-) is added to the water, the Au is readily dissolved. The cyanide ion is not an oxidant but forms a complex with the Au. In its presence, the oxidation potential for Au is reduced from −1.50 to −0.20 V and the dissolved O_2 in the water then becomes a sufficiently strong oxidant to dissolve the Au. Similarily, Au is oxidised by MnO_2 in acid chloride solutions [45]. Krauskopf [57] emphasised that the purpose of the chloride ion is to form a very stable complex ion with the dissolved Au. Gold is slowly oxidised in the same acid sulphate solutions containing bromide ions, and it is rapidly oxidised in the acid sulphate solution containing iodide ions. For example, at pH 2 in a solution of $Fe_2(SO_4)_3$ and $CuSO_4$ containing 0.05 M Cl^-, Au leaf is dissolved very slowly, if at all —<0.004 mg/l in 1 month; when 0.005 M Br^- is used, 1.2 mg/l is dissolved in 1 month; but with 0.005 M I^-, 72 mg/l of Au is dissolved in 1 week. This sequence illustrates the increasing ease of oxidation of Au with increasing strength of the complex ion. Although the ease of oxidation of Au increases in the order Cl^-, Br^-, I^-, the abundance of these halides in the Earth's crust decreases in the same order—their abundance ratio in the crust is approximately 100, 0.6, 0.06. In a carbonate environment, the oxidation of pyrite may produce sufficient $S_2O_3^{2-}$ to dissolve Au. In the oxidising environment at pH 5–8, only $S_2O_3^{2-}$, CNS^-, and CN^- may be expected to dissolve Au [45]. Thiosulphates are transient products of biological activity in soils but never reach substantial concentrations; thiocyanates are even less abundant in soils, so these forms are probably unimportant in most soils. In contrast, cyanides are produced by the hydrolysis of cyanogenic glycosides, which are abundant in plant residues and soils. More than 1000 species of plants yield HCN on hydrolysis, as do many arthropods and fungi. Macerated aqueous suspensions of 16 species of native plants collected from central areas of the USA, for example, dissolved Au leaf under experimental conditions [45].

The ability of microorganisms to solubilise elemental Au is discussed in several papers, and was summarised by Doxtader [45]. These findings were not clear cut. Doxtader performed experiments on colloidal Au with bacterial and fungal strains which had been reported by previous workers to increase the solubility of Au. His results were somewhat ambiguous. On the basis of the low levels of Au detected and the sensitivity of the chemical assay, it was difficult to draw definite conclusions regarding the relative levels of metal solubilisation by different cultures, but, apparently some solubilisation took place both in cultures and in sterile media. By comparison with the Au levels in the uninoculated media, little, if any additional Au was solubilised by the test strains. In fact, there is a slight indication that some of the cultures removed Au from solution. Doxtader suggested that the presence of significant levels of Au in solution in the uninoculated culture medium may indicate that organic solutions, such as peptone and yeast extract, may in themselves be capable of solubilising Au [45].

Baker [58, 59] presented experimental data to show that humic acids can dissolve, complex and transport Au. For example, as much as 330 μg Au/1 was taken into solution from 0.07–0.15 mm Au particles by 500 mg/1 solutions of humic acid during a 50-day period. Gold has been reported at 0.02–0.12 μg/1 in leachates extracted from Au-bearing mull [60]. Reported values for organic matter concentrations in soil solutions range from about 1 mg/1 to 1600 mg/1 [61], so the conditions used in this experiment were typical of field conditions. Gold was also mobilised in electrophoresis experiments in the presence of humic acid, whereas Au alone remained immobile. The formation of a complex between Au and humic acid was indicated by the results of polarographic, solvent extraction and X-ray diffraction investigations [58]. Preliminary high-voltage electrophoresis studies performed on equilibrated mixtures of ^{198}Au with commercially available preparations of humic acid (10 or 100 mg/1) indicated that several organically-complexed forms of Au may be present in soil extracts, each with different charge/mass characteristics [62]. However, a high proportion of the ^{198}Au appeared to be present in a similar form (presumably complexed) to that in a solution of $^{198}AuCl_4^-$. The ^{198}Au was probably also complexed with fractions of different molecular weight; 9.5% and 8.0%, respectively, of the ^{198}Au activity in the filtered solutions of 10 mg HA/1 and 100 mg HA/1 was retained on the 0.8 μm Millipore filters. Studies with size fractionation permeation gels have suggested ^{198}Au retention with components having a range of molecular weight [63]. Other workers have reported the occurrence of fine (< 0.05 μm) Au colloids in soil solutions [60, 64]. The formation of a Au-humic acid complex would clearly be important in the mobilisation of Au in the environment. It appears that once it has entered the aqueous phase (by dissolution) Au may be rapidly and readily complexed with humic and/or fulvic acids [65] and transported as an organic complex [66]. The enrichment of Au in humus mull and near-surface soil also suggests that the mobility of Au as a complex ion is transitory. The following complex ions have been suggested as playing a role in the migration of Au and are likely to be stable in soil solutions: $AuCl_4^-$, $AuBr_4^-$, AuI_2^-, $Au(CN)_2^-$, $Au(CNS)_4^-$, and $Au(S_2O_3)_2^{3-}$. Nonetheless, there is a marked difference in the likely stability of these various forms, particularly between the halide complexes and those with cyanide, thiocyanate and thiosulphate; the halide complexes are much less stable and therefore less mobile in the natural environment [45].

Adsorption and retention of Au by humates may subsequently limit its biological availability and ultimately its rate of recycling within the soil system. Strong adsorption or complexation of Au by humic colloids, with possible incorporation into the structure of humic components [66] or the formation of relatively stable organic chelate complexes would thus prevent its subsequent removal from the upper horizons, such as forest litter and mull humus. Clearly, this provides further support for the potential of assaying soil humus for Au and pathfinder elements for biogeochemical prospecting studies [39, 47, 48, 67, 68].

There is a substantial volume of literature concerning both the biogeochemical methods of exploration for Au and measurements of Au uptake by plants [66,

69–76]. Much of this in recent years has come from North America, particularly Canada, where the combination of auriferous deposits along with soils composed of deep glacial overburden and the occurrence of extensive, deep-rooting boreal forest tree species (*Alnus*, *Picea*, *Betula*, *Pinus*, etc.) has encouraged the use of biogeochemical prospecting.

The absorption of Au from solution by a given plant species is a function of the Au complex present. Because colloidal Au is not taken up by plants, and because the simple ions Au^+ and Au^{3+} cannot exist in appreciable quantities in aqueous solutions, Au must enter the plant as a soluble complex ion. Gold chloride, thiocyanate, and thiosulphate are very weakly absorbed through the roots of *Impatiens hostii*, for example, whilst Au bromide and iodide are absorbed roughly 100 times more than chloride [51].Cyanide and thiocyanate ions have the ability to complex Au in moderately oxidising environments, to remain in solution when in contact with common rocks and minerals, and to be taken up by plants. The other ions—chloride, bromide, iodide, and thiosulphate—may form complexes with Au which remain stable only long enough to permit some restricted movement of Au under special conditions. Thiocyanate occurs in appreciable concentrations in many plants, such as members of the Brassica genus (cabbages) and Umbelliferae (carrot family). Complexed forms of Au prepared with humic and fulvic acids are not as readily taken up by plants grown in hydroponic solution as other aqueous forms of Au, such as $AuCl_3$ and $AuCN_3$ [62].

Background Au concentrations are typically < 1 μg/kg in vegetation [56]. Studies in Vietnam [77] on Au content of rice seed, using Neutron Activation Analysis, indicated a range between 0.05–0.28 $\mu g\ g^{-1}$ dry wt. Studies with tree species in Canada indicate that Au concentrations vary substantially throughout the year. Data obtained from several hundred sites indicate that background levels in ashed alder twigs are 20 μg/kg in early summer, 10 μg/kg in midsummer and 15μg/kg in late summer. Background levels in April may be 30–50 μg/kg [42]. One complication as far as the biogeochemical prospector is concerned is that anomalous Au concentrations above mineral deposits may be ephemeral, apparent in the overlying vegetation only at certain times of the year. Another, emphasised by Dunn [42] is that the biogeochemical response to Au mineralisation varies between locations.

A recent study by Xu [77] describes a series of environmental problems which may be attributable to Gold contamination of vegetation in western Guandong/ Hainan, China. In this region, Au content of vegetation is elevated, some 10% above background, and is associated with some characteristic changes in physical and biochemical properties of affected plants. These include reductions in water and carotenoid content, reductions in leaf temperature and increase in spectral reflectance. Identification of such features in native vegetation are considered to aid in the detection of Au-rich anomalies by satellite surveys. Xu claims to have identified two potential mine sites in densely-vegetated terrain using this system.

References: Gold

35. Minski, M.J., Girling, C.A. and Peterson, P.J., *Radiochem. Radioanal. Lett.* **30** (1977), 179–186.
36. Tjioe, P.S., Vokers, K.J., Kroon, J.J., ed Goei, J.J.M. and The, S.K., *Internat. J. Environ. Anal. Chem.* **17** (1984), 13–24.
37. Jones, K.C., *J. Geochem. Explor.* **24** (1985), 237–246.
38. Hoffman, E.L. and Booker, E.J., in *Mineral Exploration: Biological Systems and Organic Matter*, eds. Carlisle, D., Berry, W.L., Kaplan, I.R. and Watterson, J.R., Rubey Series Vol. 5, Prentice-Hall, Englewood Cliffs, NJ. (1985), 160–169.
39. Dunn, C.E., in *Mineral Exploration: Biological Systems and Organic Matter*, eds. Carlisle, D., Berry, W.L., Kaplan, I.R. and Watterson, J.R., Rubey Series Vol. 5, Prentice-Hall, Englewood Cliffs, NJ. (1985), 134–139.
40. Brooks, R.R. and Naidu, S.D., *Anal. Chim. Acta.* **170** (1985), 325–329.
41. Dunn, C.E., *Summary of Investigations 1983*. Saskatchewan Geological Survey, Miscellaneous Report 83–4, 106–122.
42. Dunn, C.E., *Summary of Investigations 1985*. Saskatchewan Geological Survey, Miscellaneous Report 85–4, 37–49.
43. Gregoire, D.C., *J. Geochem. Explor.* **23** (1985), 299–313.
44. Clifton, H.E., Hunter, R.E., Swanson, F.J. and Phillips, R.L., *US Geological Survey Prof. Paper* 625-C (1969).
45. Lakin, H.W., Curtin, G.C. and Hubert, A.E., *US Geological Survey Bull. 1330* (1974).
46. Curtin, G.C., Lakin, H.W., Hubert, A.E., Mosier, E.L. and Watts, K.C., *US Geological Survey Bull.* (1971), 1278-B.
47. Curtin, G.C. and King, H.D., in *Mineral Exploration: Biological Systems and Organic Matter*, eds. Carlisle, D., Berry, W.L., Kaplan, I.R. and Watterson, J.R., Rubey Series Vol. 5, Prentice-Hall, Englewood Cliffs, NJ. (1985), 357–375.
48. Dunn, C.E., *J. Geochem. Explor.* **25** (1986), 21–40.
49. Goldschmidt, V.M., *Chem. Soc. (London) Journal* (1937), 655–673.
50. Goldschmidt, V.M., *Geochemistry*. Blackwell, Oxford (1954).
51. Korbushkina, E.D., Chernyak, A.S. and Mineev, G.G., *Mikrobiologiya* **43** (1974), 49–54.
52. Korbushkina, E.D., Karavaiko, G.I., and Korobushkin, I.M., in *Environmental Biogeochemistry, Ecol. Bull.* (Stockholm) **35** (1983), 325–333.
53. Fetzer, W.G., *Econ. Geol.* **41** (1946), 47–56.
54. Fetzer, W.G., *Econ. Geol.* **26** (1931) , 421–431.
55. Girling, C.A., Peterson, P.J. and Warren, H.V., *Econ. Geol.* **74** (1979), 902–907.
56. Girling, C.A. and Peterson, P.J., in *Trace Substance in Environmental Health*—XII, ed. Hemphill, D.D. University of Missouri, Columbia (1978).
57. Krauskopf, K.B., *Econ. Geol.* **46** (1951), 858–870.
58. Baker, W.E., *Geochim, Cosmochim. Acta.* **42** (1978), 645–649.
59. Baker, WE., in *Mineral Exploration: Biological Systems and Organic Matter*, eds. Carlisle, D., Berry, W.L., Kaplan, I.R. and Watterson, J.R., Rubey Series Vol. 5, Prentice-Hall, Englewood Cliffs, NJ. (1985), 378–407.
60. Curtin, G.C., Lakin, H.W and Hubert, A.E., *US Geological Survey Prof. Paper* 700-C (1970).
61. Burch, R.S., Langford, C.H. and Gamble, D.S., *Can. J. Chem.* **56** (1978), 1196–1201.
62. Jones, K.C. and Peterson, P.J., *Biogeochemistry* **7** (1989), 3–10.
63. Jones, K.C., unpublished data.
64. Ong, H.L. and Swanson, V.E., *Colorado School of Mines Quarterly* **69**, 395–425.
65. Kerndorff, H. and Schnitzer, M., *Geochem. Cosmochim. Acta.* **44** (1980), 1701–1708.
66. Boyle, R.W., *Canadian Geological Survey Bull.* **280** (1979).
67. Banister, D.P., *U.S. Bureau of Mines. Report of Investigations* 7417 (1970).
68. Schnitzer, M., in *Mineral Exploration: Biological Systems and Organic Matter*, eds. Carlisle, D., Berry, W.L., Kaplan, I.R. and Watterson, J.R., Rubey Series Vol. 5, Prentice-Hall, Englewood Cliffs, NJ. (1985), 409–427.
69. Kovalevskii, A.L., *Biogeochemical Exploration for Mineral Deposits*. Amerind Publ. Co., New Delhi.
70. Brooks, R.R., *J. Geochem. Explor.* **17** (1982), 109–122.
71. Baker, W.E., in *Mineral Exploration: Biological Systems and Organic Matter*, eds. Carlisle,

D., Berry, W.L., Kaplan, I.R. and Watterson, J.R., Rubey Series Vol. 5, Prentice-Hall, Englewood Cliffs, NJ. (1985), 151–158.
72. Schacklette, H.T., Lakin, H.W., Hubert, A.E. and Curtin, G.C., *US Geological Survey Bull. 1314-B* (1970), 1–23.
73. Warren, H.V. and Delavault, R.E., *Bull. Geol. Soc. America* **61** (195), 123–128.
74. Warren, H.V., Towers, G.H.N., Horsky, S.J., Kruckeberg, A. and Lipp, C., *Western Miner* (1983), 19–25.
75. Malyuga, D.P., *Biogeochemical Methods of Prospecting*. Consultants Bureau, New York, (1964).
76. Jones, R.S., *U.S. Geol. Surv. Circ.* **625** (1970), 1–15.
77. Van Tran, L. and Teherani, .K., *J. Radioanal. Nuc. Chem.*, **128** (1988), 43–52.
78. Xu, R., *Acta. Geologica Sinica*, **66** (1992) 170–180.

14.3 Molybdenum

Molybdenum is unusual amongst plant nutrients because it exists as an anion in soil and consequently shows an increased solubility with increasing pH. It is mainly associated with hydrous Fe and Al oxides, and with organic substances, and has significant interaction with other soil elements in an antagonistic or synergistic way. Molybdenum is used primarily in steel manufacturing and in alloys for improving strength, corrosion resistance, and preventing metal fatigue. The element occurs in fossil fuels and is mobilised by combustion processes which can result in soil contamination via atmospheric deposition. Soil contamination can also result from soil amendments such as fly-ash or sewage sludge with consequent effects on plant and animal nutrition. Molybdenum has an essential role in plant nitrogen metabolism, and high levels in herbage can cause molybdenosis—a Mo-induced Cu deficiency in livestock [79].

Molybdenum exists in the lithosphere at oxidation states ranging from Mo^{3+} to Mo^{6+}, with Mo^{4+} and Mo^{6+} predominating under reducing and oxidising conditions respectively. It has its main association with acidic and basic igneous rocks and particularly organic-rich argillaceous sediments. The primary chalcophilic Mo mineral, molybdenite (MoS_2, Mo^{4+}) is recovered commercially from magmatic, metasedimentary, and metasomatic deposits [80]. It is often mined from quartz veinlets in granites (often pegmatites) and is associated with scheelite, wolframite, topaz, fluorite, and also with contact metamorphic deposits with lime silicates, scheelite, or chalcopyrite [81]. Molybdenite is also associated with minerals of W, Fe, Sn and Ti. Other ores of commercial importance include molybdite (MoO_3), wulfenite ($PbMoO_4$) and powellite ($CaMoO_4$). Molybdenum also displays lithophilic behaviour as the Mo^{4+} ion can substitute for Al^{3+} and other elements in micas and feldspars.

The Mo content of parent rocks is usually reflected in overlying soils developed *in situ* [82, 83]. Typical Mo concentrations in parent materials are 1.4 mg/kg in basic rocks, 1 mg/kg in acidic rocks and 2 mg/kg in sedimentary rocks [84]. Non-pyritic shales low in organic content have similar Mo concentrations to igneous rocks [85] but black bituminous shales have concentrations up to 70 mg/kg [86] or higher [87]. Concentrations of molybdate in carbonate rocks such as dolomites or limestones may be enhanced by the presence of organic substances or other foreign components in the rock matrix. Average Mo soil concentrations at a global scale are given as about 1–2 mg/kg [80, 88, 89]. A figure for British soils is 1 mg/kg [90] with a range of 1–5 mg/kg for Scottish soils [91]. Total Mo concentrations in Malaysian soils vary from 1.13 to 10.53 mg/kg [92]. Ranges for US soils are of the order of 0.08–30 mg/kg [93] to 1–40 mg/kg [94] with median values of 1mg/kg and 1.2–1.3 mg/kg respectively. Soils developing on a range of marine black shales were found to have a Mo concentration as high as 100 mg/kg, with an average of 7.1 mg/kg for British soils of this type [95]. In peat, up to 4000 mg/kg Mo has been reported for many reclaimed peatland areas in the Irish Midlands [9, 6].

Molybdenum may be used as a fertiliser in its own right (e.g. ammonium molybdate, sodium molybdate, molybdenum trioxide) or be incorporated into NPK fertilisers to obtain satisfactory distribution of the low quantities usually required [94]. As indigenous soil Mo is strongly influenced by soil conditions, the addition of fertilisers can have a direct effect on Mo geochemistry via synergistic/antagonistic interactions. Copper, Mn, and Al, for example, are all known to have an antagonistic effect on Mo availability [89, 97]. Addition of sulphate fertilisers also generally has an antagonistic effect due to soil acidification, whilst addition of gypsum has in some cases been shown to enhance availability by decreasing soluble carbonate and hydroxyl ions [98]. Addition of superphosphate usually increases available Mo by the substitution of phosphoric acid ions for molybdate (MoO_4^{2-}) on the exchange complex and by increased solubility of ammonium phosphomolybdate [97, 98]. However, this effect is not always clear as the presence of sulphate in super-phosphate fertilisers can lead to reduced Mo uptake [99]. Nitrogen fertilisers have a variable effect, depending on whether ammonium or nitrate ions are predominating. A collection of data on Mo in agricultural materials [97] showed that phosphate fertilisers contain on average 0.1–60 mg/kg Mo, limestone 0.1–15 mg/kg, nitrate fertilisers 1.7 mg/kg, and manures 0.5–3 mg/kg.

As already indicated, Mo availability is strongly pH-dependent with anionic availability increasing at higher pH due to hydroxyl ion competition for specific absorption sites. Addition of lime to soil has therefore been used to increase indigenous Mo availability on the assumption that sufficient MoO_4^{2-} is present for exchange with hydroxyl ions. However, at high liming rates care must be taken that excessive Mo absorption does not occur due to high concentrations of $CaCO_3$ present [100].

Atmospheric deposition of Mo to soil can occur through melting and smelting operations, manufacturing and oil refining, and the burning of fossil fuels. Although mobilisation of Mo by combustion processes in comparison to natural weathering is low (64000t/yr as river load versus 2300t/yr from coal and oil [101]), soil contamination can be expected at a local scale. Molybdenum is concentrated in fly-ash by the combustion of coal (typical values for US coal 0.2–50 mg/kg [102] with ranges of 7–160 mg/kg) [103]. The actual concentration is dependent on coal type and shows highest affinity for small ash particulates [103]. The deposition of fine particulates onto soil or the application of fly-ash as a soil amendment is particularly significant because of the alkaline pH of ashes. Atmospheric concentrations of Mo in the UK have been shown to range from 0.29–1.29 ng/m^3 in rural areas to 2–18 ng/m^3 inside steelworks [104]. Atmospheric deposition adjacent to Al alloy factories, steelworks and oil refineries has produced high pasture Mo concentrations and associated molybdenosis in livestock [104]. A causal effect of atmospheric pollution on soil Mo behaviour has been found in relation to sulphate deposition where soil acidification in forest stands has resulted in reduced Mo uptake by trees [105].

Sewage sludge addition to soils is another source of localised soil contamination of Mo together with cationic metals. A survey of Mo concentrations in sewage sludge from a variety of industrial and non-industrial sources in England and Wales ranged from 2–30 mg/kg [106] (median 5 mg/kg) and for the USA 1.2–40 mg/kg [107, 108] (median 8.1 mg/kg), although larger concentrations have been reported [109]. A pH increase which is normally associated with sludge addition due to the presence of lime residues can be expected to enhance indigenous and sludge derived Mo availability. Mandatory regulations [110] to maintain a high soil pH following sludge addition in order to reduce cationic toxicity will effectively increase Mo availability. A possible benefit of the addition of Cu-enriched sludge exists through the increased herbage uptake of Cu and reduction of molybdenosis, although this is usually limited.

Oxidation of Mo minerals during weathering yields mainly molybdate (MoO_4^{2-}) ions in the soil solution via soluble intermediates, e.g. MoO_2SO_4. At lower pH (< 4.2), molybdate ions become protonated forming monomeric $HMoO_4^-$ and H_2MoO_4 followed by condensation to give polymolybdate [111]. Mobilised anions are readily co-precipitated by organic matter and cations, or are adsorbed by sesquioxides. Studies of soil surface reactions have shown that Mo is most strongly adsorbed with hydrous ferric oxides (e.g. goethite, haematite) where Mo_4^{4+} substitutes for Fe^{3+} [112, 113]. Studies have also shown that the amount of reactive surface rather than the residual amount of Fe was most important in adsorption processes [114]. Molybdate may also be held as an anion on clay surfaces or secondary minerals complexed with organic substances [91].

The net balance of sorbed- solution phase equilibria is fundamentally dependent on soil chemical conditions, notably pH and E_h. Availability is subject to modification by organic substances and, as already noted, by the presence of other elements. The equation relating to the effect of pH on Mo availability [115] is

$$\text{soil} + MoO_4^{2-} \rightarrow \text{soil–}MoO_4 + 2OH^-$$

It has been shown that the increase in Mo solubility below pH 7 is as high as 100-fold, which may be due to the presence of wulfenite ($PbMoO_4$). The proposed exchange mechanism at low pH involves the protonation of molybdic acid and conversion of a hydroxyl group on the exchange surface to a water molecule which is displaced by the anion. The reaction is exothermic and involves chemical bonding where the adsorption surface has a specific affinity for Mo, and an increased negative charge on adsorption [116]. The concentration of Mo in soil solution as a consequence of strong sesquioxide absorption assumes significance only in alkaline soils of impeded drainage and is normally in the range 2–8 × 10^{-3} m/kg. Solid-solution phase equilibria may also change over time through the mineral diagenesis of amorphous Fe oxides and fixation of Mo via the formation of ferro-molybdite ($Fe_2(MoO_4)_3 8H_2O$) and other semicrystalline forms [117, 118]. This has been shown to occur in stored acidic soils that have been subjected to extensive leaching [119], and fixation has also been correlated to soil C/N ratio indicating the importance of soil organic constituents [120].

Organic complexation of Mo is not yet fully understood, and its effect in soil may be obscured by pH or E_h as a consequence of poor drainage. Although Mo exists in soil mainly in a complex anionic state, enhanced fixation at low pH can be explained by reduction of molybdate to cationic form (Mo^{5+}) by humic acid and polysaccharide components [121, 122]. Molybdenum has been shown to be mobilised as anions by aerobically decomposing plant material with subsequent fixation by organic-colloidal complexes at low pH [123]. Anaerobic conditions result in the persistence of the anionic form. Variable responses to organic matter addition to soil have been noted under experimental conditions, depending on whether it was accompanied by the addition of Mo [124]. Addition of organic matter alone increased indigenous Mo availability whilst addition with Mo had the opposite effect. Prolonged organic addition has been found to increase ammonium oxalate-soluble Mo and plant availability [125]. The possibility of organic matter enhancing Mo availability over a range of pH is possible by organic chelation of Mo preventing diagenesis [81]. Molybdenum has been found to accumulate in lower A or upper B horizons in association with humus and where pH reaches a minimum in freely drained soils [86].

Poorly drained soils often have high levels of available Mo; these soils are often of high pH and organic status, which favours the mobilisation of anions. Conditions of impeded drainage have been found to result in greater herbage Mo concentrations due to higher levels in soil solution [127, 128]. A change from Fe^{3+} to Fe^{2+} iron, accompanied by the increased dissolution of ferric molybdate compounds or complexes, is probably involved [128]. The high activity of molybdate under alkaline conditions and its ability to form soluble thiomolybdates under reducing conditions also contributes. In the presence of organic matter, redox potentials may be lowered more rapidly on wetting [128]. In such circumstances, a reduction in soil pH will reduce Mo availability. Studies have shown that a reduction in soil pH by half a unit reduced plant Mo concentration by approximately one half [129].

Plant tissue concentrations of Mo at about 0.03–0.15 mg/kg are in most cases adequate for plant physiological requirements—where normal leaf tissue concentrations are in the order of about 1 mg/kg, below 0.5 mg/kg symptoms such as chlorosis, necrosis and downward curling of leaves indicate Mo deficiency [130]. Legumes have a higher demand for Mo because it is a biochemical component of the metalloenzyme nitrogenase [131] which fixes N in the symbiotic rhizobial nodule (typical values 0.13–2.3 mg/kg). Molybdenum deficiency therefore leads to poor N fixation, leading to symptoms consistent with plant N deficiency. Molybdenum is also involved with nitrate reductase in plant N metabolism, acting as a redox electron carrier in the reduction of nitrate to plant-utilisable ammonia. The sensitivity of Mo metalloenzymes has been used to detect Mo deficiency in soils and plant tissue [132, 133]. Low Mo levels in plants are also associated with an altered P metabolism [134], and reduced levels of sugar, ascorbic acid and certain amino acids [135]. Brassicas are often sensitive to lack of Mo, showing specific deficiency symptoms. Cauliflower 'whiptail', for example is associated with leaf deformation and biochemical and microstructural changes to the photosynthetic system [136]. Severe deficiency of Mo has been shown to reduce 5-methylcytosine levels in tomato and cauliflower plants. This causes disturbances in the replication and transcription of DNA which may lead to damage of the plants' reproductive structures [137]. Addition of Mo or lime to increase pH levels have been shown to increase legume nodulation, nitrogenase activity and leghaemoglobin levels [138, 139] and can also be expected to raise nitrate reductase activity in crop plants.

Molybdate is the dominant form of Mo taken up by plants from soil solution, so soil physicochemical factors, particularly pH, will strongly affect plant tissue concentrations. Direct evidence for active plant uptake is lacking, although the mechanism should be comparable to other anions such as sulphate, nitrate and phosphate. The form of Mo in translocation is unknown, although organic compexing is a possibility [140]. Elements such as S have an inhibitory effect on plant uptake of Mo, due to competition for root absorption sites as a function of similar ionic size and charge. Phosphorus is known to enhance translocation, which may be due to the formation of soluble phosphomolybdate complexes readily assimilable by roots or to biochemical changes in the plant releasing organic P into the translocation system [125].

Soil microorganisms are known to benefit from Mo additions to soil. Microbial responses have been shown to include increased decomposition, higher non-symbiotic N fixation rates and numbers of actinomycetes, and greater soil enzymic activity [141]. Foliar spraying of rice with Mo has been shown to stimulate a range of rhizospheral bacteria [142], and rhizosphere microflora of radish plants have been shown to accumulate up to 55 mg/kg of Mo when grown in liquid medium containing soil extract [143]. Plants with mycorrhizal infection are known to benefit from a greater Mo uptake [144]. A study of Mo immobilised in a range of soils revealed that up to 0.16% of Mo occurred in the microbial biomass 145], and other figures suggest that up to 148 g/ha of Mo (top 20 cm soil) may be within the biomass [146]. This immobilisation may act as a source–sink for plant-available Mo through microbial nutrient turnover processes [147]. Organic complexes formed by the soil microflora may also prevent fixation of Mo by other soil components in acid soils [127].

Molybdenum becomes toxic in mature leaf tissue at concentrations of 10–50 mg/kg [93] or lower, and this can lead to molybdenosis in livestock grazing high Mo herbage. Ruminants are particularly sensitive to Mo toxicity which is expressed in the form of a Cu deficiency with an added interaction with sulphate [79, 147]. Ratios of Cu/Mo in the forage below 2:1 have been quoted as being potentially toxic, but consideration of the total concentrations of Cu and Mo as well as sulphate intake is needed to accurately assess the risk associated with dietary intake of Mo [147]. Applications of sewage sludge at 0.41 mg Mo/ha at pH 7.2 resulted in Mo concentrations of 94 mg/kg in white clover and 20.4 mg/kg in ryegrass [148], whilst applications of up to 17 kg Mo/ha in another study resulted in 31 mg/kg of Mo in clover at pH 8.0 [149]. Addition of fly-ash at 8% by weight to soil has resulted in 44.6 mg/kg Mo in white clover [107]. Plants growing adjacent to a Mo processing plant have been shown to have concentrations as high as 1016 mg/kg [150].

Molybdenum contamination can therefore assume local significance as a consequence of soil additions and atmospheric deposition with effects in terms of animal and plant nutrition. Careful consideration is needed in controlling the addition of this metal to soil with respect to its contrasting chemical behaviour compared to cationic elements. Knowledge of its interaction with soil physico-chemical factors increase its potential to be exploited as a beneficial micronutrient in terms of microbial activity and crop production.

References: Molybdenum

79. Underwood, E.J., *Trace Elements in Human and Animal Nutrition*, 4th edn. Academic Press, New York (1977).
80. Ronov, A.B. and Yarovshevskii, A.A., in *Encyclopedia of Geochemistry and Environmental Sciences* 4A, ed. Fairbridge, R.W. Van Nostrand-Reinhold, New York (1972).
81. Mason, B. and Berry, L.G., *Elements of Mineralogy*, W.H. Freeman, San Francisco (1968).
82. Massey, H.F. and Lowe, R.H., *Soil Sci. Soc. Amer. Proc.* **25** (1961), 161.
83. Gorbacheva, A.E., *Sov. Soil Sci.* **8** (1976), 129.
84. Vinogrador, A.P., *Geochemistry* 7 (1962), 641.
85. Kuroda, P.K. and Sandell, E.B., *Geochim. Cosmochim. Acta* **8** (1954), 213.
86. Wedepohl, K.H., *Handbook of Geochemistry*, II-4/42. Springer-Verlag, Berlin (1978).
87. Manskaya, S.M. and Drozdova, T.V., *Geochemistry of Organic Substances*, Pergamon Press, New York (1968).
88. Taylor, S.R., *Geochim. Cosmochim. Acta* **28** (1964), 1273.
89. Ure, A.M. and Berrow, M.L., in *Environmental Chemistry*, Vol 2 (H.J.M. Bowen, Senior Reporter). The Royal Society of Chemistry, London (1980).
90. Swaine, D.J. and Mitchell, R.L., *J. Soil Sci.* **11** (1966), 347.
91. Mitchell, R.L., in *Trace Elements in Soils and Crops*. Technical Bulletin 21, Ministry of Agriculture Fisheries and Food HMSO, London (1971).
92. Lau, C.H. and Lim, T.S., *J. Nat. Rubber. Res.* **7** (1992), 60.
93. Kubota, J., in *Molybdenum in the Environment*, Vol. 2., eds., Chappell, W.R. and Peterson, K.K. Marcel Dekker, New York (1985),
94. Tisdale, S.L., Nelson, W.L. and Beaton, J.D., *Soil Fertility and Fertilisers*. Macmillian, New York.
95. Thornton, I. and Webb, J.S., in *Copper in Farming Symposium*. Copper Development Association, Potters Bar, UK (1976).
96. Talbot, V. and Ryan, P., *Sci. Total Environ.* **76** (1988), 217.
97. Barshad, I., *Soil Sci.* **71** (1951), 297.
98. Adams, F., in Proceedings of Symp. on *The Role of Phosphorus in Agriculture*, ASA, CSSA and SSSA. Madison, Wisc., USA (1976).
99. Takkar, P.N., in *Abstr. 12th Int. Soil Sci. Cong. Part 1*, New Delhi (1982).
100. Kabata-Pendias, A. and Pendias, H., *Trace Elements in Soils and Plants*. CRC Press, Boca Raton (1984).

101. Bertine, K.K. and Goldberg, E.D., *Science* **173** (1971), 233.
102. Swanson, V.E., Medlin, J.M., Hatch, J.R., Coleman, S.L., Wood, G.H., Jr., Woodruff, S.D. and Hildebrand, R.T., *Collection, Chemical Analysis and Evalution of Coal Samples in 1975*. U.S. Dep. of the Int. Geological Survey, Open File Rep 76–468. U.S. Geological Survey, Reston, New Jersey (1976).
103. Page, A.L., Elseewi, A.A. and Straughan, I.R., *Resid. Rev.* **71** (1979), 83.
104. Thornton, I., in *Molybdenum in the Environment*, Vol. 1, eds. Chappell, W.R. and Peterson, K.K. Marcel Dekker, New York (1977).
105. Fiedler, H.J., Ilgen, G. and Hoffman, W., *Arch. Naturschutz Landschaftsforsch.* **27** (1987), 177.
106. Berrow, M.L. and Burriedge, J.C., in *Inorganic Pollution and Agriculture*. MAFF Reference Book 326 (1980).
107. Jarrel, W.M., Page, A.L. and Elseewi, A.A., *Resid. Rev.* **74** (1980), 1.
108. Lisk, D.J., Gutenman, W.H., Rutzke, M., Kuntz, H.T. and Chu, G., *Arch. Environ. Contam. Toxicol.* **22** (1992), 190.
109. Lahann, R.W., *Water, Air and Soil Pollut.* **6** (1976), 3.
110. Council Directive on the protection of the environment and in particular of the soil when sewage sludge is used in agriculture. *Off. J. Europ. Comm. No.* 181/6 (1986).
111. Reisenbauer, H.M., Tabilch, A.A. and Stout, P.R., *Soil Sci. Soc. Amer. Proc.* **26** (1962), 23.
112. Davies, E.B., *Soil Sci.* **81** (1965), 209.
113. Jones, L.J., J. *Soil Sci.* **8** (1957), 313.
114. Jones, C.H.P. and Smith, B.H., *Plant and Soil* **37** (1972), 649.
115. Vlek, P.L.G. and Lindsay, W.L., *J. Soil Sci. Soc. Am.* **41** (1977), 42.
116. Barrow, N.J., in *Molybdenum in the Environment*, Vol. 1, eds. Chappell, W.R. and Peterson, K.K. Marcel Dekker, New York (1977).
117. Barrow, N.J., *Soil Sci.* **116** (1973), 423.
118. Barrow, N.J. and Shaw, T.C., *Soil Sci.* **119** (1975), 301.
119. Smith, B.H. and Leeper, G.W., *J. Soil Sci.* **20** (1969), 246.
120. Lal, S., De, S.K. and Shukla, R.K., *Anales de Telafologia y Agrobiologia* **30** (1971), 423.
121. Szalay, A. and Szilagyi, M., *Plant and Soil* **29** (1968), 219.
122. Goodman, B.A. and Cheshire, M.V., *Nature* **299** (1962), 618.
123. Bloomfield, C. and Kelso, W.I., *J. Soil Sci.* **24** (1973), 368.
124. Gupta, U.C., *Plant and Soil* **34** (1971), 249.
125. Selevtsova, G.A., *Agrokhimiya* **9** (1969), 74.
126. Mitchell, R.L., in *Chemistry of the Soil*, 2nd edn. ed. Bear, F.E. Reinhold, New York (1964).
127. Kubbota, J., Lemon, E.R. and Allaway, W.H., *Soil Sci. Amer. Proc.* **27** (1963), 679.
128. Ferguson, W.S.A., Lewis, A.J. and Watson, S.J., *J. Agric. Sci. Camb.* **33** (1943), 44.
129. Feely, L., *Irish J. Agricl. Res.* **29** (199), 129.
130. Cox, A., *Hortscience*, **27** (1992), 894.
131. Schrauzer, G.N., in *Molybdenum in the Environment*, Vol. 1, eds. Chappell, W.R. and Peterson, K.K. Marcel Dekker, New York (1977).
132. Peres, J.R.R., Nery, M. and Fraco. A., In *Annals of the Fifteenth Brazilian Congress of Soil Science*, Campurias, Brazil. Brazilian Society of Soil Science (1976).
133. Witt, H.H. and Jungle, A., *A. Pflanzener-nahrung Bodenkunde* **140** (1977), 209.
134. Follet, R.H., Murphy, L.S. and Donahue, R.L., *Fertilizers and Soil Amendments*. Prentice-Hall, Englewood Cliffs, NJ. (1981).
135. Epstein, E., *Mineral Nutrition of Plants: Principles and Perspectives*. John Wiley, New York (1972).
136. Fido, R.J., Gundry, C.S., Hewitt, G.J. and Notton, B.A., *Aust. J. Plant Physiol.* **4** (1977), 675.
137. Bozhenko, V.P. and Balycieva, V.N., *Sov. Plant Physiol.* **24** (1977), 281.
138. Fedorova, E.E. and Potatueva, Y.A., *Sov. Plant Phsiol.* **31** (1984), 876.
139. Duval, L., More, E. and Sicot, A., *Comptus Rendus de l'Academie d'Agriculture de France*, **78** (1992), 27.
140. Tiffin, L.O., in *Micronutrients in Agriculture*, eds. Mortvedt, J.J., Giordano, P.M. and Lindsay, W.L. Soil Sci. Soc. Amer., Madison, Wisconsin (1972).
141. Krasinskaya N.P. and Letunovea, S.V., *Agrokhimiya* **9** (1982), 108.
142. Dey, B.K. and Ghosh, A., *J. Indian Soc. Soil Sci.* **34** (1986), 264.
143. Loutit, M.W. and Loutit, J.S., *Plant and Soil* **27** (1967), 335.

144. Mosse, B., *Nature* **179** (1957), 922.
145. Ledtunova, S.V. and Gribovskaya, I.F., *Agrokhimiya* **3** (1975), 123.
146. Domsch, K.H., Jagnow, J. and Anderson, T.H., *Resid. Rev.* **86** (1983), 65.
147. Beeson, K.C. and Matrone, G., *The Soil Factor in Nutrition: Animal and Human*. Marcel Kekker, New York (1976).
148. Williams, J.H. and Gogna, J.C., in *Proc. Int. Conf. on Heavy Metals in The Environment*, Amsterdam. CEP Consultants, Edinburgh (1981).
149. Davis, R.A., in *Proc. Int. Conf. on Heavy Metals in The Environment*, Amsterdam. CEP Consultants, Edinburgh (1981).
150. Hornick, S.B., Baker, D.E. and Guss, S.B., in *Symposium on Molybdenum in the Environment*, Denver, Colorado (1975).

14.4 Silver

Silver is a precious metal of considerable economic and historical importance. Reference is made to Ag in many ancient manuscripts, and Ag artefacts have been found at the site of the Great Chaldean temple at Ur and in the tombs of Pharaohs dating back to 6000 BC. In the Codes of Menes, who reigned in Egypt to about 5500 BC, it was decreed that one part of Au was equal to two and one-half parts of Ag in value. By 4000 BC the Assyrians had established a system of trade based on Ag. Other uses mentioned in old manuscripts include the fabrication of idols, shrines, bowls, vases, flasks and jewellery. Major contemporary uses include the production of very thin films for electroplating; manufacture of reflecting mirrors, electrical contacts, alloys, coinage, and jewellery; and various uses in dentistry, optics, photography and medicine. The largest world producers of Ag, each yielding over 1000 tonnes/yr, are Canada, the USA, Mexico, Nicaragua and the USSR. In Europe, Sweden, Spain and Yugoslavia are the largest contributors.

Silver, like Cu and Au, is a member of Group 1B of the Periodic Table. It is the most reactive of the noble metals, forming three cationic species, Ag^+, Ag^{2+} and Ag^{3+}, of which only the monovalent form is environmentally significant. Silver as Ag^+ is among the most toxic of heavy metals to a range of life forms (notably microorganisms, algae and fish) and metabolic processes [151–156]. Silver has been used industrially because of its great potency as a microbial poison, although it is relatively harmless to higher life forms. Silver is almost unique in combining very low solubility of most of its compounds with exceedingly high toxicity of the soluble fraction [157]. Because Ag^+ is very easily reduced it is not readily accessible to living organisms in the natural environment. Silver sulphide, Ag_2S, is relatively soluble but has a low ionisation potential. The toxic action of Ag is related to the binding potential of the Ag^+ ion to enzymes and other active molecules at cell surfaces. Binding of –SH groups in the formation of mercaptides is the principal mechanism of enzyme inhibition.

Laboratory experiments have shown Ag to be one of the most toxic elements to microorganisms and biochemical processes in the soil. For example, 6×10^{-9} g Ag^+/ml kills *Escherichia coli* in 2–24 h, depending on the numbers of bacteria [158], and the heterotropic activity of bacteria is reduced by 0.1×10^{-9} g Ag/ml [159]. Silver nitrate at a concentration of 10×10^{-9} g Ag/ml killed 50% of the fungus *Alternatia tenuis* and of the metals tested only Os was more toxic [160]. Consequently, concern has been expressed about the possible harmful effect of Ag^+ on the soil microbial population and its subsequent influence on soil biochemical processes. Drucker *et al.* [161] showed that 1 mg/kg Ag decreased the total number of aerobic soil bacteria and the soil respiration rate of cultures; Ag was the most toxic of the 17 metals investigated. Hoffman and Hendrix [162] showed that soluble Ag had an adverse effect on *Thiobacillus ferrooxidans* at 0.1 μg/ml, and growth was restricted in solutions containing 1.0 μg Ag/ml or more. Adverse effects of Ag on various microbial metabolic processes, such as glucose respiration [163], CO_2 release [164], N mineralisation [155] and arylsulphatase activity [156] have been reported for spiked soils under laboratory conditions, but no attempt has been made to assess the importance of Ag on metabolic processes under field conditions.The impact of Ag on soil microbial functioning will, however, be limited by a number of factors. Firstly, only a fraction of the total Ag is biologically available (see below); secondly, Ag^+ is readily reduced to form insoluble compounds such as AgCl, or compounds with a low ionisation potential such as Ag_2S, which will restrict the presence of the Ag ion in some soils (see below). Thirdly, other authors have demonstrated the ability of some fungal and bacterial species to accumulate Ag without any apparent adverse effect. Some bacterial communities have been isolated from soils, for example, which are able to tolerate very high concentrations of Ag (100 mM Ag^+) [165]. These soil bacteria, particularly *Pseudomonas* and *Thiobacillus* have a great capacity for Ag accumulation and tolerance, probably due to reduction of the Ag ion to metallic Ag or Ag_2S [166].

Until recently there has been a lack of information on the environmental chemistry of Ag, largely due to analytical limitations imposed by the low concentrations present. Nowadays, however, Ag can be analysed routinely using flameless atomic absorption techniques or INAA, either by the conventional lengthy, single irradiation that generates ^{110m}Ag (half-life 253 days), or by a cyclic activation programme which alternates the sample between the irradiation source and a detector, optimised for the short-lived isotope ^{110}Ag (half-life 24 days) [167–169]. Varying estimates of the terrestrial abundance of Ag have been reported. Goldschmidt [171] estimated 0.2 mg/kg in the lithosphere, and later suggestions are close to that figure [172–174]. Silver occurs in massive sulphide ores [175], in mineralised veins as the native ore [176] and in various natural alloys. However, it occurs most frequently as a secondary component of a great variety of minerals. Argentiferous galena (PbS) is the most common of these, but associations with Sb, As, Te and Se also occur. Galena and sphalerite are often enriched in Ag, following its isomorphous substitution into the crystal lattice [177]. Halide minerals are less common sources of Ag, found in oxidised zones of ore deposits and at shallow surface depths. Originally Ag was recovered by reduction of Pb–Ag sulphides and the subsequent desilverisation of the crude Pb. By the time of the Greek civilisation, cupellation was extensively employed in the separation process. Today Ag is frequently recovered as a by-product of the base metals Ni, Pb and Zn, but also from the Cu, Pt and Au ores by a number of processes, including cyanidation, smelting and electrolysis.

Igneous rocks contain on average 0.1 mg Ag/kg, sedimentary rocks 0.05–0.25 mg/kg, organic-rich shales up to 1 mg/kg. Sediments may contain slightly higher concentrations than soils and are prone to enrichment near mineralised zones or anthropogenic sources; the highest reported value is 154 mg/kg for a grossly polluted section of the river Rhine [178]. Elevated Ag levels have also been reported in anthropogenically derived media such as sewage sludges (up to 960 mg/kg [179, 180]) and municipal wastes [181]. It should be stressed, however, that this extraordinarily high value for one sewage sludge sample is influenced by discharges from a photographic works in the catchment area of the sewage treatment plant.

The normal range of Ag in soils varies from < 0.01 to 5 mg/kg with an average of 0.1 mg/kg [173]. Soils derived from black shales (typically 0.5 mg/kg), or high in organic matter (0.5 mg/kg), are inherently richer in Ag than the sandstone (0.05 mg/kg) or limestone (0.07 mg/kg)-derived soils [182]. In some instances the distribution of Ag in soils has been shown to accurately reflect bedrock Ag distributions and to delineate lithologic and ore boundaries [183, 184]. This has obvious implications for the use of soil analyses as a rapid geochemical prospecting reconnaissance technique. However, caution will need to be exercised, because presumably soils composed of surficial deposits not derived from underlying bedrock, or very young soils, will give a false picture of the bedrock's Ag content. Contaminated soils from areas of mineralisation may contain up to 10 mg/kg [168].

Silver can form several ionic species (0, +1, +2, +3), although Lindsay and co-workers have shown by a consideration of chemical thermodynamics that only the 0 and +1 states are important in soils [185, 186]. These studies have also shown that in highly reduced soils Ag may be expected to precipitate in sulphide minerals, and that all Ag halide minerals (except AgF) are sufficiently insoluble that they must be considered as possible stable minerals. In highly oxidised soils Ag(c) cannot exist because it is too soluble. Instead one of the halide minerals will most likely control Ag^+ solubility. Ag_2CO_3(c) and Ag_3PO_4(c) are both too soluble to be of importance in soils [185, 186].

The behaviour of Ag in soils is strongly influenced by the prevailing pH and redox conditions, and by interactions with soil organic matter. In field soils Ag tends to accumulate in the surface, organic-rich horizons [187, 188]. Presant and Tupper [187], performed a detailed study on the distribution of Ag in the horizons of a range of Canadian podzolic soils and noted the following range of values for each horizon (in mg/kg): L-H, 0.2–7.8mg/kg, A_e, 0.1–1.7; B1, 0.2–0.6; B2, 0.1–0.7; C, 0.1–0.9 mg/kg. Other studies indicate that, once accumulated in the soil surface layers, Ag is persistent and leaching is slow. Leaching experiments with ^{110m}Ag and stable Ag salts, for example indicate that Ag is highly immobile in soils relative to other metals [189, 190]. Strong binding by soil is evident at locations in the USA where AgI has been used in cloud seeding experiments to promote precipitation locally. Sokol and Klein [191] analysed soil surrounding a seedling generator site and found that virtually all of the deposited Ag remained in the top 20 mm of the soil and concentrations decreased below this by an order of magnitude for each 20 mm increase in depth. The mobility of Ag is increased by humus decomposition or by lowering pH [189, 190]. Sorption of Ag to a variety of mineral substrates has been modelled successfully using the Freundlich equation, including kaolinite [192], hydrous ferric oxides [191], Mn oxides [192] and a range of soils [189, 194].

Various studies with soil extractants have been conducted which provide further evidence for the comparative insolubility/immobility of Ag. Sequential and selective extraction procedures also provide evidence for chemical affinities between Ag and the soil organic matter, and Fe/Mn oxides. Jones *et al.* [168] tested a range of chemical extractants on soils from a mineralised area of Wales, and found that no single extractant removed more than 10% of the total soil Ag (as determined by a concentrated HNO_3 digestion) for any of the soil samples. Levels of soluble plus exchangeable Ag were of the order 0.4–40 μg/kg, a small proportion of the total soil Ag (< 1%). Biologically available Ag measured with 0.005 M DTPA or 1 M HNO_3 was in the range 3–540 μg/kg (< 5% of total). This implies that biologically available Ag exists predominantly as complexed or adsorbed forms. Because the total Ag levels are over an order of magnitude greater than amounts extractable by DTPA or 1 M HNO_3 acid, it appears that Ag is predominantly occluded or co-precipitated and/or in the mineral fraction. A chemical fractionation procedure was also applied to the Welsh soils. In soils contaminated with Ag and other heavy metals from metalliferous mining activity between 80 and 150 years ago there were negligible amounts of readily exchangeable Ag; nearly half the total Ag was 'residual' (i.e. solubilised only in concentrated acid) [195, 196]. Intact soil cores with high organic matter contents and spiked with ^{110m}Ag still contained a considerable proportion of the ^{110m}Ag in a readily exchangeable form after one year, but most was bound in 'acid-reducible' or 'oxidisable-organic' forms [195]. The evidence for a strong association between soil organic matter and Ag indicates that humus may control the availability of Ag in the short term and that mineral prospecting for Ag by the chemical analysis of soil humus may be a useful reconnaissance technique. Fixation in the residual fraction may be a mechanism which reduces the bioavailability of the element in the long term. Silver enrichment in ferromanganese nodules has been demonstrated in the marine environment [197], so this interaction may be important in certain soils [193, 194]. Organically-bound soil Ag may represent a significant proportion of the total Ag component and play an important part in controlling the cycling, mobility and behaviour of Ag in soils. Silver becomes enriched in the humus and organic matter of soils and is bound as humic complexes in Ag-organo ligands [173]. Strong adsorption of Ag by humic components, or the formation of relatively stable organic complexes would thus prevent its subsequent removal from the upper horizons [189, 190]. Complexes may not be limited to humic acids, since fulvic acids also show strong retentive capacities and have been reported to contain up to 30 mg/kg in some uncontaminated soils [198, 199].

Silver has been reported as a trace constituent in many plant forms, for example, fungi [169, 200–204], bryophytes [169, 203, 204] and higher plants [169, 173, 205]. Concentrations are very variable, but usually < 1 mg/kg ash [173]. Data on Ag in plants have mainly been gathered to aid biogeochemical exploration for ores [173, 206–208] and are consequently often limited to information on relatively few species; little 'baseline' data exist for Ag concentrations in plants.

Byrne *et al.* [202] studied Ag accumulation by a range of fungi growing in Yugoslavia or Germany on sites free from pollution. From a soil with < 0.1 mg/kg a specimen of *Agaricus campestris* accumulated as much as 133 mg/kg with a median value of 30 mg/kg for a range of six species of fungi. *Boletus* and members of the Lycoperdaceae can also contain appreciable amounts of Ag, although the ability to accumulate Ag appears to vary considerably within a particular genus [169]. Sporophores of several species are capable of accumulating Ag relative to the soil concentration, which suggests that fungal hyphae (including mycorrhizal activity) may play a major role in the enrichment and subsequent retention of Ag in the rhizosphere.

Various workers have studied the uptake of Ag by higher plants [209–211]. Wallace and co-workers [211] studied uptake in nutrient solution and found Ag was lethal to bush beans (*Phaseolus vulgaris*) at 10^{-4} M $AgNO_3$; at 10^{-5} M yields were greatly decreased. Leaf, stem and root concentrations for this treatment were 5.8, 5.1 and 1760 mg/kg weight respectively. Other workers have subsequently confirmed the accumulation of Ag in roots [209, 210, 212]. Autoradiograms of bean and corn plants, using ^{110m}Ag, showed that Ag was uniformly deposited in the bean shoot, but corn tips had regions of high activity along the leaf margins and at the tips where guttation had occurred. Roots were heavily labelled and shoots (especially the new growth) continued to accumulate Ag even after the intact plant was returned to Ag-free solution. When present at high levels in plant tissue, Ag^+ may be precipitated as Ag_2S or reduced to metallic Ag; this may be visible as a darkening of the tissue. In some instances, Ag appears to pre-dispose plants to infection by microorganisms. Hausebeck *et al.* [213] reported increased susceptibility of Ag thiosulphate-treated pelargoniums to infection by the soil-borne pathogenic fungus *Pythium ultimum*, frequently followed by death. Applications of Ag thiosulphate are used to suppress ethylene production in plants; this could indicate that ethylene production in response to fungal infection is a part of the plant's defence strategy, and that elevated Ag levels in plants may be acting to suppress this mechanism. The continuing horticultural use of Ag thiosulphate may reveal further examples of such interaction in the future.

References: Silver

151. Chamber, C.W., Proctor, C.M. and Kabler, P.W., *J. Am. Water Works Assoc.* **54** (1962), 208–216.
152. Horsfall, J.G., *Principles of Fungicidal Action.* Chronica Botanica Co., Waltham, Mass. (1956).
153. Jones, J.R.E., *J. Exp. Biol.* **16** (1939), 425–437.
154. Shaw, W.H.R., *Science* **120** (1954), 361–363.
155. Liang, C.N. and Tabatabai, M.A., *Environ. Pollut.* **12** (1977), 141–147.

156. Al-Khafaji, A.A. and Tabatabai, M.A., *Soil Sci.* **127** (1979), 129–133.
157. Copper, C.F. and Jolly, W.C., *Water Resources Res.* **6** (1970), 88–98.
158. Lawrence, C.A. and Block, S.S., *Disinfection, Sterilisation and Preservation.* Lea and Febiger, Philadelphia (1968).
159. Albright, L.J., Wentworth, J.W. and Wilson, E.M., *Water Res.* **6** (1972), 1589–1596.
160. Somers, E., *Ann. Appl. Biol.* **49** (1961), 246–253.
161. Drucker, H., Garland, T.R. and Wildung, R.E., in *Trace Metals in Health and Disease*, ed. Kharasck, N. Raven Press, New York (1979).
162. Hoffman, L.E. and Hendrix, J.L., *Biotechnol. Bioengng* **18** (1976), 1161–1165.
163. Molise, E.M. and Klein, D.A., *Ann. Meeting Am. Soc. Microbiol.* **74** (1974), 2.
164. Cornfield, A.H., *Geoderma* **19** (1977), 199–203.
165. Charley, R.C. and Bull, A.T., *Arch. Microbiol.* **123** (1979), 239–244.
166. Pooley, F.D., *Nature* **296** (1982), 642–643.
167. Steinnes, E., *J. Radioanal. Chem.* **58** (1980), 387–391.
168. Jones, K.C., Peterson, P.J. and Davies, B.E., *Geoderma* **33** (1984), 157–168.
169. Jones, K.C., Peterson, P.J., Davies, B.E. and Minski, M.J., *Intern. J. Environ. Anal. Chem.* **21** (1985), 23–32.
170. Goldschmidt, V.M., *Geochemistry*. Blackwells, Oxford (1954).
171. Green, J., *Bull. Geol. Surv. Soc. Amer.* **70** (1959), 1127–1184.
172. Taylor, S.R., *Geochim. Cosmochim. Acta* **28** (1964), 1273.
173. Boyle, R.W., *Bull. Geol. Surv. Can.* **160** (1968).
174. Smith, I.C. and Carson, B.L., *Trace Metals in the Environment*, Vol. II. *Silver*. Ann Arbor Science Pub. (1977).
175. Amcoff, O., *Mineral. Deposita* **19** (1984), 63–69.
176. Watson, P.H., Godwin, C.I. and Christopher, P.A., *Can. J. Earth. Sci.* **19** (1982), 1264–1274.
177. El Shazly, E.M., Webb, J.S. and Williams, D., *Trans. Inst. Min. Metall.* **66** (1956), 241–271.
178. Dissanayake, C.B., Kritsotakis, K. and Tobschall, H.J., *Internat. J. Environ. Stud.* **22** (1984), 109–119.
179. Beckett, P.H.T., *Wat. Pollut. Controls* (1978), 539–546.
180. Gerstle, R.W. and Albrinck, D.N., *J. Air Pollut. Control Assoc.* **32** (1982), 1119–1123.
181. Greenberg, R.R., Zoller, W.H. and Gordon, G.E., *Environ. Sci. Technol.* **12** (1978), 566–573.
182. Jones, K.C., Peterson, P.J. and Davies, B.E., *Minerals in the Environ.* **5** (1983), 122–127.
183. Cox, R., Curtis, R., *J. Geochem. Explor.* **8** (1977), 189–202.
184. Leavitt, S.W. and Goodell, H.G., *J. Geochem. Explor.* **11** (1979), 89–100.
185. Lindsay, W.L., *Chemical Equilibria in Soils.* John Wiley, New York (1974).
186. Lindsay, W.L. and Sadiq, M., in *The Environmental Impact of Ice Nucleating Agents*, ed. Klein, D.A. Dowden, Hutchinson and Ross, Stroudsburg, Pa. (1978), 25–40.
187. Presant, E.W. and Tupper, W.M., *Can J. Soil Sci.* **45** (1963), 305–310.
188. Romney, E.M., Wallace, A., Wood, R., El-Gazzar, A.M., Childress, J.V. and Alexander, J.V., *Commun. Soil Sci. Plant Anal.* **8** (1979), 719–725.
189. Cameron, R.D., Unpublished PhD. Thesis, Colorado State University (1973).
190. Khan, S., Nadan, D. and Khan, N.N., *Environ. Pollut.* (Ser. B) **4** (1982), 119–125.
191. Sokol, R.A. and Klein, D.A., *J. Environ. Qual.* **4** (1975), 211–214.
192. Daniels, E.A. and Rao, S.M., *Int. J. Appl. Radiat. Isot.* **34** (1983), 981–984.
193. Dyck, W., *Can. J. Chem.* **46** (1968), 1441–1444.
194. Anderson, B.J., Jenne, E.A. and Chao, T.T., *Geochim. Cosmochim. Acta* **37** (1973), 611–622.
195. Jones, K.C., Davies, B.E. and Peterson, P.J., *Geoderma* **37** (1986), 157–174.
196. Jones, K.C., *Environ. Pollut.* (Ser. B) **12** (1986), 249–263.
197. Bolton, B.R., Ostwald, J. and Monzier, M., *Nature* **320** (1986), 518–520.
198. Chen, T., Senesi, N. and Schnitzer, M., *Geoderma* **20** (1978), 87–104.
199. Jones, K.C. and Peterson, P.J., *Plant and Soil* **95** (1986), 3–8.
200. Schmitt, J.A., Meisch, H.U. and Reinle, W., *Z. Naturforsch*, **33c** (1978), 608–615.
201. Allen, R.W. and Steinnes, E., *Chemosphere* **4** (1978), 371–378.
202. Byrne, A.R., Dermelj, M. and Vakselj, T., *Chemosphere* **10** (1979), 815–821.
203. Shacklette, H.T., *U.S. Geol. Surv. Bull.* 1198-D (1965).
204. Jones, K.C., Peterson, P.J. and Davies, B.E., *Water, Air and Soil Pollut.* **24** (1985), 329–338.
205. Horovitz, C.T., Schock, H.H. and Horovitz-Kisimova, L.A., *Plant and Soil* **40** (1974), 397–403.

206. Warren, H.V., Delavault, R.E., *Bull. Geol Soc. Amer.* **61** (1950), 123–128.
207. Webb, J.S. and Millman, A.P., *Trans. Inst. Min. and Metall.* **60** (1950), 473–504.
208. Warren, H.V., Towers, G.H.N., Horsky, S.J., Kruckeberg, A. and Lipp, C., *Western Miner* (1983), 19–25.
209. Koontz, H.V. and Berle, K.L., *Plant Physiol.* **65** (1980), 336–339.
210. Ward, N.I., Roberts, E. and Brooks, R.R., *New Zealand J. Sci.* **22** (1979), 129–132.
211. Wallace, A., Alexander, G.V. and Chaudhry, F.M., *Commun. Soil Sci. Plant Anal.* **8** (1977), 751–756.
212. Ward, N.I., Brooks, R.R. and Roberts, E., *Environ. Pollut.* **13** (1977), 269–280.
213. Hausebeck, M.K., Stephens, C.T. and Heins, R.D., *Plant Disease*, **73** (1989) 627–630.

14.5 Thallium

Thallium is a rare and dispersed element in the lithosphere. Estimates of its mean crustal abundance range from 0.5–1.0 mg/kg [214]. Known deposits of Tl minerals are rare; these are chiefly sulphides and occur in Switzerland, Yugoslavia, the Caucasus and Soviet Central Asia (Table 14.3). Most Tl in the earth's crust is dispersed, mainly as a random isomorphous replacement for K in minerals such as micas, felspars and clays [215, 16].

There is little detailed information on the Tl status of soils. Values quoted include 0.06–0.5 mg/kg dry weight [214], 0.1 mg/kg [217], 2.5–4.0 mg/kg [218], 5 mg/kg [219]; 0.5–10 mg/kg [220] and 100–350 μg/kg, measured by pulse anodic stripping voltametry [221]. Table 14.4 gives details of Tl levels found in a range of British soils by Logan [222]; values, even from the metal anomaly at Shipham, fall around the average values quoted by Smith and Carson [214]. It is assumed that literature references refer to 'total' soil Tl; there is an absence of information relating to 'extractable' Tl, or indeed to what extractants might be used.

Industrial uses of Tl are varied, but all are, at present, small scale. Its use as a catalyst for organic synthesis is increasing [214, 224], and its use in the semiconductor and electrical engineering industries is also expanding [225]. It may also be used in specialist glass production , and as a component of low-melting -point alloys [214]. However, its major use was as a pesticide and rodenticide, until banned in many countries in the mid 1970s [214]. World production was estimated at 10–12 t/yr in 1970 [216], but with the ban on Tl pesticides, this has fallen to 0.5 t/yr [214]. However, the amount of Tl recoverable from industrial processes is much higher than the production figures; Zitco [227] considers that

Table 14.3 Naturally occurring Tl minerals

Lorandite	$TlAsS_2$
Urbaite	$Hg_3Tl_4As_8Sb_2S_20$
Hutchinsonite	$(Pb, Tl)_2 (Cu, Ag)As_5S_{10}$
Crooksite	$(Cu, Tl, Ag)_2Se$
Hatchite	$PbTlAgAs_2S_5$

Table 14.4 HNO_3-extractable Tl in some British soils (mg/kg). Each sample mean of triplicate determinations. Figures in parentheses represent numbers of samples. Data from Logan [222]

Location	Tl content (range)
Kent (4)	0.06–0.18
Berkshire (2)	0.035–0.15
London (7)	0.12–0.29
New Forest, Hants (3)	0.04–0.35
Shipham, Somerset (5)	0.42–0.99
Bangor, Gwynedd (3)	0.04–0.09
Prestatyn, Clwyd (2)	0.05–0.13
Hesketh Bank, Lancs. (3)	0.04–0.08
Perthshire (3)	0.03–0.32

the pollution potential of Tl should not be based on industrial production, but on the quantity released into the atmosphere in waste products.

Tl is mainly produced from flue dusts associated with smelting of sulphate ores of Pb and Zn. Dusts may contain up to 5300 mg/kg Tl [214]. Thallium may be released by combustion of fossil fuels; US coal has an average Tl content of 0.7 mg/kg [214], with at least 50% released into the atmosphere associated with smaller respirable particles which may contain up to 76 mg/kg [228]. Based on these data, annual total Tl emissions from coal combustion have been estimated at 180 t/yr. Thallium content of oils has received little investigation. Yen [229] lists it as one of 24 metals found by spectrographic analysis of US crude oils. Some Russian crude oils contain 0.4–0.5 μg/kg Tl [230]. As at least 90% of the metal content of oil concentrates in asphalts and residual fuel oils, analysis of the Tl content of these would seem to be prudent. Certain oil shales in the US contain up to 25.8 mg/kg Tl [214].

Metal processing represents another major environmental Tl source. Due to its volatility, the temperature used in primary ore treatment and subsequent metal refinement (900–1400°C) would be expected to volatilise most of the Tl present.

The amount of Tl emitted from Zn smelting is difficult to predict; Zitco [227] estimated that up to 48 t/yr Tl may be present in Zn wastes. Lead smelting may be an important environmental Tl source. Sintering dusts are recycled until the Cd content becomes elevated; Tl content increases parallel with those of Cd, and there is a greater likelihood of 'uncontrolled' Tl emission to the atmosphere at Pb smelters [214]. Smith and Carson [214] suggest that Mn production could be an important source of Tl emissions. US Mn ores contain several thousand mg/kg Tl and up to 140 t of Tl may be lost annually in uncontrolled dust emissions. Manganese nodules recovered from the ocean floor contain Tl at levels of percent [214]. Thallium is also found in elevated concentrations in Au tailings and concentrates, and is associated with Au-bearing ores. At one waste dump in Utah, USA, it was estimated that over 2000 t of Tl are present in the Au wastes. In the USA it is probable that up to 350 t of Tl are released annually into the environment, mainly as wastes and slags.

There is little information on the chemical form taken by Tl in soils. It is generally believed [215, 231] that Tl is present as Tl^+, and is transferred to solution in this form. It becomes enriched in sediments, especially in strongly reducing environments where organic matter is accumulating under anaerobic conditions. In strongly oxidising conditions, Tl^+ will be removed from solution as Tl^{3+} by precipitation with Mn or Fe [214].

Crafts [232, 333] found Tl_2SO_4 to be strongly fixed in the upper 10 cm of various soil types, and to strongly resist leaching to lower horizons. McCool [234] suggested that Tl was removed from solution by base (anion) exchange. Recent studies [235] on alluvial soils with and without Tl enrichment indicate the fractionation of natural and anthropogenic Tl. Greatest enrichment of Tl in the polluted soil was found in the fractions extractable in acetic acid and ammonium chloride/ammonium nitrate, indicative of adsorption. Crops grown on the contaminated soil removed a significant proportion of the soil Tl pool in their biomass.

Detailed investigation of plant uptake dates from the use of Tl ground baits for rodent control. Brooks [236] noted sterilisation of soils adjacent to baits; controlled experiments revealed complete destruction of vegetation around such baits, and an absence of regeneration over a 2 yr period following their application. McMurty [237] found that at soil Tl levels of 35 mg/kg and 75 mg/kg tobacco plants were severely damaged or killed, and that 1 mg/kg Tl caused toxicity to the same species in nutrient solution. McCool [234], adding Tl(1) sulphate to sandy loam soil, found slight growth reductions in crop plants at 2.1 mg/kg added Tl, and severe injury at 8.5 mg/kg. Crafts [232] found Tl toxicity to be greatest in soils of low fertility, and also found that clay loam soils were capable of rendering Tl innocuous, up to levels of 10 000 mg/kg. From a series of pot trials involving Tl incorporation, Horn *et al.* [238] concluded that additions of 5 to 10 mg/kg were beneficial to plant growth, with toxic effects only evident at 100 to 1000 mg/kg.

More recent studies on Tl uptake by plants indicate that (i) it is a mobile element with ready movement from root to shoot and (ii) Tl uptake is profoundly affected by the presence of K [235, 239–242]. Tl^+ ions are actively accumulated in excised *Hordeum* roots, where Tl^{3+} ions are absorbed passively, probably by cation exchange and diffusion, in the same system [222]. Experiments with whole plants (*Lycopersicon esculentum*) supplied with either Tl^+ or Tl^{3+} produced results contrasting to the excised root studies. There was little difference in uptake and growth reduction between either form of Tl, and it was concluded that Tl^{3+} was undergoing reduction to Tl^+ prior to uptake. The location of this step was not clear [222]. Tl uptake by soyabean has been studied in hydroponic culture and in field conditions, following soil amendment. Low (1 μM) Tl concentrations, applied to hydroponically grown plants, dramatically reduced both biomass and nutrient contents. These treatments also caused reductions in node number in stems of treated plants, together with stunting of roots and development of chlorosis in mature iron-sufficient leaves. Increases in plant Tl levels were accompanied by increases in K levels, but levels of Ca, Mg and Mn decreased to deficiency status [243].

The influence of K^+ over Tl^+ uptake is not surprising. Both elements have similar ionic radii, and Tl^+ can replace K^+ in several enzyme systems, notably Na-K-ATPases [244]. Other systems similarly affected include pyruvate kinase [245] and phosphatases [246]. Thallium can be up to 10 times more effective than K at activating certain mammalian ATPases [244, 247].

Reduction of K content of nutrient solution enhances Tl transport and toxicity in both *Zea* and *Lycopersicon* [240], but root levels remain essentially similar. Plants supplied with only 25% of 'normal' K still contained 100–200 times more K than Tl, so explanations of Tl toxicity based on K replacement are perhaps somewhat facile. Tl^+ appears to be absorbed by roots via the K^+ system [241], based on interpretation of competitive uptake studies; however, Tl has a lower affinity for this carrier than K^+. Studies on Tl uptake in the floating aquatic plant *Lemna minor* demonstrated an apparent saturation of uptake at higher levels of applied Tl (5–10 μM), with steady-state Tl concentration reached after 140 h continuous exposure. It was concluded that Tl was predominantly accumulated in the cell vacuoles, as up to 80% of the orginal steady-state Tl concentration was retained by treated plants after 140 h exposure to a Tl-free solution. Tl uptake was almost entirely active [248].

Thallium effects on plant metabolism include reductions in photosynthesis and transpiration [249, 250]. Thallium interfered with stomatal function, a 2 mg/l solution causing reduction in stomatal aperture of up to 90%. Seed germination of *Plantago maritima* was adversely affected by Tl(I) nitrate, an effect reduced by K^+ ions [251]. Similar inhibitions of seed germination have been reported for a range of agricultural species; turnip and lettuce showed greatest sensitivity to applied Tl, but cereals (wheat and millet) were more resistant [252]. The upper critical level for Tl in reducing growth of spring barley by 10% was 0.5 mg/kg in solution culture [252]. Studies on Tl binding and speciation in plants [254, 255] agree that the majority of tissue Tl is present in a free ionic form: 80% of Tl in *Lemna* was ionic, with no significant fraction associated with the cell water [254]. In Tl-treated rape, cell fractionation indicated that up to 70% of the total plant Tl is fraction was associated with the cytosol as uncomplexed ionic species. There was no evidence for the methylation of Tl. It was concluded that Tl is probably bound to peptides (m.w. 3800 g mol with no S amino acids), but that increased tissue Tl levels did not lead to increases in these natural Tl-binding agents [255].

The most widely documented case of Tl pollution was reported from West Germany in an area adjacent to a cement works at Lengerich (North Rhine-Westphalia). Here, Fe ore additives used in the production of hardened concrete contained 0.03% Tl; over several years some 3–5 kg of Tl were emitted each day [256, 257]. Surveys showed that 80% of the local population had elevated urinary Tl levels; the normal upper limits for Tl in urine are given as 0.8 μg/l, but here the values ranged up to 76.5 μg/l. The main route of Tl intake by the population was shown to be via consumption of contaminated food grown in the vicinity of the cement works. Large numbers of people were diagnosed as suffering from Tl-related health disorders including depression, insomnia and various 'nervous disorders'. Following advice to forego consumption of locally grown vegetables or offal from local domestic animals, Tl levels in the population dropped significantly over a 12-month period [257]. Hoffman *et al.* [258] cultivated agricultural crops on soils with varying natural Tl levels, and some amended with Tl(I) nitrate. Green cabbage, turnip-rooted cabbage (kohlrabi) and rape accumulated significant Tl levels even from soils with low Tl status. These authors conclude that 'a soil Tl content of 1 mg/kg should be recommended as the tolerable margin'. In Switzerland, this has been doubled, with 2 mg/kg total Tl being the recommended maxium for agricultural soils (Fluckiger, personal communication). A recent report from Russia describes an outbreak of Tl poisoning in the Ukrainian town of Chernotsky [259]. Many inhabitants are reported as suffering hair loss and hallucinations,with at least 160 undergoing treatment. Soil in the town appears heavily contaminated with Tl, the origins of which are unknown; there is some indication of inadequate industrial storage and use of Tl as a home-made fuel additive which could be implicated.

Natural Tl mineralisation in the Aslar region of southern Yugoslavia has been reported [260, 261], with indications of Tl toxicity in grazing ruminants from that region. Thallium contents of 'grab samples' of plants from this region give Tl values (mg/kg ash wt) of 10–5990 for various plant genera. However, there is no indication as to the soil Tl content at sites where material was collected, or if these were washed prior to spectrographic analysis. More detailed studies in this region could give a valuable insight into natural biogeochemical Tl cycles, and allow further predictions to be made regarding the future pollution hazards this extremely toxic element may pose.

References: Thallium

214. Smith, I.C. and Carson, B.L. *Trace Metals in the Environment, I. Thallium*. Ann Arbor Science, Ann Arbor, Michigan (1977).
215. Shaw, D.M., *Geochim. Cosmochim. Acta* **2** (1952), 118.
216. Day, F.H., *Chemical Elements in Nature*. Harrap. London (1963).
217. Bowen, H.J., *Trace Elements in Biochemistry*. Academic Press, London (1966).
218. Seeger, R. and Gross, M., *Z. Leben. Unters. Forsch.* **173** (1981), 9.
219. Reilly, C., *Metal Contamination of Our Food. Applied Science*, London (1983).
Umweltbelastung durch Thallium. Dokumentation von der Landesanstalt für Immissionsschutz (lis) des Landes Nordrhein-Westfalen (1980).
220. Lukaszewski, Z and Zembrzuski, W. *Talanta*, **39** (1992), 221.
221. Logan, P.G., Thallium uptake and transport in plants. PhD Thesis, CNAA (1985).
222. Taylor, E.C. and McKillop, A., *Acc. Chem. Res.* **3** (1970), 338.
223. Kurosawa, H., *J. Organometall. Chem.* **254** (1983), 107.
224. Izmerov, N.F. (ed.). *Scientific Reviews of the Soviet Literature on the Toxicity and Hazards of*
225. *Chemicals. Thallium*. U.N. Environmental Programme **17**. Moscow (1982).
226. Kogan, B.J., *Chem. Abs.* **76** (1970), 153.
227. Zitko, V., *Sci. Total Env.* **4** (1975), 185.
228. Natusch, D.F.S., Wallace, J.R. and Evans, C.A, *Science* **183** (1973), 76.

229. Yen, T.F., in *The Role of Trace Metals in Petroleum*, ed. Yen T.F. (1975), 1–30.
230. Nuriev, A.I. and Efendiev, G.K., *Chem. Abs.* **65** (1966), 15099.
231. Wedepohl, K.H. (ed.). *Handbook of Geochemistry*, Springer-Verlag, Berlin (1972).
232. Crafts, A.S., *Science* **79** (1934), 62.
233. Crafts, A.S., *Hilgardia* **10** (1936), 377.
234. McCool, M.M., *Contrib. Boyce Thomspon Inst.* **5** (1933), 289.
235. Lehn, H. and Schoer, J., *Plant and Soil* **97** (1987), 253.
236. Brooks, S.C., *Science* **75** (1932), 105.
237. Mcmutrey, J.E., *Science* **76** (132), 86.
238. Horn, E.E., Ward, J.C., Munch, J.C. and Garlough, F.E., *The Effect of Thallium on Plant Growth*, USDA Circular 409, Washington, DC (1936).
239. Logan P.G., Lepp, N.W. and Phipps, D.A., *Trace Substance in Env. Health* **18** (1984), 570.
240. Logan P.G., Lepp, N.W. and Phipps, D.A., *Proc. VI Int. Colloq. Optimistion of Plant Nutrition*. Montpellier (1984), 345.
241. Logan P.G., Lepp, N.W. and Phipps, D.A., *Rev. Port. Chim.* **17** (1985), 411.
242. Allus, M.A., Martin, M.H. and Nickless, G., *Ghemosphere* **16** (1987), 929.
243. Kaplan, D.I., Adrinao, D.C. and Sajwan, K.S., *J. Env. Qual.* **19** (1990), 359.
244. Britten, J.S. and Blank, M., *Biochim. Biophys. Acta.* **159** (1968), 160.
245. Reuben, J. and Kane, F.J., *J. Biol. Chem.* **20** (1971), 6227.
246. Inturrisi, C.E., *Biochim. Biophys. Acta* **179** (1969), 630.
247. Mullins, L.J. and Moore, P.D., *J. Gen. Physiol.* **43** (1960), 759.
248. Kwan, K.M.H. and Smith, S., *New Phytol.* **117** (1990), 91.
249. Bazzaz, F.A., Carlson, R.W. and Rolfe, G.L., *Environ. Pollut.* **7** (1974), 241.
250. Carlson, R.W., Bazzaz, F.A. and Rolfe, G.L., *Environ. Res.* **10** (1975), 113.
251. Siegel, B.Z. and Siegel, S.M., *Bioinorganic Chim.* **4** (1974), 93.
252. Carlson, C.L., Adriano, D.C., Sajwan, K.S., Abels, S.L. and Driver, J.T., *Water, Air and Soil Pollution* **59** (1991), 231.
253. Davis, R.D., Beckett, P.H.T. and Wollan, E., *Plant and Soil* **49** (1978), 395.
254. Kwan, K.M.H. and Smith, S., *Chem. Spec. Bioavail.* **2** (1990), 77.
255. Guenther, K. and Umland, F., *J. Inorg. Biochem.* **36** (1989), 63.
256. Prinz, V.B., Krause, G.H.M. and Stratmann, H., *Staub.* **39** (1979), 457.
257. Brockhaus, A., Dolgner, R., Ewers, U., Weigand, H., Freir, I., Jermann, E. and Kramer, U., in *Mechanisms of Toxicity and Hazard Evaluation*, eds. Holmstedt, B. Lanwerys, R., Mercier, M. and Roberfroid, M. Elsevier/North Holland, Amsterdam.
258. Hoffmann, V.G.G., Schweiger, P. and Scholl, W., *Landwirtsch, Forschung* **35** (1982), 45.
259. *New Scientist* **1649** (1989), 28.
260. Zyka, V., *Sb. Geol. Ved. Technol. Geochem.* **10** (1970), 91.
261. Zyka, V., *Sb. Geol. Ved. Technol. Geochem.* **12** (1974), 157.

14.6 Tin

Tin was one of the first metals used in antiquity, known as early as 3000 BC and preceded only by Cu, Au and Ag. The Cu–Sn alloy, bronze, was invented about 2500 BC. Elemental tin has two crystallite modifications, α- or grey tin has the diamond structure and is stable only below 13.2°C. Above this transition temperature, the metallic form β- or white tin is stable [262]. Tin finds numerous applications today, as a protective coating, in solders, bearings and other alloys. The worldwide production of organotin compounds has risen as a consequence of their use as stabilisers in polyvinyl chloride, homogeneous catalysts for vulcanisation and as biocidic agents. The biocidal preparations include agrochemicals, fungicides, preservatives and antifouling agents in marine paints [263, 264].

Tin is ubiquitous in nature; its concentration in the geosphere is low and uniform.

It is classed as a trace element in soil and Sn is usually present at 1 to 10 mg/kg, but enrichment factors of three or four orders of magnitude are possible in tin-rich soils and mineral deposits [265]. The concentation of Sn in soils is of the same order of magnitude as that of cobalt and molybdenum [263].

The most important and most common of all Sn minerals is cassiterite or tinstone, SnO_2. Cassiterite occurs in pneumatolytic and high temperature hydrothermal veins or metasomatic deposits that are closely associated with highly siliceous igneous rocks, usually granite or rhyolite [266]. Cassiterite is typically a resistate, representing the chemically unchanged residue from weathering. Therefore it becomes concentrated in bands and layers of varying thickness forming economically valuable deposits such as those found in Malaya, Bolivia, Indonesia, Zaire, Thailand and Russia. Tin also occurs in many sulphide ores in the form of relatively pure sulphides or incorporated with a number of metals in more complex sulphides, stannite, $CuFeSnS_4$ being most noteworthy [265, 267].

Average concentrations of Sn in common rocks are typically reported as silicic, 3.6 mg/kg, intermediate, 1.5 mg/kg, mafic, 0.9 mg/kg and ultramafic, 0.35 mg/kg, whilst in argillaceous sediments concentrations are higher (6 to 10 mg/kg) [268, 269]. In soils, tin is largely derived from the bedrock, although all soil surface horizons show similar abundances of Sn [270]. Tin concentrations in soils have been reported in the range of 1 to 4.6 mg/kg [271, 272] and 1 to 10 mg/kg in non-contaminated soils [273, 274]. The most comprehensive survey of the concentration of Sn in soils in the USA has been carried out by Shacklette and Boerngen who reported a mean concentration of 1.3 mg/kg Sn [275]. In mineralised and contaminated soils, Sn concentrations above 250 mg/kg are not uncommon [276–278].

Tin may reach agricultural soils by a number of different routes such as the addition of soil amendments including sewage sludge, fly ash and organo-tin pesticides, or else it can be leached from mine spoils and Sn-containing products by biological, chemical and mechanical processes. No reliable data have been published for Sn levels in sewage sludge, but it is known that Sn has a high affinity for organic fractions, and Sn is known to be concentrated in humus-rich and organic-rich sediments at 80–100 mg/kg and 239 mg/kg respectively [279, 280]. Average Sn levels in fly ash have been reported as 4.8 mg/kg [281].

The transport of Sn from the continent to the hydrosphere takes place primarily via the atmosphere. Tin is one of the three most highly enriched metals in atmospheric particulate matter as compared to the earth's crust, along with Pb and Tl [282], with concentrations of 340 mg/kg Sn in airborne particulates reported [283]. Total Sn concentrations in air vary considerably, depending on the origin of the sample. Typical atmospheric concentrations of Sn are 300 ng/m^3 in heavily industrialised areas, approximately 10 ng/m^3 in urban and city areas whereas rural areas are reported to have less than 3 ng/m^3 of Sn [277, 282, 284]. In the vicinity of a Sn smelter the dry weight Sn content of the leaves of *Ehretia microphylla* was 2165 mg/kg, which reflects aerial deposition of Sn-containing particulates trapped by the epicuticular waxes and leaf hairs of the leaf surface rather than accumulated

Sn from the soil [285, 286]. The mean concentration of Sn in airborne particulates in Northern Hemisphere marine air is estimated at 12–800 ng/m^3, with the concentration decreasing with distance from land. This value is about one to two orders of magnitude less in Southern Hemisphere marine air [281]. It is estimated that anthropogenic emissions of Sn are dominated by waste incineration and non-ferrous metal production (in high temperature processes tin is very volatile), whilst natural inputs arising from sources such as sea spray, soil dust, volcanoes, forest fires and biomethylation, are relatively unimportant (with the exception of the latter) compared to anthropogenic sources [282].

There is little information about the geochemical behaviour of Sn during weathering processes or its mobility in groundwaters and leachate. Tin forms simple compounds with oxides, halides, sulphates, phosphates and carbonates in oxidation states +2 and +4. The standard electrode potential for the couple Sn^{4+}/Sn^{2+} is −0.15 V [263]. Tin(II) species are readily oxidised to Sn(IV) species by relatively mild conditions such as dissolved atmospheric oxygen. The rate of oxidation of Sn(II) can be reduced particularly in solutions which contain sufficiently strong electron donor species such as fluorides and chlorides. In solutions above pH 6, the oxidation is more rapid, and Sn(II) compounds can function as relatively strong reducing agents.

The mobility of Sn in natural waters is very low and is strongly pH-dependent. Tin concentrations in saline and seawaters are approximately 0.01 to 0.3 ng/g, with values higher than 1 mg/kg indicative of pollution [288–290]. In rivers, typical concentrations are 0.3 to 17 ng/g [282, 291] and concentrations in excess of 17 ng/g have been recorded for groundwaters leaching through mine tailings in Malaya [285]. The relatively low concentrations of Sn in the hydrosphere are due to the poor solubilities of the Sn oxides and their ready adsorption and coprecipitation on iron and manganese oxides, whilst in sediments and anoxic conditions Sn sulphides tend to be the limiting factor [292]. The major species for the transport of Sn in the hydrosphere is $SnO(OH)_3^-$ and other oxhydroxy species as both Sn^{2+} and Sn^{4+} tend to hydrolyse in solution [293]. The concentration of dissolved and suspended tin in rivers vary greatly depending upon river type and location. Byrd and Andrea carried out a comprehensive survey of 39 rivers worldwide and found that Sn concentrations vary from a few pM in pristine mountain rivers to 500 ng/g in polluted rivers such as the Rio Tinto, Spain. The arithmetic mean of the concentration of dissolved Sn was 2 to 3 ng/g. Total particulate Sn was determined and found to be about an order of magnitude higher than dissolved Sn [294].

In addition, Sn in either oxidation state forms soluble coordination complexes with many naturally occurring compounds, for example, amino acids and proteins [295, 296]. Such compounds may prove to be essential in the translocation mechanism of Sn in plants, although no such compounds have been isolated. Curtin *et al.* have identified an organic-Sn compound expired from the needles of a coniferous tree and have tentatively suggested its role in the movement of Sn throughout the tree [297].

Measureable amounts of Sn are only rarely found in native plant species and are usually in concentrations of 20–30 mg/kg ash weight [270]. In cultivated plants, levels between 1.6 and 7.9 mg/kg dry weight [298, 299] and 15 mg/kg ash weight Sn have been reported for corn and wheat [300], and for food plants such as carrot and sugar beet levels reported are 0.04 to 0.1 mg/kg dry weight [301] and 15 mg/kg ash weight [300].

Tin preferentially accumulates in roots, but elevated levels of Sn have been recorded in the twigs and leaves of birch and oak at 17 mg/kg (ash weight) [302]. Millman reported a maximum of only 1 mg/kg (dry weight) from the twigs of a range of tree species growing in soils containing as much as 250 mg/kg Sn in the form of cassiterite [276], but species of fern such as *Gleichiena linearis* and *Cyclosorus unitus* growing near Malayan Sn mines and on mine spoils contained up to 326 and 127 mg/kg ash weight respectively [286]. According to Sarosiek and Klys, these would be referred to as tin accumulators, along with a number of others mentioned in their study which contained between 12 and 84 mg/kg ash weight [303]. The bioaccumulation of Sn from soils is strongly influenced by the nature and form of Sn. Uptake would be much greater in areas where stannite and other Sn sulphides are present since these minerals are much more easily oxidised than cassiterite and would liberate quantities of available Sn in the form of collidal stannates [276].

There is no evidence that Sn is either beneficial or causes adverse effects in plants. Laboratory tests with sugar beet indicated that solutions containing as much as 40 mg/kg Sn^{2+} in sand cultures did not affect growth in any way [403].

The increased use of organotin compounds as biocides and antifouling agents is reflected by increased concentrations of such species in the environment. In particular, levels of monomethyl and dimethyl Sns between 2 and 49 ng/g have been recorded in sediments [305] and up to 400 ng/dm^{-3} were found in a comprehensive study of Spanish habours by Gomez-Ariza *et al.* [306]. Other workers have reported lower concentrations in estuaries, rivers and oceans [307, 308]. Although the origin of these organotins is probably anthropogenic, Branman and Tompkins have detected up to 9.1 ng/dm^{-3} methyltin in fresh water which appears to be of natural origin [309]. The possibility of Sn methylation has been assessed by a number of workers [310–312]. and bacteria such as *Desulfovito* spp., isolated from anoxic sediments, have been suggested as the cause of Sn methylation [310], following similar mechanisms to those of arsenic. Organotins are toxic to both prokaryotes and eukaryotes, but their potency is reduced by hydrolysis in sunlight. Despite the use of organotins as biocides, analysis of plant tissues over the past few years has not shown any sizeable accumulation of Sn because of its low bioavailiability [263].

There is a dearth of data relating to the environmental impact of Sn, because it is only recently the extremely precise and advanced analytical techniques required for the detection of sub mg/kg levels of Sn in the environment have been available. Many questions remain unanswered, in particular the importance biomethylation has in the redistribution of Sn and how Sn speciation and interaction with soil and plant components affect its bioavailiabilty and distribution within the plant.

References: Tin

262. Lide, D.R., ed. *Handbook of Chemistry and Physics. 72nd Edn.* CRC Press, Boca Raton (1991), 4.
263. Tsangaris, J.M. and Williams, D.R., *Appl. Organomet. Chem.* **6** (1992), 3.
264. Weber, G., *Fresenius Z. Anal. Chem.* **321** (1985), 217.
265. Wedepohl, K.H., ed. *Tin*, in *Handbook of Geochemistry, Vol. 2/4*, Springer-Verlag, Berlin (1974).
266. Singh, D.S. and Bean, J.H., *International Tin Conference*, **2** (1967), 457.
267. Yim, W.W., *Environ. Geol.* **4** (1981), 245.
268. Hamaguchi, H., Kuroda, R., Onuma, N., Kawabuchi, K., Mitsubayashi, T. and Hosohara, K., *Geochim. et Cosmochim. Acta.* **28** (1964), 1039.
269. Onishi, H. and Sandell, E.B., *Geochim. et Cosmochim Acta.* **12** (1957), 262.
270. Kabata-Pendias, A. and Pendias, H., *Tin* in *Trace Elements in Soils and Plants*, CRC Press, Boca Raton (1984), 154.
271. Presant, E.W., *Geol. Surv. Can. Bull.* **174** (1971), 1.
272. Kick, H., Burger, H. and Sommer, K., *Landwirtsch. Forsch.* **33** (1980), 12.
273. Chapman, H.D., *Diagnostic Criteria for Plants and Soils*, Univ. California Press, Riverside, Calif. (1972), 793.
274. Schroll, E., *Analytische Geochemei*, I.F. Enke, Stuttgart, (1975).
275. Shacklette, H.T. and Boerngen, J.G., *U.S. Geol. Surv. Prof. Paper*, **1270** (1984).
276. Millman, A.P., *Geochem. et Cosmochim. Acta.* **12** (1957), 85.
277. Hutzinger, O., ed. *Handbook of Environmental Chemistry, Vol 1A*, Springer-Verlag, Berlin (1980).
278. Bowen, H.J.M., *Trace Elements in Biochemistry*, Academic Press, London (1966).
279. Dogan, S. and Haerdi, W., *Int. J. Environ. Anal. Chem.* **8** (1980), 249.
280. Hallas, L.E. and Cooney, J.J., *Appl. Environ. Microbiol.* (1981), 446.
281. Imura, H. and Suzuki, N., *Talanta*, **28** (1981), 73.
282. Byrd, J.T. and Andreae, M.O. *Science*, **218** (1982), 565.
283. Sugimae, A., *Anal. Chem.*, **46** (1974), 1123.
284. Tabor, E.C. and Warren, W.V., *Arch. Ind. Health*, **17** (1958), 145.
285. Peterson, P.J., Burton, M.A.S., Gregson, M., Nye, S.M. and Porter, E.K., *Sci Total Environ.* **11** (1979), 213.
286. Peterson, P.J., Burton, M.A.S., Gregson, M., Nye, S.M. and Porter, E.K., *Trace Substances in Environmental Health, Vol 10*, University of Missouri, Columbia, Mo (1976), 123.
287. Lantzy, R.J. and Mackenzie, F.T., *Geochim. Cosmochim. Acta.* **43** (1979), 511.
288. Clark, R.B., *Marine Pollution*, Oxford Scientific Publications, Oxford (1989), 70.
289. Greenberg, R.R. and Kingston, H.M., *Anal. Chem.* 55 (1983), 1160.
290. Hodge, V.F., Seidel, S.L. and Goldberg, E.D., *Anal. Chem.* **51** (1979), 1256.
291. Andrea, M., Bryd, J.T. and Froehlich, P.N., *Environ. Sci. Technol.*, **17** (1983), 131.
292. Sager, M., *Microchim. Acta.* (1986), 129.
293. Macchi, G. and Pettine, M., *Environ. Sci. Technol.* **14** (1980), 815.
294. Byrd, J.T. and Andrea, M.O., *Geochim. et Cosmochim. Acta*, **50** (1986), 835.
295. Cusack, P.A. and Smith, P.J., *J. Chem. Soc. Dalton Trans.* (1982), 439.
296. Rose, M.R. and Lock, E.A., *Biochem. J.* **120** (1970), 151.
297. Curtin, G.C., King, H.D. and Mosier, E.L., *Geochem. Explor.*, 3 (1974), 245.
298. Kent, N.K., *J. Soc. Ind. Chem.*, **61** (1942), 183.
299. Zook, E.G., Green, F.E. and Morris, E.R., *Cereal Chem.*, **47** (1970), 72.
300. Connor, J.J. and Shacklette, H.T., *U.S. Geol. Survey Prof. Paper*, **574** (1975).
301. Duke, J.A., *Econ. Bot.*, **23** (1970), 344.
302. Harbaugh, J.W., *Econ. Geol.* **45** (1950), 548.
303. Sarosiek, J. and Klys, J., *Acta. Soc. Botan. Polon.*, **31** (1962), 737.
304. Schroeder, H.A., Balassa, J.J. and Tippon, I.H., *J. Chronic Diseases*, **17** (1964).
305. Weber, J.H., Randall, L. and Han, J.S., *Environ. Technol. Lett.*, **7** (1985), 571.
306. Gomez-Ariza, J.L., Morales, E. and Ruiz-Beitez, M., *Analyst*, **117** (1992), 641.
307. Maguire, R.J., *Environ. Sci. Technol.*, **18** (1984), 291.
308. Maguire, R.J., Tkacz, R.J. and Chau, G.A., *Chemosphere*, **15** (1986), 253
309. Braman, R.S. and Tompkins, M.A., *Anal. Chem.* **51** (1979), 12.

310. Sigleo, A.C. and Hattori, A., *Marine and Estuarine Geochemistry*, Lewis Publishers Inc., Mich. (1985), 239.
311. Ashby, J.R. and Craig, P.J., *Sci. Total Environ.*, **73** (1988), 127.
312. McDonald, L. and Trevors, J.T., *Water, Air and Soil Pollut.* **40** (1988), 215.

14.7 Uranium

There is considerable interest in the behaviour of U in soils, and perhaps more particularly in the elements which constitute the transuranic series. This is generated because of the importance of these elements to the nuclear industry, the need to have reliable methods for prospecting, and the necessity of understanding these environmentally important elements with regard to their cycling, fate and long-term health significance. A reasonably extensive literature is accumulating on the environmental chemistry of Pu and other transuranics; the reader is referred to the radiochemistry literature such as the *Journal of Environmental Radioactivity*, and various monographs for further coverage [313–315]. This section will deal specifically with U.

The U normally found in nature consists of four isotopes of mass numbers 230, 234, 235 and 238. ^{238}U (half-life 4.5×10^9 yr), constitutes 99.28% of the total crustal burden of U, and is usually in equilibrium with ^{234}U, which is present as 0.0058%. ^{235}U (half-life 0.71×10^9 yr), the parent isotope of the actinide series, is present as 0.71%. ^{230}U, which is also a member of the ^{238}U series, has a short half-life (20.8 days). Because of these long half-lives U is conventionally determined in environmental samples by measuring the levels of radioactivity although INAA is a popular alternative [315, 316]. Consequently, U values in the literature may be expressed in terms of activity (i.e. pCi or Beq/g), or as conventional concentrations.

Uranium concentrations vary with rock type; the average crustal abundance in rocks is about 2.5 mg/kg, with rocks commonly in the range 0.05–5 mg/kg. Typical activity values for igneous rocks, sandstones, shales and limestones are 1.3, 0.4, 0.4 and 0.4 pCi/g, respectively. Sedimentary rocks high in organic matter (e.g. coals), and phosphate-rich deposits tend to display U enrichment. The process of enrichment is discussed in detail by Manskaya and Drozdova [317], but essentially it is caused by the reduction of U^{6+} to U^{4+}, and subsequent precipitation of relatively insoluble phosphate or sulphide compounds. Typical values in soils are of the same order as those in rocks—averaging *ca.* 1 mg/kg. Plant ash and fresh waters from unmineralised areas typically contain 0.6 mg/kg and 0.5 μg/l respectively [318]. Uranium is commonly enriched in the minerals zircon, pitchblende and apatite, and it is only rocks with large quantitites of these minerals that can be radiometrically dated using the U series dating procedures that measure activity generated by ^{235}U, ^{238}U, and the respective daughter products. Industrially exploited sources of U usually contain the minerals uraninite (UO_2) and coffinite ($USiO_4$), or less commonly rock phosphate or uraniferous-rich organic matter. Other elements that are frequently associated with U in various rock types are P,

Ag, Mo, Pb and F in phosphorites, Mo, Pb and F in U-rich veins, and Se, Mo, V, Cu and Pb in sandstones [319].

Uranium is mined in many areas of North America: in the USA the states of New Mexico, Wyoming, Utah and Colorado are minor contributors. However, more than 90% of U in the USA is commonly obtained as a by-product of phosphate mining in Florida, where the phosphate rocks contain as much as 120 mg U/kg (ave. 40). North African rock phosphate, the other major source of the world's phosphate fertilisers, contains around 20–30 mg U/kg. Uranium ore above 0.01% U_3O_8 is presently considered economically feasible to mine although most ore actually exploited averages between 0.04 and 0.42% [313]. The nuclear fuel cycle involves mining of U ore, milling it down to a concentration where it can be economically transported, chemical conversion, purification, enrichment of ^{235}U (the fissile isotope of U) and then finally fabrication of the fuel into an appropriate form. When the fuel elements have been used or 'spent' to a certain degree in the reactor core, they must be stored for a period to allow decay of short-lived fission products, and then either disposed of or reprocessed to separate unused fissile material from waste products that have no practical use. Consequently, localised enrichment of U in soils may arise largely as a result of the mining and milling processes. Other processes which may result in the release of U into the environment and subsequent retention by soils are coal combustion [321]—the elevated concentrations of U in coal may give rise to atmospheric releases or concentration on to fly-ashes, or the application of phosphate fertilisers to agricultural land.

Comparatively little information seems to be available with respect to U in soils. Only surface soils in the USA have been extensively studied, and the results show relatively small variation in U content with soil type. The worldwide data summarised by Kabata-Pendias and Pendias [318] gives U values in a narrow range of 0.79–11 mg/kg. Meriwether *et al.* [321] determined ^{238}U and its decay products in soils throughout the state of Louisiana and found some broad relationships with soil taxonomic groups. They observed, for example, that the lowest values were consistently found in the Udults of the upland, northern portion of the state and that the highest concentrations were in the Aquepts of the floodplains and the Hemists of the coastal marshes. This was interpreted as indicating that processes such as weathering, transport of solids and dissolution may be fractionating these elements into soils of water environments.

Compared to other cations, U has been classified as being moderately mobile in oxidising conditions across the whole pH range and immobile in reducing conditions [319]. The formation of the oxycation UO_2^{2+} is probably reponsible for the solubility of U over a broad range of pH. Also, several organic acids may increase the solubility of U in soils [322, 323]. Its solubility may be limited by the formation of slightly soluble precipitates (e.g. phosphates, oxides) and adsorption to clays and organic matter. The dominant aqueous species are thought to be UO_2^{2+}, $UO_2(CO_3)_3^{4-}$, $UO_2(CO_2)_2^{2-}$, $UO_2(HPO_4)_2^{2-}$, with $UO_2(CO_3)_3^{4-}$ or phosphate complexes forming the dominant species in alkaline waters [334].

Continued application of phosphate fertilisers to soil over a period of many years can result in increases in the U content of soils. Varying amounts can be added to soils in this way, depending on the rock phosphate source, the processing method used to produce the fertiliser, and the application rate to agricultural land [325–327]. Tetravalent U replaces Ca^{2+} present in the apatite crystal lattice, but U^{4+} is readily oxidised to U^{6+} by weathering. Apatite is destroyed by H_2SO_4 during the preparation of superphosphate, but the U remains as uranyl sulphate $[(UO_2)SO_2]$ and uranous sulphate $[U(SO_4)_2]$, both of which are soluble in water.

Rothbaum *et al.* [325] have given the most comprehensive and interesting description of the accumulation of U in soils from superphosphate applied annually to arable and grassland soils. They studied agricultural experimental plots at Rothamsted, UK and in New Zealand where phosphate fertilisers had been applied over many years and at known amounts. Rates of application of superphosphate were equivalent to about 33 kg P and 15 g U/ha/yr in three experiments at Rothamsted and to about 37 kg P and 16 g U/ha/yr in one experiment in New Zealand. Most of the U (about 1300 g/ha) applied in superphosphate to the clay loam soil at Rothamsted since 1889 was retained, like P, in the plough layer of arable soils or was adsorbed by the organic layers of soils under grassland. The U contents of the subsoils (23–46 cm) showed, on average, no evidence of U enrichment due to superphosphate applications. Unlike P, little U appeared to move by transport or leaching in grassland. Two soil samples were separated into 'light' and 'heavy' fractions using a bromoform/ethanol mixture of specific gravity 2.2 g/cm and analysed for U. In soils without P, the 'light' highly-organic fraction contained less U than the 'heavy' low-organic fraction; however, the U added in superphosphate in 85 years increased the U concentration of the 'light' fraction four times as much as that of the 'heavy' fraction. Total removals of U in the wheat grain and straw from Broadbalk, calculated from average yields and the U content of the crops (< 0.02 g U/t), would not exceed 10 g U/ha in 88 years. Rothbaum *et al.* [325] emphasised that their results found virtually all the U applied in superphosphate remained in the surface horizons of soils, but felt that this finding did not necessarily conflict with the work of Spalding and Sackett [328] who found significantly increased U contents in rivers draining intensively fertilised and farmed agricultural land in southwestern USA. They argued that the enrichment observed by Spalding and Sackett [328] could have been due to losses by leaching or soil erosion, and that in the absence of organic matter, U is generally considered to be mobile and is transported as a hexavalent carbonate complex likely to form in ground waters and aquifers with high concentrations of dissolved CO_2 [329], or as a divalent uranyl ion [330]. Uranium concentrations have been related to organic matter content [331], and U can be adsorbed by humic acids [323].

Ryan [332] has argued that the contribution of U to agricultural land due to the application of phosphate fertilisers does not significantly affect the radiation dose received by the general population. Nonetheless, the cumulative effect of long-continued applications at normal rates could lead to a 50% inrease in the U content of agricultural soils over a century or so. Phosphorus, in the form of mineral phosphate rock, is sometimes added to cattle feed, and this practice can result in increased levels of U (and Ra) in cows' milk [333].

The data on the extractability and chemical associations of U in soils seem remarkably limited. Only one rather specialised study appears to have been conducted, and this was by Lowson and Short [334] who developed a method for the chemical separation of the principal phases (amorphous Fe oxide, crystalline Fe oxide and resistate), and examined the distribution of ^{238}U and its decay products between these for soils collected from the Northern Territory of Australia. A selective extraction step for organically-bound material was deliberately excluded, since it was felt this would be unimportant in these arid zone soils. The major part of the U was associated with the Fe phases; in fact the concentration of ^{238}U was two orders of magnitude greater in the combined iron phases than in the resistate phase. Ion exchange, amorphous alumina and silica phases were not present in significant amounts. Near the soil surface more U was associated with the crystalline Fe phase than the amorphous Fe phase. This report is probably of limited value for inferring the partitioning and associations of U in soils from temperate lattitudes, and those generally where there is an organic matter constituent.

The need for new U deposits to be found to support the demands of the nuclear industry has generated considerable interest in biogeochemical prospecting techniques. A wide range has been used, with varying degrees of success. One which is reasonably well developed for locating buried U deposits is the use of Rn in soil gas and natural waters [335]. Extensive programmes of orientation work using this technique have been carried out by the Geological Survey of Canada [336–339] and British government agencies [340–242]. One advantage is that anomalies for Rn in soil gas can have greater contrast than those for U in the soil itself. In several areas, Rn anomalies have been detected above ores covered by up to 100 m of transported overburden. Prospecting by using the U content of soils has not been used extensively because of the greater effectiveness of radioactivity and Rn as indicators, although some work on organic soils and bogs, and particularly in organic lake sediments, has been shown to be highly effective under specialised circumstances. Michie *et al.* [341], for example, reported the identification of a hydromorphic anomaly of U in soils by a seepage area downslope from a mineralised zone in bedrock in Scotland. Soil U concentrations in the anomaly exceeded 7 mg/kg.

Geochemical prospecting and the succesful identification of anomalies have also been carried out using stream waters [343, 344]. Regional surveys in Saskatchewan, Canada have used lake waters; clusters of lakes with U contents twice the background level defined anomalous areas several kilometres in extent, with stronger anomalies within these areas apparently related to local sources [337, 345]. Stream sediment data are also available; Rose and Keith [346] advocated selectively extracting the U in organic matter and Fe oxides and leaving silicates in the residue to enhance the anomalies.

Uptake of the actinides by terrestrial plants from soil is generally considered to be low, but this varies considerably between elements. Plant/soil concentration ratios for Pu, for example, are of the order of 10^{-4} or less, although the limited evidence available suggests somewhat higher ratios for U (up to excess of 10^{-1}) [313, 347]. Tracy *et al.* [348] conducted a study of gardens contaminated by U processing wastes in the town of Port Hope, Ontario, Canada. Concentrations of total U were measured in soils and garden produce. The highest concentrations were found in root and stem vegetables, whilst fruit concentrations were generally the lowest. The mean concentration factor for U was 0.075×10^{-3}. Background soil concentrations in the area were 2.0 mg/kg in the 0–15 cm layer; the contaminated gardens ranged between 7.5–420 mg/kg. Despite this, there have been notable biogeochemical prospecting successes based on the analysis of plant tissues. This implies that the soluble fraction can be readily absorbed by plants; vegetation in mineralised areas can accumulate up to 100 times more U than plants from other areas.

Much of the recent work on U in vegetation has been conducted by Dunn [349–352] in Saskatchewan, Canada. He states [351] that 'the pristine boreal forest of central Canada provides an excellent field laboratory for biogeochemical investigations' because the natural process of element cycling through vegetation can be studied away from anthropogenic inputs. Under these conditions unusually high concentrations of U have been found in many types of vegetation, but in particular the twigs (less the needles) of black spruce (*Picea mariana*) which locally contain in excess of 1000 mg/kg dry weight; background values are in the range < 1 to 6 mg/kg. Twigs have been used to locate a massive U biogeochemical anomaly near Wollaston Lake that covers at least 3600 km^2. Dunn [351] emphasises that the vegetation should be sampled by taking similar amounts of growth, e.g. the latest 10 years from each tree, to even out seasonal differences. His data show that, in general, the order of U concentration is twigs > leaves > roots > trunk, which is in agreement with the findings of Steubing *et al.* [353]. In other words, U migrates towards the other extremities of trees. Interestingly at the Wollaston Lake anomaly there is no discernible correlation between the U content of the twigs and U in the bedrock, lodgement tills, peats, or soils—all of these media have normal background concentrations of U. However, all the other known U deposits in this province occur at depths of 60–250 m below the surface. Formation waters near the ore bodies of the Wollaston site contain elevated concentrations of U (up to about 50 μg/l). Hence it is concluded that the anomaly is the result of predominantly upward movement of slightly uraniferous formation waters from uraniferous sources, and uptake by the deeply penetrating boreal forest tree roots. Other reported data on the U content of vegetation include: up to 2.2 mg/kg ash weight in trees from a mineralised location; up to 1800 mg/kg in aquatic mosses collected in Basin Creek, central Idaho, USA, where spring water concentrations up to 6.5 μg/l were measured [354]; root concentrations, shoot concentrations and root: shoot ratios for plants from the Thompson district of Utah, USA were as follows; deep-rooting species (e.g. *Quercus, Juniperus*) 2–10, roots 140–1600, ratios 19–200; shallow rooting species 1.2–70,; 2–70;0.5–5.6 [355, 356]. There is conflicting evidence relating to the phytotoxicity of U in soils. Levels as low as 5 mg/kg, well within normal background concentrations, have been cited as toxic, whereas many studies report no toxicity at levels of 500 to 5000 mg U/kg soil. Laboratory experiments with *Brassica rapa* growing in controlled levels of U in soils showed no detrimental effects below 300 mg/kg [357]. Effects due to excess amounts of U in plants include abnormal numbers of chromosomes, unusually shaped fruits, sterile apetalous forms and stalked leaf rosettes [357]. Most of the U in plant roots apparently precipitates in the tips as autunite, $Ca(UO_2)_2PO_4$ [358]. That which reaches the plant shoot may be held as a protein complex.

References: Uranium

313. Whicker, F.W. and Schultz, V., *Radioecology: Nuclear Energy and the Environment*, Vol. I. CRC Press, Boca Raton, Fla (1982).
314. Eisenbud, M., *Environmental Radioactivity from Nature, Industrial and Military Sources*, 3rd edn. Academic Press, New York (1987).
315. Ostle, D., Coleman, R.F. and Ball, T.K., in *Uranium Prospecting Handbook*, ed. Bowie, S.H.U. Institution of Mining and Metallurgy, London (1972), 95–109.
316. Reimer, G.M., *J. Geochem. Expor.* **4** (1975), 425–431.
317. Manskaya, S.M. and Drozodova, T.V., *Geochemistry of Organic Substances*. Pergamon Press, Oxford (1968) Chapter 6, 164–172.
318. Kabata-Pendias, A. and Pendias, H., *Trace Elements in Soils and Plants*. CRC Press, Boca Raton, Fla (1984).
319. Rose, A.W., Hawkes, H.E. and Webb, J.S., *Geochemistry in Mineral Exploration*, 2nd edn. Academic Press, New York (1979).
320. Paspastefanou, C., Manolopoulou, M. and Charalambous, S., in *Seminar on the Cycling of Long-lived Radionuclides in the Biosphere: Observations and Models*. 15–19 Septmeber 1986. Madrid, Commission of the European Communities (1987).
321. Meriwether, J.R., Beck, J.N., Keeley, D.N., Langley, M.P., Thompson, R.H. and Young, J.C., J. *Environ. Qual.* **17** (1988), 562–568.
322. Szalay, A., *Acta Phys. Acad. Sci. Hungary* **8** (1957), 25–35.
323. Szalay, A., *Geochim. Cosmochim. Acta* **28** (1964), 1605–14.
324. Langmuir, D., *Geochim. Cosmochim. Acta.* **42** (1978), 547–570.
325. Rothbaum, H.P., McGaveston, D.A., Wall, T., Johnston, A.E. and Mattingly, GE.G., *J. Soil Sci.* **30** (1979), 147–153.
326. Mustonen, R., *Sci. Total Environ.* **45** (1985), 127–134.
327. Menzel, R.G., *J. Agric. Fd. Chem.* **16** (1968), 231–234.
328. Spalding, R.F. and Sackett, W., *Science* **175** (1972), 629–631.
329. Betcher, R.N., Gascoyne, M. and Brown, D., *Can. J. Earth Sci.* **25** (1988), 2089.
330. Hostetler, P.B. and Garrels, R.M., *Econ. Geol.* **57** (1962), 137–167.
331. Talibudeen, O., *Soils Fertil.* **27** (1964), 347–359.
332. Ryan, M.T., *Nucl. Saf.* **22** (1981), 70–76.
333. Reid, D.G., Sackett, W.M. and Spalding, R.F., *Health Phys.* **32** (1977), 535–540.
334. Lowson, R.T. and Short, S.A., in *Speciation of Fission and Activation Products in the Environment*, eds. Bulman, R.A. and Copper, J.R. Elsevier Applied Science Publishers, London (1985), 128–142.
335. Smith, A.Y., Barreto, P.M.C. and Pournis, S., in *Exploration for Uranium Ore Deposits*. International Atomic Energy Agency, Vienna (1976), 185–211.
336. Dyck, W. and Smith, A.Y., *Colorado School of Mines Quarterly* **64** (1969), 223–235.
337. Dyck, W., Dass, A.S. and Durham, C.C., *Toronto Symposium Volume* (1971), 132–150.
338. Dyck, W., *Geol. Survey Canada, Paper 73–28* (1973).
339. Dyck, W., Chatterjee, A.K. and Gemmrell, D.A., *J. Geochem. Explor.* **6** (1976), 139–162.
340. Bowie, S.H.U., Ostle, D. and Ball, T.K., *Toronto Symposum Volume* (1971), 103–111.
341. Michie, U.M., Gallagher, M.J. and Simson, A., *London Symposium Volume* (1973), 117–130.
342. Miller, J.M. and Ostle, D., in *Uranium Exploration Methods*. International Atomic Energy Agency, Vienna (1973), 237–247.

343. Wodzicki, A., *N.Z. J. Geol. Geophys.* **2** (1959), 602–612.
344. Denson, N.M., Zeller, H.D. and Stephens, J.G., *U.S. Geol. Survey Prof. Paper* **300** (1956) 673–680.
345. MacDonald, J.A., *Colorado School of Mines Quarterly* **64** (1969), 357–376.
346. Rose, A.W. and Keith, M.L., *J. Geochem. Explor.* **6** (1976), 119–137.
347. Vyas, B.N. and Mistry, K.V., *Plant and Soil* **59** (1981), 75.
348. Tracy, B.L., Prantl, F.A. and Quinn, J.M., *Health Phys.* **44** (1983), 469–477.
349. Dunn, C.E., *Summary of Investigations 1981*. Saskatchewan Geological Survey Miscellaneous Report 81–4, 117–126.
350. Dunn, C.E., *Summary of Investigations 1983*. Saskatchewan Geological Survey Miscellaneous Report 83/4, 106–122.
351. Dunn, C.E., in *Mineral Exploration: Biological Systems and Organic Matter*, eds. Carlisle, D., Berry, W.L., Kaplan, I.R. and Watterson, J.R. Rubey Series Vol. 5. Prentice-Hall, Englewood Cliffs, NJ, 134–149.
352. Dunn, C.E., *J. Geochem. Explor.* **15** (1981), 437–452.
353. Steubing, L., Haeke, J., Biermann, J. and Gnittke, J., *Agnew. Botanik*, **63** (1989), 361.
354. Shacklette, H.T. and Erdman, J.A., *J. Geochim. Explor.* **17** (1982), 221–236.
355. Cannon, H.L., *U.S. Geol. Survey Bull.* **1085-A** (1960), 1–50.
356. Cannon, H.L., *Science* **132** (1960), 591–598.
357. Sheppard, S.C., Evenden, W.G. and Anderson, A.J., *Environ. Toxicol. Water Qual.* 7 (1992), 275.
358. Cannon, H.L., *U.S. Geol. Survey Bull.* **1030-M** (1957), 399–516.

14.8 Vanadium

Vandium is ubiquitous in the lithosphere, with a mean crustal abundance of 150 mg/kg; this is of the same order as Ni, Cu, Zn and Pb, but V is more dispersed than these metals [359, 360]. At least 60 different V ores are known, the most abundant being the polysulphide, partronite, found in association with sulphur, Ni and Fe sulphides [361]. In igneous rocks, V is largely associated with basic magmas, especially titaniferous magnatites, and may be present in elevated concentrations; in acidic and silicic igneous rocks, V content is much lower [362]. Vanadium contents of metamorphic and sedimentary rocks are intermediate between those of basic and acidic igneous rocks. V^{3+} has a similar ionic radius to Fe3+, and may replace it in Fe minerals. There is general agreement that V is present in rocks as an insoluble salt of V^{3+}; only in rare instances (some sulphide minerals) does it occur in a bivalent form [363].

The V content of a soil depends upon the parent material and the pedogenic processes associated with its development. Composition of the parent material has less bearing on V content of mature, developed soils. Mitchell [364] reported V levels of 20–250 mg/kg from a range of Scottish soils, and related those differences to the parent material of each soil. Studies in Poland indicate a mean baseline level of 18.4 mg/kg^{-1} dry wt, with higher levels in silt and loamy soils and lower levels in sandy soils [365]. Swaine [366] gives a range of 20–500 mg/kg for the V content of 'normal' soils, and a frequently quoted mean value of 100 mg/kg is derived from Vinogradov [367] and Hopkins *et al.* [368].

The proportion of 'extractable' V in a soil is linked to extractant, soil type and drainage. Berrow and Mitchell [369] found 'extractable'V content of a range of

Scottish soils to be 0.03–26.0 mg/kg, depending upon the degree of a drainage of the soil, and the type of extractant used. EDTA was a more efficient extractant than acetic acid, indicative of an organic bound V fraction.

Vanadium has a wide and varied industrial usage, in metallurgy, electronics and dyeing, and as a catalyst. In all cases, the actual quantity of V used is small, and represents a very small potential environmental input. It appears that major anthropogenic sources of V enrichment of soil are from combustion of fossil fuels and the wastes from such processes. Fly-ash has been postulated as a major V source, based on elevated levels frequently encountered, but a series of studies [370–372] have failed to demonstrate V toxic responses in plants grown in fly-ash-amended soil. Both sewage sludges [373, 374] and phosphate fertilisers [375] have been investigated; it was concluded that neither represented a significant potential V input to agricultural soils.

Combustion of coals and oils represents the major source of V enrichment of the biosphere. Vanadium is the major trace metal in petroleum products, especially in the heavier fractions. Bertine and Goldberg [376] quote an average V content of crude oil as 50 mg/kg, with a range from 0.6–1400 mg/kg. Venezuelan oils have elevated V levels (mean 112 mg/kg, maximum 1400 mg/kg [377]). Vanadium is present as an organometallic porphyrin complex of low volatility; in consequence it becomes concentrated during distillation, giving residual oils an elevated V content (e.g. Venezuelan crude 112 mg/kg, Venezuelan Bunker 'C' no 5 and 6 residuals 870 mg/kg [378]. Such residual oils are used for domestic heating and power generation. Combustion releases V as V_2O_5 particulates; approximately 1 t of particulate V_2O_5 is released by the combustion of 1000 t crude oil [378]. Based on this calculation, it is estimated that V emissions from soil combustion equal V emissions from natural sources. Particulate V can be transported long distances [379].

Jacks [380] studied the fate of deposited V in the environment. He found that 5% was lost from his study site in runoff water, but the bulk of V was retained in the soil, mainly in association with organic matter. At the rates of deposition he observed, the author concluded that soil accumulation of V posed no environmental threat.

There is a dearth of information of the form and behaviour of V in soils. During soil formation V^{3+} in the mineral lattice is oxidised to V^{5+} [367, 381]. Oxyanions of V^{5+} are soluble over a wide range of pH, and are generally considered to be the mobile forms of V in soils [368], the degree of mobility being dependent on prevailing physical and geochemical factors. Goldschmidt [382] lists the following four groups of factors which may cause precipitation of less soluble V compounds:

(i) Presence of reducting agents,
(ii) Local concentrations of elements forming insoluble vanadates (e.g. Ca),
(iii) Precipitation in the presence of uranyl $(UO_2)^{2+}$ cations,
(iv) Presence of Al or ferric ions.

Reducing agents, such as organic matter, are important in soil V cycling. Szalay and Szilagy [383] demonstrated reduction of vanadate ions by humic acid, with subsequent geochemical enrichment.

Goodman and Cheshire [384] also found reduction of V^{5+} to V^{4+} during incubation with a peat humic preparation. Cheshire *et al.* [385] showed association of V with humic and fulvic acid fractions of a soil, and compared this with other metals. This percentage of total metals in a soil extracted by alkali (i.e. humic and fulvic acid fractions) decreased in the order:

$$Cu > Al > V > Ni = Co > Mn > Cr > Fe > Sr > Ba.$$

These authors also showed uneven distribution of V between humic and fulvic acid fractions, more being associated with the latter. Taylor and Giles [386] suggested that V could move as a vanadyl complex in certain soils, especially when associated with Fe oxides. Several V 'pools' can be envisaged in a soil system (Figure 14.1); exchange between pools will depend upon prevailing physical, chemical and biological conditions.

The few detailed surveys in the literature suggest that average plant V content associated with unpolluted, non-anomalous soils is between 0.5–2.00 mg/kg dry weight [360, 377, 387]. These data mean very little, as numerous edaphic factors, such as pH, redox potential and organic matter content, will exert a profound influence over eventual plant uptake. Vanadium accumulator species are known to exist. Cannon [388] records average plant values of 144 mg V/kg in *Astraglus confertiflorus* from vanadiferous sandstone-derived soil in Utah, USA; several other species typical of this region were found to possess elevated tissue V levels (Table 14.5).

The sporophore of the basidiomycete fungus *Amanita muscaria* is well known to contain elevated levels of V [389–393]. Vanadium exists as a unique

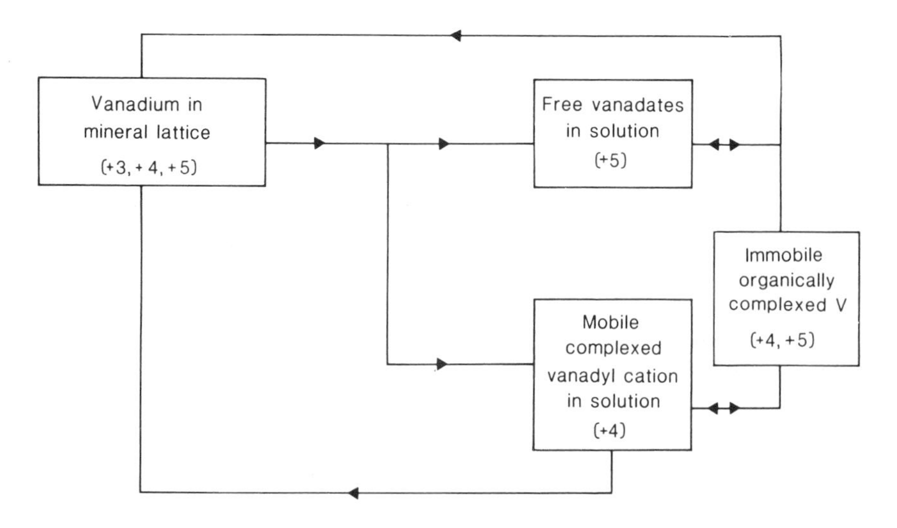

Figure 14.1 Vanadium pools in soil

Table 14.5 Tissue V content (mg/kg) of plants from vanadiferous soils in the USA (data from [368] and [388]).

Species	Tissue V content
Astraglus confertiflorus	144
Allium macropetallum	133
Astraglus preussi	67
Oenothera caespitosa	38
Castilleja angustifolia	37
Chrysothamnus viscidiflorus	37
Eriogonum inflatum	15
Lepidium montanum	11
Triticum aestivum (wheat)	5

organometallic complex, amavadin [394, 395], whose synthesis and function are at present unclear. These sporophores act as a major natural biogeochemical pathway for V [393]. Calculations based on laboratory experiments using powdered sporophore tissue incorporated into woodland soils and test plants indicate that an abundant population of *A. muscaria* sporophores could cycle 0.65% of the total V pool in the 0–5 cm soil layer of a skeletal pararendzina over a 14-day period (average lifespan of an individual sporophore). This represents a unique and rapid biological focusing of V.

In controlled laboratory or field experiments, results indicate a strong tendency for V to accumulate in roots [393, 396–399], although some data of Baisouny [400] indicate a more even distribution in pot-grown tomato. When V is supplied to whole plants as VO^{2+} or VO_3^-, there is little difference in uptake pattern and within-plant distribution between either form. ESR studies indicate the formation of VO^{2+} (V^{4+}) in roots of *Hordeum* initially supplied with VO_3^- (V^{5+}) [401]. This reduction could explain the similarities between the two forms, indicating an immobilisation of V in the root system. Experiments on the interaction between VO^{2+}, VO_3^- and Ca^{2+} ions support this hypothesis. Ca^{2+} has little influence on V uptake and transport in *Zea* when applied with VO^{2+}, but causes a significant reduction in uptake of VO_3^-; this indicates the formation of Ca vanadate prior to uptake, and would seem to preclude a major extracellular reduction of V^{5+} to V^{4+} during uptake [396]. Further experimental work is required to pinpoint the precise location of the reductive step.

There has been lengthly debate on the essentiality (or otherwise) of V for normal plant growth. It has been recognised as an essential micronutrient for certain green algae [402], and possibly for certain marine macroalgae [403]. Both Hewitt [404] and Welch and Huffman [405] grew various crop plants to full maturity in nutrient solutions containing between 2–4 μg V/l. It is probable that any obligate plant requirement for V is met at levels below these.

Vanadium is a specific catalyst for N fixation in *Azotobacter*, replacing Mo in the nitrogenase enzyme. However, it is not as effective as the Mo enzyme, possibly due to a lower stability [406–408]. N-metabolism in legumes is responsive to V treatment. Application of 8 μM V to *Phaseolus* plants resulted in increased growth and nitrate reductase activity together with an increase in leaf protein content towards the end of the experimental period [409].

Phytotoxicity attributed to V was recognised in early studies on phosphate fertilisers [410]. Warington [411, 412] found V toxicity in soya bean which could be alleviated by application of Fe. Wallace *et al.* [370] showed reductions in dry matter content of bush bean following root application of 10^{-4} M vanadate. Hara *et al.* [413] noted growth reduction in cabbage when root levels exceeded 2500 mg/kg dry weight. Lepp [414] noted little effect of $V(VO^{2+})$ on germination of lettuce (*Lettuca sativa*) seeds, but subsequent seedling growth was greatly reduced. Treatment with low concentrations of V stimulated root growth in seedlings of six agricultural crops [415]. Radicle extension of collards (*Brassica oleracea* cv *acephala*) was stimulated with treatments of < 1 mg/l^{-1} V, but severe toxicity was observed at V levels > 3 mg/l [415]. Sing and Wort [416] followed foliar application of $VOSO_4$ to sugar beet. They found significant reductions in leaf growth, concomitant with increases in sugar content of the storage roots. Further work by Singh [398] suggested injurious effects on maize at V levels above 0.25 mg/l in nutrient solution, a finding corroborated by Hidalgo *et al.* using onion roots [417].

Studies on V toxicity in whole plants [415] showed that soil type exerted a considerable influence over plant response. Collards showed reduction in dry matter at application rates of 80 mg/kg^{-1} when grown in Blanton Sand, but showed no response to 100 mg/kg^{-1} when grown in Orangeburry loamy sand. These responses could be attributed to reduced plant-available V, as a result of interactions with soil constituents. Humic and fulvic acids have been shown to reduce V toxicity in pot-grown corn (*Zea mays*) [418]. Incorporation of either humate of fulvate into V-containing solutions used to treat test plants resulted in increased growth in comparison to plants receiving V alone. Humate treatments were more effective than fulvate treatments. Aller *et al.* review recent data on the effects of V on crop growth [419]. Based on experiments with *Hordeum* seedlings Davis *et al.* [420] proposed an upper critical V level in actively growing shoots for growth reduction of 2 mg/kg, which is in good agreement with results of Hara [413] for cabbage.

Data relating to field episodes of V pollution are very rare. Despite the estimates of Vouk and Piver [421] that amounts of V released into the atmosphere from anthropogenic sources exceeds total world V production, there is little evidence for phytotoxic effects. Vaccarino *et al.* [422] reported instances of leaf and fruit necrosis associated with V-rich ash from an oil-burning power plant. A recent study in the Canary Islands [423] on V distribution from an oil-fired power plant concludes that contamination of native vegetation was restricted to the immediate (< 1 km) vicinity of the source, and could be mainly attributed to dry deposition of V-enriched fly ash. Vanadium in fly ash may represent a future pollution source. Applications of fly ash of 167 tonnes/ha^{-1} have been calculated to raise total soil V levels by up to 10% [424]. Vanadium can be leached from fly ash under both acid and alkaline conditions in the later stages of weathering [425] and up to 50% of total V can be easily released from chemically weathered alkaline fly ash [426].

Cement factories may also be sources of V pollution. A study in Cairo, Egypt [427] has identified such a factory as a significant pollution source, due to elevated levels of V in the raw materials and to the combustion of fossil fuels in the manufacturing process.

In view of the well-documented effect of VO_3^- on Na–K ATPase activity in cells [428], there is the potential for dramatic effects of V on all living organisms. However, the apparent ease of reduction of VO_3^- to VO^{2+}, which does not affect those essential systems, indicates that the environmental threat posed by V may be slight.

References: Vanadium

359. Fleischer, M., *Ann. N.Y. Acad. Sci.* **199** (1971), 6.
360. Bertrand, D., in *Survey of Contemporary Knowledge of Biochemistry*. Am. Museum of Natural History Bulletin 94 (1950).
361. Clarke, R.J.H., *The Chemistry of Titanium and Vanadium*. Elsevier, New York (1968).
362. Van Zinderen Bakker, E.M. and Jaworski, J.F., *Effects of Vanadium in the Canadian Environment*. Env. Secretariat Publication No. 18132, NRCC/CNRC, Ottawa (1980).
363. Rose, E.R., *Econ. Geology Report* No. 27 Dept. Energy, Mines and Resources, Ottawa, Canada (1973).
364. Mitchell, R.L., MAFF Technical Bulletin **21** (1971), 8.
365. Dudka, S. and Markert, B., *Sci. Tot. Env.* **122** (1992), 279.
366. Swaine, D.J., *Commonwealth Bureau of Soil Science Tech. Comm.* **48** (1955), 117.
367. Vinogradov, A.P., *The Geochemistry of Rare and Dispersed Chemical Elements in Soils*. Consultants Bureau, New York (1959).
368. Hopkins, L.L., Cannon, H.L., Meisch, A.T., Welch, R.M. and Nielson, P.M., *Geochem. Environ.* **2** (1977), 93.
369. Berrow, M.L. and Mitchell, M.L., *Trans. Roy. Soc. Edinburgh Earth Sci.* **71** (1980), 105.
370. Wallace, A., Alexander, G.V. and Chaudhry, F.M., *Commun. Soil Sci. Plant Anal.* **8** (1977), 751.
371. Adriano, D.L., Page, A.L., Elseewi, A.A., Change, A.C. and Straughan, I.L., *J. Environ. Qual.* **9** (1980), 333.
372. Jastrow, J.D., Zimmerman, C.A., Dvorak, A.J. and Hinchman, R.R., *J. Environ. Qual.* **10** (1981), 154.
373. Berrow, M.L. and Webber, J., *J. Sci. Fd Agric.* **23** (1972), 93.
374. Bradford, G.R., Page, A.L., Lund, L.J. and Olmstead, E., *J. Environ. Qual.* **4** (1975), 123.
375. Goodroad, L.L. and Caldwell, A.C., *J. Environ. Qual.* **8** (1979), 493.
376. Bertine, L. and Goldberg, E.C., *Science* **177** (1971), 233.
377. Bengtsson, S. and Tyler, G., *Vanadium in the Environment. A Technical Report*. MARC, London (1976).
378. Zoller, W.H., Gordon, G.E., Gladney, E.S. and Jones, A.G., in *Trace Elements in the Environment*, ed. Kothny, E.L., Am. Chem. Soc. (1972).
379. Brosset, C., *Ambio* **5** (1976), 157.
380. Jacks, G., *Environ. Pollut.* **11** (1976), 289.
381. Yen, T.F., *Trace Substances in Environmental Health* VI (1972), 347.
382. Goldschmidt, V.M., *Geochemistry*. Clarendon Press, Oxford (1958).
383. Szalay, A. and Szilagy, U., *Geochim. Cosmochim. Acta* **31** (1977), 1.
384. Goodman, B.A. and Cheshire, M.V., *Geochim. Cosmochim. Acta.* **39** (1975), 1111.
385. Cheshire, M.V., Berrow, M.L., Goodman, B.A. and Mundie, C.M., *Geochim. Cosmochim. Acta.* **41** (1977), 1131.
386. Taylor, R.M. and Giles, J.B., *J. Soil Sci.* **27** (1970), 203.

387. Schroedr, H.A., Balassa, J.J. and Tipton, J.H., *J. Chron. Dis.* **16** (1963), 1047.
388. Cannon, H.L., *Soil Sci.* **96** (1963), 196.
389. Ter Meulen, H., *Rec. Trans. Chem. Pays-Bas* **50** (1931), 491.
390. Bertrand, D., *Bull. Soc. Chim. Biol.* **25** (1943), 194.
391. Byrne, A.R., Ravnik, V. and Kosta, L., *Sci. Total Env.* **6** (1976), 65.
392. Tyler, G., *Trans. Br. Mycol. Soc.* **74** (1980), 41.
393. Lepp, N.W., Harrison, S.C.S. and Morrell, B.G., *Env. Geochem. Health* **9** (1987), 61.
394. Kneifel, H. and Bayer, R., *Angew. Chem.* **85** (1973), 542.
395. Lancashire, R.J., *Chem. Education* **17** (1980), 88.
396. Morrell, B.G., Lepp, N.W. and Phipps, D.A., *Min. Environ.* **5** (1983), 79.
397. Welch, R.M., *Plant Physiol.* **51** (1973), 825.
398. Singh, B.R., *Plant and Soil* **34** (1971), 209.
399. Wallace, A., *Soil Sci.* **147** (1989), 461.
400. Baisouny, F.M., *J. Plant Nutr.* **7** (1984), 1059.
401. Morrell, B.G., Lepp, N.W. and Phipps, D.A., *Env. Geochem. Health* **8** (1986), 14.
402. Arnon, D.J. and Wessel, G., *Nature* **172** (1953), 1039.
403. Fries, L., *Planta* **154** (1982), 393.
404. Hewitt, E.J., *Sand and Water Culture Methods Used in the Study of Plant Nutrition*. C.A.B., Farnham (1966).
405. Welch, R.M. and Huffman, E.W.D. Jr., *Plant Physiol.* **52** (1973), 183.
406. Mishra, D.K. and Kumar, H.D., *Biol. Plant*, **26** (1984), 448.
407. Vaishampayan, A. and Hementaranjan, A., *Pl. Cell Physiol.* **25** (1984), 845.
408. Eady, R., Robson, R. and Postgate, J., *New Scientist*, **114** (1987), 59.
409. Salo, D., Alvarez, M. and Martin, S., *Suelo y Planto*, **2** (1992), 723.
410. Brenchley, W.E., *J. Agric. Sci.* **22** (1933), 704.
411. Warington, K., *Ann. Appl. Biol.* **38** (1951), 624.
412. Warington, K., *Ann. Appl. Biol.* **38** (1954), 1.
413. Hara. T., Sonoda, Y. and Iwai, I., *Soil Sci. Pl. Nutr.* **22** (1976), 307.
414. Lepp, N.W., *Z. Pflanzenphysiol.* **83** (1977), 185.
415. Kaplan, D.I., Sajwan, K.S., Adriano, D.C. and Gettier, S., *Water, Air, Soil Pollut.* **53** (1990), 203.
416. Singh, B.R. and Wort, D.J., *Pl. Physiol.* **44** (1969), 1312.
417. Hidalgo, A., Navas, P. and Garciaherdugo, G., *Environ. Expt. Bot.*, **28** (1988), 131.
418. Ullah, S.M. and Gerzabek, M.H., *Bodenkultur*, **42** (1991), 123.
419. Aller, A.J., Bernal, J.L., Jesus de Nozal, M. and Deban, L., *J. Sci. Fd. Agric.* **51** (1990) 447.
420. Davis, R.D., Beckett, P.H.T. and Wollan, E., *Plant and Soil* **49** (1978), 395.
421. Vouk, V.B. and Piver, W.T., *Env. Hlth Perspect.* **47** (1983), 201.
422. Vaccarino, C., Gimmino, G., Tripodi, M.N., Lagona, G., LoGuicide, L. and Materese, R., *Agric. Ecosystems Environ.* **10** (1983), 275.
423. Alvarez, C.E., Fernandez, M., Perez, N., Inglesia, E. and Snelling, R., *J. Env. Sci. Hlth. Part A, Env. Sci. Eng.* **28** (1993), 269.
424. Warren, C.J., Evans, L.J. and Sheard, R.W., *Waste Man. Res.* **11** (1993), 3.
425. Texeira, E.C., Samama, J. and Brun, A., *Env. Tech.* **13** (1992), 1187.
426. Warren, C.J. and Dudas, M.J., *Sci. Tot. Env.* **76** (1988), 229.
427. Hindy, K.T., Abel-Shafy, H.I. and Farag, S.A., *Environ. Pollut.* **66** (1990), 195.
428. Macara, I.G., *TIBS* **5** (1980), 92.

Appendices

Appendix 1: Chemical properties of the heavy metals

Element	Group in Periodic Table	Atomic number	Atomic weight	Ions	Ionic radius*	Electro-negativity†	Ion potential (charge/radius)
Ag	IB	47	107.87	Ag^{+}	1.26	1.9	–
As	VA	33	74.92	As^{3+}	0.58	–	–
				As^{5+}	0.46	1.9	–
Au	IB	79	196.97	Au^{+}	1.37	2.4	–
Cd	IIB	48	122.40	Cd^{2+}	0.97	1.7	–
Co	VIII	27	58.93	Co^{2+}	0.72	1.8	2.6
Cr	VIB	24	52.00	Cr^{3+}	0.63	1.6	4.3
				Cr^{6+}	0.52	–	16.0
Cu	IB	29	63.54	Cu^{+}	0.96	1.9	–
				Cu^{2+}	0.72	2.0	2.5
Hg	IIB	80	200.59	Hg^{2+}	1.10	1.9	–
Mn	VIIB	25	54.94	Mn^{2+}	0.80	1.5	–
				Mn^{3+}	0.66	–	–
				Mn^{4+}	0.60	–	6.5
Mo	VIB	42	95.94	Mo^{4+}	0.70	–	–
				Mo^{6+}	0.62	1.8	12.0
Ni	VIII	28	59.71	Ni^{2+}	0.69	1.8	2.6
Pb	IVA	82	207.19	Pb^{2+}	1.20	1.8	1.9
Sb	VA	51	121.75	Sb^{3+}	0.76	–	–
				Sb^{5+}	0.62	1.9	–
Se	VIA	34	78.96	$Se^{=}$	[2.00]	2.4	3.7
				Se^{6+}	0.42	–	–
Sn	IVA	50	118.69	Sn^{2+}	0.93	1.8	1.5
				Sn^{4+}	0.71	1.9	–
Tl	IIIA	81	204.37	Tl^{+}	1.47	–	–
				Tl^{3+}	0.95	1.8	–
U	Actinide series	92	238.04	U^{4+}	0.97	–	–
				U^{6+}	0.80	1.7	–
V	VB	23	50.94	V^{3+}	0.74	1.6	–
				V^{4+}	[0.65]	–	–
				V^{5+}	0.59	–	11.0
W	VIB	74	183.85	W^{6+}	0.62	1.7	–
Zn	IIB	30	65.37	Zn^{2+}	0.74	1.7	2.6

*Ionic radius is for 6-coordination.
†Electronegativity values for other elements: S 2.5, O 3.5, I 2.5, Cl 3.0, F 4.0. From these numbers it can be generalised that a bond between any two atoms will be largely covalent if the electronegativties are similar and mainly ionic if they are very different.

Sources:

Kabata-Pendias, A. and Pendias, H., *Trace Elements in Soils and Plants*, 2nd edn. CRC Press, Boca Raton, Fla (1992).

Krauskopf, K.B., Introduction to Geochemistry. McGraw-Hill, New York (1967).

Appendix 2: Concentrations of heavy metals in soils and plants

Element	Normal range in soils*	Critical soil total concs† (mg/kg)	Normal range in plants*	Critical concs in plants‡	
				a	*b*
Ag	0.01–8	2	0.1–0.8	–	1–4
As	0.1–40	20–50	0.02–7	5–20	1–20
Au	0.001–0.02	–	⊂0.0017	–	<1
Cd	0.01–2.0	3–8	0.1–2.4	5–30	4–200
Co	0.5–65	25–50	0.02–1	15–50	4–40
Cr	5–1500	75–100	0.03–14	5–30	2–18
Cu	2–250	60–125	5–20	20–100	5–64
Hg	0.01–0.5	0.3–5	0.005–0.17	1–3	1–8
Mn	20–10000	1500–3000	20–1000	300–500	100–7000
Mo	0.1–40	2–10	0.03–5	10–50	–
Ni	2–750	100	0.02–5	10–100	8–220
Pb	2–300	100–400	0.2–20	30–300	–
Sb	0.2–10	5–10	0.0001–0.2	–	1–2
Se	0.1–5	5–10	0.001–2	5–30	3–40
Sn	1–200	50	0.2–6.8	60	63
Tl	0.1–0.8	1	0.03–3	20	–
U	0.7–9	–	0.005–0.06	–	–
V	3–500	50–100	0.001–1.5	5–10	1–13
W	0.5–83	–	0.005–0.15	–	–
Zn	1–900	70–400	1–400	100–400	100–900

Notes and Sources:
*Data mainly from Bowen, H.J.M., *Environmental Chemistry of the Elements.* Academic Press, London (1979).
†The critical soil total concentration is the range of values above which toxicity is considered to be possible. Data from Kabata-Pendias, A. and Pendias, H., *Trace Elements in Soils and Plants,* 2nd edn. CRC Press, Boca Raton, Fla (1992).
‡The critical concentration in plants is the level above which toxicity effects are likely. *a*, data from Kabata-Pendias and Pendias (1992); *b*, values likely to cause 10% depression in yield; data from McNichol, R.D. and Beckett, P.H.T., *Plant and Soil* **85** (1985), 107–129.

Appendix 3: Maximum metal concentrations allowed in soils treated with sewage sludge in different countries (mg/kg)

Country	Cd	Cr	Cu	Hg	Ni	Pb	Zn
European Union	1–3	100–150	50–140	1–1.5	30–75	50–300	150–300
France	2	150	100	1	50	100	300
Germany	1.5	100	60	1	50	100	200
Italy	3	150	100	–	50	100	300
UK	3	400	135	1	75	300	300
Denmark	0.5	30	40	0.5	15	40	100
Finland	0.5	200	100	0.2	60	60	150
Norway	1	100	50	1	30	50	150
Sweden	0.5	30	40	0.5	15	40	100
USA	20	1500	750	8	210	150	1400

Ref. McGrath, S.P., Chang, A.C., Page, A.L. and Witter, E. (1994) *Environ. Rev. 2* (in press).

Appendix 4: US Environmental Protection Agency Part 503 regulations for sewage sludge applied to land

Metal	Max permitted conc in sludge (mg/kg)	Max permitted conc in 'clean' sludge (mg/kg)	Max annual loading (kg/ha/yr)	Max cumulative pollutant loading (kg/ha)
As	75	41	2.0	41
Cd	85	39	1.9	39
Cr	3000	1200	150	3000
Cu	4300	1500	75	1500
Pb	840	300	15	300
Hg	57	17	0.85	17
Mo	75	18	0.90	18
Ni	420	420	21	420
Se	100	36	5.0	100
Zn	7500	2800	140	2800

Ref. US Environmental Protection Agency (1993) Standards for the use or disposal of sewage sludge. *Federal Register*, **58:47**, 210–47 238.

Appendix 5: UK Department of the Environment ICRCL 'trigger' concentrations for metals in contaminated land which is to be redeveloped for the uses specified

Metal	Proposed use	Threshold trigger concentration (mg/kg)
Contaminants which pose hazards to human health		
As	Gardens, allotments	10
	Parks, playing fields, open space	40
Cd	Gardens, allotments	3
	Parks, playing fields, open space	15
Cr (hexavalent)*	Gardens, allotments	25
	Parks, playing fields, open space	–
Cr (total)	Gardens, allotments	600
	Parks, playing fields, open space	1000
Pb	Gardens, allotments	500
	Parks, playing fields, open space	2000
Hg	Gardens, allotments	1
	Parks, playing fields, open space	20
Se	Gardens, allotments	3
	Parks, playing fields, open space	6
Phytotoxic contaminants not normally hazardous to human health		
B (water soluble)	Any uses where plants grown	3
Cu (total)	Any uses where plants grown	130
Cu (extractable)†		50
Ni (total)	Any uses where plants grown	70
Ni (extractable)†		20
Zn (total)	Any uses where plants grown	300
Zn (extractable)†		130

*Hexavalent Cr extracted by 0.1MHCl adjusted to pH at 37.5°C.
†Cu, Ni and Zn extracted in 0.05M EDTA (ethylenediaminetetraacetic acid).

Ref: Interdepartmental Committee on the Redevelopment of Contaminated Land, Guidance on the assessment and redevelopment of contaminated land, Guidance Note 59/83, Department of the Environment, London, 1987.

Appendix 6: Guide values and quality standards used in the Netherlands for assessing soil contamination ('A' and 'C'* values in current use, 'B'† value now discontinued but included for information)

Metal	Reference (A) value*	Intervention (C) value*	Test (B) value†
As	29	50	30
Ba	200	2000	400
Cd	0.8	12	5
Co	10	300	50
Cr	100	380	250
Cu	36	190	100
Hg	0.3	10	2
Mo	10	200	40
Ni	35	210	100
Pb	85	530	150
Sn	20	300	50
Zn	140	720	500

*A = Reference value based on concentrations ound in nature reserves where only the contamination is from atmospheric deposition. C = Intervention values where the soil must be cleaned-up.
†B = Test value (proposed in 1986) indicating need for further investigation – now discontinued but included for information.

Ref. Netherlands Ministry of Housing, Physical Planning and Environment, *Environmental Quality Standards for Soil and Water*, Netherlands Ministry of Housing, Physical Planning and Environment, Leidschendam, Netherlands, 1986 and 1991.

Appendix 7: Comparison between the US soil taxonomy and the FAO/UNESCO soil classification schemes

FAO/UNESCO	US soil taxonomy	Other names
Fluvisols	Fluvents	Alluvial soils
Gleysols	Haplaquents, Psammaquents	Hydromorphic soils
Regosols	Orthents, Psamments	–
Lithosols	Lithosols	–
Arenosols	Psamments	Sandy soils
Rendzinas	Rendolls	–
Rankers	Lithic Haplumbrept	–
Andosols	Andepts	Volcanic soils
Vertisols	Vertisols	–
Solonchaks	Salorthids, Salorthidic Haplustolls	Saline soils
Solonetz	Natrustalfs, Natrixeralfs, Naturgids	–
Yermosols	Typic Aridisols	–
Xerosols	Mollic Aridisols	–
Kastanozems	Ustolls	–
Chernozems	Haploborolls, Vermiborolls	–
Phaeozems	Hapludolls	Degraded Chernozerms
Greyzems	Argiborolls	–
Cambiosols	Eutrochrepts	Brown Earths
Luvisols	Hapludalfs, Haploxeralfs	Sols-Lessive
Podzoluvisols	Glossudalfs, Glossoboralfs	Brown podzolic soils
Podzols	Spodosols, Orthods	–
Planosols	Palexeralfs, Paleustalfs	Sols Lessives
Acrisols	Hapludults, Haploxerults	–
Nitosols	Tropudalfs, Paleudalfs, Rhodustalfs	Red-Brown soil
Ferralsols	Oxisols	Lateritic soil
Histosols	Histosols	Peats, Mucks

Adapted from: Clayden, B. in *Principles and Applications of Soil Geography*, eds. Bridges, E.M., and Davidson, D.A. Longman, London (1982); and Kabata-Pendias, A. and Pendias, H., *Trace Elements in Soils and Plants*, 2nd edn. CRC Press, Boca Raton, Fla (1992).

Appendix 8: Common and botanical names of cultivated plants

Common name	Botanical name
Alfalfa	*Medicago sativa* L.
Barley	*Hordeum vulgare* L.
Bean (field, kidney and French)	*Phaseolus vulgaris* cultivars
Beetroot (red beet)	*Beta vulgaris var. crassa* Alef
Cabbage	*Brassica oleracae var. capiatata*
Carrot	*Daucus carota* L.
Celery	*Apium graveiolus var.* dulce
Chard (Swiss)	*Beta vulgare* var. *cicla* L.
Chinese cabbage	*Brassica chinensis*
Corn	*Zea mais* L.
Clover	*Trifolium* spp (Leguminosae)
Cucumber	*Cucumis sativis* L.
Fescues (grasses)	*Festuca* spp.
Flax	*Linum usitassiumum* L.
Kale	*Brassica oleracae* var. *acephala*
Leek	*Allium porrum* L.
Lettuce	*Lactuca sativa* L.
Lucerne (see alfalfa)	*Medicago sativa* L.
Mangold	*Beta vulgaris* var. *macrorhiza*
Oats	*Avena sativa* L.
Onion	*Allium cepa* L.
Peanut	*Arachis hypogea* L.
Potato	*Solanum tubersum* L.
Radish	*Raphanus sativus* L.
Red fescue	*Festuca rubra*
Rice	*Oryza sativa* L.
Ryegrass	*Lolium perenne* L.
Sorghum	*Sorghum vulgare* L.
Soya bean	*Glycine max* L.
Smooth bromegrass	*Bromus racemosus* L.
Spinach	*Spinacia oleracea* L.
Squash	*Cucurbita bepo* L.
Sudan grass	*Sorghum sudanese* (Piper)
Sugar beet	*Beta vulgaris* var. *altissima*
Tobacco	*Nicotiana sinensis* L.
Tomato	*Lycopersicum esculentum* Mill
Turnip	*Brassica napus* L.
Wheat	*Triticum aestivum* L.

Appendix 9: Element concentrations in selected certified reference materials

1. Analytical Quality Control Service, International Atomic Agency Vienna, P.O. Box 100, A-1400 Vienna, Austria.

IAEA Soil-6

ELEMENT:	^{90}Sr	^{137}Gs	^{226}Ra	^{239}Pu
CONTENT:	30.34	52.65	79.92	1.04 (Bq/kg)

IAEA SOIL-7

ELEMENT:	As	Ce	Co	Cr	Cs	Cu	Dy	Eu	Hf	La	Mn
CONTENT: (μg/g)	13.	61	8.9	60	5.4	11	3.9	1.0	5.1	28	631

ELEMENT:	Nd	Pb	Rb	Sb	Sc	Sm	Sr	Ta	Tb	Th	U
CONTENT:	30	60	51	1.7	8.3	5.1	108	0.8	0.6	8.2	2.6

ELEMENT:	V	Y	Zn	Zr
CONTENT:	66	21	104	185

2. Commision of the European Communities. Community Bureau of Reference (BCR), Rue de la Loi, B 1049 Brussels, Belgium.

BCR No. 141 Calcareous loam soil

Certified content:

ELEMENT:	Cd	Cu	Hg	Pb	Zn
CONTENT: (μg/g)	0.36	32.6	56.8×10^{-3}	29.4	81.3

Indicative values:

ELEMENT:	Se	Cr	Co	Mn	Ni
CONTENT: (Mean)	0.16	75	9.2	547	30.9

Other indicative values for 20 elements.

Aqua regia-soluble contents:

ELEMENT:	Cd	Cr	Cu	Mn	Ni	Pb	Zn
CONTENT: (μg/g)	0.3	53	31.2	512	28.0	26.3	70

Aqua regia-soluble indicative values for seven other elements.

BCR No. 142. Light sandy soil.
Certified contents:

ELEMENT:	Cd	Cu	Hg	Ni	Pb	Zn
CONTENT: (μg/g)	0.25	27.5	0.104	29.2	37.8	92.1

Appendix 9: *Cont'd*

Indicative values:

ELEMENT: (mean)	Se	Cr	Co	Mn
CONTENT: (μg/g)	0.53	74.9	7.9	569

Other indicative values for 31 elements.

Aqua regia-soluble contents:

ELEMENT:	Cd	Cr	Cu	Mn	Ni	Pb	Zn
CONTENT: (μg/g)	0.22	44.4	25.3	527	28.9	30.9	79.6

Aqua regia-soluble indicative values for eight other elements.

BCR No. 143 Sewage sludge-amended soil.

Certified contents:

ELEMENT:	Cd	Cu	Hg	Ni	Pb	Zn
CONTENT: (μg/g)	31.1	236.5	3.92	99.5	1333	1272

Indicative values:

ELEMENT:	Co	Cr	Mn	Se
CONTENT: (mean) (μg/g)	11.8	228	999	0.6

Other indicative values for 34 elements.

Aqua regia-soluble contents:

ELEMENT:	Cd	Cr	Cu	Mn	Ni	Pb	Zn
CONTENT:	31.5	208	236	935	92.7	1317	1301

Aqua regia soluble indicative values for nine other elements.

BCR No. 144. Sewage sludge of domestic origin.

BCR No. 145. Sewage sludge—Domestic and some industrial origin.

BCR No. 146. Sewage sludge of mainly industrial origin.

3. Canadian Reference Materials Project, Mineral Sciences Laboratory, CANMET, 555 Booth St. Ottawa, Canada, K1A 0G1.

CCRMP Reference Soil SO1, SO2, SO3, SO4.

CCRMP SO1

Appendix 9: *Cont'd*

Recommended values:

ELEMENT:	Al	Ca	Fe	K	Mg	Mn	P	Si	Ti
CONTENT: (%)	9.38	1.80	6.00	2.68	2.31	0.089	0.062	25.72	0.53

ELEMENT:	Cr	Cu	Hg	Ni	Pb	V	Zn
CONTENT: (μg/g)	160	61	0.022	94	21	139	140

CCRMP SO2

Recommended values:

ELEMENT:	Al	Ca	Fe	K	Mg	Mn	Si	Ti
CONTENT: (%)	8.07	1.96	5.56	2.45	0.54	0.072	24.99	0.86

ELEMENT:	Cr	Cu	Hg	Pb	Sr	V	Zn
CONTENT: (μg/g)	16	7	0.082	21	340	64	124

CCRMP SO3

Recommended values:

ELEMENT:	Al	Fe	K	Mn	Na	Si
CONTENT: (%)	3.05	1.51	1.16	0.052	0.74	15.86

ELEMENT:	Cr	Cu	Hg	Ni	Pb	Sr	Zn
CONTENT:	26	17	0.017	16	14	217	52

CCRMP SO4

Recommended values:

ELEMENT:	Al	Ca	Fe	K	Mg	Mn	Si	Ti
CONTENT: (%)	5.60	1.16	2.43	1.76	0.60	0.062	0.097	0.36

ELEMENT:	Cr	Cu	Hg	Pb	Sr	V	Zn
CONTENT: (μg/g)	66	23	29	19	188	101	97

Sources:

Commission of the European Communities, Reports EHR8833EN, EHR8834EN, EHR8835EN, EHR8836EN, EHR8837EN, EHR8838EN, Brussels (1983).

Bowman, W.S., *et al.*, *Geostandards Newsletter* **3** (1979), 2.

Index